Peptides for Health Benefits 2019

Peptides for Health Benefits 2019

Volume 1

Special Issue Editors

Blanca Hernández-Ledesma
Cristina Martínez-Villaluenga

MDPI • Basel • Beijing • Wuhan • Barcelona • Belgrade • Manchester • Tokyo • Cluj • Tianjin

Special Issue Editors
Blanca Hernández-Ledesma
Institute of Food Science Research (CIAL, CSIC-UAM, CEI UAM+CSIC)
Spain

Cristina Martínez-Villaluenga
Institute of Food Science, Technology and Nutrition (ICTAN, CSIC)
Spain

Editorial Office
MDPI
St. Alban-Anlage 66
4052 Basel, Switzerland

This is a reprint of articles from the Special Issue published online in the open access journal *International Journal of Molecular Sciences* (ISSN 1422-0067) (available at: https://www.mdpi.com/journal/ijms/special_issues/Peptides_2019).

For citation purposes, cite each article independently as indicated on the article page online and as indicated below:

LastName, A.A.; LastName, B.B.; LastName, C.C. Article Title. *Journal Name* **Year**, *Article Number*, Page Range.

Volume 1
ISBN 978-3-03936-080-2 (Hbk)
ISBN 978-3-03936-081-9 (PDF)

Volume 1-2
ISBN 978-3-03936-082-6 (Hbk)
ISBN 978-3-03936-083-3 (PDF)

Contents

About the Special Issue Editors ix

Cristina Martínez-Villaluenga and Blanca Hernández-Ledesma
Peptides for Health Benefits 2019
Reprinted from: *Int. J. Mol. Sci.* **2020**, *21*, 2543, doi:10.3390/ijms21072543 1

Rosa Gaglione, Angela Cesaro, Eliana Dell'Olmo, Rocco Di Girolamo, Luca Tartaglione, Elio Pizzo and Angela Arciello
Cryptides Identified in Human Apolipoprotein B as New Weapons to Fight Antibiotic Resistance in Cystic Fibrosis Disease
Reprinted from: *Int. J. Mol. Sci.* **2020**, *21*, 2049, doi:10.3390/ijms21062049 7

Valeria Tarallo, Emanuela Iaccarino, Valeria Cicatiello, Riccardo Sanna, Menotti Ruvo and Sandro De Falco
Oral Delivery of a Tetrameric Tripeptide Inhibitor of VEGFR1 Suppresses Pathological Choroid Neovascularization
Reprinted from: *Int. J. Mol. Sci.* **2020**, *21*, 410, doi:10.3390/ijms21020410 25

Tomoko T. Asai, Fumi Oikawa, Kazunobu Yoshikawa, Naoki Inoue and Kenji Sato
Food-Derived Collagen Peptides, Prolyl-Hydroxyproline (Pro-Hyp), and Hydroxyprolyl-Glycine (Hyp-Gly) Enhance Growth of Primary Cultured Mouse Skin Fibroblast Using Fetal Bovine Serum Free from Hydroxyprolyl Peptide
Reprinted from: *Int. J. Mol. Sci.* **2020**, *21*, 229, doi:10.3390/ijms21010229 37

Ewelina Russjan and Katarzyna Kaczyńska
Beneficial Effects of Neurotensin in Murine Model of Hapten-Induced Asthma
Reprinted from: *Int. J. Mol. Sci.* **2019**, *20*, 5025, doi:10.3390/ijms20205025 47

Ewelina Russjan, Kryspin Andrzejewski, Dorota Sulejczak, Patrycja Kleczkowska and Katarzyna Kaczyńska
Endomorphin-2- and Neurotensin- Based Chimeric Peptide Attenuates Airway Inflammation in Mouse Model of Nonallergic Asthma
Reprinted from: *Int. J. Mol. Sci.* **2019**, *20*, 5935, doi:10.3390/ijms20235935 59

Olga Victorovna Pershina, Angelina Vladimirovna Pakhomova, Darius Widera, Natalia Nicolaevna Ermakova, Anton Alexandrovich Epanchintsev, Edgar Sergeevich Pan, Vyacheslav Andreevich Krupin, Olga Evgenevna Vaizova, Olesia Dmitrievna Putrova, Lubov Alexandrovna Sandrikina, et al.
Gender Differences in the Pharmacological Actions of Pegylated Glucagon-Like Peptide-1 on Endothelial Progenitor Cells and Angiogenic Precursor Cells in a Combination of Metabolic Disorders and Lung Emphysema
Reprinted from: *Int. J. Mol. Sci.* **2019**, *20*, 5414, doi:10.3390/ijms20215414 77

Adenrele Oludiran, David S. Courson, Malia D. Stuart, Anwar R. Radwan, John C. Poutsma, Myriam L. Cotten and Erin B. Purcell
How Oxygen Availability Affects the Antimicrobial Efficacy of Host Defense Peptides: Lessons Learned from Studying the Copper-Binding Peptides Piscidins 1 and 3
Reprinted from: *Int. J. Mol. Sci.* **2019**, *20*, 5289, doi:10.3390/ijms20215289 99

Francesca Caccuri, Antonella Bugatti, Silvia Corbellini, Sara Roversi, Alberto Zani, Pietro Mazzuca, Stefania Marsico, Arnaldo Caruso and Cinzia Giagulli
The Synthetic Dipeptide Pidotimod Shows a Chemokine-Like Activity through CXC Chemokine Receptor 3 (CXCR3)
Reprinted from: *Int. J. Mol. Sci.* **2019**, , 5287, doi:10.3390/ijms20215287 . . . . **113**

Mihee Jang, Jieun Kim, Yujin Choi, JeongKyu Bang and Yangmee Kim
Antiseptic Effect of Ps-K18: Mechanism of Its Antibacterial and Anti-Inflammatory Activities
Reprinted from: *Int. J. Mol. Sci.* , *20*, 4895, doi:10.3390/ijms20194895 . . . . **131**

Anna Golda, Paulina Kosikowska-Adamus, Aleksandra Kret, Olena Babyak, Kinga Wójcik, Ewelina Dobosz, Jan Potempa, Adam Lesner and Joanna Koziel
The Bactericidal Activity of Temporin Analogues Against Methicillin Resistant *Staphylococcus aureus*
Reprinted from: *Int. J. Mol. Sci.* **2019**, , 4761, doi:10.3390/ijms20194761 . . . . **149**

Tomislav Rončević, Jasna Puizina and Alessandro Tossi
Antimicrobial Peptides as Anti-Infective Agents in Pre-Post-Antibiotic Era?
Reprinted from: *Int. J. Mol. Sci.* **2019**, *20*, 5713, doi:10.3390/ijms20225713 . . . . **165**

Lucinda J. Bessa, Julia R. Manickchand, Peter Eaton, José Roberto S. A. Leite, Guilherme D. Brand and Paula Gameiro
Intragenic Antimicrobial Peptide Hs02 Hampers the Proliferation of Single- and Dual-Species Biofilms of *P. aeruginosa* and *S. aureus*: A Promising Agent for Mitigation of Biofilm-Associated Infections
Reprinted from: *Int. J. Mol. Sci.* **2019**, *20*, 3604, doi:10.3390/ijms20143604 . . . . **197**

Suhanya V. Prasad, Krzysztof Fiedoruk, Tamara Daniluk, Ewelina Piktel and Robert Bucki
Expression and Function of Host Defense Peptides at Inflammation Sites
Reprinted from: *Int. J. Mol. Sci.* **2020**, *21*, 104, doi:10.3390/ijms21010104 . . . . **209**

Josep Nácher-Juan, María Carmen Terencio, María José Alcaraz and María Luisa Ferrándiz
Osteostatin Inhibits Collagen-Induced Arthritis by Regulation of Immune Activation, Pro-Inflammatory Cytokines, and Osteoclastogenesis
Reprinted from: *Int. J. Mol. Sci.* **2019**, *20*, 3845, doi:10.3390/ijms20163845 . . . . **233**

Katarzyna Palus, Krystyna Makowska and Jarosław Całka
Alterations in Galanin-Like Immunoreactivity in the Enteric Nervous System of the Porcine Stomach Following Acrylamide Supplementation
Reprinted from: *Int. J. Mol. Sci.* **2019**, *20*, 3345, doi:10.3390/ijms20133345 . . . . **247**

Shang Eun Park, Kiumars Shamloo, Timothy A. Kristedja, Shaban Darwish, Marco Bisoffi, Keykavous Parang and Rakesh Kumar Tiwari
EDB-FN Targeted Peptide–Drug Conjugates for Use against Prostate Cancer
Reprinted from: *Int. J. Mol. Sci.* **2019**, *20*, 3291, doi:10.3390/ijms20133291 . . . . **263**

Samuel Fernández-Tomé, Fei Xu, Yanhui Han, Blanca Hernández-Ledesma and Hang Xiao
Inhibitory Effects of Peptide Lunasin in Colorectal Cancer HCT-116 Cells and Their Tumorsphere-Derived Subpopulation
Reprinted from: *Int. J. Mol. Sci.* **2020**, *21*, 537, doi:10.3390/ijms21020537 . . . . **285**

Sara María Martínez-Sánchez, Horacio Pérez-Sánchez, José Antonio Gabaldón, José Abellán-Alemán and Silvia Montoro-García
Multifunctional Peptides from Spanish Dry-Cured Pork Ham: Endothelial Responses and Molecular Modeling Studies
Reprinted from: *Int. J. Mol. Sci.* **2019**, *20*, 4204, doi:10.3390/ijms20174204 . . . . **299**

Shi-Ying Cai, Yu-Mei Wang, Yu-Qin Zhao, Chang-Feng Chi and Bin Wang
Cytoprotective Effect of Antioxidant Pentapeptides from the Protein Hydrolysate of Swim Bladders of Miiuy Croaker (*Miichthys miiuy*) against H_2O_2-Mediated Human Umbilical Vein Endothelial Cell (HUVEC) Injury
Reprinted from: *Int. J. Mol. Sci.* **2019**, *20*, 5425, doi:10.3390/ijms20215425 . . . . **315**

About the Special Issue Editors

Blanca Hernández-Ledesma, Ph.D. earned a B.S. in Pharmacy in 1998, and defended her Ph.D. thesis in Pharmacy in 2002. Her research career has focused on the biological activity of food proteins/peptides, aiming to better understand their health implications and the development of novel food ingredients. She is author of 76 JCR articles, 9 popular science articles, and 30 book chapters, with an h-index 33 (WoS). Her results have been presented in 78 international and national conferences. She has supervised 4 Doctoral theses and 14 Master theses. She has participated in more than 30 international and national research projects. She has participated as a member of the Selection Board for Tenured Scientists, Ph.D. and Masters' thesis dissertation committees, reviewer of international Ph.D. theses, and member of national and international projects evaluation panels. She is member of the Editorial Committees of 3 books and 8 journals, and collaborates as a reviewer for more than 90 journals.

Cristina Martínez-Villaluenga (Ph.D.), B.S. in Biology by University Complutense of Madrid in 2001, Ph.D. in Food Science from the University Autonoma of Madrid in 2006. She joined the Spanish Research Council (CSIC) in 2009. The long-term goal of Dr. Martinez's research program is to enhance the health of individuals by identifying and determining the benefits of the bioactive components of plant foods with special focus on bioactive peptides. Dr. Martínez's research on legumes, cereals, and pseudocereals has led to increased understanding of the anti-inflammatory, anti-hypertensive, anti-diabetic, and other physiological properties of these foods. She is the author of 94 JCR articles and 9 book chapters with an h-index 31 (WoS). Her results have been disseminated in 84 international and national conferences and social media. In the last 10 years, she has supervised a total of 9 Ph.D. theses, 5 Masters' theses, and more than 20 undergraduate students. She has participated in a total of 35 international and national R&D projects and contracts with the agri-food sector. She is the member of the Editorial Committees of 3 books and 3 journals.

International Journal of
Molecular Sciences

MDPI

Editorial

Peptides for Health Benefits 2019

Cristina Martínez-Villaluenga [1] and Blanca Hernández-Ledesma [2,*]

1 Institute of Food Science, Technology and Nutrition (ICTAN-CSIC), Juan de la Cierva 3, 28006 Madrid, Spain; c.m.villaluenga@csic.es
2 Institute of Food Science Research (CIAL, CSIC-UAM, CEI UAM+CSIC), Nicolás Cabrera, 9, 28049 Madrid, Spain
* Correspondence: b.hernandez@csic.es

Received: 31 March 2020; Accepted: 2 April 2020; Published: 6 April 2020

In recent years, peptides have received increased interest in pharmaceutical, food, cosmetics and various other fields. The high potency, specificity and good safety profile are the main strengths of bioactive peptides as new and promising therapies that may fill the gap between small molecules and protein drugs. Peptides possess favorable tissue penetration and the capability to engage into specific and high-affinity interactions with endogenous receptors. These positive attributes of peptides have driven research in evaluating peptides as versatile tools for drug discovery and delivery. In addition, among bioactive peptides, those released from food protein sources have acquired importance as active components in functional foods and nutraceuticals because they are known to possess regulatory functions that can lead to health benefits.

This Special Issue of *International Journal of Molecular Sciences* represents the second in a series dedicated to peptides. This issue includes thirty-six outstanding papers describing examples of the most recent advances in peptide research and its applicability.

The Special Issue begins with a group of papers exploring aspects of synthetic peptides that are of significance to develop novel drugs for controlling and/or managing chronic diseases. It begins with a study of Gaglione et al. [1] on the identification of three cryptides in human apolipoprotein B and evaluation of their antimicrobial and anti-biofilm properties individually or in combination with ciprofloxacin towards *Pseudomonas* and *Burkholderia* strains clinically isolated from cystic fibrosis patients. These findings will open interesting perspectives to apoB cryptides applicability in the treatment of chronic lung infections associated with cystic fibrosis disease. The issue follows with research by Tarallo et al. [2] on a new tetrameric tripeptide inhibitor of vascular endothelial growth factor receptor 1 that exerts anti-angiogenic activity at ocular level by oral delivery in a preclinical model of age-related macular degeneration. Asai et al. [3] demonstrate that Pro-Hyp and Hyp-Gly play crucial roles in proliferation of fibroblasts attached on collagen gel. Russjan and Kaczynska [4] investigate the beneficial effects of neurotensin in murine model of hapten-induced asthma. In another paper, Russjan et al. [5] investigate the anti-inflammatory potency of hybrid peptide-PK20, composed of neurotensin and endomorphin-2 pharmacophores in a mouse model of non-allergic asthma. Improved anti-inflammatory potency of the hybrid over the mixture of its moieties shows potential as a promising tool in modulating airway inflammation in asthma. Pershina et al. [6] study the gender specific effects of a pegylated glucagon-like peptide 1 (GLP-1), used in the treatment regime for metabolic disorder and chronic obstructive pulmonary disease. Oludiran et al. [7] demonstrate that potency of antimicrobial piscidin peptides depends on environmental oxygen, therefore, the development of pharmaceuticals from host-defense peptides such as piscidin will necessitate consideration of oxygen levels in the targeted tissue. The chemokine-like activity of the synthetic dipeptide pidotimod is studied by Caccuri et al. [8]. The study also defines the mechanism of action for chemokine-like activity of pidotimod and points on the possible role that this synthetic dipeptide may play in leukocyte trafficking and function. The potency, toxicity and mechanisms of action of Ps-K18 is examined by Jang et al. [9] aiming to develop antibiotics derived from bioactive peptides for the treatment of Gram-negative sepsis. Golda et al. [10] screen a library of synthetic peptides to identify those with antibacterial potential against multidrug-resistant *Staphylococcus aureus*. The bactericidal and keratinocytes cytoprotective mechanisms against invading bacteria are also elucidated. *Staphylococcus aureus* and *Pseudomonas aeruginosa*, individually or in co-occurrence, are the two main pathogens implied in multiple bacterial infections. Since their discovery, the antimicrobial peptides (AMPs) of innate defense have been considered as a potential alternative to conventional antibiotics. However, no commercial AMPs are still available. The review of Rončević et al. [11] is aimed at describing these peptides, their mechanisms of action, their biological and biophysical properties as well as the developed models to design and produce new molecules with high antimicrobial potency and low toxicity. Intragenic antimicrobial peptide Hs02 is demonstrated by Bessa et al. [12] to exert antimicrobial properties against *Pseudomonas aeruginosa* and *Staphylococcus aureus*, also hampering the proliferation of their single and dual-species biofilms. The study of Prasad et al. [13] reviews the role of host defense peptides in different inflammatory conditions and diseases, associating this role with the physicochemical properties of peptides and their interaction with various receptors that define their immunomodulatory effects. In another paper, Nácher-Juan et al. [14] investigate the role of peptide osteostatin derived from parathyroid hormone-related protein against rheumatoid arthritis. This peptide, administered to collagen-induced arthritic mice, decreases the severity of the disease through modulation of immune and inflammatory biomarkers. Palus et al. [15] report the neurothropic and/or neuroprotective properties of galanin, alone or in combination with other neuroactive substances such as vasoactive intestinal peptide, neuronal nitric oxide synthase and cocaine- and amphetamine-regulated transcript peptide in the recovery processes in the stomach enteric nervous system neurons following acrylamide intoxication. The extra domain B of fibronectin (EDB-FN) localized in the extracellular matrix can differentiate aggressive prostate cancer from benign prostatic hyperplasia. Park et al. [16] synthesize two cyclic peptides, CTVRTSADC and KTVRTSADE with ability to target EDB-FN, and develop different conjugates with anticancer drugs docetaxel and doxorubicin. The conjugates show selective cytotoxic effects against prostate cancer cells without affecting normal prostate cells.

Following, there is a short series of articles dealing with the elucidation of modes of action of known food-derived bioactive peptides. Fernández-Tomé et al. [17] provide new evidence on the chemopreventive activity of peptide lunasin, a bioactive peptide from soybean and other vegetal sources, on colorectal cancer by modulating both the parental and the tumorsphere-derived subsets of HCT-116 cells. The underlying molecular mechanisms behind the inhibitory effects of lunasin on cell cycle progress of colon cancer cells and cytotoxicity were also discussed. Martínez-Sánchez et al. [18] describe the beneficial effects of dry-cured ham peptides previously identified to prevent from endothelial dysfunction and inflammation. In silico dockings show the predicted modes of binding of four bioactive peptides with the regulatory subunit NEMO of the NF-κB transcription factor and angiotensin I converting-enzyme.

Another group of papers explores the potential of new proteins as sources of bioactive peptides. Cai et al. [19] explore the cytoprotective mechanism of antioxidant pentapeptides from a protein hydrolysate of miiuy croaker (*Miichthys miiuy*) swim bladder against oxidative damage to human umbilical vein endothelial cells. Gomez et al. [20] report on the potential bioactivities of Portuguese oyster (*Crassostrea angulata*) proteins through in silico analyses and in vitro tests. *C. angulata* proteins were proven to be sources of angiotensin I-converting enzyme and dipeptidyl peptidase IV inhibitory peptides with pharmaceutical and nutraceutical applications. Using different commercial proteases, Ding et al. [21] produce hydrolysates from velvet antler with antioxidant properties. The protective effect against oxidative stress of a tetrapeptide produced by Alcalase is investigated in Chang liver cells and a zebrafish model. León-Lopez et al. [22] describe the biochemical, structure and physico-chemical features as well as the antioxidant activity of collagen hydrolysates from sheepskins. A soybean product obtained after combined hydrolysis with Prozyme and fermentation with *Lactobacillus rhamnosus* EBD1 by Daliri et al. [23] show antihypertensive properties in both in vitro and in vivo models, without losing its activity after simulating its digestion by gastrointestinal enzymes. Peptides PPNNNPASPSFSSSS, GPKALPII and IIRCTGC, in which angiotensin-converting enzyme inhibitory activity had been previously demonstrated, are included in the soy product. Another review of Brady et al. [24] summarizes the antibacterial and anti-inflammatory activities of cecropins, a group of naturally occurring antimicrobial peptides found in insects. The strategies designed to overcome the existing limitations linked to their costly large-scale production and their use as therapeutic agents are also described.

The issue includes some studies on bioinformatic and proteomic tools useful for peptide research. Using molecular docking, Chamata et al. [25] describe the structure-activity relationships of peptide sequences present in whey/milk protein hydrolysates with high angiotensin converting enzyme inhibitory activity to a better understanding and prediction of their in vivo antihypertensive activity. Minkiewicz et al. [26] review the new opportunities offered by the BIOPEP-UWM database of bioactive peptides that include the possibility of annotating peptides containing D-enantiomers of amino acids, batch processing option, converting amino acid sequences into SMILES code, new quantitative parameters characterizing the presence of bioactive fragments in protein sequences and finding proteinases that release particular peptides. Using yeast proteome microarrays, Shah et al. [27] identify a total of 140 and 137 intracellular protein targets of antifungal peptides of Lactoferricin B and Histatin-5, respectively. The usefulness of this proteomic tool to find synergistic actions of bioactive peptides is also addressed. The in silico analysis carried out by Tejano et al. [28] reveal the role of *Chlorella sorokiniana* proteins as source of bioactive peptides. The BIOPEP's profile shows that these proteins have multiple dipeptydil peptidase IV inhibitors, glucose uptake stimulants, antioxidant, regulating, anti-amnestic and anti-thrombotic peptides. Pepsin, bromelain and papain are the main proteases responsible for the release of bioactive peptides with pharmaceutical and nutraceutical potential. The review of Bozovičar and Bratkovic [29] focuses on recombinant peptide libraries useful for pharmaceutical industry in the drug discovery and delivery. These authors discuss different platforms for the display and/or expression of bioactive peptides as well as various diversification strategies for library design.

Another group of papers explores the effects of endogenous peptides on body functions and their potential for new drug alternatives. In a glioma mouse model, Kucheryavykh et al. [30] reveal by ELISA and immunofluorescence images that innate amyloid beta (Aβ) peptide is accumulated in glioma tumors and nearby blood vessels. Interestingly, the amyloidogenic Aβ peptide is co-localized with the lipid-free apolipoprotein E (apoE) in amyloid plaques in Alzheimer's disease, where the apoE4 isoform plays a crucial role for the late onset disorder. In the study of Tsiolaki et al. [31], apoE peptide-analogues serve to predict the dynamics of apoE and apoE-Aβ complexes. The homeostasis of the organism is maintained by coordinated neuroendocrine and immune systems. Vasoactive intestinal peptide (VIP) is an endogenous neuropeptide produced by both neurons and endocrine and immune cells. Martínez et al. [32] review the biology of VIP and VIP receptor's signaling and their protective immunomodulatory effects. The current evidence on strategies improving the stability, selectivity and effectiveness of VIP receptors analogs, the advances on new routes of administration and the potential clinical benefits against inflammatory and autoimmune disorders is described. Another neuropeptide described in the Special Issue is the prolactin-releasing peptide (PrRP). The anorexigenic neuroprotective effects of this peptide are reviewed by Pražienková et al. [33]. These authors also describe its therapeutic potential mediated by its actions on cardiovascular system, pain and stress. G-protein-coupled-seven-transmembrane receptors (GPCRs) are known by their modulatory properties of myeloid cell trafficking in microbial infections, inflammation, immune response and cancer progression. The review of Krepel and Wang [34] shows the existing evidence on one of these receptors from murine origin, called Fpr2, and its endogenous agonist peptide, cathelicidin-related antimicrobial peptide. Both are implied in normal mouse colon epithelial growth, repair and protective actions against inflammation-associated tumorigenesis.

Finally, a couple of articles describe new developed techniques to investigate the response of immune system. Thus, Kametani et al. [35] describe humanized mouse systems possessing immune cells as successful models to in vivo investigate the human immunity and predict the antibody response and immune adverse effects. Similarly, immune responses can be studied using an in situ mayor histocompatibility complex tetramer staining. As described by Abdelaal et al. [36], this technique, combined with immunohistochemistry, is a valuable tool for studying the Ag-specific T cell immune response in tissues. Combined techniques enable determining the localization, abundance and phenotype of T cells and characterizing Ag-specific T cells in specific tissues. Current applications in microbial infections, cancer and autoimmunity are also reviewed.

We wish to thank the invited authors for their interesting and insightful contributions, and look forward to a new set of advances in the bioactive peptides field to be included in the following Special Issue "Peptides for Health Benefits 2020" (https://www.mdpi.com/journal/ijms/special_issues/peptides_2020).

Conflicts of Interest: The authors declare no conflict of interest.

References

1. Gaglione, R.; Cesaro, A.; Dell'Olmo, E.; Di Girolamo, R.; Tartaglione, L.; Pizzo, E.; Arciello, A. Cryptides Identified in Human Apolipoprotein B as New Weapons to Fight Antibiotic Resistance in Cystic Fibrosis Disease. *Int. J. Mol. Sci.* **2020**, *21*, 2049. [CrossRef] [PubMed]
2. Tarallo, V.; Iaccarino, E.; Cicatiello, V.; Sanna, R.; Ruvo, M.; De Falco, S. Oral Delivery of a Tetrameric Tripeptide Inhibitor of VEGFR1 Suppresses Pathological Choroid Neovascularization. *Int. J. Mol. Sci.* **2020**, *21*, 410. [CrossRef] [PubMed]
3. Asai, T.T.; Oikawa, F.; Yoshikawa, K.; Inoue, N.; Sato, K. Food-Derived Collagen Peptides, Prolyl-Hydroxyproline (Pro-Hyp), and Hydroxyprolyl-Glycine (Hyp-Gly) Enhance Growth of Primary Cultured Mouse Skin Fibroblast Using Fetal Bovine Serum Free from Hydroxyprolyl Peptide. *Int. J. Mol. Sci.* **2020**, *21*, 229. [CrossRef] [PubMed]
4. Russjan, E.; Kaczyńska, K. Beneficial Effects of Neurotensin in Murine Model of Hapten-Induced Asthma. *Int. J. Mol. Sci.* **2019**, *20*, 5025. [CrossRef]
5. Russjan, E.; Andrzejewski, K.; Sulejczak, D.; Kleczkowska, P.; Kaczyńska, K. Endomorphin-2- and Neurotensin-Based Chimeric Peptide Attenuates Airway Inflammation in Mouse Model of Nonallergic Asthma. *Int. J. Mol. Sci.* **2019**, *20*, 5935. [CrossRef]
6. Pershina, O.V.; Pakhomova, A.V.; Widera, D.; Ermakova, N.N.; Epanchintsev, A.A.; Pan, E.S.; Krupin, V.A.; Vaizova, O.E.; Putrova, O.D.; Sandrikina, L.A.; et al. Gender Differences in the Pharmacological Actions of Pegylated Glucagon-Like Peptide-1 on Endothelial Progenitor Cells and Angiogenic Precursor Cells in a Combination of Metabolic Disorders and Lung Emphysema. *Int. J. Mol. Sci.* **2019**, *20*, 5414. [CrossRef]
7. Oludiran, A.; Courson, D.S.; Stuart, M.D.; Radwan, A.R.; Poutsma, J.C.; Cotten, M.L.; Purcell, E.B. How Oxygen Availability Affects the Antimicrobial Efficacy of Host Defense Peptides: Lessons Learned from Studying the Copper-Binding Peptides Piscidins 1 and 3. *Int. J. Mol. Sci.* **2019**, *20*, 5289. [CrossRef]
8. Caccuri, F.; Bugatti, A.; Corbellini, S.; Roversi, S.; Zani, A.; Mazzuca, P.; Marsico, S.; Caruso, A.; Giagulli, C. The Synthetic Dipeptide Pidotimod Shows a Chemokine-Like Activity through CXC Chemokine Receptor 3 (CXCR3). *Int. J. Mol. Sci.* **2019**, *20*, 5287. [CrossRef]
9. Jang, M.; Kim, J.; Choi, Y.; Bang, J.; Kim, Y. Antiseptic Effect of Ps-K18: Mechanism of Its Antibacterial and Anti-Inflammatory Activities. *Int. J. Mol. Sci.* **2019**, *20*, 4895. [CrossRef]
10. Golda, A.; Kosikowska-Adamus, P.; Kret, A.; Babyak, O.; Wójcik, K.; Dobosz, E.; Potempa, J.; Lesner, A.; Koziel, J. The Bactericidal Activity of Temporin Analogues Against Methicillin Resistant *Staphylococcus aureus*. *Int. J. Mol. Sci.* **2019**, *20*, 4761. [CrossRef]
11. Rončević, T.; Puizina, J.; Tossi, A. Antimicrobial Peptides as Anti-Infective Agents in Pre-Post-Antibiotic Era? *Int. J. Mol. Sci.* **2019**, *20*, 5713. [CrossRef] [PubMed]

12. Bessa, L.J.; Manickchand, J.R.; Eaton, P.; Leite, J.R.S.A.; Brand, G.D.; Gameiro, P. Intragenic Antimicrobial Peptide Hs02 Hampers the Proliferation of Single- and Dual-Species Biofilms of *P. aeruginosa* and *S. aureus*: A Promising Agent for Mitigation of Biofilm-Associated Infections. *Int. J. Mol. Sci.* **2019**, *20*, 3604. [CrossRef] [PubMed]
13. Prasad, S.V.; Fiedoruk, K.; Daniluk, T.; Piktel, E.; Bucki, R. Expression and Function of Host Defense Peptides at Inflammation Sites. *Int. J. Mol. Sci.* **2020**, *21*, 104. [CrossRef] [PubMed]
14. Nácher-Juan, J.; Terencio, M.C.; Alcaraz, M.J.; Ferrándiz, M.L. Osteostatin Inhibits Collagen-Induced Arthritis by Regulation of Immune Activation, Pro-Inflammatory Cytokines, and Osteoclastogenesis. *Int. J. Mol. Sci.* **2019**, *20*, 3845. [CrossRef] [PubMed]
15. Palus, K.; Makowska, K.; Całka, J. Alterations in Galanin-Like Immunoreactivity in the Enteric Nervous System of the Porcine Stomach Following Acrylamide Supplementation. *Int. J. Mol. Sci.* **2019**, *20*, 3345. [CrossRef] [PubMed]
16. Park, S.E.; Shamloo, K.; Kristedja, T.A.; Darwish, S.; Bisoffi, M.; Parang, K.; Tiwari, R.K. EDB-FN Targeted Peptide–Drug Conjugates for Use against Prostate Cancer. *Int. J. Mol. Sci.* **2019**, *20*, 3291. [CrossRef]
17. Fernández-Tomé, S.; Xu, F.; Han, Y.; Hernández-Ledesma, B.; Xiao, H. Inhibitory Effects of Peptide Lunasin in Colorectal Cancer HCT-116 Cells and Their Tumorsphere-Derived Subpopulation. *Int. J. Mol. Sci.* **2020**, *21*, 537. [CrossRef]
18. Martínez-Sánchez, S.M.; Pérez-Sánchez, H.; Antonio Gabaldón, J.; Abellán-Alemán, J.; Montoro-García, S. Multifunctional Peptides from Spanish Dry-Cured Pork Ham: Endothelial Responses and Molecular Modeling Studies. *Int. J. Mol. Sci.* **2019**, *20*, 4204. [CrossRef]
19. Cai, S.Y.; Wang, Y.M.; Zhao, Y.Q.; Chi, C.F.; Wang, B. Cytoprotective Effect of Antioxidant Pentapeptides from the Protein Hydrolysate of Swim Bladders of Miiuy Croaker (*Miichthys miiuy*) against H_2O_2-Mediated Human Umbilical Vein Endothelial Cell (HUVEC) Injury. *Int. J. Mol. Sci.* **2019**, *20*, 5425. [CrossRef]
20. Gomez, H.L.R.; Peralta, J.P.; Tejano, L.A.; Chang, Y.W. In Silico and In Vitro Assessment of Portuguese Oyster (*Crassostrea angulata*) Proteins as Precursor of Bioactive Peptides. *Int. J. Mol. Sci.* **2019**, *20*, 5191. [CrossRef]
21. Ding, Y.; Ko, S.C.; Moon, S.H.; Lee, S.H. Protective Effects of Novel Antioxidant Peptide Purified from Alcalase Hydrolysate of Velvet Antler Against Oxidative Stress in Chang Liver Cells In Vitro and in a Zebrafish Model In Vivo. *Int. J. Mol. Sci.* **2019**, *20*, 5187. [CrossRef] [PubMed]
22. León-López, A.; Fuentes-Jiménez, L.; Hernández-Fuentes, A.D.; Campos-Montiel, R.G.; Aguirre-Álvarez, G. Hydrolysed Collagen from Sheepskins as a Source of Functional Peptides with Antioxidant Activity. *Int. J. Mol. Sci.* **2019**, *20*, 3931. [CrossRef] [PubMed]
23. Daliri, E.B.M.; Ofosu, F.K.; Chelliah, R.; Park, M.H.; Kim, J.H.; Oh, D.H. Development of a Soy Protein Hydrolysate with an Antihypertensive Effect. *Int. J. Mol. Sci.* **2019**, *20*, 1496. [CrossRef] [PubMed]
24. Brady, D.; Grapputo, A.; Romoli, O.; Sandrelli, F. Insect Cecropins, Antimicrobial Peptides with Potential Therapeutic Applications. *Int. J. Mol. Sci.* **2019**, *20*, 5862. [CrossRef]
25. Chamata, Y.; Watson, K.A.; Jauregi, P. Whey-Derived Peptides Interactions with ACE by Molecular Docking as a Potential Predictive Tool of Natural ACE Inhibitors. *Int. J. Mol. Sci.* **2020**, *21*, 864. [CrossRef]
26. Minkiewicz, P.; Iwaniak, A.; Darewicz, M. BIOPEP-UWM Database of Bioactive Peptides: Current Opportunities. *Int. J. Mol. Sci.* **2019**, *20*, 5978. [CrossRef]
27. Shah, P.; Wu, W.S.; Chen, C.S. Systematical Analysis of the Protein Targets of Lactoferricin B and Histatin-5 Using Yeast Proteome Microarrays. *Int. J. Mol. Sci.* **2019**, *20*, 4218. [CrossRef]
28. Tejano, L.A.; Peralta, J.P.; Yap, E.E.S.; Panjaitan, F.C.A.; Chang, Y.W. Prediction of Bioactive Peptides from *Chlorella sorokiniana* Proteins Using Proteomic Techniques in Combination with Bioinformatics Analyses. *Int. J. Mol. Sci.* **2019**, *20*, 1786. [CrossRef]
29. Bozovičar, K.; Bratkovič, T. Evolving a Peptide: Library Platforms and Diversification Strategies. *Int. J. Mol. Sci.* **2020**, *21*, 215. [CrossRef]
30. Kucheryavykh, L.Y.; Ortiz-Rivera, J.; Kucheryavykh, Y.V.; Zayas-Santiago, A.; Diaz-Garcia, A.; Inyushin, M.Y. Accumulation of Innate Amyloid Beta Peptide in Glioblastoma Tumors. *Int. J. Mol. Sci.* **2019**, *20*, 2482. [CrossRef]
31. Tsiolaki, P.L.; Katsafana, A.D.; Baltoumas, F.A.; Louros, N.N.; Iconomidou, V.A. Hidden Aggregation Hot-Spots on Human Apolipoprotein E: A Structural Study. *Int. J. Mol. Sci.* **2019**, *20*, 2274. [CrossRef] [PubMed]

32. Martínez, C.; Juarranz, Y.; Gutiérrez-Cañas, I.; Carrión, M.; Pérez-García, S.; Villanueva-Romero, R.; Castro, D.; Lamana, A.; Mellado, M.; González-Álvaro, I.; et al. A Clinical Approach for the Use of VIP Axis in Inflammatory and Autoimmune Diseases. *Int. J. Mol. Sci.* **2020**, *21*, 65. [CrossRef] [PubMed]
33. Pražienková, V.; Popelová, A.; Kuneš, J.; Maletínská, L. Prolactin-Releasing Peptide: Physiological and Pharmacological Properties. *Int. J. Mol. Sci.* **2019**, *20*, 5297. [CrossRef] [PubMed]
34. Krepel, S.A.; Wang, J.M. Chemotactic Ligands that Activate G-Protein-Coupled Formylpeptide Receptors. *Int. J. Mol. Sci.* **2019**, *20*, 3426. [CrossRef] [PubMed]
35. Kametani, Y.; Ohno, Y.; Ohshima, S.; Tsuda, B.; Yasuda, A.; Seki, T.; Ito, R.; Tokuda, Y. Humanized Mice as an Effective Evaluation System for Peptide Vaccines and Immune Checkpoint Inhibitors. *Int. J. Mol. Sci.* **2019**, *20*, 6337. [CrossRef] [PubMed]
36. Abdelaal, H.M.; Cartwright, E.K.; Skinner, P.J. Detection of Antigen-Specific T Cells Using In Situ MHC Tetramer Staining. *Int. J. Mol. Sci.* **2019**, *20*, 5165. [CrossRef] [PubMed]

International Journal of
Molecular Sciences

MDPI

Article

Cryptides Identified in Human Apolipoprotein B as New Weapons to Fight Antibiotic Resistance in Cystic Fibrosis Disease

Rosa Gaglione [1,3], Angela Cesaro [1], Eliana Dell'Olmo [1], Rocco Di Girolamo [1], Luca Tartaglione [1], Elio Pizzo [2] and Angela Arciello [1,3,*]

1 Department of Chemical Sciences, University of Naples Federico II, Via Cintia 21, I-80126 Naples, Italy; rosa.gaglione@unina.it (R.G.); angela.cesaro@unina.it (A.C.); eliana.dellolmo@unina.it (E.D.); rocco.digirolamo@unina.it (R.D.G.); luca.tartaglione.lt@gmail.com (L.T.)

2 Department of Biology, University of Naples Federico II, Via Cintia 21, I-80126 Naples, Italy; elipizzo@unina.it

3 Istituto Nazionale di Biostrutture e Biosistemi (INBB), I-00136 Rome, Italy

* Correspondence: anarciel@unina.it

Received: 29 February 2020; Accepted: 13 March 2020; Published: 17 March 2020

Abstract: Chronic respiratory infections are the main cause of morbidity and mortality in cystic fibrosis (CF) patients, and are characterized by the development of multidrug resistance (MDR) phenotype and biofilm formation, generally recalcitrant to treatment with conventional antibiotics. Hence, novel effective strategies are urgently needed. Antimicrobial peptides represent new promising therapeutic agents. Here, we analyze for the first time the efficacy of three versions of a cryptide identified in human apolipoprotein B (ApoB, residues 887-922) towards bacterial strains clinically isolated from CF patients. Antimicrobial and anti-biofilm properties of ApoB-derived cryptides have been analyzed by broth microdilution assays, crystal violet assays, confocal laser scanning microscopy and scanning electron microscopy. Cell proliferation assays have been performed to test cryptide effects on human host cells. ApoB-derived cryptides have been found to be endowed with significant antimicrobial and anti-biofilm properties towards *Pseudomonas* and *Burkholderia* strains clinically isolated from CF patients. Peptides have been also found to be able to act in combination with the antibiotic ciprofloxacin, and they are harmless when tested on human bronchial epithelial mesothelial cells. These findings open interesting perspectives to cryptide applicability in the treatment of chronic lung infections associated with CF disease.

Keywords: antibiotic resistance; cystic fibrosis; antimicrobial peptides; host defense peptides; cryptides; anti-biofilm peptides; synergistic effects

1. Introduction

Cystic fibrosis (CF) is a rare autosomal recessive disease affecting 1 in 2500 newborns in Europe [1]. More than 2000 mutations have been identified in the Cystic Fibrosis Transmembrane conductance Regulator (CFTR) gene and have been associated with the disease. CFTR gene encodes a chloride ion channel whose malfunctioning causes the production of viscous secretions coating the airway epithelia [2,3]. This phenomenon is responsible for the accumulation of trapped microbes, including *Pseudomonas aeruginosa*, with consequent deterioration of lung tissue and impairment of respiratory functions [4]. Indeed, chronic respiratory infections and inflammation are the main causes of death in CF [5]. Despite aggressive antibiotic treatments, *Pseudomonas* strains often grow in CF lungs and lead to chronic and recalcitrant infections characterized by a robust host inflammatory response [6,7]. Pulmonary infections due to the Gram-negative *P. aeruginosa* strain are the main cause of lung decline and death in patients suffering from CF [8–10]. *P. aeruginosa* colonization of host tissues is mediated by an initial attachment of bacteria to epithelial cells [11,12], followed by internalization into cells [13–16]. This phenomenon protects bacteria from host defense mechanisms and from the killing action of conventional antibiotics that hardly enter epithelial cells [17]. This is generally responsible for systemic diffusion of bacteria and for the consequent chronic nature of *P. aeruginosa* lung infections [18]. Moreover, chronic inflammation and mucus provide an environment favorable to the development of resistance phenotype for bacteria in biofilms, thus hampering antibiotic efficacy [19]. *Burkholderia* species also cause serious challenges in CF patients, even if infections associated with these strains are relatively rare [20]. Indeed, a main post-transplant complication is represented by infections caused by multidrug resistant (MDR) bacteria, with the *Burkholderia* species recognized as significant contributors to CF morbidity and mortality associated with increased post-transplant death rate [21,22]. Conventional antibiotics generally appear ineffective and their prolonged use is responsible for the development of the MDR phenotype. The concomitant decrease in the pharmaceutical industry research pipeline for novel antimicrobial agents during the last three decades has, thus, resulted in an urgent need for the discovery of novel effective antimicrobial strategies [23]. In this scenario, naturally occurring antimicrobial peptides (AMPs), or their derivatives, represent an appealing source for the generation of new therapeutic agents able to treat chronic MDR bacterial infections [24,25]. AMPs are produced by all living organisms as the first line of defense against invading microorganisms, and the majority of them are characterized by net positive charge at neutral pH and by the tendency to form amphipathic structures in a hydrophobic environment [26,27]. So far, hundreds of naturally occurring AMPs have been isolated and characterized as highly efficacious, safe, and tolerable antimicrobials [28,29]. Being able to selectively interact with bacterial cytoplasmic membranes in a manner not dependent upon specific receptors, AMPs are generally endowed with broad-spectrum antimicrobial activity [30,31], and several of them have been reported to combat biofilms because of their potent bactericidal activity and their ability to first penetrate and then to disorganize biofilm structures [32]. Furthermore, AMPs frequently synergize with antimicrobial compounds to repress molecular pathways leading to biofilm development [32]. Here, we analyze for the first time the antimicrobial and anti-biofilm properties of two recently characterized AMPs [33] towards *Pseudomonas* and *Burkholderia* strains clinically isolated from CF patients. AMPs under test have been identified in human apolipoprotein B (ApoB) by using a bioinformatic method developed by our research group [33–41]. Indeed, it is increasingly evident that eukaryotic proteins, with functions not necessarily related to host defense, act as sources of "cryptic" bioactive peptides released upon proteolytic processing by bacterial and/or host proteases [42–44]. We previously characterized two variants of the cryptide identified in human ApoB (residues 887–922), i.e., peptides ApoB887–923 and ApoB887–911 [33]. These two host defense peptides (HDPs), recombinantly produced in bacterial cells, have been here named r(P)ApoB$_L^{Pro}$ and r(P)ApoB$_S^{Pro}$ because of the presence of a Pro residue becoming the N-terminus of the peptides released by the acidic cleavage of an Asp-Pro bond [33,36]. Here, we also characterized a further peptide, i.e., a version of the longest peptide characterized by the presence of an Ala residue instead of a Pro residue in position six of peptide sequence, here named r(P)ApoB$_L^{Ala}$. Peptides r(P)ApoB$_L^{Pro}$ and r(P)ApoB$_S^{Pro}$ have been previously found to be endowed with antimicrobial, anti-biofilm, wound healing and immunomodulatory properties, and are able to synergistically act in combination with either conventional antibiotics or EDTA [33]. On the other hand, peptides have been found to be neither toxic nor hemolytic towards mammalian cells [33]. It has been also demonstrated that electrostatic interactions between negatively charged bacterial membranes and positively charged ApoB-derived AMPs play a key role in mediating peptide toxicity, although they are strongly influenced by the composition of negatively charged bacterial surfaces and by defined extracellular microenvironments [35]. Here, we demonstrate that the three ApoB-derived cryptides exert significant antimicrobial and anti-biofilm effects towards *Pseudomonas* and *Burkholderia* strains clinically isolated from CF patients and that they are able to act in combination with the ciprofloxacin antibiotic, widely used to treat chronic lung infections in CF patients [45]. Furthermore, ApoB-derived cryptides have been found to be not toxic when tested on human bronchial epithelial mesothelial cells. Altogether, these findings open interesting perspectives to peptide applicability, suggesting the possibility to develop in the future successful combinatorial therapeutic approaches, based on the concomitant administration of AMPs and conventional antibiotics, with a consequently very low potential to induce a resistance phenotype.

2. Results

2.1. *Evaluation of ApoB-Derived Peptide Effects on Clinically Isolated Baterial Strains*

In order to evaluate the ApoB-derived peptide ability to counteract microbial infections in CF, their effects were tested on six clinically isolated bacterial strains, i.e., *P. aeruginosa* RP 73, *P. aeruginosa* KK 27, *P. aeruginosa* 14, *P. aeruginosa* AA2, *Burkholderia multivorans* LMG 17582, and *Burkholderia cenocepacia* LMG 18863. To this purpose, the susceptibility of planktonic bacteria to ApoB-derived peptides was examined by using broth microdilution method [33] that allows the measurement of minimum inhibitory concentration (MIC) values. As reported in Table 1, the three ApoB-derived peptides under test were found to exert antimicrobial effects on three out of six bacterial strains tested. In particular, bacterial strains *P. aeruginosa* RP 73, *P. aeruginosa* KK 27, and *B. multivorans* LMG 17582 were found to be susceptible to ApoB-derived peptide antimicrobial activity, with MIC_{100} values ranging from 5 to 40 μM (Table 1). Peptide r(P)ApoB$_L$^{Ala} was found to be the most active in directly killing bacterial cells (Table 1).

Table 1. Minimum inhibitory concentration (MIC) values determined for r(P)ApoB$_L$^{Pro}, r(P)ApoB$_L$^{Ala} and r(P)ApoB$_S$^{Pro} tested on clinically isolated bacterial strains.

	MIC_{100} (μM)		
	r(P)ApoB$_L$^{Pro}	r(P)ApoB$_L$^{Ala}	r(P)ApoB$_S$^{Pro}
P. aeruginosa RP 73	10–20	5–10	20–40
P. aeruginosa 14	>40	>40	>40
P. aeruginosa AA2	>40	>40	>40
P. aeruginosa KK 27	20–40	10–20	20–40
Burkholderia cenocepacia LMG 18863	>40	>40	>40
Burkholderia multivorans LMG 17582	10–20	10–20	20–40

2.2. *Evaluation of ApoB-Derived Peptide Anti-biofilm Activity on Clinically Isolated Baterial Strains*

2.2.1. Evaluation of ApoB-Derived Peptide Anti-Biofilm Activity by Microtiter Plate Assay

To evaluate whether recombinant ApoB-derived peptides are endowed with anti-biofilm activity, analyses were performed on clinically isolated bacterial strains *P. aeruginosa* RP 73, *P. aeruginosa* KK 27, *P. aeruginosa* 14, *P. aeruginosa* AA2, *B. multivorans* LMG 17582, and *B. cenocepacia* LMG 18863 in 0.5X Mueller Hinton Broth (MHB). By following different experimental approaches, peptide effects were tested on the three main stages of biofilm development, such as attachment, formation and detachment. To test peptide effects on biofilm attachment, following overnight growth, a bacterial culture was diluted into MHB medium containing increasing concentrations of the peptide under test (0–40 μM), and incubated for 4 h at 37 °C [33]. When, instead, peptide effects were tested on biofilm formation, the experimental procedure described above was followed with the only exception that bacterial cells were incubated with increasing concentrations of peptides for 24 h at 37 °C [33]. Finally, the effects of ApoB-derived peptides were tested on biofilm detachment [33]. In each case, following incubation with peptides, biofilm was analyzed by staining with crystal violet. As shown in Figure 1, ApoB-derived peptides have been found to be effective on biofilm attachment, with the greatest effects obtained in the case of *P. aeruginosa* KK 27 and *P. aeruginosa* 14 bacterial strains for all the three peptides under test. In the case of biofilm formation, the greatest effects were found to be exerted by r(P)ApoB$_L$^{Ala} and r(P)ApoB$_S$^{Pro} on *P. aeruginosa* 14 (~50% inhibition) (Figure 1). Even more interestingly, about 30%–40% biofilm eradication was observed in the case of *B. cenocepacia* LMG 18863 upon treatment with 2.5 μM r(P)ApoB$_L$^{Ala} (Figure 1). A similar effect was obtained upon treatment of *P. aeruginosa* KK 27 preformed biofilm with 2.5 μM r(P)ApoB$_S$^{Pro} (Figure 1). Moreover, about 20% biofilm eradication was observed upon treatment of *P. aeruginosa* RP 73 with very low concentrations (1.25 μM) of r(P)ApoB$_S$^{Pro}.

Altogether, obtained data indicate that peptides exert anti-biofilm effects even on bacterial strains not sensitive to their direct antimicrobial activity. In most of the cases, significant anti-biofilm effects were detected at peptide concentrations (1.25–2.5 μM) lower than those required to directly kill planktonic cells (Table 1 and Figure 1). Data reported in Figure 1 represent the mean ± standard deviation (SD) of at least three independent experiments.

2.2.2. Evaluation of ApoB-Derived Peptides Anti-Biofilm Activity by Laser Scanning Confocal Microscopy

In order to further investigate anti-biofilm properties of ApoB-derived peptides, analyses were also performed by confocal laser scanning microscopy (CLSM). For this approach, we selected two bacterial strains not responsive to ApoB-derived peptides direct antimicrobial activity, such as *B. cenocepacia* LMG 18863 and *P. aeruginosa* 14. Peptide effects on biofilm attachment, formation and detachment were evaluated upon sample staining with LIVE/DEAD BacLight bacterial viability kit. Analyses revealed that all three peptides are able to affect biofilm attachment and formation in the case of *B. cenocepacia* LMG 18863 (Figure 2). Even more interestingly, peptides are able to affect pre-formed biofilm with the strongest effects observed in the presence of r(P)ApoB$_L$^{Ala} (Figure 2). By staining bacterial biofilm with SYPRO® ruby dye, which is able to specifically stain biofilm extracellular matrix, the appearance of highly fluorescent aggregates is clearly evident upon treatment with peptides (Figure 2), thus indicating that peptides induce strong alterations of biofilm matrix architecture, as previously reported for different anti-biofilm agents [46]. Similar results were obtained also in the case of *P. aeruginosa* 14 (Figure 3). These findings are also supported by biofilm biovolume determinations by CLSM reported in Figure 4, that indicate a strong and significant effect of r(P)ApoB$_S$^{Pro} peptide on biofilm eradication in the case of both bacterial strains (Figure 4c,f). Furthermore, peptides have been found to exert significant effects on biofilm biovolume when attachment is tested (Figure 4a,d), except for r(P)ApoB$_L$^{Ala} for which no significant reduction in biovolume is observed, although a disaggregating effect is clearly evident (Figures 2 and 3). This might be due to the fact that, upon treatment with r(P)ApoB$_L$^{Ala} peptide, planktonic cells escape from biofilm by floating, with a consequent significant contribution to biovolume. Altogether, these findings confirm that peptides are able to exert significant anti-biofilm effects even on bacterial strains not sensitive to their direct antimicrobial activity.

2.2.3. Evaluation of ApoB-Derived Peptides Anti-Biofilm Activity by Scanning Electron Microscopy

To analyze morphological modifications of bacterial biofilm upon treatment with peptides, scanning electron microscopy (SEM) analyses were also performed on bacterial strains not responsive to peptide direct antimicrobial activity, such as *B. cenocepacia* LMG 18863 and *P. aeruginosa* 14. In untreated samples, bacteria present smooth and intact surfaces and appear embedded into the extracellular biofilm matrix in the case of both bacterial strains (Figure 5). When bacteria are treated with peptides, instead, a significant decrease or disappearance of biofilm matrix is clearly evident with a concomitant decrease of cell density.

P. aeruginosa RP 73

Assay	Peptide (µM)	$r(P)ApoB_L^{Pro}$ % of biofilm growth	SD	$r(P)ApoB_L^{Ala}$ % of biofilm growth	SD	$r(P)ApoB_S^{Pro}$ % of biofilm growth	SD
Attachment	0	100	0	100	0	100	0
	0.612	85	20	77	3	78	3
	1.25	74	16	67	11	72	3
	2.5	71	9	65	12	65	4
	5	71	9	72	15	65	12
	10	74	9	65	17	66	8
	20	73	7	64	13	66	8
	40	72	10	75	18	65	17
Formation	0	100	0	100	0	100	0
	0.612	72	6	94	14	75	2
	1.25	73	7	95	19	77	8
	2.5	68	0	90	1	85	9
	5	82	5	100	1	78	4
	10	77	10	95	5	89	6
	20	82	7	93	10	89	5
	40	36	5	37	11	79	10
Detachment	0	100	0	100	0	100	0
	0.612	85	4	95	0	76	10
	1.25	82	6	84	7	75	3
	2.5	76	17	82	1	66	11
	5	89	2	92	4	80	16
	10	98	5	84	3	71	13
	20	99	1	88	8	70	25
	40	95	4	83	12	69	14

P. aeruginosa KK 27

Assay	Peptide (µM)	$r(P)ApoB_L^{Pro}$ % of biofilm growth	SD	$r(P)ApoB_L^{Ala}$ % of biofilm growth	SD	$r(P)ApoB_S^{Pro}$ % of biofilm growth	SD
Attachment	0	100	0	100	0	100	0
	0.612	92	4	95	23	101	23
	1.25	75	10	93	14	85	18
	2.5	74	11	67	24	62	8
	5	53	4	46	4	45	0
	10	22	8	20	2	20	4
	20	18	6	21	3	14	4
	40	20	9	20	1	17	7
Formation	0	100	0	100	0	100	0
	0.612	81	4	93	5	84	10
	1.25	75	13	84	18	80	11
	2.5	70	18	80	17	74	11
	5	72	8	79	14	80	4
	10	64	13	69	25	74	21
	20	58	15	67	13	71	6
	40	64	3	29	7	27	3
Detachment	0	100	0	100	0	100	0
	0.612	87	3	91	3	87	9
	1.25	80	15	87	14	78	1
	2.5	88	5	88	9	75	7
	5	80	11	87	8	80	13
	10	79	18	83	7	71	16
	20						
	40						

P. aeruginosa 14

Assay	Peptide (µM)	$r(P)ApoB_L^{Pro}$ % of biofilm growth	SD	$r(P)ApoB_L^{Ala}$ % of biofilm growth	SD	$r(P)ApoB_S^{Pro}$ % of biofilm growth	SD
Attachment	0	100	0	100	0	100	0
	0.612	77	12	98	5	98	5
	1.25	73	20	83	8	83	8
	2.5	75	14	85	5	85	5
	5	70	20	69	17	69	17
	10	55	19	50	18	50	18
	20	40	20	37	12	37	12
	40	33	17	29	8	29	8
Formation	0	100	0	100	0	100	0
	0.612	75	8	81	14	81	14
	1.25	74	7	75	20	75	20
	2.5	71	7	71	26	71	26
	5	76	18	66	28	66	28
	10	62	20	54	28	54	28
	20	33	13	54	32	54	32
	40	35	15	46	26	46	26
Detachment	0	100	0	100	0	100	0
	0.612	94	8	96	6	96	6
	1.25	96	5	95	6	95	6
	2.5	93	4	86	13	86	13
	5	100	1	93	2	93	2
	10	95	7	92	8	92	8
	20	105	13	85	17	85	17
	40	94	3	107	10	107	10

P. aeruginosa AA2

Assay	Peptide (µM)	$r(P)ApoB_L^{Pro}$ % of biofilm growth	SD	$r(P)ApoB_L^{Ala}$ % of biofilm growth	SD	$r(P)ApoB_S^{Pro}$ % of biofilm growth	SD
Attachment	0	100	0	100	0	100	0
	0.612	94	12	95	10	91	13
	1.25	95	11	93	12	88	14
	2.5	99	3	92	14	89	14
	5	98	7	90	16	88	11
	10	97	5	88	19	78	21
	20	94	12	87	19	83	17
	40	87	22	68	21	75	17
Formation	0	100	0	100	0	100	0
	0.612	90	12	88	10	94	5
	1.25	85	10	79	12	94	12
	2.5	86	15	85	2	95	5
	5	84	15	77	11	89	17
	10	85	16	89	6	90	7
	20	87	12	86	9	90	9
	40	100	7	91	0	95	3
Detachment	0	100	0	100	0	100	0
	0.612	103	7	103	2	100	1
	1.25	102	12	100	1	96	5
	2.5	93	1	92	7	96	1
	5	104	8	97	2	97	3
	10	96	3	91	6	94	1
	20	100	4	99	11	102	4
	40	100	3	95	13	101	6

Burkholderia multivorans LMG 17582

Assay	Peptide (µM)	$r(P)ApoB_L^{Pro}$ % of biofilm growth	SD	$r(P)ApoB_L^{Ala}$ % of biofilm growth	SD	$r(P)ApoB_S^{Pro}$ % of biofilm growth	SD
Attachment	0	100	0	100	0	100	0
	0.612	86	8	86	12	80	6
	1.25	74	5	80	20	76	2
	2.5	79	2	77	25	75	5
	5	79	2	77	22	73	5
	10	77	3	80	23	71	11
	20	82	1	79	22	65	11
	40	82	6	81	19	68	7
Formation	0	100	0	100	0	100	0
	0.612	77	16	79	9	77	6
	1.25	67	13	75	7	68	9
	2.5	65	18	75	10	71	17
	5	68	18	73	11	66	19
	10	69	17	75	14	74	21
	20	71	11	77	16	75	19
	40	80	5	83	13	78	11
Detachment	0	100	0	100	0	100	0
	0.612	85	23	86	21	85	25
	1.25	88	17	91	15	82	31
	2.5	100	2	96	7	84	29
	5	92	12	96	7	89	17
	10	93	8	104	6	94	8
	20	93	4	101	4	95	4
	40	93	9	92	9	92	7

Burkholderia cenocepacia LMG 18863

Assay	Peptide (µM)	$r(P)ApoB_L^{Pro}$ % of biofilm growth	SD	$r(P)ApoB_L^{Ala}$ % of biofilm growth	SD	$r(P)ApoB_S^{Pro}$ % of biofilm growth	SD
Attachment	0	100	0	100	0	100	0
	0.612	71	7	63	2	85	4
	1.25	65	5	61	9	64	8
	2.5	67	13	61	11	68	15
	5	71	11	66	7	73	15
	10	75	2	78	17	69	14
	20	69	13	73	5	68	18
	40	69	12	78	3	69	12
Formation	0	100	0	100	0	100	0
	0.612	74	14	68	13	73	8
	1.25	68	13	75	1	66	8
	2.5	71	18	61	6	66	11
	5	78	13	74	0	64	13
	10	74	15	83	6	63	15
	20	73	13	83	3	78	10
	40	67	8	78	11	79	15
Detachment	0	100	0	100	0	100	0
	0.612	92	10	89	15	94	9
	1.25	87	18	87	18	98	3
	2.5	90	8	75	10	105	15
	5	89	17	84	14	89	7
	10	101	2	94	5	89	5
	20	91	7	85	18	86	13
	40	108	17	104	14	84	10

Figure 1. Anti-biofilm activity of $r(P)ApoB_L^{Pro}$, $r(P)ApoB_L^{Ala}$, and $r(P)ApoB_S^{Pro}$ peptides on *P. aeruginosa* RP 73, *P. aeruginosa* KK 27, *P. aeruginosa* 14, *P. aeruginosa* AA2, *B. multivorans* LMG 17582, and *B. cenocepacia* LMG 18863 in MHB medium.

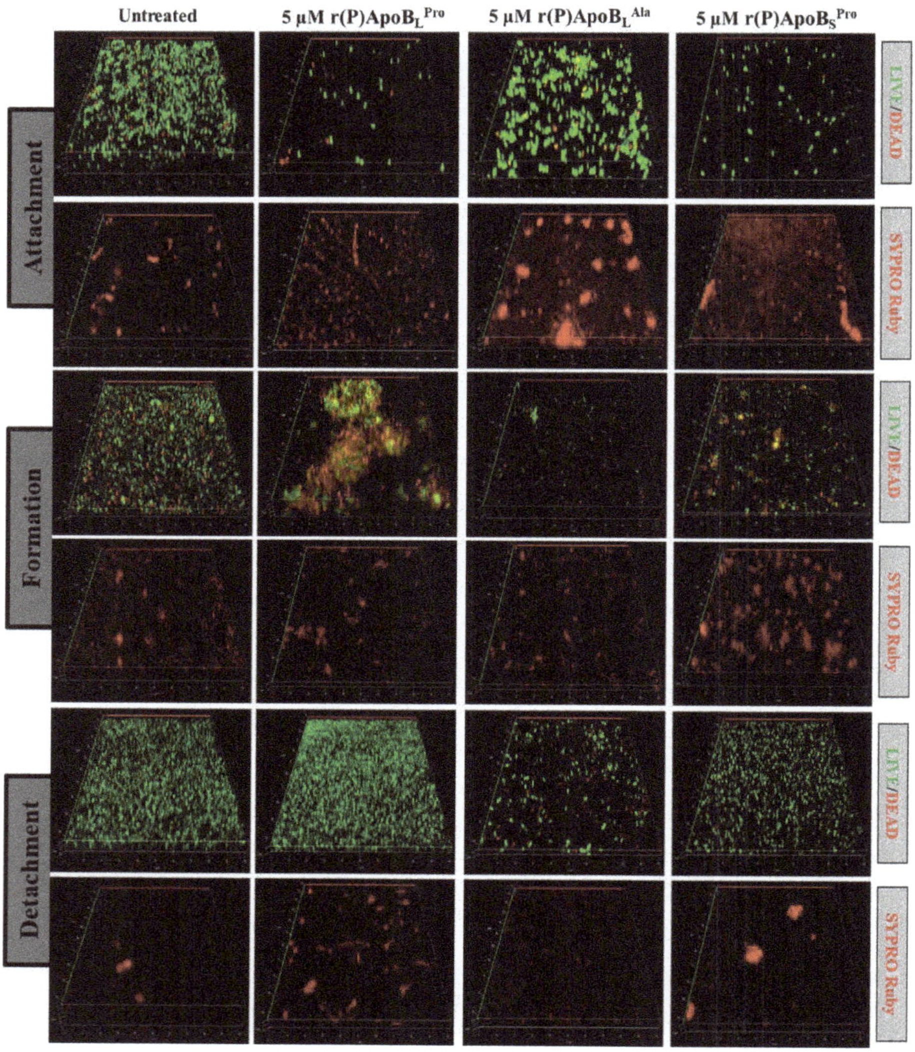

Figure 2. Effects of r(P)ApoB$_L$^{Pro}, r(P)ApoB$_L$^{Ala}, and r(P)ApoB$_S$^{Pro} peptides on *B. cenocepacia* LMG 18863 biofilm attachment, formation and detachment. Biofilm cells were stained by using LIVE/DEAD BacLight bacterial viability kit (Molecular Probes, Eugene, OR, USA) containing 1:1 ratio of Syto-9 (green fluorescence, all cells) and propidium iodide (PI, red fluorescence, dead cells) and FilmTracer™ SYPRO® Ruby biofilm matrix staining (Invitrogen™, F10318). Images are 3D projections of biofilm structure obtained by laser scanning confocal z-stack using Zen Lite 2.3 software. All images were taken under identical conditions.

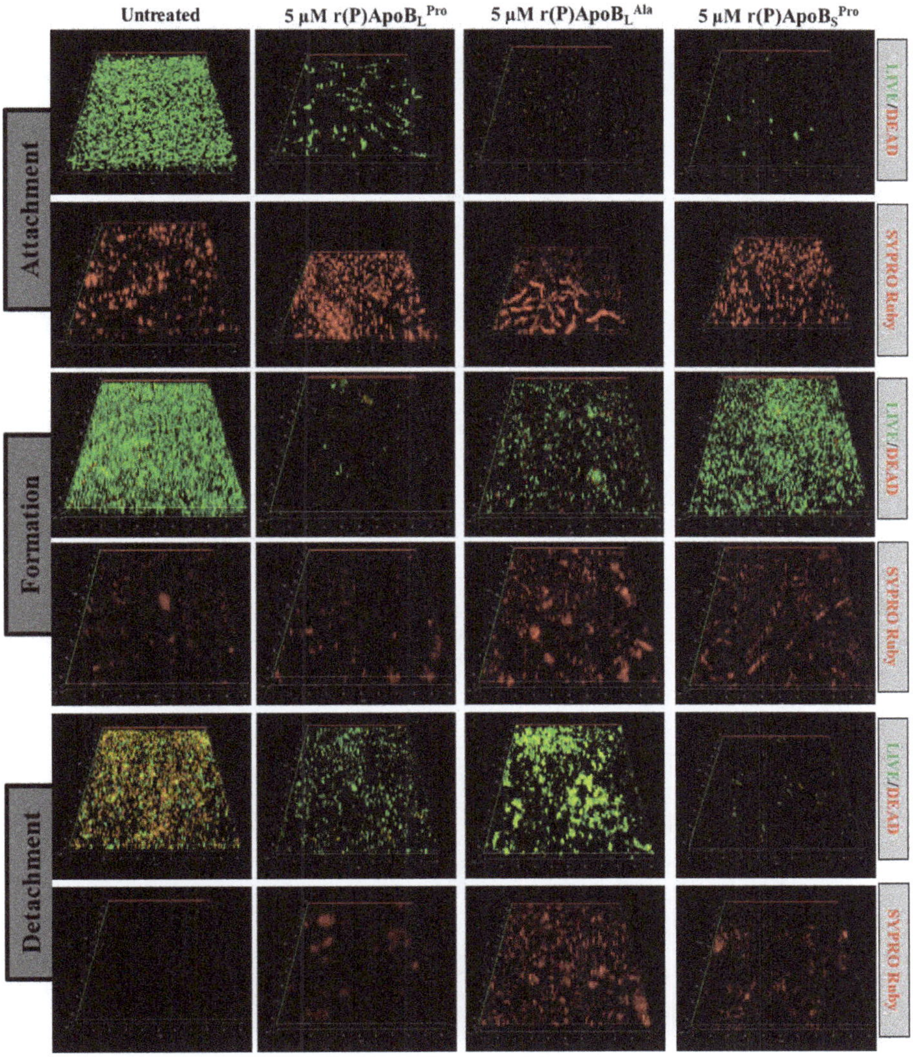

Figure 3. Effects of r(P)ApoB$_L^{Pro}$, r(P)ApoB$_L^{Ala}$ and r(P)ApoB$_S^{Pro}$ peptides on *P. aeruginosa* 14 biofilm attachment, formation and detachment. Biofilm cells were stained by using LIVE/DEAD BacLight bacterial viability kit (Molecular Probes, Eugene, OR, USA) containing 1:1 ratio of Syto-9 (green fluorescence, all cells) and propidium iodide (PI, red fluorescence, dead cells) and FilmTracer™ SYPRO® Ruby biofilm matrix staining (Invitrogen™, F10318). Images are 3D projections of biofilm structure obtained by laser scanning confocal z-stack using Zen Lite 2.3 software. All images were taken under identical conditions.

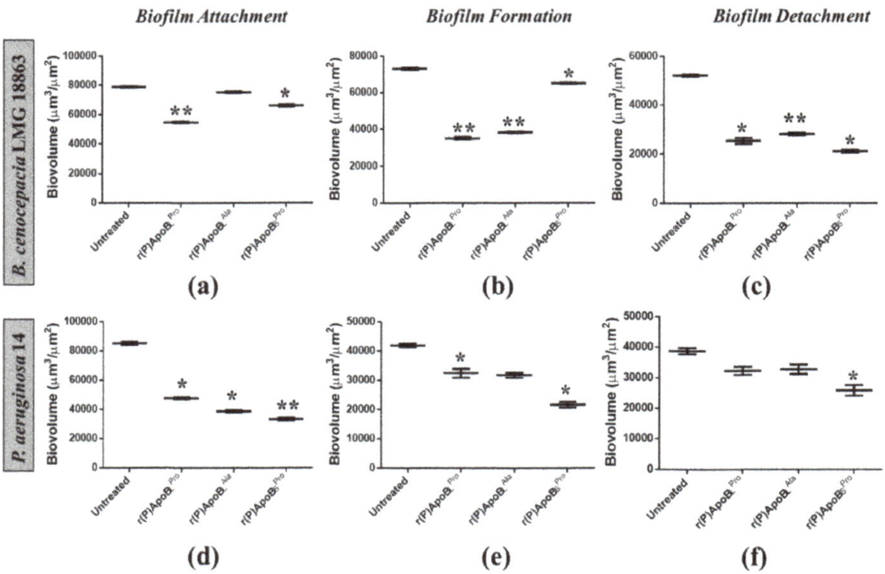

Figure 4. Analysis of the effects of r(P)ApoB$_L$^{Pro}, r(P)ApoB$_L$^{Ala} and r(P)ApoB$_S$^{Pro} peptides on biofilm attachment (**a**,**d**), formation (**b**,**e**) and detachment (**c**,**f**) in the case of *B. cenocepacia* LMG 18863 (**a**–**c**) and *P. aeruginosa* 14 (**d**– **f**). Biovolume (μm^3/μm^2) was measured by using Zen Lite 2.3 software. Significant differences were indicated as * $p < 0.05$ or ** $p < 0.01$ for treated versus control samples.

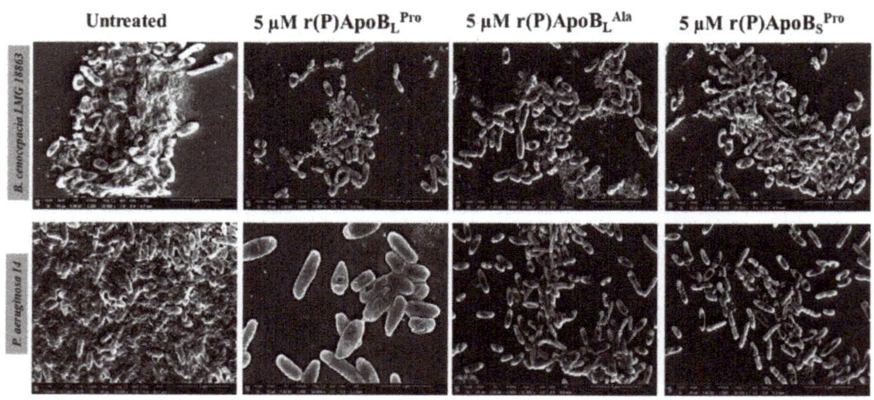

Figure 5. Morphological analyses of *B. cenocepacia* LMG 18863 (top panel) and *P. aeruginosa* 14 (lower panel) preformed biofilms by SEM. Representative images are shown upon treatment of bacterial biofilm with 5 μM r(P)ApoB$_L$^{Pro}, r(P)ApoB$_L$^{Ala}, and r(P)ApoB$_S$^{Pro}. Bars 5 μm.

2.3. Combinatorial Therapeutic Approach

To verify whether ApoB-derived peptides are able to synergistically act in combination with conventional antibiotics to counteract bacterial infections associated with biofilm in CF, CLSM analyses were performed to evaluate the effects of combinations of r(P)ApoB$_L$^{Pro} or r(P)ApoB$_L$^{Ala} and ciprofloxacin on preformed biofilm. Analyses were performed on *B. cenocepacia* LMG 18863 bacterial strain, since chronic lung infections associated with this strain strongly contribute to CF morbidity and mortality and are generally recalcitrant to conventional antibiotics [21,22]. Effects of r(P)ApoB$_L$^{Pro} or r(P)ApoB$_L$^{Ala} peptide in combination with ciprofloxacin were tested on preformed biofilm, in order to better simulate clinical conditions.

As shown in Figure 6, by comparing the effects of combinations of r(P)ApoB$_L$^{Pro} and ciprofloxacin with the effects of single agents on preformed biofilm, a significantly greater reduction of biofilm biovolume is observed in the case of the sample treated with the compound mixture together with a concomitant increase of the number of dead cells embedded into the biofilm matrix (Figure 6). Similarly, about the effects of combinations of r(P)ApoB$_L$^{Ala} and ciprofloxacin on preformed biofilm, a significantly greater reduction of biofilm biovolume is observed in the presence of compounds combination (Figure 7).

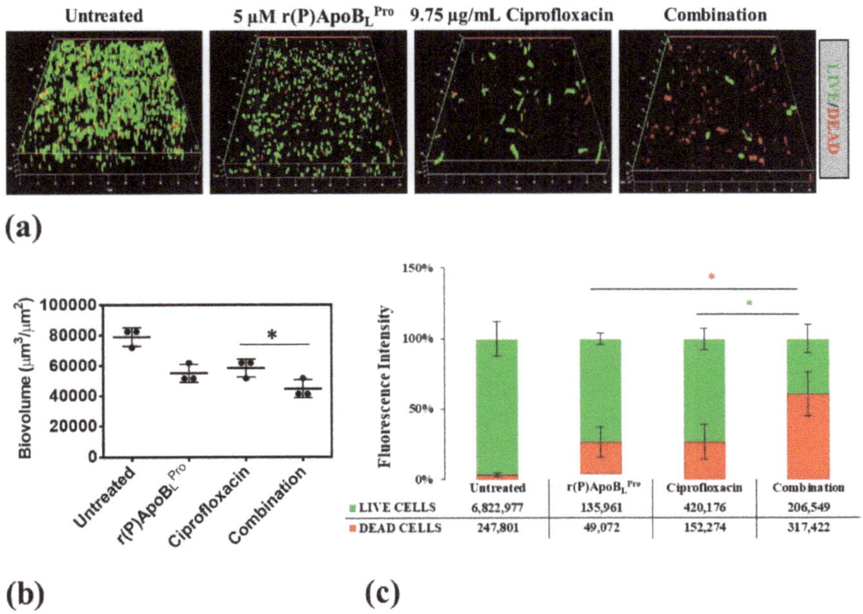

Figure 6. Effects of r(P)ApoB$_L$^{Pro}, ciprofloxacin and a combination of the two compounds on preformed biofilm (**a**). Biofilm cells were stained by using LIVE/DEAD BacLight bacterial viability kit (Molecular Probes, Eugene, OR) containing 1:1 ratio of Syto-9 (green fluorescence, all cells) and propidium iodide (PI, red fluorescence, dead cells). Images are 3D projections of biofilm structure obtained by laser scanning confocal z-stack using Zen Lite 2.3 software. All images were taken under identical conditions. Biovolume (µm^3/µm^2) was measured by using Zen Lite 2.3 software. Significant differences were indicated as * $p < 0.05$ for treated versus control samples (**b**). Numbers of live and dead cells were evaluated by using Zen Lite 2.3 software (**c**).

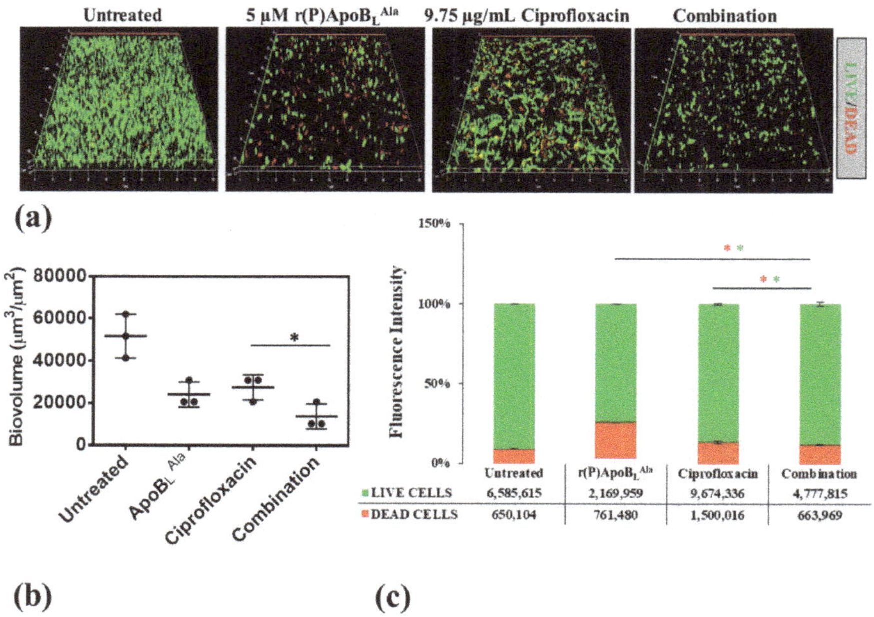

Figure 7. Effects of r(P)ApoB$_L$^{Ala}, ciprofloxacin and a combination of the two compounds on preformed biofilm (**a**). Biofilm cells were stained by using LIVE/DEAD BacLight bacterial viability kit (Molecular Probes, Eugene, OR) containing 1:1 ratio of Syto-9 (green fluorescence, all cells) and propidium iodide (PI, red fluorescence, dead cells). Images are 3D projections of biofilm structure obtained by laser scanning confocal z-stack using Zen Lite 2.3 software. All images were taken under identical conditions. Biovolume (µm^3/µm^2) was measured by using Zen Lite 2.3 software. Significant differences were indicated as * $p < 0.05$ for treated versus control samples (**b**). Numbers of live and dead cells were evaluated by using Zen Lite 2.3 software (**c**).

2.4. Evaluation of Peptide Biocompatibility

Peptide applicability in therapeutic approaches aimed at counteracting bacterial infections associated with CF is strongly dependent on the absence of any toxic effect towards host cells. For this reason, biocompatibility assays were performed to test ApoB-derived peptide effects on immortalized human bronchial epithelial mesothelial (BEAS) cells. As shown in Figure 8, only slight toxic effects were detected and the most biocompatible peptide was found to be r(P)ApoB$_L$^{Ala}. Indeed, in the presence of this peptide, only slight toxic effects were detected upon 72 h treatment and at the highest peptide concentrations tested (Figure 8).

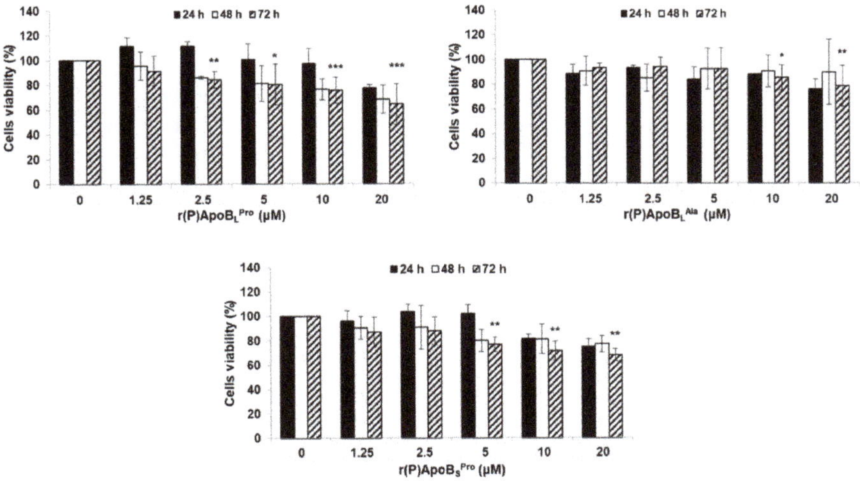

Figure 8. Effects of r(P)ApoB$_L$^{Pro}, r(P)ApoB$_L$^{Ala} and r(P)ApoB$_S$^{Pro} peptides on the viability of BEAS cells. Cell viability was assessed by MTT assays, and expressed as the percentage of viable cells with respect to controls (untreated cells). Error bars indicate standard deviations obtained from at least three independent experiments, each one carried out with triplicate determinations. Significant differences were indicated as * $p < 0.05$, ** $p < 0.01$ or *** $p < 0.001$ for treated versus control samples.

3. Discussion

AMPs represent novel promising effective alternative agents to counteract chronic bacterial infections affecting CF patients. Indeed, as these infections are generally recalcitrant to conventional antibiotics because of the development of the MDR phenotype and of biofilm formation, the development of novel therapeutic strategies is strongly necessary. To this purpose, three versions of a cryptide identified in human ApoB [33,35] have been here tested towards six bacterial strains clinically isolated from CF patients, such as *P. aeruginosa* RP 73, *P. aeruginosa* KK 27, *P. aeruginosa* 14, *P. aeruginosa* AA2, *B. multivorans* LMG 17582, and *B. cenocepacia* LMG 18863. ApoB-derived cryptides have been found to exert direct antimicrobial activity towards three out of six bacterial strains tested. Indeed, ApoB-derived AMPs have been found to be active on *P. aeruginosa* RP 73, *P. aeruginosa* KK 27, and *B. multivorans* LMG 17582, with MIC_{100} values ranging from 5 to 40 μM. This is in agreement with recent findings indicating that ApoB-derived cryptides direct antimicrobial activity, although mediated by electrostatic interactions between cationic peptides and negatively charged bacterial membranes, is strongly influenced by chemical composition of LPS molecules exposed on the surface of different strains of *P. aeruginosa* [35]. Indeed, although several bacterial resistance components against antimicrobial peptides have been reported [47], LPS chemical composition has been proposed to play a key role in the case of ApoB-derived cryptide antimicrobial activity [35]. It has been reported that different *Burkholderia* strains present different LPS chemotypes, such as rough, partial rough, or smooth [48]. In particular, in the case of *B. cenocepacia* LMG 18863, the LPS chemotype has been identified as smooth [48]. These differences in LPS chemotype might be responsible for the different susceptibility of *Burkholderia* strains to ApoB-derived cryptides' direct antimicrobial activity. It also has to be highlighted that several *B. cenocepacia* strains have been reported to be naturally resistant to different classes of antibiotics and even to several antimicrobial peptides [49]. This is probably due to the ability of *B. cenocepacia* strains to acquire a resistance phenotype by modifying the LPS chemical composition by substituting a phosphate group with a cationic charged residue of 4-amino-4-deoxy-L-arabinose (L-Ara4N), with a consequent reduction in membrane negative potential, that plays a key role in the interaction between bacterial membranes and antimicrobial peptides [50,51]. Based on these observations, ApoB-derived cryptides direct antimicrobial activity towards *B. multivorans* LMG 17582 appears to be really interesting. It also has to be considered that, in the case of chronic infections affecting CF patients, the mucus phenotype favors bacterial biofilm formation [20]. Bacteria embedded into biofilm matrix are more resistant to conventional antibiotics for several reasons: i) low antibiotic diffusion rate inside biofilm matrix; ii) bacteria metabolic changes due to nutrients missing, with a consequently lower susceptibility to antibiotics; and iii) appearance of persisted cells recalcitrant to conventional antibiotics and playing a key role in long-term infections [52]. Anti-biofilm cationic amphipathic peptides represent an alternative promising approach to treat infections associated with biofilm formation, since peptides act on biofilm specific targets, such as matrix components and/or highly conserved regulatory mechanisms [53]. Here, we tested the ability of r(P)ApoB$_L$^{Pro}, r(P)ApoB$_L$^{Ala} and r(P)ApoB$_S$^{Pro} to affect the biofilm of bacterial strains clinically isolated from CF patients. In particular, we analyzed the ability of ApoB-derived cryptides to interfere with the three main stages of biofilm development, i.e., attachment, formation and detachment [33]. We found that all the three ApoB-derived cryptides are able to exert significant effects on biofilm attachment and formation. Even more interestingly, ApoB-derived cryptides have been found to exert significant anti-biofilm effects even on bacterial strains not sensitive to their direct antimicrobial activity. In particular, ApoB-derived AMPs have been found to affect *P. aeruginosa* 14 biofilm attachment and formation and *B. cenocepacia* LMG 18863 preformed biofilm at a very low concentration (2.5 μM). As reported for different AMPs, obtained data allow us to exclude any correlation between peptide direct antimicrobial activity and their anti-biofilm properties. Indeed, peptide IDR-1018 has been reported to be endowed with strong anti-biofilm activity towards a pool of *P. aeruginosa* and *Burkholderia* strains in the absence of any direct antimicrobial effect [54]. To deeply characterize ApoB-derived cryptides anti-biofilm activity, we also performed analyses by confocal laser scanning microscopy (CLSM) and scanning electron microscopy (SEM) on *P. aeruginosa 14* and *B. cenocepacia* LMG 18863, two strains not sensitive to ApoB-derived AMP's direct antimicrobial activity. CLSM analyses revealed the ability of ApoB-derived cryptides to alter biofilm architecture, as indicated by the appearance of highly fluorescent aggregates only in treated samples upon staining with SYPRO® Ruby, a dye able to specifically label biofilm extracellular matrix. Accordingly, a significant reduction in biofilm biovolume has been evaluated in the case of samples treated with ApoB-derived cryptides. Furthermore, scanning electron microscopy analyses clearly indicate the ability of ApoB-derived cryptides to disrupt the biofilm matrix of bacterial strains not responsive to the peptides' direct antimicrobial activity. These observations are in perfect agreement with data reported for peptide 6K-F17, which is able to strongly affect *P. aeruginosa* biofilm by disrupting the extracellular matrix, thus determining a significant decrease of biofilm biovolume [55]. However, CLSM and SEM analyses indicate only a slight increase of bacterial cells death upon treatment with 6K-F17 [55]. Based on obtained results, we also evaluated the possibility to set up effective combinatorial therapeutic approaches by concomitantly administrating ApoB-derived cryptides and conventional antibiotics to bacterial cells. To this purpose, we analyzed the anti-biofilm properties of combinations of r(P)ApoB$_L$^{Pro} or r(P)ApoB$_L$^{Ala} and the antibiotic ciprofloxacin, which is widely used to treat bacterial infections in CF patients [45]. The effects of compound mixtures have been tested on *P. aeruginosa 14* or *B. cenocepacia* LMG 18863 preformed biofilm. Effects of combinations of r(P)ApoB$_L$^{Pro} or r(P)ApoB$_L$^{Ala} and ciprofloxacin on preformed biofilm have been found to be stronger than those of single agents, with more severe effects on biofilm biovolume. It has been previously reported that, upon biofilm treatment with the ciprofloxacin antibiotic, a deep alteration of the matrix structure and a strong decrease of biofilm biovolume are immediately observed, probably associated with a high killing rate of bacterial cells embedded into the biofilm matrix [56]. However, upon a prolonged exposure to ciprofloxacin, the activation of specific mechanisms leading to a variation of biofilm phenotype makes the antibiotic ineffective [56]. This phenomenon might be overcome by the development of successful combinatorial therapeutic approaches, which present several advantages over conventional therapeutic treatments based on the administration of single agents. Indeed, several anti-biofilm peptides have been reported to be able to act in synergism with a broad range of conventional antibiotics [57]. This allows us to significantly reduce the effective dose of antibiotics up to 64-fold, with a consequent lower possibility to induce MDR phenotype and to simultaneously reduce effective peptide concentrations [57]. In the case of CAMA peptides, synergism with the conventional antibiotics tobramycin, ciprofloxacin and colistin has been demonstrated in the treatment of *P. aeruginosa* biofilm, with the consequent possibility of reducing antibiotics doses up to 8-fold and peptide concentrations up to 10-fold [58]. Since one of the bottlenecks for the development of successful peptide-based therapies is peptide cytotoxicity, we also tested ApoB-derived cryptides effects on immortalized human bronchial epithelial mesothelial (BEAS) cells, and found that peptides are biocompatible, since slight toxic effects are detected only upon 72 h cell treatment and at the highest peptide concentrations tested. Altogether, obtained findings open interesting perspectives to the applicability of ApoB-derived cryptides in the treatment of bacterial chronic infections associated with biofilm formation and characterized by MDR phenotype, such as those affecting CF patients, and to the development in the future of successful combinatorial therapeutic approaches based on the concomitant administration of peptides and conventional antibiotics.

4. Materials and Methods

4.1. Materials

All the reagents were purchase from Sigma-Aldrich (Milan, Italy), unless differently specified.

4.2. Recombinant Production of ApoB-Derived Peptides

Expression and isolation of recombinant ApoB-derived peptides was carried out as previously described [33,35]. Pro → Ala substitution in position six of the longest peptide was obtained by QuikChange II site-directed mutagenesis performed by using the following primers: primer forward 5′-CATTTTACCCGCTTTCAGCGCAACATGCGGGTG-3′ and primer reverse 5′-GATCCGCATGTTGCGCTGAAAGCGGGTAAACTG-3′.

4.3. Bacterial Strains and Growth Conditions

Bacterial strains *P. aeruginosa* RP 73, *P. aeruginosa* KK 27, *P. aeruginosa* 14, *P. aeruginosa* AA2, *B. multivorans* LMG 17582, and *B. cenocepacia* LMG 18863 were kindly provided by Dr. Alessandra Bragonzi (Infection and CF Unit, San Raffaele Scientific Institute, Milan, Italy). Bacterial strains were grown in MHB (Becton Dickinson Difco, Franklin Lakes, NJ, USA) and on Tryptic Soy Agar (TSA; Oxoid Ltd., Hampshire, UK). In all the experiments, bacteria were inoculated and grown overnight in MHB at 37 °C.

4.4. Eukaryotic Cells and Growth Conditions

Immortalized human bronchial epithelial mesothelial cells (BEAS) were cultured in high-glucose Dulbecco's modified Eagle's medium (DMEM) supplemented with 10% fetal bovine serum and 1% penicillin-streptomycin at 37 °C in the presence of 5% carbon dioxide (CO_2).

4.5. Cell Viability Assays

Peptide effects on eukaryotic cell viability was evaluated by seeding cells in 96-well plates (100 μL/well) at a density of 3×10^3 cells/well. Upon 24 h, cells were incubated with increasing peptide concentrations (0–40 μM), for 24, 48 and 72 h. At the end of the treatment, cell viability was assessed by the 3-(4,5-dimethylthiazol-2-yl)-2,5-diphenyltetrazolium bromide (MTT) assay. MTT reagent, dissolved in DMEM without phenol red, was added to the cells (100 μL/well) at a final concentration of 0.5 mg/mL. After 4 h at 37 °C, the culture medium was removed and the resulting formazan salts were dissolved by the addition of isopropanol containing 0.1 N HCl (100 μL/well) [41]. Absorbance values of blue formazan were determined at 570 nm by using an automatic plate reader (Synergy™ H4 Hybrid Microplate Reader, BioTek Instruments, Inc., Winooski, VT, USA). Cell survival was expressed as the percentage of viable cells in the presence of the peptide under test, with respect to control cells grown in the absence of the peptide.

4.6. Antimicrobial Activity Assays

To test the antimicrobial activity of ApoB-derived peptides, a previously described experimental procedure was used [33]. MIC_{100} values correspond to the lowest concentration of peptide associated with no detectable bacterial growth.

4.7. Anti-Biofilm Activity by Crystal Violet Assay

ApoB-derived peptides effects on biofilm attachment, formation and detachment were evaluated as previously described [33]. Optical densities at 595 nm of biofilm stained biomasses were measured by using a microtiter plate reader (Synergy™ H4 Hybrid Microplate Reader, BioTek Instruments, Inc., Winooski, VT, USA).

4.8. Anti-Biofilm Activity by CLSM Analyses

Bacterial biofilm was grown on glass cover slips in 24-well plates in 0.5X MHB in static conditions. In particular, bacterial cells from an overnight culture were diluted to about 1×10^8 CFU/mL and then seeded into wells for 4 or 24 h at 37 °C in the presence of the peptide under test, in order to evaluate biofilm attachment and formation, respectively. When effects on preformed biofilm were evaluated, bacterial biofilms were formed for 24 h at 37 °C, and then treated with peptides under test for further 24 h to evaluate their ability to eradicate preformed biofilm. Afterwards, non-adherent bacteria were removed by gently washing samples with sterile phosphate buffer and viability of cells embedded into biofilm structure was determined by sample staining with LIVE/DEAD® BacLight™ Bacterial Viability kit (Molecular Probes, Thermo Fisher Scientific, Waltham, MA, USA), while FilmTracer™ SYPRO® Ruby biofilm matrix dye has been used to stain matrices of biofilms (Invitrogen, Carlsbad, CA, USA). Staining was performed accordingly to manufacturer instructions. Biofilm images were captured by using a confocal laser scanning microscopy (Zeiss LSM 710, Zeiss, Germany) and a 63X objective oil immersion system. Biofilm architecture was analyzed by using the Zen Lite 2.3 software package (Zeiss, Germany). Each experiment was performed in triplicate. All images were taken under identical conditions.

4.9. Anti-Biofilm Activity by Scanning Electron Microscopy

To perform scanning electron microscopy (SEM) analyses, *B. cenocepacia* LMG 18863 and *P. aeruginosa 14* cells were incubated with 5 μM r(P)$ApoB_L^{Pro}$, r(P)$ApoB_L^{Ala}$ or r(P)$ApoB_S^{Pro}$ peptides for 24 h at 37 °C. Following incubation, bacterial biofilms were fixed in 2.5% glutaraldehyde. Following overnight incubation, bacterial biofilms were washed three times in distilled water and then dehydrated with a graded ethanol series: 25% ethanol (1 × 10 min); 50% ethanol (1 × 10 min); 75% ethanol (1 × 10 min); 95% ethanol (1 × 10 min); 100% anhydrous ethanol (3 × 30 min). Bacterial biofilms deposited onto glass substrate were sputter coated with a thin layer of Au-Pd (Sputter Coater Denton Vacuum DeskV) to allow subsequent morphological characterization using a FEI Nova NanoSEM 450 at an accelerating voltage of 5 kV with Everhart Thornley Detector (ETD) and Through Lens Detector (TLD) at high magnification.

4.10. Statistical Analysis

Statistical analysis was performed using a Student's t-test. Significant differences were indicated as * $p < 0.05$, ** $p < 0.01$ or *** $p < 0.001$.

Author Contributions: Conceptualization, R.G. and A.A.; methodology, R.G., A.C., E.D.O., L.T. and R.D.G.; validation, R.G.; formal analysis, R.G.; investigation, R.G., A.C., E.D.O. and R.D.G.; resources, E.P. and A.A.; data curation, R.G., A.C., L.T. and E.D.O.; writing—original draft preparation, R.G., A.C., E.D.O. and A.A.;

writing—review & editing, all authors; supervision, A.A.; project administration, A.A.; funding acquisition, E.P. and A.A. All authors have read and agreed to the published version of the manuscript.

Funding: This research was funded by the Italian Cystic Fibrosis Foundation (grant number FFC#16/2017).

Conflicts of Interest: The authors declare no conflict of interest.

Abbreviations

CF	Cystic Fibrosis
MDR	Multidrug Resistance
ApoB	Apolipoprotein B
HDPs	Host Defense Peptides
CFTR	Cystic fibrosis transmembrane conductance regulator
AMPs	Antimicrobial peptides
MIC	Minimum inhibitory concentration
MHB	Mueller Hinton Broth
CLSM	Confocal laser scanning microscopy
SEM	Scanning electron microscopy
PI	Propidium iodide
LPS	Lipopolysaccharide

References

1. Farrell, P.M. The prevalence of cystic fibrosis in the European Union. *J. Cyst. Fibros.* **2008**, *7*, 450–453. [CrossRef]
2. Evans, C.M.; Koo, J.S. Airway mucus: The good, the bad, the sticky. *Pharmacol Ther.* **2009**, *121*, 332–348. [CrossRef]
3. Chace, K.V.; Flux, M.; Sachdev, G.P. Comparison of physicochemical properties of purified mucus glycoproteins isolated from respiratory secretions of cystic fibrosis and asthmatic patients. *Biochemistry* **1985**, *24*, 7334–7341. [CrossRef] [PubMed]
4. Knowles, M.R.; Durie, P.R. What is cystic fibrosis? *N. Engl. J. Med.* **2002**, *347*, 439–442. [CrossRef] [PubMed]
5. Ciofu, O.; Hansen, C.R.; Høiby, N. Respiratory bacterial infections in cystic fibrosis. *Curr. Opin. Pulm. Med.* **2013**, *19*, 251–258. [CrossRef] [PubMed]
6. Rudkjøbing, V.B.; Thomsen, T.R.; Alhede, M.; Kragh, K.N.; Nielsen, P.H.; Johansen, U.R.; Givskov, M.; Høiby, N.; Bjarnsholt, T. The microorganisms in chronically infected end-stage and non-end-stage cystic fibrosis patients. *FEMS Immunol. Med. Microbiol.* **2012**, *65*, 236–244. [CrossRef]
7. Bjarnsholt, T.; Jensen, P.Ø.; Fiandaca, M.J.; Pedersen, J.; Hansen, C.R.; Andersen, C.B.; Pressler, T.; Givskov, M.; Høiby, N. Pseudomonas aeruginosa biofilms in the respiratory tract of cystic fibrosis patients. *Pediatr. Pulmonol.* **2009**, *44*, 547–558. [CrossRef] [PubMed]
8. Parker, D.; Ahn, D.; Cohen, T.; Prince, A. Innate Immune Signaling Activated by MDR Bacteria in the Airway. *Physio. Rev.* **2016**, *96*, 19–53. [CrossRef]
9. Blanc, D.S.; Petignat, C.; Janin, B.; Bille, J.; Francioli, P. Frequency and molecular diversity of *Pseudomonas aeruginosa* upon admission and during hospitalization: A prospective epidemiologic study. *Clin. Microbiol. Infect.* **1998**, *4*, 242–247. [CrossRef]
10. Sadikot, R.T.; Blackwell, T.S.; Christman, J.W.; Prince, A.S. Pathogen–host interactions in *Pseudomonas aeruginosa* pneumonia. *Am. J. Respir. Crit. Care Med.* **2005**, *171*, 1209–1223. [CrossRef]
11. Larrosa, M.; Truchado, P.; Espin, J.C.; Tomas-Barberan, F.A.; Allende, A.; Garcia-Conesa, M.T. Evaluation of Pseudomonas aeruginosa (PAO1) adhesion to human alveolar epithelial cells A549 using SYTO 9 dye. *Mol. Cell Probes* **2012**, *26*, 121–126. [CrossRef] [PubMed]
12. Hirakata, Y.; Izumikawa, K.; Yamaguchi, T.; Igimi, S.; Furuya, N.; Maesaki, S.; Tomono, K.; Yamada, Y.; Kohno, S.; Yamaguchi, K.; et al. Adherence to and penetration of human intestinal Caco-2 epithelial cell monolayers by Pseudomonas aeruginosa. *Infect. Immun.* **1998**, *66*, 1748–1751. [CrossRef]
13. Pier, G.B.; Grout, M.; Zaidi, T.S. Cystic fibrosis transmembrane conductance regulator is an epithelial cell receptor for clearance of Pseudomonas aeruginosa from the lung. *Proc. Natl. Acad. Sci. USA* **1997**, *94*, 12088–12093. [CrossRef] [PubMed]

14. Pier, G.B.; Grout, M.; Zaidi, T.S.; Olsen, J.C.; Johnson, L.G.; Yankaskas, J.R.; Goldberg, J.B. Role of mutant CFTR in hypersusceptibility of cystic fibrosis patients to lung infections. *Science* **1996**, *271*, 64–67. [CrossRef]
15. Esen, M.; Grassme, H.; Riethmuller, J.; Riehle, A.; Fassbender, K.; Gulbins, E. Invasion of human epithelial cells by Pseudomonas aeruginosa involves src-like tyrosine kinases p60Src and p59Fyn. *Infect. Immun.* **2001**, *69*, 281–287. [CrossRef] [PubMed]
16. Darling, K.E.; Evans, T.J. Effects of nitric oxide on Pseudomonas aeruginosa infection of epithelial cells from a human respiratory cell line derived from a patient with cystic fibrosis. *Infect. Immun.* **2003**, *71*, 2341–2349. [CrossRef]
17. Brinch, K.S.; Frimodt-Moller, N.; Hoiby, N.; Kristensen, H.H. Influence of antidrug antibodies on plectasin efficacy and pharmacokinetics. *Antimicrob. Agents Chemother.* **2009**, *53*, 4794–4800. [CrossRef]
18. Chi, E.; Mehl, T.; Nunn, D.; Lory, S. Interaction of Pseudomonas aeruginosa with A549 pneumocyte cells. *Infect. Immun.* **1991**, *59*, 822–828. [CrossRef]
19. Drenkard, E.; Ausubel, F.M. Pseudomonas biofilm formation and antibiotic resistance are linked to phenotypic variation. *Nature* **2002**, *416*, 740–743. [CrossRef]
20. Scoffone, V.C.; Chiarelli, L.R.; Trespidi, G.; Mentasti, M.; Riccardi, G.; Buroni, S. Burkholderia cenocepacia Infections in Cystic Fibrosis Patients: Drug Resistance and Therapeutic Approaches. *Front. Microbiol.* **2017**, *8*, 1592. [CrossRef]
21. Alexander, B.D.; Petzold, E.W.; Reller, L.B.; Palmer, S.M.; Davis, R.D.; Woods, C.W.; Lipuma, J.J. Survival after lung transplantation of cystic fibrosis patients infected with Burkholderia cepacia complex. *Am. J. Transplant.* **2008**, *8*, 1025–1030. [CrossRef] [PubMed]
22. Chaparro, C.; Keshavjee, S. Lung transplantation for cystic fibrosis: An update. *Expert Rev. Respir. Med.* **2016**, *10*, 1269–1280. [CrossRef] [PubMed]
23. Mercer, D.K.; O'Neil, D.A. Peptides as the next generation of anti-infectives. *Future Med. Chem.* **2013**, *5*, 315–337. [CrossRef] [PubMed]
24. Haney, E.F.; Hancock, R.B. Peptide design for antimicrobial and immunomodulatory applications. *Biopolymers* **2013**, *100*, 572–583. [CrossRef]
25. Cruz, J.; Ortiz, C.; Guzman, F.; Fernandez-Lafuente, R.; Torres, R. Antimicrobial peptides: Promising compounds against pathogenic microorganisms. *Curr. Med. Chem.* **2014**, *21*, 2299–2321. [CrossRef]
26. Epand, R.M.; Vogel, H.J. Diversity of antimicrobial peptides and their mechanisms of action. *Biochim. Biophys. Acta* **1999**, *1462*, 11–28. [CrossRef]
27. Shai, Y. Mode of action of membrane active antimicrobial peptides. *Biopolymers* **2002**, *66*, 236–248. [CrossRef]
28. Cao, H.; Ke, T.; Liu, R.; Yu, J.; Dong, C.; Cheng, M.; Huang, J.; Liu, S. Identification of a Novel Proline-Rich Antimicrobial Peptide from Brassica napus. *PLoS ONE* **2015**, *10*, e0137414. [CrossRef]
29. Fosgerau, K.; Hoffmann, T. Peptide therapeutics: Current status and future directions. *Drug Discov. Today* **2015**, *20*, 122–128. [CrossRef]
30. Yeaman, M.R.; Yount, N.Y. Mechanisms of antimicrobial peptide action and resistance. *Pharmacol. Rev.* **2003**, *55*, 27–55. [CrossRef]
31. Vineeth Kumar, T.V.; Sanil, G. A Review of the Mechanism of Action of Amphibian Antimicrobial Peptides Focusing on Peptide-Membrane Interaction and Membrane Curvature. *Curr. Protein. Pept. Sci.* **2017**, *18*, 1263–1272. [PubMed]
32. Shahrour, H.; Ferrer-Espada, R.; Dandache, I.; Bárcena-Varela, S.; Sánchez-Gómez, S.; Chokr, A.; Martinez-de-Tejada, G. AMPs as Anti-biofilm Agents for Human Therapy and Prophylaxis. *Adv. Exp. Med. Biol.* **2019**, *1117*, 257–279. [PubMed]
33. Gaglione, R.; Dell'Olmo, E.; Bosso, A.; Chino, M.; Pane, K.; Ascione, F.; Itri, F.; Caserta, S.; Amoresano, A.; Lombardi, A.; et al. Novel human bioactive peptides identified in Apolipoprotein B: Evaluation of their therapeutic potential. *Biochem. Pharmacol.* **2017**, *130*, 34–50. [CrossRef] [PubMed]
34. Pane, K.; Durante, L.; Crescenzi, O.; Cafaro, V.; Pizzo, E.; Varcamonti, M.; Zanfardino, A.; Izzo, V.; Di Donato, A.; Notomista, E. Antimicrobial potency of cationic antimicrobial peptides can be predicted from their amino acid composition: Application to the detection of "cryptic" antimicrobial peptides. *J. Theor. Biol.* **2017**, *419*, 254–265. [CrossRef]
35. Gaglione, R.; Cesaro, A.; Dell'Olmo, E.; Della Ventura, B.; Casillo, A.; Di Girolamo, R.; Velotta, R.; Notomista, E.; Veldhuizen, E.J.A.; Corsaro, M.M.; et al. Effects of human antimicrobial cryptides identified in apolipoprotein B depend on specific features of bacterial strains. *Sci. Rep.* **2019**, *9*, 6728. [CrossRef]

36. Gaglione, R.; Pane, K.; Dell'Olmo, E.; Cafaro, V.; Pizzo, E.; Olivieri, G.; Notomista, E.; Arciello, A. Cost-effective production of recombinant peptides in Escherichia coli. *N. Biotechnol.* **2019**, *51*, 39–48. [CrossRef]
37. Pane, K.; Cafaro, V.; Avitabile, A.; Torres, M.T.; Vollaro, A.; De Gregorio, E.; Catania, M.R.; Di Maro, A.; Bosso, A.; Gallo, G.; et al. Identification of Novel Cryptic Multifunctional Antimicrobial Peptides from the Human Stomach Enabled by a Computational-Experimental Platform. *ACS Synth. Biol.* **2018**, *7*, 2105–2115. [CrossRef]
38. Zanfardino, A.; Bosso, A.; Gallo, G.; Pistorio, V.; Di Napoli, M.; Gaglione, R.; Dell'Olmo, E.; Varcamonti, M.; Notomista, E.; Arciello, A.; et al. Human apolipoprotein E as a reservoir of cryptic bioactive peptides: The case of ApoE 133-167. *J. Pept. Sci.* **2018**, *24*, e3095. [CrossRef]
39. Pizzo, E.; Pane, K.; Bosso, A.; Landi, N.; Ragucci, S.; Russo, R.; Gaglione, R.; Torres, M.D.T.; de la Fuente-Nunez, C.; Arciello, A.; et al. Novel bioactive peptides from PD-L1/2, a type 1 ribosome inactivating protein from Phytolacca dioica L. Evaluation of their antimicrobial properties and anti-biofilm activities. *Biochim. Biophys. Acta Biomembr.* **2018**, *1860*, 1425–1435. [CrossRef]
40. Pizzo, E.; Cafaro, V.; Di Donato, A.; Notomista, E. Cryptic Antimicrobial Peptides: Identification Methods and Current Knowledge of their Immunomodulatory Properties. *Curr. Pharm. Des.* **2018**, *24*, 1054–1066. [CrossRef]
41. Gaglione, R.; Pirone, L.; Farina, B.; Fusco, S.; Smaldone, G.; Aulitto, M.; Dell'Olmo, E.; Roscetto, E.; Del Gatto, A.; Fattorusso, R.; et al. Insights into the anticancer properties of the first antimicrobial peptide from Archaea. *Biochim. Biophys. Acta Gen. Subj.* **2017**, *1861*, 2155–2164. [CrossRef] [PubMed]
42. Kasetty, G.; Papareddy, P.; Kalle, M.; Rydengård, V.; Walse, B.; Svensson, B.; Mörgelin, M.; Malmsten, M.; Schmidtchen, A. The C-terminal sequence of several human serine proteases encodes host defense functions. *J. Innate Immun.* **2011**, *3*, 471–482. [CrossRef] [PubMed]
43. Lee, D.Y.; Huang, C.M.; Nakatsuji, T.; Thiboutot, D.; Kang, S.A.; Monestier, M.; Gallo, R.L. Histone H4 is a major component of the antimicrobial action of human sebocytes. *J. Invest. Dermatol.* **2009**, *129*, 2489–2496. [CrossRef] [PubMed]
44. Beck, W.H.; Adams, C.P.; Biglang-Awa, I.M.; Patel, A.B.; Vincent, H.; Haas-Stapleton, E.J.; Weers, P.M. Apolipoprotein A-I binding to anionic vesicles and lipopolysaccharides: Role for lysine residues in antimicrobial properties. *Biochim. Biophys. Acta* **2013**, *1828*, 1503–1510. [CrossRef] [PubMed]
45. Langton Hewer, S.C.; Smyth, A.R. Antibiotic strategies for eradicating Pseudomonas aeruginosa in people with cystic fibrosis. *Cochrane Database Syst. Rev.* **2014**, *11*, CD004197.
46. Sanchez, Z.; Tani, A.; Kimbara, K. Extensive reduction of cell viability and enhanced matrix production in Pseudomonas aeruginosa PAO1 flow biofilms treated with a D-amino acid mixture. *Appl. Environ. Microbiol.* **2013**, *79*, 1396–1399. [CrossRef]
47. Bechinger, B.; Gorr, S.U. Antimicrobial Peptides: Mechanisms of Action and Resistance. *J. Dent. Res.* **2017**, *96*, 254–260. [CrossRef]
48. De Soyza, A.; Ellis, C.D.; Khan, C.M.A.; Corris, P.A.; Demarco De Hormaeche, R. Burkholderia cenocepacia lipopolysaccharide, lipid A, and proinflammatory activity. *Am. J. Respir. Crit. Care Med.* **2004**, *170*, 70–77. [CrossRef]
49. Mahenthiralingam, E.; Urban, T.A.; Goldberg, J.B. The multifarious, multireplicon Burkholderia cepacia complex. *Nat. Rev. Microbiol.* **2005**, *3*, 144–156. [CrossRef]
50. Raetz, C.R.H.; Reynolds, C.M.; Trent, M.S.; Bishop, R.E. Lipid A Modification Systems in Gram-Negative Bacteria. *Annu. Rev. Biochem.* **2007**, *76*, 295–329. [CrossRef]
51. Olaitan, A.O.; Morand, S.; Rolain, J.M. Mechanisms of polymyxin resistance: Acquired and intrinsic resistance in bacteria. *Front. Microbiol.* **2014**, *5*, 29–30. [CrossRef] [PubMed]
52. Mulcahy, L.R.; Burns, J.L.; Lory, S.; Lewis, K. Emergence of Pseudomonas aeruginosa strains producing high levels of persister cells in patients with cystic fibrosis. *J. Bacteriol.* **2010**, *192*, 6191–6199. [CrossRef] [PubMed]
53. Pletzer, D.; Hancock, R.E.W. Antibiofilm peptides: Potential as broadspectrum agents. *J. Bacteriol.* **2016**, *198*, 2572–2578. [CrossRef] [PubMed]
54. Mansour, S.C.; de la Fuente-Núñez, C.; Hancock, R.E. Peptide IDR-1018: Modulating the immune system and targeting bacterial biofilms to treat antibiotic-resistant bacterial infections. *J. Pept. Sci.* **2015**, *21*, 323–329. [CrossRef] [PubMed]

55. Beaudoin, T.; Stone, T.A.; Glibowicka, M.; Adams, C.; Yau, Y.; Ahmadi, S.; Bear, C.E.; Grasemann, H.; Waters, V.; Deber, C.M. Activity of a novel antimicrobial peptide against Pseudomonas aeruginosa biofilms. *Sci. Rep.* **2018**, *8*, 14728. [CrossRef] [PubMed]
56. Soares, A.; Roussel, V.; Pestel-Caron, M.; Barreau, M.; Caron, F.; Bouffartigues, E.; Chevalier, S.; Etienne, M. Understanding Ciprofloxacin Failure in *Pseudomonas aeruginosa* Biofilm: Persister Cells Survive Matrix Disruption. *Front. Microbiol.* **2019**, *10*, 2603. [CrossRef]
57. Reffuveille, F.; De La Fuente-Núñez, C.; Mansour, S.; Hancock, R.E.W. A broad-spectrum antibiofilm peptide enhances antibiotic action against bacterial biofilms. *Antimicrob. Agents Chemother.* **2014**, *58*, 5363–5371. [CrossRef]
58. Dosler, S.; Karaaslan, E. Inhibition and destruction of Pseudomonas aeruginosa biofilms by antibiotics and antimicrobial peptides. *Peptides* **2014**, *62*, 32–37. [CrossRef]

International Journal of
Molecular Sciences

MDPI

Communication

Oral Delivery of a Tetrameric Tripeptide Inhibitor of VEGFR1 Suppresses Pathological Choroid Neovascularization

Valeria Tarallo [1], Emanuela Iaccarino [2], Valeria Cicatiello [1], Riccardo Sanna [3], Menotti Ruvo [2,4] and Sandro De Falco [1,4,*]

1 Istituto di Genetica e Biofisica 'Adriano Buzzati-Traverso'—CNR, 80131 Napoli, Italy; valeria.tarallo@igb.cnr.it (V.T.); valeria.cicatiello@igb.cnr.it (V.C.)

2 Istituto di Biostrutture e Bioimmagini—CNR, 80134 Napoli, Italy; emanuela.iaccarino@gmail.com (E.I.); menotti.ruvo@unina.it (M.R.)

3 BIOVIIIx s.r.l., Department of R&D, 80142 Napoli, Italy; riccardo.sanna@bioviiix.com

4 ANBITION s.r.l., Department of R&D, 80128 Napoli, Italy

* Correspondence: sandro.defalco@igb.cnr.it or sandro.defalco@anbition.it; Tel.: +39-081-6132-354

Received: 13 December 2019; Accepted: 7 January 2020; Published: 9 January 2020

Abstract: Age-related macular degeneration (AMD) is the primary cause of blindness in advanced countries. Repeated intravitreal delivery of anti-vascular endothelial growth factor (VEGF) agents has represented an important advancement for the therapy of wet AMD with significative results in terms of blindness prevention and partial vision restore. Nonetheless, some patients are not responsive or do not attain significant visual improvement, intravitreal injection may cause serious complications and important side effects have been reported for the prolonged block of VEGF-A. In order to evaluate new anti-angiogenic strategies, we focused our attention on VEGF receptor 1 (VEGFR1) developing a specific VEGFR-1 antagonist, a tetrameric tripeptide named inhibitor of VEGFR 1 (iVR1). We have evaluated its anti-angiogenic activity in the preclinical model of AMD, the laser-induced choroid neovascularization (CNV). iVR1 is able to potently inhibit CNV when delivered by intravitreal injection. Surprisingly, it is able to significantly reduce CNV also when delivered by gavage. Our data show that the specific block of VEGFR1 in vivo represents a valid alternative to the block of VEGF-A and that the inhibition of the pathological neovascularization at ocular level is also possible by systemic delivery of compounds not targeting VEGF-A.

Keywords: AMD; CNV; VEGFR1; multimeric peptides; oral delivery

---

1. Introduction

Aberrant ocular neovascularization is involved in many vision-threatening diseases including age-related macular degeneration (AMD), diabetic retinopathy (DR), central retinal vein occlusion (CRVO), retinopathy of prematurity (ROP) and corneal neovascularization. Among them, wet AMD accounts for about eight percent of all blindness worldwide and is the primary cause of blindness among the elderly in industrialized nations [1,2].

In the last years, anti-angiogenesis agents have revolutionized the treatment of ocular neovascular diseases [3,4]. Three anti-VEGF agents are currently available for therapy: ranibizumab and bevacizumab that specifically neutralize VEGF-A [5], and aflibercept, able to block VEGF-A, VEGF-B and placental growth factor (PlGF) [6]. Their delivery by repeated intravitreal injections blocks the growth of pathological vessels preventing blindness and, in many cases, restores vision.

Despite this significant clinical success, many patients do not attain significant visual improvement [7,8]. It has been shown that expression of VEGF in eyes is regulated by advanced glycation end products generated by protracted diabetes symptoms and that this can be causally associated with diabetic retinopathy, characterized by an increased retinal neovascularization due to the action of VEGF [9]. Adverse effects of VEGFA neutralization on multiple retinal cell types, widely reported in animal models [10,11], are observed in patients who have been treated with anti-VEGF drugs for several years [5,12,13]. Moreover, drug delivery by repeated intravitreal injections may generate devastating ocular complications. The most frequent are infectious endophthalmitis and intraocular inflammation. Ocular hemorrhage, intraocular pressure elevation and rhegmatogenous retinal detachment have also been observed [14], therefore, alternative or additional therapeutic anti-angiogenic strategies possibly coupled to different routes of administration are continuously sought.

VEGFR1 is the common receptor of the pro-angiogenic members of the VEGF family: VEGF-A, VEGF-B and PlGF [15]. It is also known as the high affinity receptor for VEGF-A since VEGF-A also recognizes VEGFR2, but with a KD increased by one order of magnitude [16]. The VEGF-A/VEGFR2 axis activates the main signaling pathway for the formation of new blood vessels from the pre-existing ones both in physiological and pathological conditions [17]. In parallel, several reports have highlighted the crucial function of VEGFR1 activation, mainly in pathological angiogenesis. Genetic ablation of VEGFR1 TK domain [18] or of VEGFR1 specific ligands, PlGF [19,20] and VEGF-B [21], gives rise to normal mice that show an inhibited pathological angiogenesis. Mirroring these studies, biochemical inhibition of VEGFR1 or PlGF, by the use of neutralizing monoclonal antibodies [22,23] or by small peptides [24,25] is effective in inhibiting the angiogenesis associated to several disease states, such as cancer, ocular neovascular diseases, inflammatory diseases, atherosclerosis and obesity [26].

To match the need of new therapeutic anti-angiogenic strategies, we developed a tetrameric tripeptide inhibitor of VEGF receptor 1 (VEGFR1), named iVR1 [27].

iVR1 binds to VEGFR-1, inhibiting the interaction with all three natural ligands with a half maximal inhibitory concentration (IC_{50}) close to 8–10 μM. It is composed by unnatural amino acids that, together with the multimeric structure, confer high resistance to the degradation in biological fluids. iVR1 activity has been already fully characterized in vitro and in vivo. It specifically binds VEGFR1 and does not interfere with VEGFR2 activity, is able to prevent VEGFR1 phosphorylation and capillary-like tube formation of human primary endothelial cells, and also blocks neovascularization of chicken embryo chorioallantoic membrane induced by PlGF or VEGF-A [27]. In vivo, iVR1 suppresses tumor growth and neoangiogenesis in xenograft models of colorectal cancer to an extent similar to that exhibited by bevacizumab. It is able to synergize with the chemotherapeutic agent irinotecan, inducing a significant prolongation of survival similar to that observed with the combination of bevacizumab and irinotecan. Moreover, iVR1 delivered by intravitreal injection is also able to inhibit pathological angiogenesis in the preclinical model of wet AMD, the laser-induced choroid neovascularization (CNV) [28].

We have here evaluated the favorable option of administering iVR1 by a less traumatic route of administration for the therapeutic treatment of wet AMD. In the perspective of a systemic delivery of iVR1, we first performed a counter-ion exchange from trifluoroacetate (TFA), deriving from the chemical synthesis of the peptide (named here iVR1-TFA), to acetate, obtaining a new compound named iVR1-Ac. Replacement of TFA with acetic acid has also entailed a potency gain evaluated as a reduction of the IC_{50} of the inhibition of VEGF-A/VEGFR1 interaction in vitro of about five times. We have confirmed the inhibitory activity of iVR1-Ac in the preclinical model of wet AMD after intravitreal delivery and we have investigated whether the oral delivery by gavage could provide effective inhibition of laser-induced CNV.

2. Results

2.1. Synthesis of iVR1

iVR1 is a tetrameric tripeptide composed by unnatural amino acids having the following sequence and structure: [D-Glu-L-Cys(Bzl)-L-Cha]4-Lys2-Lys-Gly, whereby Cys(Bzl) indicates S-benzylated

L-cysteine and Cha indicates cyclohexylalanine (Figure 1A,B). The peptide has been assembled by solid phase synthesis and removed from the resin using a mixture of TFA containing suitable scavengers, as previously reported [29]. The peptide has been obtained in a good yield (about 60% after purification) and in high purity (>95% as determined by HPLC analysis at 214 nm). Peptide identity has been assessed by LC–MS using an electrospray ionization time-of-flight (ESI-TOF) mass spectrometer coupled to HPLC. By virtue of the final TFA treatment and reverse phase purification the peptide is obtained as trifluoroacetate salt [29]. Since we have observed several times in previous studies [27,28] that TFA-peptide compounds are partially irritating the skin of the animal on the site of injection and given the known much higher toxicity of TFA salts compared to acetic acid, we have thus evaluated the option of using the peptide bearing the acetate as counter-ion instead of TFA, also considering that acetate is largely used in the formulation of therapeutic peptides.

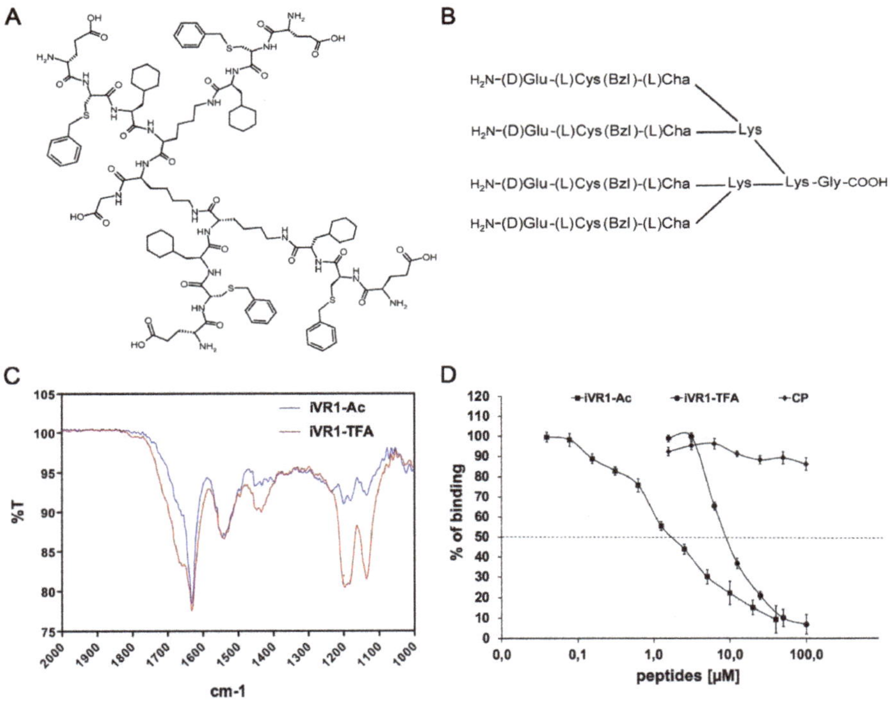

Figure 1. iVR1-Ac shows an increased inhibitory activity compared to iVR1-TFA. (**A**) Chemical structure of iVR1 tetrameric tripeptide that has a calculated molecular weight of 2362.02 g/mol. (**B**) Schematic representation of the iVR1. L-Cys(Bzl), L-cysteine(S-benzyl); L-Cha, L-cyclohexylalanine. (**C**) Overlay of FT-IR spectra in the spectral range between 1000 and 2000 cm^{-1} of the peptide before (red line) and after treatment with acetic acid (blue line). The spectrum collected on the acetic acid treated peptide shows that the bands characteristic of TFA at about 1145 and 1200 cm^{-1} are drastically reduced following repeated lyophilization in 0.1 M acetic acid. Some bands at around 1666 cm^{-1} were also strongly suppressed as consequence of the TFA removal. (**D**) Competitive ELISA for the binding of VEGF-A to immobilized VEGFR1. iVR1-Ac showed a decreased IC_{50} as compared to iVR1-TFA (1.94 and 9.35 µM, respectively) estimated as the concentration at 50% inhibition. Control peptide (CP) was inactive in the concentration range tested (up to 100 µM). Data are presented as the mean ± SEM of two independent experiments performed in triplicate.

2.2. FT-IR Characterization of Peptides Before and after Counter-Ion Exchange

Trifluoroacetate to acetate exchange is generally not a trivial operation. However, we tried to replace TFA by a triple step of peptide dissolution in acetic acid solutions and lyophilization. TFA removal was assessed by FT-IR spectroscopy comparing specific TFA IR bands before and after the treatment. We referred in particular to bands appearing as a shoulder at about 1670 cm^{-1} and two strong absorption bands at about 1135 and 1200 cm^{-1} [30]. These bands in proteins and peptides appear well above 1200 cm^{-1} [31], therefore they could be used as a good fingerprint to determine TFA amounts by comparing relative intensities. The data in Figure 1C shows that more than about 90% of the initial TFA content was removed by our treatment [30], suggesting that the method, with some improvements, could be used for replacing TFA with acetate in peptide samples.

2.3. iVR1Ac Shows Increased Inhibitory Activity Compared to iVR1-TFA

To verify whether the exchange of TFA with acetic acid could interfere with the inhibitory activity of iVR1, we performed a competitive ELISA using both compounds and as further control an unrelated peptide. iVR1-TFA, iVR1-Ac and the control peptide (CP) were used in competition for the binding of VEGF-A to VEGFR1. As shown in Figure 1D, iVR1-TFA was able to inhibit the binding of VEGF-A to VEGFR1 with an IC_{50} of 9.35 μM, whereas iVR1-Ac showed a greater competing activity with an IC_{50} close to 1.94 μM. The CP was unable to inhibit VEGF-A/VEGFR1 interaction. Thereby the counter-ion exchange determined an increase in terms of inhibitory activity of about five times, likely due to a partial improved solubility.

2.4. Intravitreal Delivery of iVR1-Ac Potently Inhibit Laser-Induced CNV

Laser-induced CNV experiments were performed to verify the ability of iVR1-Ac to inhibit pathological neovascularization in vivo. Immediately after the induction of laser damage, single intravitreal injections of iVR1-Ac or of the vehicle (DMSO) and of anti-mouse VEGF-A polyclonal antibody used as positive controls, were performed. After seven days, CNV volume was evaluated by immunofluorescence analysis of retinal pigment epithelium (RPE) choroid flat mounts. Anti-mouse VEGF-A induced a strong and significant inhibition compared to PBS (−54.3%, $p = 0.001$ vs. PBS). iVR1-Ac was able to induce a dose-dependent inhibition of CNV. Indeed, at 10 μg it already induced a significant reduction of CNV (−37.8%, $p = 0.0464$ vs. DMSO) and become even stronger at the highest quantity delivered of 50 μg (−73.9%, $p = 0.0002$ vs. DMSO; Figure 2).

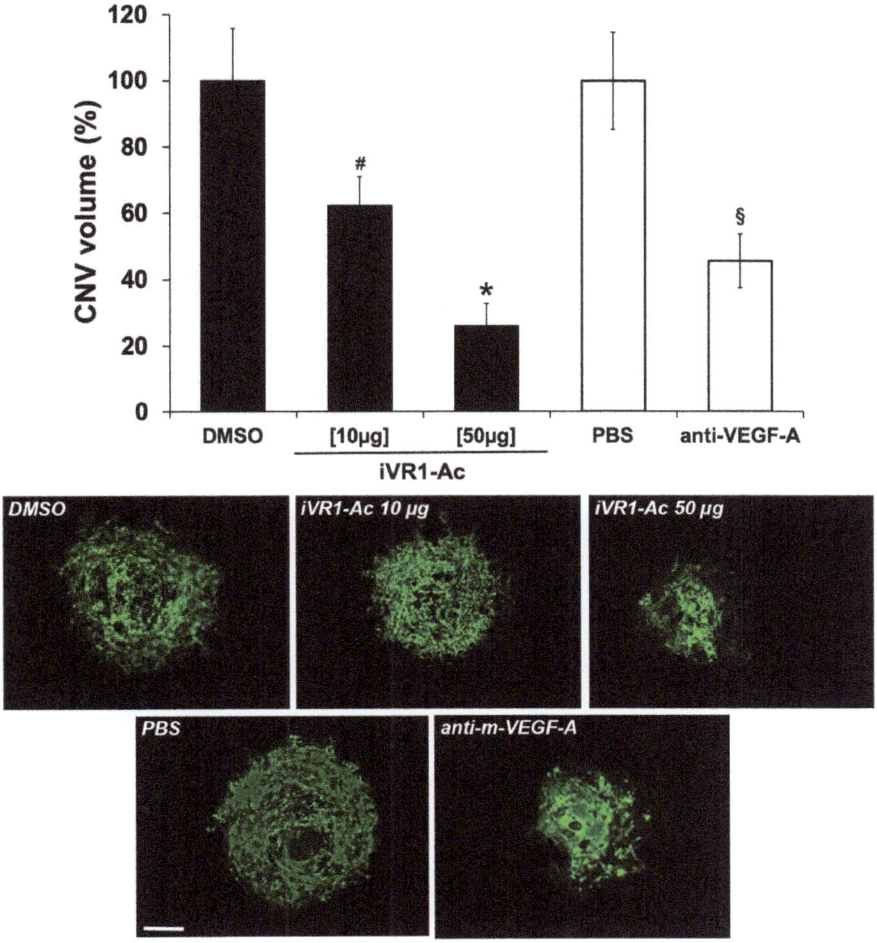

Figure 2. iVR1-Ac inhibits laser-induced choroid neovascularization (CNV) in a dose-dependent manner after intravitreal delivery. After 7 days from laser-induced damage, CNV volumes were measured by Isolectin B4 staining of RPE-choroid flat mounts. $N = 5$ mice per group. The following number of spots were analyzed: DMSO = 14, iVR1-Ac [10 µg] = 12, iVR1-Ac [50 µg] = 15; PBS = 10, anti-m-VEGF-A = 8. Data are presented as the mean ± SEM. * $p = 0.0002$ and # $p = 0.0464$ vs. DMSO; § $p = 0.001$ vs. PBS. On the bottom, representative pictures of CNV are shown. Scale bar: 100 µm.

2.5. iVR1-Ac Delivered by Gavage Provides Effective CNV Inhibition

In order to look for alternative route of administration for treating wet AMD, we evaluated whether systemic delivery of iVR1-Ac by gavage was similarly effective. The administration of the peptide, and as control of the vehicle (200 µL each dose), started 12 h after the damage induced by laser and was performed over 7 days, two times per day, at 50 mg/Kg. This dosage was chosen based on previous data obtained in in vivo experiments for tumor studies [28].

iVR1-Ac suppressed CNV by of about 50%, inhibiting pathological neovascularization ($p = 0.001$ vs. vehicle; Figure 3).

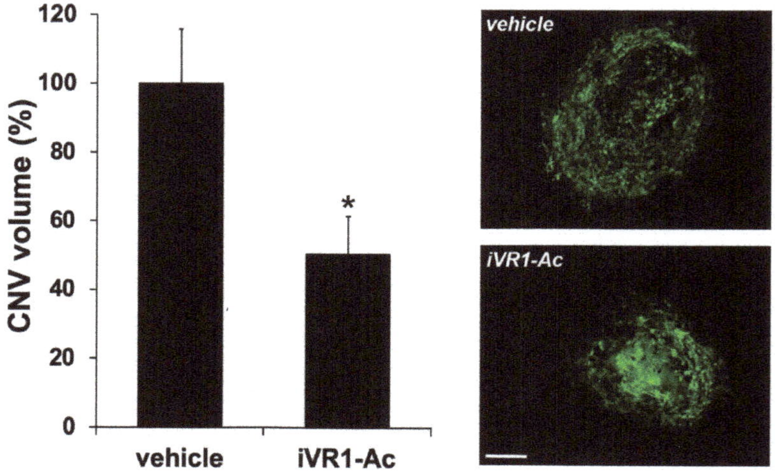

Figure 3. iVR1-Ac inhibited laser-induced CNV when delivered by gavage. After 7 days from laser-induced damage, CNV volumes were measured by Isolectin B4 staining of RPE-choroid flat mounts. $N = 5$ mice per group. The following number of spots were analyzed: vehicle = 10, iVR1-Ac = 20. Data are presented as the mean ± SEM. * $p = 0.001$ vs. vehicle. On the right, representative pictures of CNV are shown. Scale bar: 100 μm.

3. Discussion

Two main concerns affect the current anti-angiogenic therapies for ocular neovascular diseases: the side effects deriving from the prolonged block of VEGF-A and the tedious and the potentially dangerous practice of intravitreal injection. This last concern is also associated with the general reluctance of patients to be submitted to intravitreal punctures, most often accepted with worries and fright.

Several data from preclinical models and patients show how detrimental can be the block of VEGF-A, and consequently of VEGF-A/VEGFR2 signaling, given its involvement also in physiological settings. VEGFR1 is also deeply involved in neoangiogenesis, however its activity is mostly restricted to pathological conditions. On this basis, we chose it as a privileged and more selective therapeutic target for angiogenesis inhibition. If the VEGF-A/VEGFR2 pathway is crucial for the stimulation, differentiation and migration of endothelial cells, as well as for the physiological homeostasis of vessels [32], the ability of VEGFR1 to drive neo-angiogenesis depends essentially on its wide pattern of expression and on its ability to drive survival, migratory, and cells recruitment signals [15,33]. Indeed, it is expressed on endothelial cells, where it has a role in vessel sprouting and growth [34], in mural cells, where drives their recruitment essential to stabilize nascent vessels [32], and in inflammatory cells that are recruited and activated at neo-angiogenic sites where they result crucial to further foster the vessel growth and stabilization [35].

The importance of VEGFR1 in pathological neoangiogenesis was also confirmed by our previous data. In the preclinical model of colorectal cancer [28], iVR1 was able to inhibit tumor growth and vascularization with an extent similar to that induced by bevacizumab, despite an IC_{50} close to 10 μM. It potently inhibited VEGFR1 phosphorylation in vivo and induced a strong decrease of monocyte-macrophages and mural cells recruitment. Here, we reported on a further advancement of knowledge on iVR1. We had indeed observed a considerable improvement of the peptide ability to inhibit VEGFR1 activity when the TFA counter-ion was removed or reduced from the formulation and replaced with acetate. Such improvement is reflected by the IC_{50} passing from about 10 μM to less than 2 μM. Intravitreal delivery of this new version of iVR1 in the preclinical model of wet AMD determines an inhibition of neovascularization to an extent never observed with anti-mouse VEGF-A antibodies. Together, these results demonstrated that in these in vivo preclinical models of diseases, the prevention of VEGFR1 activation was a valid and effective alternative to the block of VEGF-A.

As anticipated, a major concern associated to the anti-angiogenic therapy for eye diseases is the delivery achieved by intravitreal injection of the drugs. Very few reports have so far described the delivery of anti-angiogenic drugs by oral administration for wet AMD. They are all referred to multitarget tyrosine kinase inhibitors (TKi; i.e., lenvatinib, pazopanib) mainly selected for their activity on VEGFR2 [36–38]. It must be underlined that the use of TKi substantially mirrors the block of VEGF-A achieved with current therapeutic agents but add further complications deriving from their multitargeting properties. It is therefore expected that this therapeutic strategy may have the dual side effects of prejudicing the physiological role of VEGF-A/VEGFR2 axis and suppressing the potentially beneficial activity of many other unrelated receptors targeted by the TKi, both at ocular and systemic levels.

Here we demonstrated that iVR1 was able to inhibit the CNV by 50% when delivered by gavage, displaying, for the first time at the best of knowledge, that inhibition of pathological neovascularization at ocular level was also feasible by systemic delivery of an anti-angiogenic peptide not inhibiting the VEGF-A/VEGFR2 axis. The dosage of 50 mg/kg was based on previous results obtained in tumor models, in which iVR1 was delivered by intraperitoneal injection every other day at 50 mg/kg for 14 days or at 25/mg/kg for 80 days [28]. This dosage regimen had not evidenced any gross side effects or sign of toxicity. These previous data and those we reported here allowed us to speculate that the systemic block of VEGFR1 could determine low or negligible side effects deriving from its minor role in physiological processes.

The prevalence of wet AMD is destined to increase in the next years as a consequence of exponential population ageing [1]. Accordingly, the request of new anti-angiogenic drugs, even better if delivered by routes alternative to the intravitreal injection, will consequently increase expanding enormously the interest for this class of molecules. Clearly, to propose iVR1 for further therapeutic studies, iVR1 toxicity, pharmacokinetic and pharmacodynamic studies at ocular and systemic levels are needed. In this perspective, analogues of iVR1 obtained from the iVR1 chemical structure have already been identified and will be soon characterized for their anti-angiogenic properties to assess the therapeutic potential.

4. Materials and Methods

4.1. Peptide Synthesis

All chemicals for peptide synthesis were from commercial sources and used without further purification unless otherwise stated. Solvents, including acetonitrile (CH_3CN), dimethylformamide (DMF) were purchased from Sigma Aldrich (Milano, Italy). Acetic acid, *N*,*N*-Diisopropylethylamine (DIPEA), piperidine, and TFA were also from Sigma-Aldrich. Protected amino acids and the coupling agents OxymePure® (ethyl 2-cyano-2-(hydroxyimino)acetate, called Oxyma Pure) and *N*-*N*′ diisopropylcarbodiimide (DIC) used for peptide synthesis were from IRIS Biotech (Marktredwitz, Germany). The peptide was prepared as previously reported [27,28], with some modifications. In particular as coupling and deprotection agents we used Oxyma/DIC/DIPEA and 40% piperidine in DMF to improve the final yield and to obtain purer products. The crude material was characterized by LC–MS [29] and purified by preparative RP-HPLC using a gradient from 20% solvent B (CH_3CN, 0.1% TFA) to 80% solvent B over solvent A (H_2O, 0.1% TFA) in 20 min, monitoring the eluate at 214 nm. The purified material was collected, lyophilized, and characterized for purity and identity by LC–MS using the conditions previously reported [29].

4.2. FT-IR Characterization of Peptides Before and After Counter-Ion Exchange

To assess the removal of TFA, purified peptides were characterized by FT-IR spectroscopy. To remove the TFA the peptide samples were dissolved in 0.1 M acetic acid in water and lyophilized for at least 24 h. The treatment was repeated at least three times. To assess removal of TFA, solid samples were analyzed before and after treatment by FT-IR spectroscopy using a Jasco FT/IR 4100 spectrometer (Jasco Europe, Cremella, Italy). Samples were ground into a fine powder and analyzed by FT-IR following the attenuated total reflection (ATR) technique. In these experiments a dry peptide film, from which most of the free solvent water was removed, was prepared on the ATR crystal. Then the characteristic peaks of IR transmission spectra were recorded at a resolution of 4 cm^{-1} over the wavenumber region of 400–4000 cm^{-1}.

4.3. Competitive ELISA for VEGF-A/VEGFR-1 Interaction

The competitive ELISA based assay was performed by coating on 96-well plates a recombinant form of VEGFR-1 (R&D Systems, Minneapolis, MN, USA) at 0.5 μg/mL in PBS, 100 μL/well (the same volume was used for all subsequent steps), 16 h at room temperature. The plate was then blocked for 3 h at RT with 1% bovine serum albumin (BSA). Recombinant form of VEGF-A (R&D Systems) at 5 ng/mL in PBS containing 0.1% BSA, 5 mM EDTA, 0.004% Tween 20 (PBET) and 4% DMSO, was added alone or mixed with peptides at a concentration ranging between 0.04 and 100 μM, and incubated for 1 h at 37 °C followed by 1 h at RT. A biotinylated anti VEGF-A polyclonal antibody (R&D Systems) diluted in PBET at 150 ng/mL, was added to the wells and incubated for 1 h at 37 °C followed by 1 h at RT. A solution containing an avidin and biotinylated horseradish peroxidase (HRP) macromolecular complex was prepared as suggested by the manufacturer (Vector Laboratories, Burlingame, CA, USA) and added to the wells and incubated for 1 h at RT followed by the HRP substrate composed of 1 mg/mL of ortho-phenylenediamine in 50 mM citrate phosphate buffer pH 5, 0.006% of H_2O_2, incubated for 40 min in the dark at RT. The reaction was blocked by adding 30 μL/well of 4 N H_2SO_4 and the absorbance measured at 490 nm on a microplate reader (BenchMark, Biorad, Hercules, CA, USA).

4.4. Animals

C57Bl6/J were purchased from Charles River. Animal experiments were run in accordance with European directives no. 2010/63/UE and Italian directives D.L. 26/2014, and were approved by the Italian Ministry of Health (authorization no. 695/2015-PR of 17 July 2015). For laser-induced CNV, anesthesia was performed by intraperitoneal injection of 100 mg/kg ketamine hydrochloride and 10 mg/kg xylazine. Pupils were dilated with topical tropicamide (1%, Visupharma, Rome, Italy).

4.5. Choroidal Neo-Vascularization Model: Intravitreal Delivery

Laser photocoagulation was performed on 6–8 weeks old C57Bl6/J mice (n = 5 per group), using a 532-nm laser (Meridian, Thun, Switzerland) connected to the Micron IV apparatus (Phoenix Research Labs), Pleasanton, CA, USA). Eyes in which a massive sub-retinal hemorrhage developed after laser induction of CNV were excluded from the analysis. Immediately after the laser application, 10 or 50 μg of iVR1 in 1 μL of vehicle (DMSO) were intravitreally injected with a microsyringe (Hamilton Italy, Agrate Brianza, Italy) carrying a 33-gauge needle. As positive control of inhibition, 2 ng of anti-mouse VEGF-A polyclonal antibody (R&D Systems) in 1 μL of PBS were intravitreally delivered. Of vehicles (DMSO or PBS) 1 μL was delivered to the contralateral eyes.

4.6. Choroidal Neo-Vascularization Model: Oral Delivery by Gavage

Laser photocoagulation was performed on 6–8 weeks old C57Bl6/J mice (n = 5 per group). iVR1 was dissolved in DMSO and daily mixed with food thickener Nutilis (Nutricia, Hoofddorp, The Netherlands) to have a final concentration of 6.25 mg/mL of iVR1 in 4.25% Nutilis, 10% DMSO. Of peptide preparation (50 mg/Kg) 200 μL was administered by gavage using a syringe carrying a rounded tip 20-gauge needle, starting 12 h after laser damage, over 7 days, two times per day. To control animals, the same volume of vehicle was delivered.

4.7. Choroidal Neo-Vascularization Volume Quantification

Seven days after laser injury, eyes were enucleated and processed to isolate eye-cups that were stained with 0.7% FITC-conjugated *Griffonia simplicifolia* Isolectin B4 (Vector Laboratories). Afterwards, retinae were removed and RPE-choroid were flat mounted by four incisions under dissecting microscope and then mounted with Vectashield. CNV were visualized under Leica DM6000 fluorescent microscope and horizontal optical sections were obtained at every 1-µm step from the surface to the deepest focal plane. The CNV volume was measured summing the whole fluorescent area of each optical section with ImageJ software.

4.8. Statistical Analysis

Results are expressed as mean ± SEM, with p values < 0.05 considered statistically significant. Differences among groups were compared by the Student's t test (two-tailed) or one-way ANOVA.

5. Patents

Patents resulting from the work reported in this manuscript: Italian patent n. 102018000008507 filed on 11 November 2018, entitled: "Peptidi ed usi medici correlati".

Author Contributions: Conceptualization: V.T. and S.D.F.; investigation, V.T., E.I., V.C., R.S.; formal analysis, V.T., M.R., S.D.F.; writing: M.R. and S.D.F.; supervision, S.D.F.; funding acquisition, M.R., S.D.F. All authors have read and agreed to the published version of the manuscript.

Funding: This research was funded by: Regione Campania, Project SATIN—POR Campania FESR 2014/2020 to M.R. and S.D.F. and Project Campania Oncotherapies, to M.R.; MIUR (Italian Ministry of Instruction, University and Research) project PRIN 2017XJ38A4, to S.D.F.

Acknowledgments: The authors thank the IGB integrated microscopy and animal house facilities, Anna Maria Aliperti for manuscript editing, and Rosario Chiusolo for continuous support.

Conflicts of Interest: M.R. and S.D.F. are coinventors of the Italian patent n. 102018000008507 and scientific founders of ANBITION s.r.l. The funders had no role in the design of the study; in the collection, analyses, or interpretation of data; in the writing of the manuscript, or in the decision to publish the results.

Abbreviations

AMD	age-related macular degeneration
CNV	choroid neovascularization
CRVO	central retinal vein occlusion
DR	diabetic retinopathy
iVR1	inhibitor of VEGF receptor 1
ROP	retinopathy of prematurity
RPE	retinal pigment epithelium
TFA	trifluoroacetate
TKi	tyrosine kinase inhibitors
VEGF	vascular endothelial growth factor
VEGFR1	vascular endothelial growth factor receptor 1
VEGFR2	vascular endothelial growth factor receptor 2

References

1. Wong, W.L.; Su, X.; Li, X.; Cheung, C.M.; Klein, R.; Cheng, C.Y.; Wong, T.Y. Global prevalence of age-related macular degeneration and disease burden projection for 2020 and 2040: A systematic review and meta-analysis. *Lancet Glob. Health* **2014**, *2*, e106–e116. [CrossRef]
2. Gelfand, B.D.; Ambati, J. A Revised Hemodynamic Theory of Age-Related Macular Degeneration. *Trends Mol. Med.* **2016**, *22*, 656–670. [CrossRef] [PubMed]
3. Ferrara, N.; Mass, R.D.; Campa, C.; Kim, R. Targeting VEGF-A to treat cancer and age-related macular degeneration. *Annu. Rev. Med.* **2007**, *58*, 491–504. [CrossRef] [PubMed]

4. Tarallo, V.; De Falco, S. The vascular endothelial growth factors and receptors family: Up to now the only target for anti-angiogenesis therapy. *Int. J. Biochem. Cell Biol.* **2015**, *64*, 185–189. [CrossRef] [PubMed]
5. Group, C.R.; Martin, D.F.; Maguire, M.G.; Ying, G.S.; Grunwald, J.E.; Fine, S.L.; Jaffe, G.J. Ranibizumab and bevacizumab for neovascular age-related macular degeneration. *N. Engl. J. Med.* **2011**, *364*, 1897–1908.
6. Schmidt-Erfurth, U.; Kaiser, P.K.; Korobelnik, J.F.; Brown, D.M.; Chong, V.; Nguyen, Q.D.; Ho, A.C.; Ogura, Y.; Simader, C.; Jaffe, G.J.; et al. Intravitreal aflibercept injection for neovascular age-related macular degeneration: Ninety-six-week results of the VIEW studies. *Ophthalmology* **2014**, *121*, 193–201. [CrossRef]
7. Lux, A.; Llacer, H.; Heussen, F.M.; Joussen, A.M. Non-responders to bevacizumab (Avastin) therapy of choroidal neovascular lesions. *Br. J. Ophthalmol.* **2007**, *91*, 1318–1322. [CrossRef]
8. Krebs, I.; Glittenberg, C.; Ansari-Shahrezaei, S.; Hagen, S.; Steiner, I.; Binder, S. Non-responders to treatment with antagonists of vascular endothelial growth factor in age-related macular degeneration. *Br. J. Ophthalmol.* **2013**, *97*, 1443–1446. [CrossRef]
9. Treins, C.; Giorgetti-Peraldi, S.; Murdaca, J.; Van Obberghen, E. Regulation of vascular endothelial growth factor expression by advanced glycation end products. *J. Biol. Chem.* **2001**, *276*, 43836–43841. [CrossRef]
10. Nishijima, K.; Ng, Y.S.; Zhong, L.; Bradley, J.; Schubert, W.; Jo, N.; Akita, J.; Samuelsson, S.J.; Robinson, G.S.; Adamis, A.P.; et al. Vascular endothelial growth factor-A is a survival factor for retinal neurons and a critical neuroprotectant during the adaptive response to ischemic injury. *Am. J. Pathol.* **2007**, *171*, 53–67. [CrossRef]
11. Kurihara, T.; Westenskow, P.D.; Bravo, S.; Aguilar, E.; Friedlander, M. Targeted deletion of Vegfa in adult mice induces vision loss. *J. Clin. Investig.* **2012**, *122*, 4213–4217. [CrossRef] [PubMed]
12. Rofagha, S.; Bhisitkul, R.B.; Boyer, D.S.; Sadda, S.R.; Zhang, K.; SEVEN-UP Study Group. Seven-year outcomes in ranibizumab-treated patients in ANCHOR, MARINA, and HORIZON: A multicenter cohort study (SEVEN-UP). *Ophthalmology* **2013**, *120*, 2292–2299. [CrossRef] [PubMed]
13. Grunwald, J.E.; Daniel, E.; Huang, J.; Ying, G.S.; Maguire, M.G.; Toth, C.A.; Jaffe, G.J.; Fine, S.L.; Blodi, B.; Klein, M.L.; et al. Risk of geographic atrophy in the comparison of age-related macular degeneration treatments trials. *Ophthalmology* **2014**, *121*, 150–161. [CrossRef] [PubMed]
14. Falavarjani, K.G.; Nguyen, Q.D. Adverse events and complications associated with intravitreal injection of anti-VEGF agents: A review of literature. *Eye* **2013**, *27*, 787–794. [CrossRef]
15. Fischer, C.; Mazzone, M.; Jonckx, B.; Carmeliet, P. FLT1 and its ligands VEGFB and PlGF: Drug targets for anti-angiogenic therapy? *Nat. Rev. Cancer* **2008**, *8*, 942–956. [CrossRef]
16. Waltenberger, J.; Claesson-Welsh, L.; Siegbahn, A.; Shibuya, M.; Heldin, C.H. Different signal transduction properties of KDR and Flt1, two receptors for vascular endothelial growth factor. *J. Biol. Chem.* **1994**, *269*, 26988–26995.
17. Koch, S.; Claesson-Welsh, L. Signal transduction by vascular endothelial growth factor receptors. *Cold Spring Harb. Perspect. Med.* **2012**, *2*, a006502. [CrossRef]
18. Hiratsuka, S.; Maru, Y.; Okada, A.; Seiki, M.; Noda, T.; Shibuya, M. Involvement of Flt-1 tyrosine kinase (vascular endothelial growth factor receptor-1) in pathological angiogenesis. *Cancer Res.* **2001**, *61*, 1207–1213.
19. Carmeliet, P.; Moons, L.; Luttun, A.; Vincenti, V.; Compernolle, V.; De Mol, M.; Wu, Y.; Bono, F.; Devy, L.; Beck, H.; et al. Synergism between vascular endothelial growth factor and placental growth factor contributes to angiogenesis and plasma extravasation in pathological conditions. *Nat. Med.* **2001**, *7*, 575–583. [CrossRef]
20. Apicella, I.; Cicatiello, V.; Acampora, D.; Tarallo, V.; De Falco, S. Full Functional Knockout of Placental Growth Factor by Knockin with an Inactive Variant Able to Heterodimerize with VEGF-A. *Cell Rep.* **2018**, *23*, 3635–3646. [CrossRef]
21. Bellomo, D.; Headrick, J.P.; Silins, G.U.; Paterson, C.A.; Thomas, P.S.; Gartside, M.; Mould, A.; Cahill, M.M.; Tonks, I.D.; Grimmond, S.M.; et al. Mice lacking the vascular endothelial growth factor-B gene (Vegfb) have smaller hearts, dysfunctional coronary vasculature, and impaired recovery from cardiac ischemia. *Circ. Res.* **2000**, *86*, e29–e35. [CrossRef] [PubMed]
22. Wu, Y.; Zhong, Z.; Huber, J.; Bassi, R.; Finnerty, B.; Corcoran, E.; Li, H.; Navarro, E.; Balderes, P.; Jimenez, X.; et al. Anti-vascular endothelial growth factor receptor-1 antagonist antibody as a therapeutic agent for cancer. *Clin. Cancer Res.* **2006**, *12*, 6573–6584. [CrossRef] [PubMed]
23. Fischer, C.; Jonckx, B.; Mazzone, M.; Zacchigna, S.; Loges, S.; Pattarini, L.; Chorianopoulos, E.; Liesenborghs, L.; Koch, M.; De Mol, M.; et al. Anti-PlGF inhibits growth of VEGF(R)-inhibitor-resistant tumors without affecting healthy vessels. *Cell* **2007**, *131*, 463–475. [CrossRef] [PubMed]

24. Bae, D.G.; Kim, T.D.; Li, G.; Yoon, W.H.; Chae, C.B. Anti-flt1 peptide, a vascular endothelial growth factor receptor 1-specific hexapeptide, inhibits tumor growth and metastasis. *Clin. Cancer Res.* **2005**, *11*, 2651–2661. [CrossRef]
25. Taylor, A.P.; Goldenberg, D.M. Role of placenta growth factor in malignancy and evidence that an antagonistic PlGF/Flt-1 peptide inhibits the growth and metastasis of human breast cancer xenografts. *Mol. Cancer Ther.* **2007**, *6*, 524–531. [CrossRef]
26. De Falco, S. Antiangiogenesis therapy: An update after the first decade. *Korean J. Intern. Med.* **2014**, *29*, 1–11. [CrossRef]
27. Ponticelli, S.; Marasco, D.; Tarallo, V.; Albuquerque, R.J.; Mitola, S.; Takeda, A.; Stassen, J.M.; Presta, M.; Ambati, J.; Ruvo, M.; et al. Modulation of angiogenesis by a tetrameric tripeptide that antagonizes vascular endothelial growth factor receptor 1. *J. Biol. Chem.* **2008**, *283*, 34250–34259. [CrossRef]
28. Cicatiello, V.; Apicella, I.; Tudisco, L.; Tarallo, V.; Formisano, L.; Sandomenico, A.; Kim, Y.; Bastos-Carvalho, A.; Orlandi, A.; Ambati, J.; et al. Powerful anti-tumor and anti-angiogenic activity of a new anti-vascular endothelial growth factor receptor 1 peptide in colorectal cancer models. *Oncotarget* **2015**, *6*, 10563–10576. [CrossRef]
29. Caporale, A.; Doti, N.; Monti, A.; Sandomenico, A.; Ruvo, M. Automatic procedures for the synthesis of difficult peptides using oxyma as activating reagent: A comparative study on the use of bases and on different deprotection and agitation conditions. *Peptides* **2018**, *102*, 38–46. [CrossRef]
30. Valenti, L.E.; Paci, M.B.; De Pauli, C.P.; Giacomelli, C.E. Infrared study of trifluoroacetic acid unpurified synthetic peptides in aqueous solution: Trifluoroacetic acid removal and band assignment. *Anal. Biochem.* **2011**, *410*, 118–123. [CrossRef]
31. Barth, A. Infrared spectroscopy of proteins. *Biochim. Biophys. Acta* **2007**, *1767*, 1073–1101. [CrossRef] [PubMed]
32. Lee, S.; Chen, T.T.; Barber, C.L.; Jordan, M.C.; Murdock, J.; Desai, S.; Ferrara, N.; Nagy, A.; Roos, K.P.; Iruela-Arispe, M.L. Autocrine VEGF signaling is required for vascular homeostasis. *Cell* **2007**, *130*, 691–703. [CrossRef] [PubMed]
33. Dewerchin, M.; Carmeliet, P. PlGF: A multitasking cytokine with disease-restricted activity. *Cold Spring Harb. Perspect. Med.* **2012**, *2*, a011056. [CrossRef] [PubMed]
34. Jakobsson, L.; Franco, C.A.; Bentley, K.; Collins, R.T.; Ponsioen, B.; Aspalter, I.M.; Rosewell, I.; Busse, M.; Thurston, G.; Medvinsky, A.; et al. Endothelial cells dynamically compete for the tip cell position during angiogenic sprouting. *Nat. Cell Biol.* **2010**, *12*, 943–953. [CrossRef] [PubMed]
35. Clauss, M.; Weich, H.; Breier, G.; Knies, U.; Rockl, W.; Waltenberger, J.; Risau, W. The vascular endothelial growth factor receptor Flt-1 mediates biological activities. Implications for a functional role of placenta growth factor in monocyte activation and chemotaxis. *J. Biol. Chem.* **1996**, *271*, 17629–17634. [CrossRef] [PubMed]
36. McLaughlin, M.M.; Paglione, M.G.; Slakter, J.; Tolentino, M.; Ye, L.; Xu, C.F.; Suttle, A.B.; Kim, R.Y. Initial exploration of oral pazopanib in healthy participants and patients with age-related macular degeneration. *JAMA Ophthalmol.* **2013**, *131*, 1595–1601. [CrossRef]
37. Meredith, E.L.; Mainolfi, N.; Poor, S.; Qiu, Y.; Miranda, K.; Powers, J.; Liu, D.; Ma, F.; Solovay, C.; Rao, C.; et al. Discovery of Oral VEGFR-2 Inhibitors with Prolonged Ocular Retention That Are Efficacious in Models of Wet Age-Related Macular Degeneration. *J. Med. Chem.* **2015**, *58*, 9273–9286. [CrossRef]
38. Wei, X.; Zhang, T.; Yao, Y.; Zeng, S.; Li, M.; Xiang, H.; Zhao, C.; Cao, G.; Li, M.; Wan, R.; et al. Efficacy of Lenvatinib, a multitargeted tyrosine kinase inhibitor, on laser-induced CNV mouse model of neovascular AMD. *Exp. Eye Res.* **2018**, *168*, 2–11. [CrossRef]

International Journal of
Molecular Sciences

MDPI

Article

Food-Derived Collagen Peptides, Prolyl-Hydroxyproline (Pro-Hyp), and Hydroxyprolyl-Glycine (Hyp-Gly) Enhance Growth of Primary Cultured Mouse Skin Fibroblast Using Fetal Bovine Serum Free from Hydroxyprolyl Peptide

Tomoko T. Asai [1,2], Fumi Oikawa [3], Kazunobu Yoshikawa [1], Naoki Inoue [4] and Kenji Sato [1,*]

[1] Division of Applied Biosciences, Graduate School of Agriculture, Kyoto University, Kyoto 606 8502, Japan; asai@cc.nara-wu.ac.jp (T.T.A.); kazunobu2017@gmail.com (K.Y.)

[2] Department of Food Science and Nutrition, Faculty of Human Life and Environment, Nara Women's University, Nara 630 8506, Japan

[3] Division of Applied Life Sciences, Graduate School of Life and Environment Sciences, Kyoto Prefectural University, Kyoto 606 8522, Japan; 23fumi.oioi@gmail.com

[4] Peptide Division, Nitta Gelatin Inc., 2-22 Futamata, Yao, Osaka 581 0024, Japan; na-inoue@nitta-gelatin.co.jp

* Correspondence: kensato@kais.kyoto-u.ac.jp; Tel.: +81-75-753-6444

Received: 30 November 2019; Accepted: 26 December 2019; Published: 28 December 2019

Abstract: Prolyl-hydroxyproline (Pro-Hyp) and hydroxyprolyl-glycine (Hyp-Gly) appear in human blood after ingestion of collagen hydrolysate and trigger growth of fibroblasts attached on collagen gel, which has been associated with beneficial effects upon ingestion of collagen hydrolysate, such as improvement of skin and joint conditions. In the present study, inconsistent results were obtained by using different lots of fetal bovine serum (FBS). Fibroblasts proliferated in collagen gel without adding Pro-Hyp and Hyp-Gly and did not respond to addition of Pro-Hyp and Hyp-Gly, which raises doubts about conclusions from prior research. Unexpectedly high levels of hydroxyprolyl peptides, including Pro-Hyp, however, were present in the FBS (approximately 100 μM), and also in other commercially available forms of FBS (70–80 μM). After removal of low molecular weight (LMW, < 6000 Da) compounds from the FBS by size exclusion chromatography, Pro-Hyp and Hyp-Gly again triggered growth of fibroblasts attached on collagen and increased the number of fibroblasts migrated from mouse skin. These results indicate the presence of bioactive hydroxyprolyl peptides in commercially available FBS, which can mask effects of Pro-Hyp and Hyp-Gly supplementation; our work confirms that Pro-Hyp and Hyp-Gly do play crucial roles in proliferation of fibroblasts.

Keywords: prolyl-hydroxyproline (Pro-Hyp); hydroxyprolyl-glycine (Hyp-Gly); collagen peptide; fibroblasts; fetal bovine serum (FBS)

1. Introduction

Collagen is the main protein in the extracellular matrix and has a triple-helical structure. Collagen has two specific post-translationally modified amino acids: hydroxyproline (Hyp) and hydroxylysine (Hyl). Heat treatment converts the triple-helical structure of collagen into a globular structure, which is referred to as gelatin. The protease digest of gelatin is referred to as collagen hydrolysate, gelatin hydrolysate, or collagen peptide. Collagen hydrolysate is prepared from skin, bones, and tendons of animals, or the skin and scales of fish. In human trials with placebo controls, ingestion of collagen hydrolysate (2.5–10 g/day) suppresses transepidermal water loss, reduces wrinkle volume, and increases elasticity of skin [1–4]. Furthermore, ingestion of collagen hydrolysate moderates the symptoms of osteoarthritis [5,6] and enhances healing of pressure ulcers [7–9]. Our previous work has shown that ingestion of collagen hydrolysate (2–20 g) increases the peptide forms of Hyp (i.e., hydroxyprolyl peptide or collagen peptide) in human peripheral blood plasma to 20–100 μM [10–13]. To date, the presence of more than ten food-derived hydroxyprolyl peptides in human blood has been reported, including prolyl-hydroxyproline (Pro-Hyp), hydroxyprolyl-glycine (Hyp-Gly), alanyl-hydroxyproline (Ala-Hyp), isoleucyl-hydroxyproline (Ile-Hyp), leucyl-hydroxyproline (Leu-Hyp), phenylalanyl-hydroxyproline (Phe-Hyp), glutamyl-hydroxyproline (Glu-Hyp), prolyl-hydroxyprolyl-glycine (Pro-Hyp-Gly), glycyl-prolyl-hydroxyproline (Gly-Pro-Hyp), alanyl-hydroxyprolyl-glycine (Ala-Hyp-Gly), and serinyl-hydroxyprolyl-glycine (Ser-Hyp-Gly). Pro-Hyp and Hyp-Gly are the main hydroxyprolyl peptides found in human blood after ingestion of collagen hydrolysate [10–13]. Pro-Hyp is also generated by the degradation of endogenous collagen in tissue undergoing inflammation [14] and at wound healing sites in the skin [15].

Mouse skin fibroblasts attached on collagen gel stopped growing without addition of Pro-Hyp, even in the presence of fetal bovine serum (FBS), whereas fibroblasts grew on plastic plates in the presence of FBS [11,16,17]. We previously reported that Pro-Hyp and Hyp-Gly triggered the growth of fibroblasts attached on collagen gel [11,17], which has been associated with biological responses upon ingestion of collagen hydrolysate. However, some researchers have obtained results inconsistent with our findings (personal communication). In the present study, we also found that fibroblasts attached on collagen gel grew without adding Pro-Hyp when we used different lots of FBS than those used in previous studies. The objectives of the present study were to solve this problem and confirm the effects of food-derived hydroxyprolyl peptides on the growth of fibroblasts.

2. Results and Discussion

2.1. Growth of Fibroblasts on Collagen Gel in Medium Containing FBS-1

Mouse skin fibroblasts were cultivated on collagen gel in a medium containing a commercially available lot of FBS (FBS-1). As shown in Figure 1, fibroblasts grew on the collagen gel even without addition of Pro-Hyp and Hyp-Gly. Addition of mixture of Pro-Hyp (100 μM) and Hyp-Gly (100 μM) did not significantly enhance growth of fibroblasts. These results are inconsistent with previous studies using different lots of the same brand of FBS [11,16,17], in which fibroblasts grew on collagen gel only after adding Pro-Hyp and Hyp-Gly. It has been demonstrated that some FBS lots contain significant amounts of free Hyp, while the presence of hydroxyprolyl peptide has not been examined [18]. We assumed that FBS might contain different levels of hydroxyprolyl peptides depending on lot number and brand, which might explain the inconsistent results.

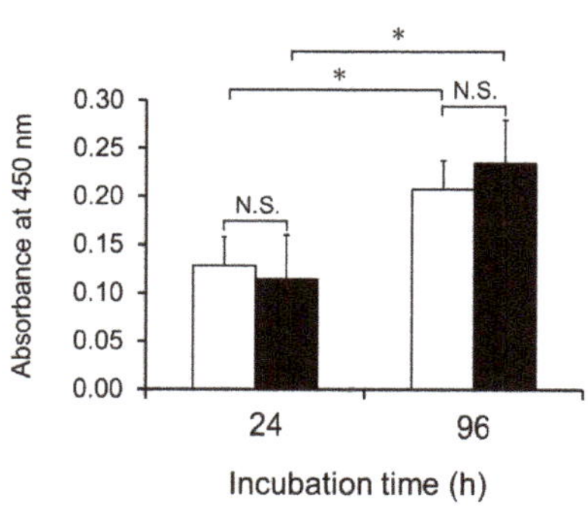

Figure 1. Effect of a mixture of Pro-Hyp and Hyp-Gly on the growth of fibroblasts on collagen gel in the presence of 10% FBS-1; (□), control; (■), medium containing Pro-Hyp and Hyp-Gly at 100 μM, respectively. Data are shown as mean ± standard deviation (SD) ($n = 5$). Asterisks indicate significant differences ($p < 0.05$; Tukey's test). N.S. indicates results that are not significantly different.

2.2. Presence of Hydroxyprolyl Peptides in FBS

Amino acid analysis revealed the presence of hydroxyprolyl peptides in the type of FBS used in the present study. As shown in Figure 2A, FBS contained unexpectedly higher levels of hydroxyprolyl peptide (approximately 70–100 µM) than adult bovine serum (ABS) and human plasma before ingestion of collagen hydrolysate [10,11,13]. As shown in Figure 2B, Pro-Hyp accounted for 37%–70% of total hydroxyprolyl peptides in FBS. These values are similar to those in human plasma after ingestion of collagen hydrolysate [10,11,13]. On the other hand, only negligible amounts of Hyp-Gly were present in the FBS. It has been demonstrated that constituents in FBS differ between lots, even from the same brand [18]. Thus, different lots of FBS might contain different levels of Pro-Hyp and other hydroxyprolyl peptides.

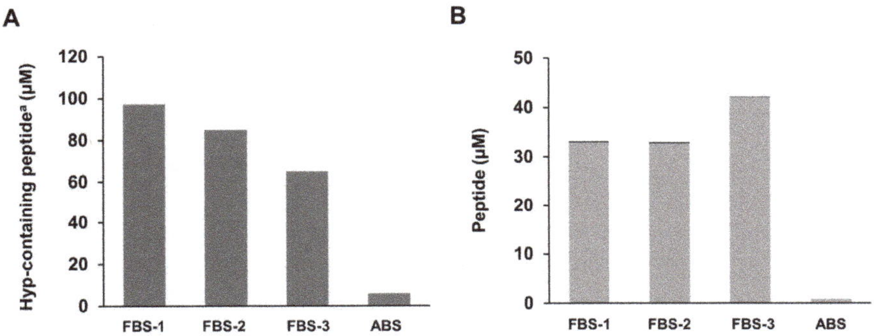

Figure 2. Contents of hydroxyprolyl peptide in commercially available fetal bovine serum (FBS) and adult bovine serum (ABS): (**A**) Hyp-containing peptide; (**B**) content of Pro-Hyp (■) and Hyp-Gly (■); [a] peptide form of Hyp.

2.3. Removal of Hydroxyprolyl Peptides from FBS-1

As shown in Figure 3, the protein in FBS-1 was eluted in fractions 4–7 by size-exclusion chromatography (SEC). Hydroxyprolyl peptides were eluted in fractions 6–10. According to the instructions from the supplier, peptides larger than 6000 Da were eluted in fractions 4–7. Thus, hydroxyprolyl peptides in fractions 6 and 7 were larger than 6000 Da. Based on these facts, fractions 4–7 were collected and used as FBS-1 free from low molecular weight (LMW) compounds. Fractions 8–13 were collected and used as LMW fractions.

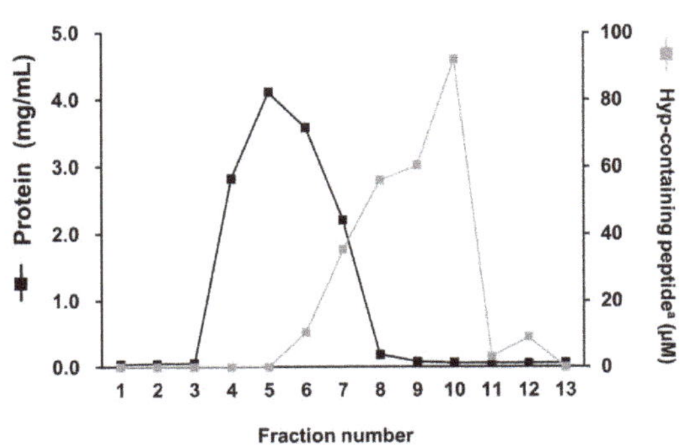

Figure 3. Elution of protein and hydroxyprolyl peptide in FBS-1 from the Econo-Pac 10DG column; (■), protein; (■), Hyp-containing peptide; [a] peptide form of Hyp.

The different fractions (FBS-1, FBS-1 free from LMW fractions, and LMW fractions) were each added to the medium of fibroblasts attached on plastic plates. FBS-1 free from LMW fractions caused fibroblast proliferation equivalent to proliferation with non-purified FBS-1 (Figure 4), while LMW fractions caused little fibroblast proliferation. These facts indicate that protein growth factors play a significant role in proliferating fibroblasts on plastic plates, as compared to LMW compounds in FBS.

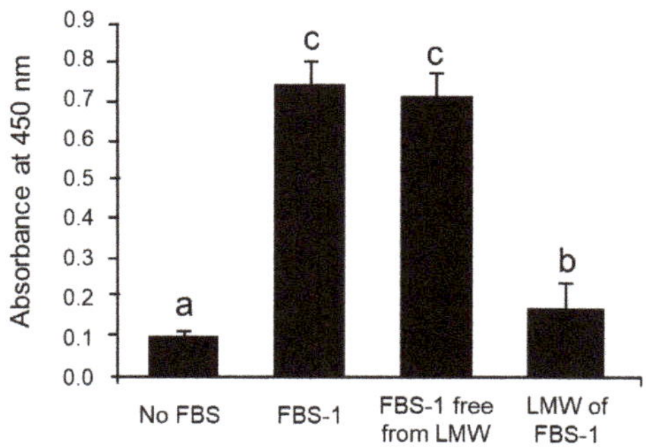

Figure 4. Effect of FBS-1 and its fractions on fibroblast growth on plastic plates. FBS-1 free from LMW, SEC Fr. 4–7; LMW of FBS-1, SEC Fr. 8–13. Data are shown as the mean ± SD ($n = 5$). Different letters indicate significant differences ($p < 0.05$, Tukey's test).

2.4. *Effect of Hydroxyprolyl Peptides on Growth of Fibroblasts on Collagen Gel*

As shown in Figure 5, fibroblasts attached on collagen gel stopped growing in the presence of FBS-1 free from LMW compounds. The addition of Pro-Hyp (100 μM) and Hyp-Gly (100 μM) triggered growth of the fibroblasts. These results are consistent with our previous studies [11,17]. However, when the present FBS-1 was used without purification, entirely different results were obtained (Figure 1) due to presence of hydroxyprolyl peptides in the FBS-1.

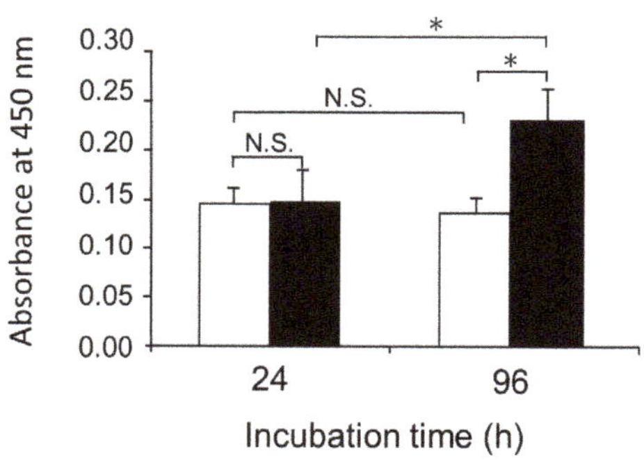

Figure 5. Effect of mixture of Pro-Hyp and Hyp-Gly on the growth of fibroblasts on collagen gel in the presence of 10% FBS-1 free from LMW hydroxyprolyl peptides; (□), control; (■), medium containing Pro-Hyp and Hyp-Gly at 100 μM. Data are shown as mean ± SD ($n = 5$). Asterisks indicate significant differences ($p < 0.05$, Tukey's test). N.S. indicates results that are not significantly different.

2.5. *Effect of Pro-Hyp on Number of Fibroblasts Migrated from Mouse Skin*

In our previous research, we demonstrated that Pro-Hyp (200 μM) increased the number of fibroblasts that migrated from skin in the absence of FBS [17]. However, the addition of FBS removed this effect [17]. The FBS-1 free from LMW compounds was added to the skin culture system to give 2% and 10% concentrations. As shown in Figure 6, the number of fibroblasts that migrated from skin was significantly increased by the addition of Pro-Hyp (200 μM) in absence of FBS after 72 h, which is consistent with previous results [17]. Even in the presence of FBS-1 free from LMW compounds at 2% and 10%, Pro-Hyp also increased the number of fibroblasts migrated from skin after 48 and 24 h, respectively. In the previous study [17], hydroxyprolyl peptides in FBS might have partially masked the effect of Pro-Hyp on the number of fibroblasts migrated from mouse skin.

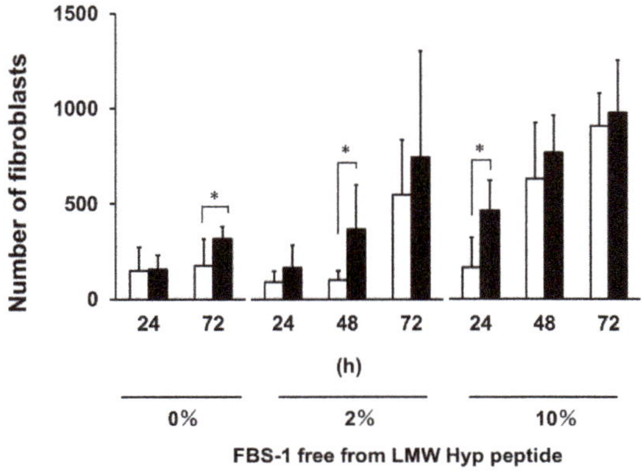

Figure 6. Effect of Pro-Hyp on the number of fibroblasts migrated from mouse skin in the absence or presence of FBS-1 free from LMW hydroxyprolyl peptides. Open column, control; closed column, medium containing Pro-Hyp at 200 μM. Data are shown as the mean ± SD (n = 6). Asterisks indicate significant differences ($p < 0.05$, Student's t-test).

3. Materials and Methods

3.1. Bovine Sera

Three different brands of FBS (FBS-1–3) were commercially obtained. One brand of ABS was commercially obtained.

3.2. Chemicals

The amino acid standard mixture (Type H), acetonitrile (for high performance liquid chromatography: HPLC), trimethylamine, and phenyl isothiocyanate were all purchased from Wako Chemicals (Osaka, Japan). Hydroxyproline and 0.25% trypsin-ethylenediaminetetraacetic acid (EDTA) solution were purchased from Nacalai Tesque (Kyoto, Japan). Pro-Hyp and Hyp-Gly were obtained from Bachem (Bubendorf, Switzerland). Dulbecco's phosphate-buffered saline (D-PBS) and gentamicin were purchased from Invitrogen (Carlsbad, CA, USA). Dulbecco's modified Eagle medium (DMEM) supplemented with L-glutamine (584 mg/L) was from Sigma-Aldrich (St. Louis, MO, USA). Cell Counting Kit-8 was purchased from Dojin Glocal (Kumamoto, Japan). Calf acid-soluble type I collagen solution (0.5%; IAC-50) was purchased from Koken (Tokyo, Japan). Here, 6-Aminoquinolyl-N-hydroxy succinimidyl carbamate (AccQ) was obtained from Waters Corporation (Milford, MA, USA). All other reagents were of analytical grade or better.

3.3. Removal of Hydroxyprolyl Peptides from FBS-1

An Econo-Pac 10DG column (Bio-Rad Laboratories, Hercules, CA, USA) was pre-equilibrated with the DMEM. In total, 3 mL of FBS-1 was loaded onto the column. After elution of the first 3 mL, 13 mL DMEM was loaded onto the column. Effluent was collected in 1 mL quantities. Elution of protein was monitored by the Bradford method using a Protein Assay Kit (Bio-Rad Laboratories). Hydroxyproline and hydroxyprolyl peptides were detected by amino acid analysis, as described previously [10].

3.4. Animals

All experiments were conducted according to the ethical guidelines of the Kyoto University Animal Research Committee. The protocol was approved by the Kyoto University Animal Research Committee (permission number: 2014–45, 2015–38). Five-week-old male Balb/c mice were purchased from Japan SLC (Shizuoka, Japan). Mice were sacrificed by cervical dislocation under deep isoflurane anesthesia. Abdominal skin was sterilized with 70% ethanol, shaved using a razor, and then stripped for use in these experiments.

3.5. Estimation of the Number of Cells Migrated from Mouse Skin

The skin was rinsed with D-PBS and DMEM to remove ethanol and placed on sterilized rubber plates. Disks measuring 4 mm in diameter were punched out using a Dermal Punch (Nipro, Tokyo, Japan). The skin disks were then placed on 12-well plastic plates (Falcon BD, Lakes, NJ, USA). DMEM was supplemented with gentamicin (0.01 mg/mL), FBS-1 (0%, 2%, and 10%) free from LMW hydroxyprolyl peptides, and Pro-Hyp (0 and 200 μM), respectively. Then, 1 mL of each mixture was added to the wells. The 12-well plastic plates were placed in a humidified incubator at 37 °C under 5% CO_2. After incubation at suitable intervals, cells were fixed with 4% paraformaldehyde and observed using a phase-contrast microscope. The number of cells attached on the plate was directly counted.

3.6. Cell Proliferation Assay

Pieces of mouse abdominal skin (approximately 6–7 mm in width) were prepared using scissors and placed on a culture dish (90 mm i.d.). DMEM supplemented with gentamicin (0.01 mg/mL) and 10% FBS-1 (5 mL) were added into dish. The skin pieces were cultured in a humidified incubator at 37 °C under 5% CO_2. During cultivation, the medium was changed every 2–3 days. After incubation for 2 weeks, the skin disks were removed; the fibroblasts migrated from skin were washed with PBS and treated with 1 mL of a 0.25% trypsin-EDTA solution at 37 °C for 10 min. Here, 9 mL of medium with 10% FBS-1 was poured into the plate to inactivate trypsin. Fibroblasts were collected by centrifugation at 3000 × g for 10 min. The pellets were suspended in each media: DMEM only, medium containing FBS-1, FBS-1 free from LMW hydroxyprolyl peptides, or LMW fraction of FBS-1 (5×10^4 cells/mL), with or without the addition of Pro-Hyp (final concentration 100 μM) and Hyp-Gly (100 μM). Fibroblasts (5×10^3 cells/100 μL) were cultured on 96-well plastic plates or collagen gel-coated plates in each media: DMEM only, medium containing FBS-1, FBS-1 free from LMW hydroxyprolyl peptides, or LMW fraction of FBS-1, with and without the addition of Pro-Hyp and Hyp-Gly. The collagen solution (0.5%) and the same volume of double-concentrated DMEM medium were mixed, and 100 μL of mixture was added into each well of the 96-well plastic plate. The plate was then incubated in a humidified incubator for 1 h at 37 °C under 5% CO_2 to allow gelation. Cell proliferation was monitored using a Cell Counting Kit-8 instrument.

3.7. Amino Acid Analysis

FBS, ABS, and SEC fractions of FBS-1 were mixed with three volumes of ethanol and centrifuged at 1000 × g for 10 min. The supernatant was used as the 75% ethanol-soluble fraction. Next, 100 μL of the 75% ethanol-soluble fraction was dried in a glass tube (5 × 60 mm) and hydrolyzed by 6 M HCl vapor at 150 °C for 1 h, as described previously [10]. The Hyp contents in the non-hydrolysate and HCl hydrolysate were determined according to the method of Bidlingmeyer et al. [19], with slight modifications [10].

3.8. Determination of Pro-Hyp and Hyp-Gly

Pro-Hyp and Hyp-Gly in FBS were derivatized with AccQ and then determined by the liquid chromatography tandem mass spectrometry (LC-MS/MS) in multireaction monitoring (MRM) mode, using an LCMS-8040 (Shimadzu, Kyoto, Japan) and high-pressure binary gradient HPLC (LC20 system, Shimadzu), as previously reported [13].

3.9. Statistical analysis

Differences between means were evaluated using one-way analysis of variance, followed by Tukey's multiple comparison test for post hoc analysis using GraphPad Prism Version 6.04 (GraphPad Software, San Diego, CA, USA). Differences between the two groups were compared using Student's *t*-tests.

4. Conclusions

The present study demonstrates that some commercially available FBS contain high levels of LMW hydroxyprolyl peptides (70–100 µM), including Pro-Hyp. These values are higher than those in ABS and human plasma without ingestion of collagen hydrolysate. By using FBS that is free from LMW hydroxyprolyl peptides, the present study clearly confirms that Pro-Hyp and Hyp-Gly play crucial roles in proliferation of fibroblasts attached on collagen gel. It has been demonstrated that hydroxyprolyl peptides exert many functions, such as anti-hypertention [20], anti-inflammation [21], and improvement of glucose tolerance [22], in addition to improving skin and joint conditions. Therefore, researchers who use cell culture systems to evaluate biological activities should be aware of the presence of bioactive hydroxyprolyl peptides in FBS.

Author Contributions: Conceptualization, K.S.; investigation, T.T.A., F.O. and K.Y.; writing—original draft preparation, T.T.A.; writing—review and editing, K.S.; funding acquisition, N.I. All authors have read and agreed to the published version of the manuscript.

Funding: This work was supported by Commissioned Research between Kyoto University and Nitta Gelatin Inc. (Project number 150141000045).

Acknowledgments: The authors would like to thank Enago (www.enago.jp) for the English language review.

Conflicts of Interest: The authors declare no conflict of interest.

Abbreviations

Pro-Hyp	prolyl-hydoxyproline
Hyp-Gly	hydroxyprolyl-glycine
Ala-Hyp	alanyl-hydroxyproline
Ile-Hyp	isoleucyl-hydroxyproline
Leu-Hyp	leucyl-hydroxyproline
Phe-Hyp	phenylalanyl-hydroxyproline
Glu-Hyp	glutamyl-hydroxyproline
Pro-Hyp-Gly	prolyl-hydroxyprolyl-glycine
Gly-Pro-Hyp	glycyl-prolyl-hydroxyproline
Ala-Hyp-Gly	alanyl-hydroxyprolyl-glycine
Ser-Hyp-Gly	serinyl-hydroxyprolyl-glycine
FBS	fetal bovine serum
LMW	low molecular weight
ABS	adult bovine serum
SEC	size-exclusion chromatography
HPLC	high performance liquid chromatography
EDTA	ethylenediaminetetraacetic acid
DMEM	Dulbecco's Modified Eagle Medium
D-PBS	Dulbecco's phosphate-buffered saline
AccQ	6-Aminoquinolyl-N-hydroxy succinimidyl carbamate
LC-MS/MS	liquid chromatography tandem mass spectrometry
MRM	multiple reaction monitoring

References

1. Proksch, E.; Schunck, M.; Zague, V.; Segger, D.; Degwert, J.; Oesser, S. Oral intake of specific bioactive collagen peptides reduces skin wrinkles and increases dermal matrix synthesis. *Skin Pharmacol. Physiol.* **2014**, *27*, 113–119. [CrossRef]
2. Proksch, E.; Segger, D.; Degwert, J.; Schunck, M.; Zague, V.; Oesser, S. Oral supplementation of specific collagen peptides has beneficial effects on human skin physiology: a double-blind, placebo-controlled study. *Skin Pharmacol. Physiol.* **2014**, *27*, 47–55. [CrossRef]
3. Sugihara, F.; Inoue, N.; Wang, X. Clinical effects of ingesting collagen hydrolysate on facial skin properties: A randomized, placebo-controlled, double-blind trial. *Jpn. Pharmacol. Ther.* **2015**, *43*, 67–70.
4. Inoue, N.; Sugihara, F.; Wang, X. Ingestion of bioactive collagen hydrolysates enhance facial skin moisture and elasticity and reduce facial ageing signs in a randomised double-blind placebo-controlled clinical study. *J. Sci. Food Agric.* **2016**, *96*, 4077–4081. [CrossRef] [PubMed]
5. Benito-Ruiz, P.; Camacho-Zambrano, M.M.; Carrillo-Arcentales, J.N.; Mestanza-Peralta, M.A.; Vallejo-Flores, C.A.; Vargas-López, S.V.; Villacís-Tamayo, R.A.; Zurita-Gavilanes, L.A. A randomized controlled trial on the efficacy and safety of a food ingredient, collagen hydrolysate, for improving joint comfort. *Int. J. Food Sci. Nutr.* **2009**, *60*, 99–113. [CrossRef] [PubMed]
6. Kumar, S.; Sugihara, F.; Suzuki, K.; Inoue, N.; Venkateswarathirukumara, S. A double-blind, placebo-controlled, randomised, clinical study on the effectiveness of collagen peptide on osteoarthritis. *J. Sci. Food Agric.* **2015**, *95*, 702–707. [CrossRef]
7. Lee, S.K.; Posthauer, M.E.; Dorner, B.; Redovian, V.; Maloney, M.J. Pressure ulcer healing with a concentrated, fortified, collagen protein hydrolysate supplement: a randomized controlled trial. *Adv. Skin Wound Care* **2006**, *19*, 92–96. [CrossRef]
8. Yamanaka, H.; Okada, S.; Sanada, H. A multicenter, randomized, controlled study of the use of nutritional supplements containing collagen peptides to facilitate the healing of pressure ulcers. *J. Nutr. Intermed. Metab.* **2017**, *8*, 51–59. [CrossRef]
9. Sugihara, F.; Inoue, N.; Venkateswarathirukumara, S. Ingestion of bioactive collagen hydrolysates enhanced pressure ulcer healing in a randomized double-blind placebo-controlled clinical study. *Sci. Rep.* **2018**, *8*, 11403. [CrossRef]
10. Iwai, K.; Hasegawa, T.; Taguchi, Y.; Morimatsu, F.; Sato, K.; Nakamura, Y.; Higashi, A.; Kido, Y.; Nakabo, Y.; Ohtsuki, K. Identification of food-derived collagen peptides in human blood after oral ingestion of gelatin hydrolysates. *J. Agric. Food Chem.* **2005**, *53*, 6531–6536. [CrossRef]
11. Shigemura, Y.; Akaba, S.; Kawashima, E.; Park, E.Y.; Nakamura, Y.; Sato, K. Identification of a novel food-derived collagen peptide, hydroxyprolyl-glycine, in human peripheral blood by pre-column derivatisation with phenyl isothiocyanate. *Food Chem.* **2011**, *129*, 1019–1024. [CrossRef] [PubMed]
12. Shigemura, Y.; Kubomura, D.; Sato, Y.; Sato, K. Dose-dependent changes in the levels of free and peptide forms of hydroxyproline in human plasma after collagen hydrolysate ingestion. *Food Chem.* **2014**, *159*, 328–332. [CrossRef] [PubMed]
13. Asai, T.; Takahashi, A.; Ito, K.; Uetake, T.; Matsumura, Y.; Ikeda, K.; Inagaki, N.; Nakata, M.; Imanishi, Y.; Sato, K. Amount of collagen in the meat contained in Japanese daily dishes and the collagen peptide content in human blood after ingestion of cooked fish meat. *J. Agric. Food Chem.* **2019**, *67*, 2831–2838. [CrossRef] [PubMed]
14. Kusubata, M.; Koyama, Y.; Tometsuka, C.; Shigemura, Y.; Sato, K. Detection of endogenous and food-derived collagen dipeptide prolylhydroxyproline (Pro-Hyp) in allergic contact dermatitis-affected mouse ear. *Biosci. Biotechnol. Biochem.* **2015**, *79*, 1356–1361. [CrossRef] [PubMed]
15. Jimi, S.; Sato, K.; Kimura, M.; Suzumiya, J.; Hara, S.; De Francesco, F.; Ohjimi, H. G-CSF administration accelerates cutaneous wound healing accompanied with increased Pro-Hyp production in db/db mice. *Clin. Res. Dermatol.* **2017**, *4*, 1–9. [CrossRef]
16. Kono, T.; Tanii, T.; Furukawa, M.; Mizuno, N.; Kitajima, J.; Ishii, M.; Hamada, T.; Yoshizato, K. Cell cycle analysis of human dermal fibroblasts cultured on or in hydrated type I collagen lattices. *Arch. Dermatol. Res.* **1990**, *282*, 258–262. [CrossRef]

17. Shigemura, Y.; Iwai, K.; Morimatsu, F.; Iwamoto, T.; Mori, T.; Oda, C.; Taira, T.; Park, E.Y.; Nakamura, Y.; Sato, K. Effect of Prolyl-hydroxyproline (Pro-Hyp), a food-derived collagen peptide in human blood, on growth of fibroblasts from mouse skin. *J. Agric. Food Chem.* **2009**, *57*, 444–449. [CrossRef]
18. Price, P.J.; Gregory, E.A. Relationship between in vitro growth promotion and biophysical and biochemical properties of the serum supplement. *In Vitro* **1982**, *18*, 576–584. [CrossRef]
19. Bidlingmeyer, B.A.; Cohen, S.A.; Tarvin, T.L. Rapid analysis of amino acids using pre-column derivatization. *J. Chromatogr.* **1984**, *336*, 93–104. [CrossRef]
20. Kouguchi, T.; Ohmori, T.; Shimizu, M.; Takahata, Y.; Maeyama, Y.; Suzuki, T.; Morimatsu, F.; Tanabe, S. Effects of a chicken collagen hydrolysate on the circulation system in subjects with mild hypertension or high-normal blood pressure. *Biosci. Biotechnol. Biochem.* **2013**, *77*, 691–696. [CrossRef]
21. Zhang, Y.; Kouguchi, T.; Shimizu, K.; Sato, M.; Takahata, Y.; Morimatsu, F. Chicken collagen hydrolysate reduces proinflammatory cytokine production in C57BL/6.KOR-ApoE^{shl} mice. *J. Nutr. Sci. Vitaminol.* **2010**, *56*, 208–210. [CrossRef] [PubMed]
22. Iba, Y.; Yokoi, K.; Eitoku, I.; Goto, M.; Koizumi, S.; Sugihara, F.; Oyama, H.; Yoshimoto, T. Oral administration of collagen hydrolysates improves glucose tolerance in normal mice through GLP-1-dependent and GLP-1-independent mechanisms. *J. Med. Food* **2016**, *19*, 836–843. [CrossRef] [PubMed]

International Journal of *Molecular Sciences*

MDPI

Article

Beneficial Effects of Neurotensin in Murine Model of Hapten-Induced Asthma

Ewelina Russjan and Katarzyna Kaczyńska *

Department of Respiration Physiology, Mossakowski Medical Research Centre, Polish Academy of Sciences, 5 Pawińskiego Street, 02-106 Warsaw, Poland; erussjan@imdik.pan.pl
* Correspondence: kkaczynska@imdik.pan.pl

Received: 20 August 2019; Accepted: 9 October 2019; Published: 11 October 2019

Abstract: Neurotensin (NT) demonstrates ambiguous activity on inflammatory processes. The present study was undertaken to test the potential anti-inflammatory activity of NT in a murine model of non-atopic asthma and to establish the contribution of NTR1 receptors. Asthma was induced in BALB/c mice by skin sensitization with dinitrofluorobenzene followed by intratracheal hapten provocation. The mice were treated intraperitoneally with NT, SR 142948 (NTR1 receptor antagonist) + NT or NaCl. Twenty-four hours after the challenge, airway responsiveness to nebulized methacholine was measured. Bronchoalveolar lavage fluid (BALF) and lungs were collected for biochemical and immunohistological analysis. NT alleviated airway hyperreactivity and reduced the number of inflammatory cells in BALF. These beneficial effects were inhibited by pretreatment with the NTR1 antagonist. Additionally, NT reduced levels of IL-13 and TNF-α in BALF and IL-17A, IL12p40, RANTES, mouse mast cell protease and malondialdehyde in lung homogenates. SR 142948 reverted only a post-NT TNF-α decrease. NT exhibited anti-inflammatory activity in the hapten-induced asthma. Reduced leukocyte accumulation and airway hyperresponsiveness indicate that this beneficial NT action is mediated through NTR1 receptors. A lack of effect by the NTR1 blockade on mast cell activation, oxidative stress marker and pro-inflammatory cytokine production suggests that other pathways can be involved, which requires further research.

Keywords: neurotensin; asthma; inflammation

---

1. Introduction

Neurotensin (NT), regarded as a hormone, paracrine factor or neurotransmitter, regulates various physiological functions. It was first isolated from bovine hypothalamus by Carraway and Leeman in 1973 and named because of its neuronal localization and hypotensive activity [1]. In the brain, NT controls anterior pituitary hormone secretion [2], hypothermia [3] and antinociceptive activity [4]. In the periphery, NT is widely distributed in the gastrointestinal tract, where it influences gut motility. It has also been identified in other peripheral organs: lung, liver, heart, pancreas and spleen [5].

NT is an endogenous tridecapeptide (pGlu-Leu-Tyr-Glu-Asn-Lys-Pro-Arg-Arg-Pro-Tyr-Ile-Leu). The C-terminal part (precisely NT 8-13) is recognized by all known NT receptors and demonstrates activity similar to an entire NT molecule [6,7]. To date, four subtypes of cell surface receptors that can mediate NT responses have been described. The high-affinity NTR1 receptor, responsible for the most physiological effects of NT, is a member of the G protein-coupled receptors family, with seven transmembrane domains [8,9]. The G protein-linked second messenger system is also affected by the low-affinity NTR2 receptor, which presents 64% amino acid homology with NTR1. In contrast to NTR1, NTR2 is characterized by the capability to recognize levocabastine, a histamine H1 receptor antagonist. This sensitivity feature allows for the distinction between NTR1 and NTR2 [10]. However, the role of NTR2 in NT signaling remains controversial. Different results suggest that NT can be treated as an agonist, inverse agonist or antagonist at this binding site [11]. Some similarities have been described between mosaic protein NTR3 and the last identified NTR4 receptor. However, their role in the physiological properties of NT is unclear [12,13].

Neurotensin plays a part as an important regulator of immune cells: lymphocytes, mastocytes and macrophages, as well as vascular, epithelial and connective tissue cells [14,15]. Similarly to other neuropeptides, NT can act as a link between the immune and nervous systems. Neurogenic stimulation is involved in the processes of inflammation, proliferation, cytokine and growth factor production [16] and this neuroimmune interaction can be observed in asthma [17]. Asthmatic airway hyperreactivity is modulated by the neurons innervating lungs [18,19]. NT receptors have been synthesized and transported within a subpopulation of afferent and efferent components of the vagus nerve [20]. Therefore, a possibility exists that, in asthma, NT might exert its effects via the modulation of both inflammatory cells and lung nerves.

Despite considerable advances in understanding NT properties and its function in physiological and pathological phenomena, its exact role in the inflammation process still remains controversial. There are several studies, demonstrating contradicting results, which point to either the pro- or anti-inflammatory activity of NT. As an example, it has been demonstrated that NT promotes an acute inflammatory response in the experimental model of colon inflammation [21], while in another study, NT augmented the healing process of colonic mucosa in the chronic phase of disease [22].

The controversies on NT involvement in the inflammatory processes and the possibility of modulatory action on lung nerves prompted us to investigate its potential anti-inflammatory activity in a murine model of non-atopic asthma induced by a low molecular compound (hapten), specifically dinitrofluorobenzene (DNFB). A simultaneous purpose is establishing the contribution of the NTR1 receptor via pretreatment with the nonpeptide NT antagonist SR 142948.

2. Results

2.1. Airway Hyperresponsiveness (AHR)

There was no significant difference between all tested groups after NaCl and the lowest dose of methacholine (5 mg/mL) nebulization. However, an intratracheal dinitrobenzene sulphonic acid (DNS) challenge in DNFB-sensitized mice increased airway hyperresponsiveness to methacholine in doses of 10, 20 and 40 mg/mL, which was reflected in increased Penh (Figure 1). Intraperitoneal injection of NT after the hapten challenge significantly reduced Penh compared to the positive control group (PC). Administration of the NT antagonist SR 142948 limited the ameliorative activity of NT on airway hyperresponsiveness (Figure 1). Although there was no meaningful difference in Penh values between NT and SR 142948 + NT treated groups ($p = 0.3$, 0.09 and 0.15 for the three highest methacholine doses), antagonist pretreatment counteracted NT-induced AHR reduction in comparison to the PC group ($p = 0.07$, 0.11 and 0.15 for 10, 20 and 40 mg/mL of methacholine).

2.2. Leukocyte Accumulation in BALF and Lung Tissue

The cell influx in the airway lumen during the course of inflammation consisted primarily of macrophages and neutrophils, and, to a lesser degree, lymphocytes. The higher percentage of neutrophils is a hallmark of non-atopic asthma. DNFB-sensitized/DNS-challenged mice (PC) demonstrated an increased number of total inflammatory cells and neutrophils compared to a vehicle-sensitized/DNS-challenged group (NC) (Figure 2A,B). The administration of NT significantly reduced leukocyte accumulation in BALF. The beneficial effect of NT was reduced by pretreatment with SR 142948 (Figure 2A,B). A histological evaluation showed similar changes—apparent peribronchial leukocyte accumulation in the PC group (Figure 2C), which was reflected by an increased inflammation score (Figure 2D). NT treatment significantly attenuated the inflammation score, while pretreatment with the NTR1 antagonist eliminated this effect (Figure 2D).

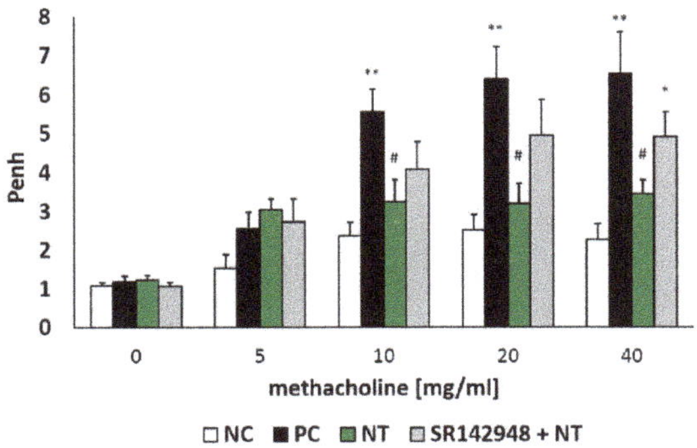

Figure 1. Airway responsiveness (Penh) to increasing concentrations of methacholine (5, 10, 20, 40 mg/mL). Effect of treatment with Neurotensin (NT) alone and NT pretreated with SR 142948 in dinitrofluorobenzene (DNFB)-sensitized/ dinitrobenzene sulphonic acid (DNS)-challenged mice. Comparison to DNFB-sensitized/DNS-challenged group (positive control; PC) and vehicle-sensitized/DNS-challenged group (negative control; NC) treated with NaCl. The results are presented as means ± SEM; n = 6–7 mice per group. * $p < 0.05$, ** $p < 0.01$ compared with the NC group, # $p < 0.05$ compared with the PC group.

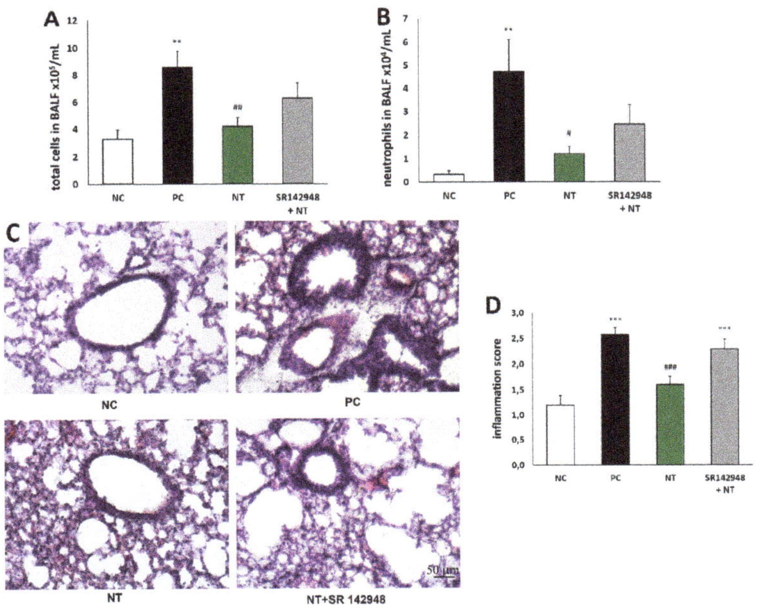

Figure 2. Reduction of the total number of inflammatory cells (**A**) and neutrophils (**B**) in bronchoalveolar lavage fluid (BALF) and effect of NT on lung tissue inflammatory cell infiltration; hematoxylin- and eosin-stained lung sections (**C**) and inflammation score (**D**); see Materials and Methods 4.9. Effect of treatment with NT alone and NT pretreated with SR 142948 in comparison to DNFB-sensitized/DNS-challenged group (positive control; PC) and vehicle-sensitized/DNS-challenged group (negative control; NC) treated with NaCl. The pictures were made at magnification × 10. The results are presented as means ± SEM (n = 7–8 in BALF and n = 4 in histology study). ** $p < 0.01$, *** $p < 0.001$ compared with the NC group, # $p < 0.05$, ## $p < 0.01$, ### $p < 0.001$ compared with the PC group.

2.3. *Pro-Inflammatory Cytokine Production*

An increased concentration of IL-17A, IL-12p40 and RANTES in lung tissue homogenates and IL-13 and TNF-α in BALF was found in the PC group (Figure 3A–E). DNFB-sensitized/DNS-challenged mice treated with NT were characterized as having a much lower level of cytokines compared to the group treated with NaCl. Pretreatment with SR 142948 did not antagonize the anti-inflammatory activity of NT on interleukin production (Figure 3A–E). The only exception was the concentration of TNF-α, which remained enhanced after pretreatment with the NT antagonist (Figure 3E).

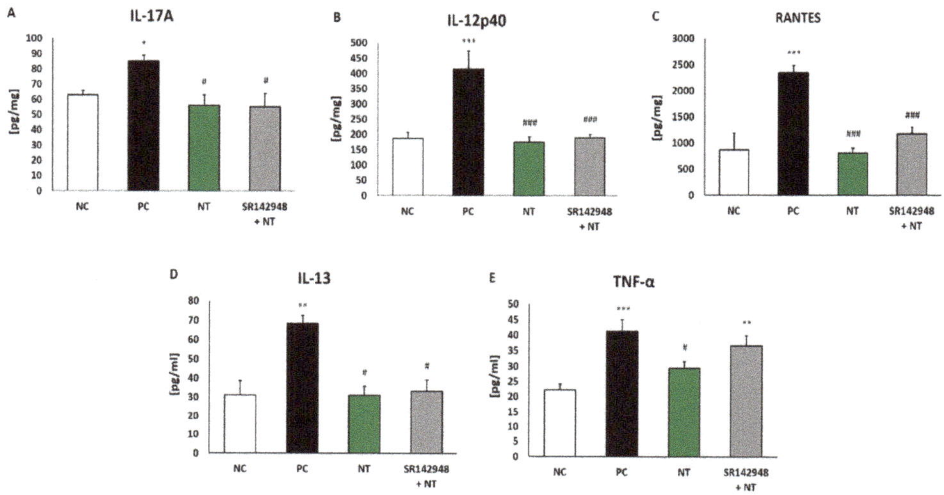

Figure 3. Pro-inflammatory cytokine concentrations in lung tissue homogenates: IL-17A (**A**), IL-12p40 (**B**), RANTES (**C**) and in BALF: IL-13 (**D**) and TNF-α (**E**) after treatment with NT alone and NT pretreated with SR 142948. Comparison to DNFB-sensitized/DNS-challenged group (positive control; PC) and vehicle-sensitized/DNS-challenged group (negative control; NC) treated with NaCl. The results are presented as means ± SEM; n = 5–7 mice per group. * $p < 0.05$, ** $p < 0.01$, *** $p < 0.001$ compared with the NC group; # $p < 0.05$, ### $p < 0.001$ compared with the PC group.

2.4. *The Level of Mouse Mast Cell Protease (MCPT 1)*

The injection of DNS into the trachea of DNFB-sensitized mice significantly increased MCPT 1 concentration compared to the same in vehicle-sensitized animals. The administration of NT alone and NT pretreatment with SR 142948 reduced levels of protease in lung tissue homogenates. However, a significant difference was only obtained in the group without the application of an antagonist (Figure 4A).

2.5. *Oxidative Stress Marker Production*

Malondialdehyde (MDA) concentration in lung homogenates of DNFB-sensitized/DNS-challenged mice was almost three times that of the negative control (Figure 4B). NT treatment completely inhibited oxidative stress marker production, while the administration of the SR 142948 antagonist had no influence on NT activity.

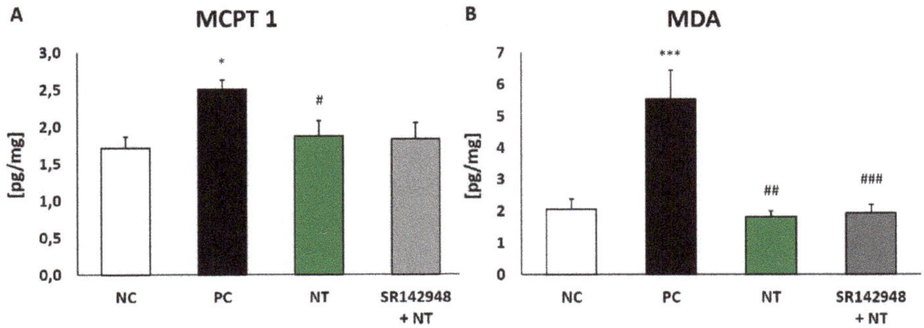

Figure 4. Mouse mast cell protease-1 (MCPT 1) (**A**) and malondialdehyde (MDA) (**B**) concentration in lung tissue homogenates after treatment with NT alone and NT pretreated with SR 142948. Comparison to DNFB-sensitized/DNS-challenged group (positive control; PC) and vehicle-sensitized/DNS-challenged group (negative control; NC) treated with NaCl. The results are presented as means ± SEM; n = 4–7 mice per group. * $p < 0.05$, *** $p < 0.001$ compared with the NC group; # $p < 0.0$, ## $p < 0.01$, ### $p < 0.001$ compared with the PC group.

3. Discussion

Although NT was isolated and described almost fifty years ago, its role in the regulation of an inflammatory response is not completely elucidated. Some studies indicate the pro-inflammatory activity of NT. It has been evidenced that plasma and intraperitoneal concentrations of NT are elevated in the murine model of sepsis, whereas the administration of the NT antagonist or using NT-deficient mice resulted in a decreased mortality rate [23]. The latest study suggested that the plasma level of the stable precursor of NT may be a biomarker of visceral adipose tissue inflammation [24]. The expression of NT and NTR1 was upregulated in animal models of colitis, suggesting that in the gastrointestinal system NT is involved in pro-inflammatory signaling [25].

On the other hand, there is evidence pointing to an ameliorative effect of NT in the development of inflammatory responses in the colon, confirmed by the repair and regeneration of colonic mucosa [22,26]. NT signaling was upregulated in experimental colitis as a part of the adaptive mechanism involved in the healing process, while the administration of the NT antagonist increased the severity of mucosal injuries. The beneficial influence of NT on wound repair has been associated with epithelial cell migration mediated through COX-2 activation and PGE_2 production [27]. In another study the neuropeptide stimulated intestinal mucosa regeneration and exhibited anti-inflammatory activity through the reduction of IL-6 and TNF-α levels in serum, and myeloperoxidase, malondialdehyde and caspase-3 in colonic tissue [28]. The healing properties of NT have been confirmed in a murine model of diabetes, where collagen matrices, loaded with NT-enhanced wound healing, decreased the expression of TNF-α and IL-1β and reduced the infiltration of inflammatory cells into the wound [29]. Furthermore, a crucial role of NT in the wound healing process was evidenced by a decrease in the pro-inflammatory features of skin dendritic cells and fibroblasts and an increase in the epidermal growth factor expression [30,31].

The discrepant data on NT's role in inflammation and the possibility of modulatory action on lung nerves prompted us to investigate its potential in a murine model of non-atopic asthma, in which an inflammatory reaction has been elicited by dermal sensitization with DNFB and an intratracheal challenge with cognate compound DNS, as was previously described by van der Kleij [32].

In our study, NT administration markedly reduced airway responsiveness to methacholine provocation and the effect was blocked by pretreatment with the NTR1 antagonist. In fact, widespread NT in components of the central and peripheral nervous systems has been postulated to regulate breathing in the physiological state and in the disease [33]. The neuropeptide and its receptors have been identified in the airway mucosa and lungs [34], and in the presynaptic cholinergic terminals and post-synaptic smooth muscles of the bronchi [35], indicating potential sites of NT action and the possibility of the modulation of airway hyperresponsiveness.

Yet, an in vitro study on guinea pig and rat tracheal preparations demonstrated the contractile properties of NT on airway smooth muscle [35,36]. Contrarily another subsequent finding showed that NT inhibited cholinergic and noncholinergic contractions of guinea pig bronchial rings stimulated by an electrical field. The effect was reversed after treatment with a selective NTR1 receptor antagonist, which corresponds to our AHR results [37]. We cannot rule out that reduced airway hyperresponsiveness induced by NT was the consequence of attenuated inflammation reflected in the decreased inflammatory cell influx in BALF, blocked by the NTR1 antagonist, as well. We used SR 142948, which is not as selective an antagonist as SR 48692 but presents approximately ten times the affinity for NTR1 than it does for NTR2 [38]. SR 48692, although characterized by high selectivity, failed to block many physiological effects of NT, such as hypothermia and analgesia [39,40]. Its newer potent antagonist, SR 142948, has the ability to antagonize a wider spectrum of NT-mediated effects [41,42].

The contribution of NTR2 in these phenomena remains unclear when taking into consideration the fact that several cell line studies indicate that both NT antagonists, SR 48962 and SR 142948 may activate, instead of block, NTR2 [43]. The ability to examine NTR2 involvement in the inflammatory process is limited due to the lack of a selective NTR2 antagonist. Levocabastine, which is recognized by NTR2, exhibits species-dependent activity—it acts as an agonist in mice and as an antagonist in humans [44].

The alleviating properties of NT have been further confirmed in our study in cytokine assays, where NT diminished the concentration of IL-12p40, RANTES and IL-17A in lung homogenates, and IL-13 and TNF-α in BALF. Many interleukins and chemokines have been described as being involved in the development of asthma. IL-12p40, which is produced by activated monocytes, neutrophils, macrophages and dendritic cells, acts as an inducer of Th1 cell differentiation [45]. Yet, anti-IL-12 treatment during the airway challenge was able to reduce the main symptoms of ovalbumin-induced asthma, suggesting that IL-12 plays a pro-inflammatory role in the effector phase of allergic airway inflammation [46]. Pro-inflammatory activity has also been demonstrated in chemokine RANTES, which is considered to be a chemoattractant for eosinophils, monocytes, macrophages and T lymphocytes. An elevated level of RANTES mRNA was observed in the respiratory tract of patients suffering from mild asthma [47]. RANTES concentration was also markedly higher in the exhaled breath condensate of asthmatics, particularly in patients with unstable asthma [48].

Another interleukin playing an important role in pulmonary inflammation is IL-17A, which is secreted by distinctive T cells of the Th17 subtype and belongs to a larger IL-17 cytokine family [49]. IL-17A is expressed in the BALF, sputum and bronchial biopsies of asthmatic patients and is closely linked to neutrophilic influx into the airways [50]. Neutrophil-mediated inflammation is connected with the stimulating activity of IL-17A, which induces the release of the potent neutrophil chemotactic factor CXCL8 from airway smooth muscle cells and airway epithelial cells, and in this way mobilizes neutrophil recruitment [51,52]. These results are in line with the present study, where an increased number of neutrophils in the BALF of DNFB-sensitized/DNS-challenged mice was accompanied by an elevated concentration of IL-17A in lung tissue homogenates.

Another cytokine whose concentration in BALF was reduced by NT treatment is IL-13, regarded as one of the most crucial cytokines in the initiation and exacerbation of asthma in humans [53]. This Th2 cytokine plays an important role in allergic inflammation, yet acts probably via the pathway independent of immunoglobulin E and eosinophils. IL-13 gene polymorphism might have an impact on susceptibility to the disease [54].

It is noteworthy that pretreatment with the NT antagonist SR 142948 did not inhibit the effect of NT on the majority of cytokines examined, suggesting that pathways for a receptor other than NTR1 might be involved. The exception was TNF-α, which was attenuated by NT, and the effect was reversed after the NTR1 blockade. Similarly to our study, NT reduced the level of TNF-α in the serum of rats with experimentally induced colitis [28]. It has been previously demonstrated that the application of TNF-α to healthy and asthmatic subjects led to airway hyperresponsiveness and neutrophil infiltration [55,56]. Increased levels of TNF-α were confirmed in bronchoalveolar lavage in the same DNFB-induced model of non-allergic asthma [32]. The main source of TNF-α in asthmatic lungs is mast cells, contributing to the development of late AHR [57]. The mast cells are known to store a large number of different mediators, including TNF-α, that might be released at the same time as other preformed granule contents, such as histamine, tryptase or chymase [58]. One of those compounds is MCPT 1, classified as a β-chymase and predominantly produced by mucosal mast cells [59]. In the present study, the DNS challenge in DNFB-sensitized mice increased the level of MCPT 1 in lung tissue and the effect was markedly reduced after NT administration. The key role of mast cells was previously established in a DNFB-induced model for non-atopic asthma [60].

We also demonstrated that NT significantly reduced the level of MDA, which corresponds to results obtained by Akcan et al. [28]. MDA, regarded as a marker of oxidative stress reaction, is formed as a consequence of the peroxidation of polyunsaturated fatty acids, which form the lipids in the cell membrane. The concentration of MDA is increased both in the exhaled breath condensate and the serum of asthmatic patients [61]. This is in line with animal studies—MDA was elevated in murine lung tissue homogenates in a model of ovalbumin-induced allergic asthma [62], as well as in occupational asthma in rats provoked by toluene diisocyanate [63], which supports the hypothesis that an antioxidative–prooxidative imbalance is one of the characteristic features of the inflammatory process in lungs.

Summarizing, the present study demonstrated NT potency in modulating the inflammatory response and airway hyperreactivity in a non-atopic asthma model. The effect of NT on leukocyte accumulation, airway hyperreactivity and the TNF-α level was mediated through NTR1 receptors. On the contrary, the lack of impact of SR 142948 on the beneficial effect of NT on pro-inflammatory cytokine production and oxidative stress indicates that another receptor pathway might be involved in the reaction. Further experiments investigating the detailed mechanism of NT ameliorative activity are required.

4. Materials and Methods

4.1. Drugs and Reagents

Neurotensin (NT) and neurotensin antagonist SR 142948 were purchased from Tocris Bioscience (Bristol, UK). 1-fluoro-2,4-dinitrobenzene (DNFB), dinitrobenzene sulfphonic acid (DNS), TX-100 and protease inhibitor cocktail were obtained from Sigma-Aldrich (Poznań, Poland). Pentobarbital sodium, ketamine and xylazine were acquired from Biowet (Puławy, Poland).

4.2. Animals and Hapten-Induced Experimental Asthma Model

Male BALB/c mice 7–8 weeks of age were obtained from the Animal House of Mossakowski Medical Research Centre, Polish Academy of Sciences. All animal experiments were conducted in accordance with the guidelines approved (15 April 2014) by the IV Warsaw Ethics Commission for the Care and Use of Laboratory Animals (permit number: 15/2014).

On day 0, mice were sensitized dermally with 0.5% DNFB (dissolved in acetone and olive oil (ratio 4:1) or vehicle control onto shaved thorax (50 μL) and on paws (50 μL on four paws). On day 1, DNFB/vehicle was only applied to the thorax. On day 5, mice were challenged with 50 μL of 0.6% DNS (dissolvable in water cognate hapten of DNFB) by intratracheal application. All procedures were performed under light anesthesia using ketamine (70 mg/kg IM) and xylazine (10 mg/kg IM). Twenty-four hours later, the airway responsiveness to methacholine exposure was measured. After that, the mice were sacrificed by intraperitoneal injection of pentobarbital sodium (150–200 mg/kg). Bronchoalveolar lavage fluid (BALF) and lung tissue were collected in order to perform further assays.

4.3. Experimental Design

The study was performed in four experimental groups (n = 8 in each group):

- I vehicle-sensitized, DNS-challenged mice treated with physiological saline (NaCl) as the negative control (NC);
- II DNFB-sensitized, DNS-challenged mice treated with NaCl as the positive control (PC);
- III DNFB-sensitized, DNS-challenged mice treated with neurotensin alone (NT);
- IV DNFB-sensitized, DNS-challenged mice pretreated with SR 142948 and treated with NT (SR 142948 + NT).

The doses of NT and SR 142948 were 7.4 μmol/kg and 2.9 μmol/kg, respectively, and were taken from previous studies [42,64]. Compounds were dissolved in NaCl (0.9%). All groups were treated twice—two and eight hours after DNS challenge by the intraperitoneal administration of equivalent volumes of 100 μL of NaCl (I, II), NT (III) or SR 142948 + NT (IV). The interval between the administration of the antagonist and the NT in the last group was thirty minutes.

4.4. Airway Hyperresponsiveness

The measurements of bronchial hyperresponsiveness were performed in the murine whole body plethysmograph chamber (Buxco Electronics, Inc., Wilmington, NC, USA) twenty-four hours after the intratracheal DNS challenge. The animals were placed in the chamber and after ten minutes of adaptation they were exposed to the NaCl aerosol control. Then, every twenty minutes, increasing concentrations of methacholine (5, 10, 20 and 40 mg/mL) were administered. Airway hyperresponsiveness was reflected by increasing values of enhanced pause (Penh), which were recorded for three minutes after each nebulization.

4.5. Leukocyte Accumulation in BALF

After sacrificing each mouse, a cannula was placed into the trachea and four aliquots of the phosphate buffered saline (PBS) solution were inserted and withdrawn (4 × 1 mL). The bronchoalveolar lavages collected were centrifuged (1500 rpm, 10 min) in order to separate the BALF cells from the supernatant. The cell pellet was then re-suspended in 150 μL of PBS. Türk reagent (100 μL) was added to 50 μL of the suspension and the total number of inflammatory cells was determined using a Bürker counting chamber. The remaining cell suspension was centrifuged with a cytocentrifuge (700 rpm for 10 min, Thermo Shandon, Cambridge, UK). The microscopic slides obtained were left to dry and then stained with a fast staining kit based on eosin and azure solutions (Hemastain, Analab, Warsaw, Poland). The leukocyte differentiation was determined using a light microscope and by counting the number of mononuclear cells and neutrophils per 500 consecutive cells in each sample. Results are expressed as the number of cells per 1 mL of BALF.

4.6. Pro-Inflammatory Cytokine Production

The first 1 mL of BALF, containing the cocktail of protease inhibitors, was centrifuged (1500 rpm, 10 min) and the supernatant obtained was collected and stored at −80 °C. After collecting the BALF, the lungs were isolated and also transferred to −80 °C. For further testing, the lungs were homogenized in liquid nitrogen and suspended in 500 μL of PBS, with the addition of a protease inhibitor cocktail and 1% Triton X-100. The sample was then centrifuged (14,000 rpm, 10 min, 4 °C) and the supernatant was used to analyze cytokine levels. Measurements of IL-13, TNF-α in BALF and IL-17A, IL-12p40 and RANTES in lung tissue homogenates were taken using the Bio-Plex Pro Mouse Cytokine panel (Bio-Rad, Warsaw, Poland) with the Bio-Plex 200 platform (Luminex, Bio-Rad, Warsaw, Poland).

4.7. The Level of Mouse Mast Cell Protease (MCPT 1)

The level of mouse mast cell protease (MCPT 1), a type of serine protease (β-chymase) stored and secreted in a tissue-specific manner by mucosal mast cells, was determined in lung tissue homogenates using the Mouse MCP-1 ELISA Ready-SET kit (eBioscience, San Diego, CA, USA) according to the manufacturer's guidelines.

4.8. Oxidative Stress Marker Production

Malondialdehyde (MDA) is a product of lipid peroxidation and a marker of oxidative stress. Determination of MDA concentration in the lung tissue homogenates was performed by the ELISA enzyme assay using a commercially available analysis kit (OxiSelect MDA, Cell Biolabs, Inc., San Diego, CA, USA).

4.9. Lung Histology

Sections of lung tissue, 20 μm in thickness, were stained with eosin and hematoxylin (Merck Millipore, Poland) and examined under a light microscope (Nikon, Japan). To assess the severity of immune cell infiltration, peribronchial and perivascular cell counts were performed based on the four-point scoring system (from 0 to 3). Briefly, the four-point scoring system described by Tourney et al. [65] was: no inflammation detectable (0), occasional cuffing with inflammatory cells (1), most bronchi or vessels surrounded by a thin ring (one to five cells) of inflammatory cells (2), most bronchi or vessels surrounded by a thick layer (more than five cells) of inflammatory cells (3).

4.10. Statistical Analysis

The results are presented as mean ± standard error of mean (SEM). The concentration of MCPT 1, MDA and pro-inflammatory cytokines in lung tissue homogenates is expressed per mg of total protein in the sample. Statistical analysis was performed using a one-way ANOVA test with a Newman–Keuls post-hoc test. The results were considered significant when the significance level of p was less than 0.05. Analyses were performed using the STATISTICA software 12 (StatSoft, Kraków, Poland).

Author Contributions: Conceptualization, methodology, formal analysis and visualization were performed by K.K. and E.R.; original draft preparation and experimental investigation were performed by E.R.; supervision, review, editing and funding acquisition were performed by K.K.

Funding: The study was supported by a grant from National Science Centre, Poland, no. 2014/13/B/NZ7/02247.

Conflicts of Interest: The authors declare no conflicts of interest with respect to the publication of this manuscript.

References

1. Carraway, R.; Leeman, S.E. The isolation of a new hypotensive peptide, neurotensin, from bovine hypothalami. *J. Biol. Chem.* **1973**, *248*, 6854–6861.
2. Bello, A.R.; Reyes, R.; Hernández, G.; Negrín, I.; González, M.; Tramu, G.; Alonso, R. Developmental expression of neurotensin in thyrotropes and gonadotropes of male and female rats. *Neuroendocrinology* **2004**, *79*, 90–99. [CrossRef]
3. Bissette, G.; Nemeroff, C.B.; Loosen, P.T.; Prange, A.J., Jr.; Lipton, M.A. Hypothermia and intolerance to cold induced by intracisternal administration of the hypothalamic peptide neurotensin. *Nature* **1976**, *262*, 607–609. [CrossRef] [PubMed]
4. Clineschmidt, B.V.; McGuffin, J.C. Neurotensin administered intracisternally inhibits responsiveness of mice to noxious stimuli. *Eur. J. Pharmacol.* **1977**, *46*, 395–396. [CrossRef]
5. Boules, M.; Li, Z.; Smith, K.; Fredrickson, P.; Richelson, E. Diverse roles of neurotensin agonists in the central nervous system. *Front. Endocrinol. (Lausanne)* **2013**, *4*, 36. [CrossRef] [PubMed]
6. Lambert, P.D.; Gross, R.; Nemeroff, C.B.; Kilts, C.D. Anatomy and mechanisms of neurotensin-dopamine interactions in the central nervous system. *Ann. N.Y. Acad. Sci.* **1995**, *757*, 377–389. [CrossRef] [PubMed]

7. Vincent, J.P.; Mazella, J.; Kitabgi, P. Neurotensin and neurotensin receptors. *Trends Pharmacol. Sci.* **1999**, *20*, 302–309. [CrossRef]
8. Tanaka, K.; Masu, M.; Nakanishi, S. Structure and functional expression of the cloned rat neurotensin receptor. *Neuron* **1990**, *4*, 847–854. [CrossRef]
9. Vita, N.; Laurent, P.; Lefort, S.; Chalon, P.; Dumont, X.; Kaghad, M.; Gully, D.; Le Fur, G.; Ferrara, P.; Caput, D. Cloning and expression of a complementary DNA encoding a high affinity human neurotensin receptor. *FEBS Lett.* **1993**, *317*, 139–142. [CrossRef]
10. Chalon, P.; Vita, N.; Kaghad, M.; Guillemot, M.; Bonnin, J.; Delpech, B.; Le Fur, G.; Ferrara, P.; Caput, D. Molecular cloning of a levocabastine-sensitive neurotensin binding site. *FEBS Lett.* **1996**, *386*, 91–94. [CrossRef]
11. Kleczkowska, P.; Lipkowski, A.W. Neurotensin and neurotensin receptors: Characteristic, structure-activity relationship and pain modulation—A review. *Eur. J. Pharmacol.* **2013**, *716*, 54–60. [CrossRef] [PubMed]
12. Mazella, J.; Zsürger, N.; Navarro, V.; Chabry, J.; Kaghad, M.; Caput, D.; Ferrara, P.; Vita, N.; Gully, D.; Maffrand, J.P.; et al. The 100-kDa neurotensin receptor is gp95/sortilin, a non-G-protein-coupled receptor. *J. Biol. Chem.* **1998**, *273*, 26273–26276. [CrossRef] [PubMed]
13. Jacobsen, L.; Madsen, P.; Jacobsen, C.; Nielsen, M.S.; Gliemann, J.; Petersen, C.M. Activation and functional characterization of the mosaic receptor SorLA/LR11. *J. Biol. Chem.* **2001**, *276*, 22788–22796. [CrossRef] [PubMed]
14. Moura, L.I.; Cruz, M.T.; Carvalho, E. The effect of neurotensin in human keratinocytes-implication on impaired wound healing in diabetes. *Exp. Biol. Med. (Maywood)* **2014**, *239*, 6–12. [CrossRef] [PubMed]
15. Katsanos, G.S.; Anogianaki, A.; Castellani, M.L.; Ciampoli, C.; De Amicis, D.; Orso, C.; Pollice, R.; Vecchiet, J.; Tetè, S.; Salini, V.; et al. Biology of neurotensin: Revisited study. *Int. J. Immunopathol. Pharmacol.* **2008**, *21*, 255–259. [CrossRef]
16. Schäffer, M.; Beiter, T.; Becker, H.D.; Hunt, T.K. Neuropeptides: Mediators of inflammation and tissue repair? *Arch. Surg.* **1998**, *133*, 1107–1116. [CrossRef]
17. Voisin, T.; Bouvier, A.; Chiu, I.M. Neuro-immune interactions in allergic diseases: Novel targets for therapeutics. *Int. Immunol.* **2017**, *29*, 247–261. [CrossRef]
18. Barnes, P.J. Asthma as an axon reflex. *Lancet* **1986**, *1*, 242–245. [CrossRef]
19. Tränkner, D.; Hahne, N.; Sugino, K.; Hoon, M.A.; Zuker, C. Population of sensory neurons essential for asthmatic hyperreactivity of inflamed airways. *Proc. Natl. Acad. Sci. USA.* **2014**, *111*, 11515–11520. [CrossRef]
20. Kessler, J.P.; Beaudet, A.J. Association of neurotensin binding sites with sensory and visceromotor components of the vagus nerve. *J. Neurosci.* **1989**, *9*, 466–472. [CrossRef]
21. Castagliuolo, I.; Wang, C.C.; Valenick, L.; Pasha, A.; Nikulasson, S.; Carraway, R.E.; Pothoulakis, C.J. Neurotensin is a proinflammatory neuropeptide in colonic inflammation. *Clin. Investig.* **1999**, *103*, 843–849. [CrossRef] [PubMed]
22. Margolis, K.G.; Gershon, M.D. Neuropeptides and inflammatory bowel disease. *Curr. Opin. Gastroenterol.* **2009**, *25*, 503–511. [CrossRef]
23. Piliponsky, A.M.; Chen, C.C.; Nishimura, T.; Metz, M.; Rios, E.J.; Dobner, P.R.; Wada, E.; Wada, K.; Zacharias, S.; Mohanasundaram, U.M.; et al. Neurotensin increases mortality and mast cells reduce neurotensin levels in a mouse model of sepsis. *Nat. Med.* **2008**, *14*, 392–398. [CrossRef] [PubMed]
24. Barchetta, I.; Cimini, F.A.; Capoccia, D.; Bertoccini, L.; Ceccarelli, V.; Chiappetta, C.; Leonetti, F.; Di Cristofano, C.; Silecchia, G.; Orho-Melander, M.; et al. Neurotensin Is a Lipid-Induced Gastrointestinal Peptide Associated with Visceral Adipose Tissue Inflammation in Obesity. *Nutrients* **2018**, *10*, 526. [CrossRef] [PubMed]
25. Law, I.K.; Bakirtzi, K.; Polytarchou, C.; Oikonomopoulos, A.; Hommes, D.; Iliopoulos, D.; Pothoulakis, C. Neurotensin–regulated miR-133α is involved in proinflammatory signalling in human colonic epithelial cells and in experimental colitis. *Gut* **2015**, *64*, 1095–1104. [CrossRef] [PubMed]
26. Zhao, D.; Pothoulakis, C. Effects of NT on gastrointestinal motility and secretion, and role in intestinal inflammation. *Peptides* **2006**, *27*, 2434–2444. [CrossRef] [PubMed]
27. Brun, P.; Mastrotto, C.; Beggiao, E.; Stefani, A.; Barzon, L.; Sturniolo, G.C.; Palù, G.; Castagliuolo, I. Neuropeptide neurotensin stimulates intestinal wound healing following chronic intestinal inflammation. *Am. J. Physiol. Gastrointest. Liver Physiol.* **2005**, *288*, G621–G629. [CrossRef] [PubMed]

28. Akcan, A.; Muhtaroglu, S.; Akgun, H.; Akyildiz, H.; Kucuk, C.; Sozuer, E.; Yurci, A.; Yilmaz, N. Ameliorative effects of bombesin and neurotensin on trinitrobenzene sulphonic acid-induced colitis, oxidative damage and apoptosis in rats. *World J. Gastroenterol.* **2008**, *14*, 1222–1230. [CrossRef] [PubMed]
29. Moura, L.I.; Dias, A.M.; Suesca, E.; Casadiegos, S.; Leal, E.C.; Fontanilla, M.R.; Carvalho, L.; de Sousa, H.C.; Carvalho, E. Neurotensin-loaded collagen dressings reduce inflammation and improve wound healing in diabetic mice. *Biochim. Biophys. Acta* **2014**, *1842*, 32–43. [CrossRef] [PubMed]
30. da Silva, L.; Neves, B.M.; Moura, L.; Cruz, M.T.; Carvalho, E. Neurotensin downregulates the pro-inflammatory properties of skin dendritic cells and increases epidermal growth factor expression. *Biochim. Biophys. Acta* **2011**, *1813*, 1863–1871. [CrossRef]
31. Pereira da Silva, L.; Miguel Neves, B.; Moura, L.; Cruz, M.T.; Carvalho, E. Neurotensin decreases the proinflammatory status of human skin fibroblasts and increases epidermal growth factor expression. *Int. J. Inflam.* **2014**, *2014*, 248240. [CrossRef] [PubMed]
32. Van der Kleij, H.P.; Kraneveld, A.D.; van Houwelingen, A.H.; Kool, M.; Weitenberg, A.C.; Redegeld, F.A.; Nijkamp, F.P. Murine model for non-IgE-mediated asthma. *Inflammation* **2004**, *28*, 115–125. [CrossRef]
33. Kaczyńska, K.; Zając, D.; Wojciechowski, P.; Kogut, E.; Szereda-Przestaszewska, M. Neuropeptides and breathing in health and disease. *Pulm. Pharmacol. Ther.* **2018**, *48*, 217–224. [CrossRef] [PubMed]
34. Robbins, R.A.; Nelson, K.J.; Gossman, G.L.; Rubinstein, I. Neurotensin stimulates neutrophil adherence to bronchial epithelial cells in vitro. *Life Sci.* **1995**, *56*, 1353–1359. [CrossRef]
35. Aas, P.; Helle, K.B. Neurotensin receptors in the rat bronchi. *Regul. Pept.* **1982**, *3*, 405–413. [CrossRef]
36. Djokic, T.D.; Dusser, D.J.; Borson, D.B.; Nadel, J.A. Neutral endopeptidase modulates neurotensin-induced airway contraction. *J. Appl. Physiol.* **1989**, *66*, 2338–2343. [CrossRef]
37. Martin, C.A.; Gully, D.; Naline, E.; Advenier, C. Neurotensin modulates cholinergic and noncholinergic neurotransmission in guinea-pig main bronchi in vitro. *Neuropeptides* **1994**, *26*, 159–166. [CrossRef]
38. Tyler-McMahon, B.M.; Boules, M.; Richelson, E. Neurotensin: Peptide for the next millennium. *Regul. Pept.* **2000**, *93*, 125–136. [CrossRef]
39. Gully, D.; Canton, M.; Boigegrain, R.; Jeanjean, F.; Molimard, J.C.; Poncelet, M.; Gueudet, C.; Heaulme, M.; Leyris, R.; Brouard, A.; et al. Biochemical and pharmacological profile of a potent and selective nonpeptide antagonist of the neurotensin receptor. *Proc. Natl. Acad. Sci. USA* **1993**, *90*, 65–69. [CrossRef]
40. Dubuc, I.; Costentin, J.; Terranova, J.P.; Barnouin, M.C.; Soubrié, P.; Le Fur, G.; Rostène, W.; Kitabgi, P. The nonpeptide neurotensin antagonist, SR 48692, used as a tool to reveal putative neurotensin receptor subtypes. *Br. J. Pharmacol.* **1994**, *112*, 352–354. [CrossRef]
41. Schaeffer, P.; Laplace, M.C.; Bernat, A.; Prabonnaud, V.; Gully, D.; Lespy, L.; Herbert, J.M. SR142948A is a potent antagonist of the cardiovascular effects of neurotensin. *J. Cardiovasc. Pharmacol.* **1998**, *31*, 545–550. [CrossRef] [PubMed]
42. Kaczyńska, K.; Szereda-Przestaszewska, M. Cardio-respiratory effects of systemic neurotensin injection are mediated through activation of neurotensin NTS_1 receptors. *Eur. J. Pharmacol.* **2012**, *691*, 245–250. [CrossRef] [PubMed]
43. Dobner, P.R. Multitasking with neurotensin in the central nervous system. *Cell. Mol. Life Sci.* **2005**, *62*, 1946–1963. [CrossRef] [PubMed]
44. St-Gelais, F.; Jomphe, C.; Trudeau, L.E. The role of neurotensin in central nervous system pathophysiology: What is the evidence? *J. Psychiatry Neurosci.* **2006**, *31*, 229–245. [PubMed]
45. Trinchieri, G.; Pflanz, S.; Kastelein, R.A. The IL-12 family of heterodimeric cytokines: New players in the regulation of T cell responses. *Immunity* **2003**, *19*, 641–644. [CrossRef]
46. Meyts, I.; Hellings, P.W.; Hens, G.; Vanaudenaerde, B.M.; Verbinnen, B.; Heremans, H.; Matthys, P.; Bullens, D.M.; Overbergh, L.; Mathieu, C.; et al. IL-12 contributes to allergen-induced airway inflammation in experimental asthma. *J. Immunol.* **2006**, *177*, 6460–6470. [CrossRef] [PubMed]
47. Berkman, N.; Krishnan, V.L.; Gilbey, T.; Newton, R.; O'Connor, B.; Barnes, P.J.; Chung, K.F. Expression of RANTES mRNA and protein in airways of patients with mild asthma. *Am. J. Respir. Crit. Care Med.* **1996**, *154*, 1804–1811. [CrossRef] [PubMed]
48. Zietkowski, Z.; Tomasiak, M.M.; Skiepko, R.; Bodzenta-Lukaszyk, A. RANTES in exhaled breath condensate of stable and unstable asthma patients. *Respir. Med.* **2008**, *102*, 1198–1202. [CrossRef] [PubMed]
49. Agache, I.; Ciobanu, C.; Agache, C.; Anghel, M. Increased serum IL-17 is an independent risk factor for severe asthma. *Respir. Med.* **2010**, *104*, 1131–1137. [CrossRef]

50. Bullens, D.M.; Truyen, E.; Coteur, L.; Dilissen, E.; Hellings, P.W.; Dupont, L.J.; Ceuppens, J.L. IL-17 mRNA in sputum of asthmatic patients: Linking T cell driven inflammation and granulocytic influx? *Respir. Res.* **2006**, *7*, 135. [CrossRef]
51. Dragon, S.; Rahman, M.S.; Yang, J.; Unruh, H.; Halayko, A.J.; Gounni, A.S. IL-17 enhances IL-1beta-mediated CXCL-8 release from human airway smooth muscle cells. *Am. J. Physiol. Lung Cell. Mol. Physiol.* **2007**, *292*, L1023–L1029. [CrossRef] [PubMed]
52. Fossiez, F.; Djossou, O.; Chomarat, P.; Flores-Romo, L.; Ait-Yahia, S.; Maat, C.; Pin, J.J.; Garrone, P.; Garcia, E.; Saeland, S.; et al. T cell interleukin-17 induces stromal cells to produce proinflammatory and hematopoietic cytokines. *J. Exp. Med.* **1996**, *183*, 2593–2603. [CrossRef] [PubMed]
53. Izuhara, K.; Arima, K.; Kanaji, S.; Ohta, S.; Kanaji, T. IL-13: A promising therapeutic target for bronchial asthma. *Curr. Med. Chem.* **2006**, *13*, 2291–2298. [PubMed]
54. Wills-Karp, M. Interleukin-13 in asthma pathogenesis. *Immunol. Rev.* **2004**, *202*, 175–190. [CrossRef] [PubMed]
55. Thomas, P.S.; Yates, D.H.; Barnes, P.J. Tumor necrosis factor-alpha increases airway responsiveness and sputum neutrophilia in normal human subjects. *Am. J. Respir. Crit. Care Med.* **1995**, *152*, 76–80. [CrossRef] [PubMed]
56. Thomas, P.S.; Heywood, G. Effects of inhaled tumour necrosis factor alpha in subjects with mild asthma. *Thorax* **2002**, *57*, 774–778. [CrossRef] [PubMed]
57. Kim, Y.S.; Ko, H.M.; Kang, N.I.; Song, C.H.; Zhang, X.; Chung, W.C.; Kim, J.H.; Choi, I.H.; Park, Y.M.; Kim, G.Y.; et al. Mast cells play a key role in the development of late airway hyperresponsiveness through TNF-alpha in a murine model of asthma. *Eur. J. Immunol.* **2007**, *37*, 1107–1115. [CrossRef] [PubMed]
58. Thomas, P.S. Tumour necrosis factor-alpha: The role of this multifunctional cytokine in asthma. *Immunol. Cell. Biol.* **2001**, *79*, 132–140. [CrossRef]
59. Dai, H.; Korthuis, R.J. Mast Cell Proteases and Inflammation. *Drug Discov. Today Dis. Models* **2011**, *8*, 47–55. [CrossRef]
60. Kraneveld, A.D.; van der Kleij, H.P.; Kool, M.; van Houwelingen, A.H.; Weitenberg, A.C.; Redegeld, F.A.; Nijkamp, F.P. Key role for mast cells in nonatopic asthma. *J. Immunol.* **2002**, *169*, 2044–2053. [CrossRef]
61. Sadowska-Woda, I.; Bieszczad-Bedrejczuk, E. Rola stresu oksydacyjnego w patogenezie astmy oskrzelowej. *Alergia Astma Immunologia* **2011**, *16*, 80–89.
62. Abdelaziz, R.R.; Elmahdy, M.K.; Suddek, G.M. Flavocoxid attenuates airway inflammation in ovalbumin-induced mouse asthma model. *Chem. Biol. Interact.* **2018**, *292*, 15–23. [CrossRef] [PubMed]
63. Muti, A.D.; Pârvu, A.E.; Muti, L.A.; Moldovan, R.; Mureşan, A. Vitamin E effect in a rat model of toluene diisocyanate-induced asthma. *Clujul. Med.* **2016**, *89*, 499–505. [CrossRef] [PubMed]
64. Kogut, E.; Kaczyńska, K.; Lipkowski, A.; Kleczkowska, P. Opioid-neurotensin hybrid influences pulmonary inflammatory process in murine model of non-atopic asthma. *Eur. Respir. J.* **2017**, *50*, PA1407.
65. Tournoy, K.G.; Kips, J.C.; Schou, C.; Pauwels, R.A. Airway eosinophilia is not arequirement for allergen-induced airway hyperresponsiveness. *Clin. Exp. Allergy* **2000**, *30*, 79–85. [CrossRef] [PubMed]

International Journal of
Molecular Sciences

MDPI

Article

Endomorphin-2- and Neurotensin- Based Chimeric Peptide Attenuates Airway Inflammation in Mouse Model of Nonallergic Asthma

Ewelina Russjan [1], Kryspin Andrzejewski [1], Dorota Sulejczak [2], Patrycja Kleczkowska [3] and Katarzyna Kaczyńska [1,*]

[1] Department of Respiration Physiology, Mossakowski Medical Research Centre, Polish Academy of Sciences, 5 Pawińskiego Str., 02-106 Warsaw, Poland; erussjan@imdik.pan.pl (E.R.); kandrzejewski@imdik.pan.pl (K.A.)

[2] Department of Experimental Pharmacology, Mossakowski Medical Research Centre, Polish Academy of Sciences, 5 Pawińskiego Str., 02-106 Warsaw, Poland; dsulejczak@imdik.pan.pl

[3] Department of Pharmacodynamics, The Centre for Preclinical Research (CBP), Medical University of Warsaw, 1B Banacha Str., 02-097 Warsaw, Poland; hazufiel@wp.pl

* Correspondence: kkaczynska@imdik.pan.pl

Received: 30 October 2019; Accepted: 22 November 2019; Published: 26 November 2019

Abstract: We examined anti-inflammatory potency of hybrid peptide-PK20, composed of neurotensin (NT) and endomorphin-2 (EM-2) pharmacophores in a murine model of non-atopic asthma induced by skin sensitization with 2,4-dinitrofluorobenzene and intratracheal challenge of cognate hapten. Mice received intraperitoneally PK20, equimolar mixture of its structural elements (MIX), dexamethasone (DEX), or NaCl. Twenty-four hours following hapten challenge, the measurements of airway responsiveness to methacholine were taken. Bronchoalveolar lavage (BALF) and lungs were collected for further analyses. Treatment with PK20, similarly to dexamethasone, reduced infiltration of inflammatory cells, concentration of mouse mast cell protease, IL-1β, IL-12p40, IL-17A, CXCL1, RANTES in lungs and IL-1α, IL-2, IL-13, and TNF-α in BALF. Simple mixture of NT and EM-2 moieties was less potent. PK20, DEX, and MIX significantly decreased malondialdehyde level and secretory phospholipase 2 activity in lungs. Intensity of NF-κB immunoreactivity was diminished only after PK20 and DEX treatments. Neither PK20 nor mixture of its pharmacophores were as effective as DEX in alleviating airway hyperresponsiveness. PK20 effectively inhibited hapten-induced inflammation and mediator and signaling pathways in a manner seen with dexamethasone. Improved anti-inflammatory potency of the hybrid over the mixture of its moieties shows its preponderance and might pose a promising tool in modulating inflammation in asthma.

Keywords: airway inflammation; non-atopic asthma; hybrid peptide; pro-inflammatory cytokines; $sPLA_2$; MDA; NF-κB

---

1. Introduction

Asthma is a chronic disease characterized by airway inflammation, remodeling, and bronchial hyperresponsiveness with usually reversible airflow limitation [1,2]. According to the Global Asthma Report, the prevalence of asthma has increased lately, and over 334 million people worldwide have been reported to be affected [3]. Treatment of both controlled and severe uncontrolled asthma patients poses a substantial economic burden for developed and developing countries [4–6]. Inhaled bronchodilators and corticosteroids are currently used to improve asthma condition. However, most of these medications produce unwanted adverse effects, resistance with long-term use, and unresponsiveness [7,8]. The disease is also difficult to treat due to its heterogeneity and different phenotypes involving a multitude of inflammatory mediators [9]. Therefore, there is an urgent need to develop novel anti-inflammatory therapies for the treatment of asthma.

In the last few years, a strong interest has been raised in hybrid compounds encompassing at least two various biologically active fragments, which may simultaneously reach several targets [10–12]. One of the hybrid peptides, designated PK20, comprises modified endomorphin-2 (EM-2) and neurotensin (NT) pharmacophores, and it has been shown previously as a highly potent analgesic, as well as a neuroprotective agent [13,14]. Detailed information on the hybrid structure, amino acid modification, and an improved stability can be found elsewhere [13,15]. Both elements of the ligand seem to possess anti-inflammatory potency. NT attenuated oxidative stress and apoptosis in the model of inflammatory bowel disease [16]. The latest study showed the ameliorating action of NT on inflammatory processes and airway hyperreactivity in an experimental asthma model [17]. Likewise, stimulation of opioid receptors resulted in the decrease of edema, plasma extravasation, immune mediators' level, and tissue damage in various inflammation models [18].

With respect to airways and their innervation, an excitation of opioid receptors by endomorphins has been shown to inhibit cholinergic [19], non-cholinergic, and non-adrenergic bronchoconstriction [20] and the release of pro-inflammatory sensory neuropeptides from capsaicin-sensitive nerves [21] and tachykinergic airway smooth muscle constriction [20]. Therefore, we think that a new therapy with PK20-hybrid comprising EM-2 and NT in the structure offers not only ant-inflammatory potency but also promise of improved bronchodilator activity.

Considering the abovementioned, the aim of the present study was to evaluate the PK20-exerted effects on airway hyperreactivity and inflammation in a mouse model of dinitrofluorobenzene (DNFB) induced non-atopic asthma [22,23]. The effect of the tested hybrid was also compared with glucocorticosteroid dexamethasone and with co-administration of simple mixture of EM-2 and NT (8–13) pharmacophores. The latter was performed to see whether connection of two different elements into one molecule provides further beneficial properties. For these purposes, bronchoalveolar lavage fluid (BALF) and lung tissue were collected for biochemical, immunohistochemical, and histological analyses.

2. Results

2.1. Effect of PK20 on DNFB-Induced Inflammatory Cell Infiltration in BALF and Lung Tissue

Compared with vehicle-sensitized/DNS-challenged group (negative control; NC), DNFB-sensitized/DNS-challenged (positive control; PC) mice showed in BALF a significant increase in the infiltrate of total cells, which consisted of an augmented cell number of macrophages (94%), neutrophils (5%), and lymphocytes (0.8%). The total cell number was significantly lowered only in PK20 and dexamethasone (DEX) groups (Figure 1A). The neutrophil number was significantly reduced in all three groups treated with PK20, mixture of its pharmacophores (MIX), and DEX (Figure 1B). These changes were parallel to histological evaluation, which showed marked peribronchial leukocyte accumulation in the PC group (Figure 1C). The inflammation score was significantly attenuated in all treated groups; however, MIX mice showed significantly higher value in comparison to the PK20 and DEX groups (Figure 1D).

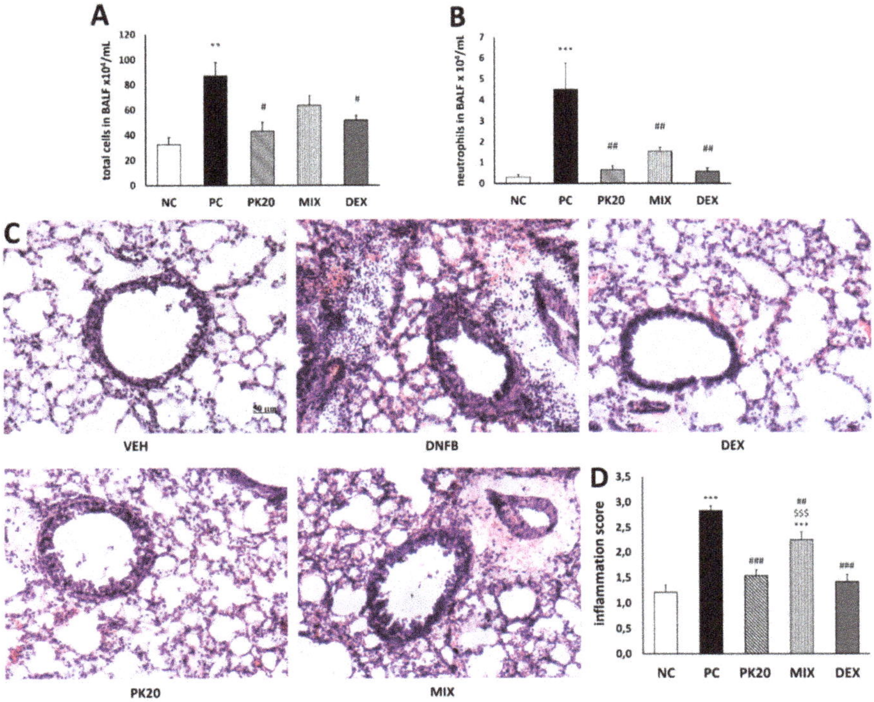

Figure 1. Effect of PK20 on the total number of inflammatory cells (**A**) and neutrophils (**B**) in BALF and effect of PK20 on lung-tissue inflammatory cell infiltration; hematoxylin-and-eosin stained lung sections (**C**) and inflammation score (**D**). Comparison to DNFB-sensitized/DNS-challenged group (positive control; PC) and vehicle-sensitized/DNS-challenged group (negative control; NC) treated with NaCl and to DNFB-sensitized/DNS-challenged groups treated with dexamethasone (DEX) and equimolar mixture of hybrid's structural elements (MIX). The pictures were made at magnification 10×. All values are the mean ± SEM (n = 6–9 in BALF and n = 4 in histology study). ** $p < 0.01$, *** $p < 0.001$ vs. NC, # $p < 0.05$, ## $p < 0.01$, ### $p < 0.001$ vs. PC, $$$ $p < 0.001$ vs. PK20 and DEX groups.

2.2. PK20 Reduces Airway Hyperresponsiveness (AHR)

To assess whether PK20 had a beneficial effect on AHR during an inflammatory response in lungs in DNFB-induced asthma, we exposed mice to the increasing doses of methacholine (MCh) aerosol in whole-body plethysmograph. Penh (enhanced pause) was measured in this noninvasive method at 24 h post-challenge. DNFB-sensitized/DNS-challenged mice displayed increased Penh to growing doses of MCh as compared to mice in the vehicle-sensitized/DNS-challenged group (Figure 2). Treatment with the hybrid peptide PK20 and the mixture of its pharmacophores significantly reduced AHR in DNFB-sensitized/DNS-challenged mice at 20 mg/mL of inhaled MCh. DEX-treated mice exhibited lowered Penh at all higher doses of MCh in comparison to the PC group (Figure 2).

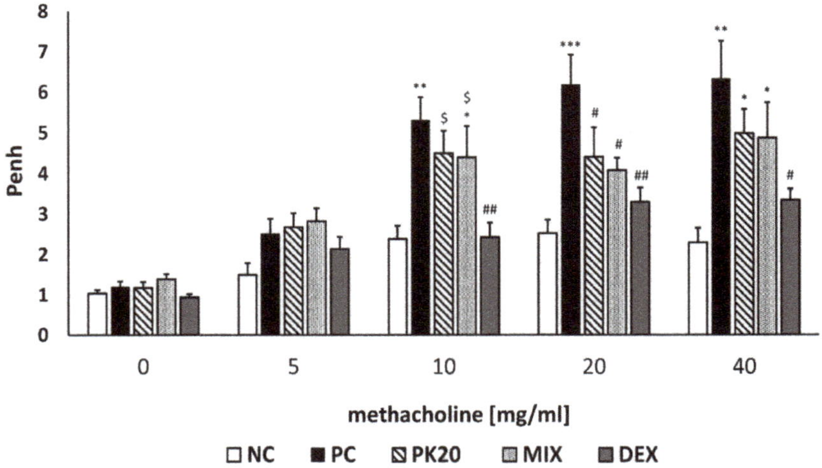

Figure 2. Effect of PK20 on the development of airway hyperreactivity in non-atopic asthma model. Penh responses to increasing concentrations of aerosolized methacholine in DNFB-sensitized/DNS-challenged group (positive control; PC) and vehicle-sensitized/DNS-challenged group (negative control; NC) treated with NaCl and in DNFB-sensitized/DNS-challenged groups treated with PK20, dexamethasone (DEX), and equimolar mixture of hybrid's structural elements (MIX). All values are the mean ± SEM (n = 7–8). * $p < 0.05$, ** $p < 0.01$, *** $p < 0.001$ vs. corresponding NC group. # $p < 0.05$, ## $p < 0.01$, vs. corresponding PC group and \$ $p < 0.05$, \$\$ $p < 0.01$ vs. corresponding DEX group.

2.3. Effect of PK20 on DNFB-Induced Pro-Inflammatory Cytokine and Chemokine Production

Measurement of pro-inflammatory cytokine content was performed 24 h after intratracheal DNS challenge in BALF and lung homogenates of DNFB or vehicle-sensitized mice, to determine whether treatment with PK20 is able to influence their production. The levels of IL-1α, IL-2, IL-13, and TNF-α were significantly increased in BALF of DNFB-sensitized/DNS-challenged mice (PC) compared to NC group (Figure 3). The levels of all cytokines were significantly decreased after PK20 and DEX treatment, whereas the co-administration of PK20's opioid- and NT-like pharmacophores resulted in decreased content of IL-2, solely (Figure 3B).

In lung homogenates PK20 and DEX in similar degree reduced levels of IL-1β, IL-17A, IL-12p40, CXCL1 (KC), and RANTES in comparison to the PC group (Figure 4). Treatment with the mixture of PK20 pharmacophores was effective only in decreasing content of IL-12p40 and RANTES (Figure 4D,E).

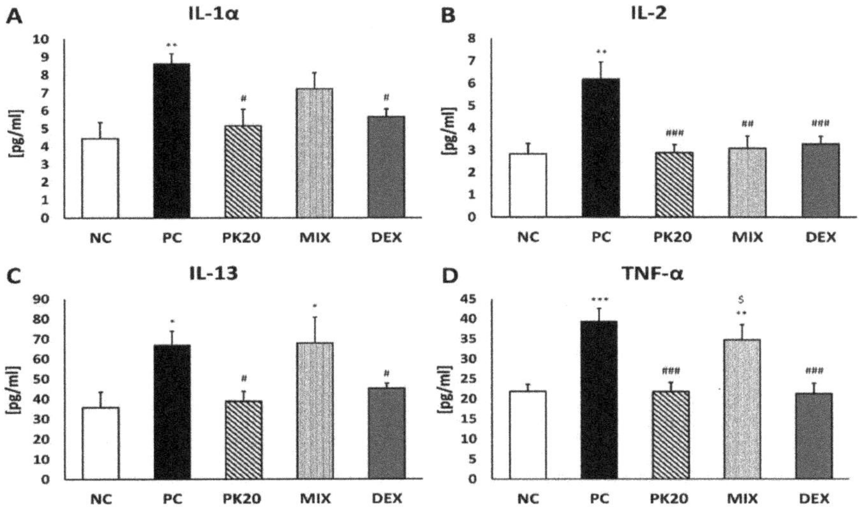

Figure 3. Concentration of pro-inflammatory cytokines in BALF: IL-1α (**A**), IL-2 (**B**), IL-13 (**C**), and TNF-α (**D**) in DNFB-sensitized/DNS-challenged mice after treatment with PK20, mixture of its structural elements (MIX), and dexamethasone (DEX). Comparison to DNFB-sensitized/DNS-challenged (positive control; PC) and vehicle-sensitized/DNS-challenged group (negative control; NC) treated with NaCl. All values are the mean ± SEM (n = 5–9). * $p < 0.05$, ** $p < 0.01$, *** $p < 0.001$ compared with NC group, # $p < 0.05$, ## $p < 0.01$, ### $p < 0.001$ compared with the PC group, \$ $p < 0.001$ vs. PK20 and DEX groups.

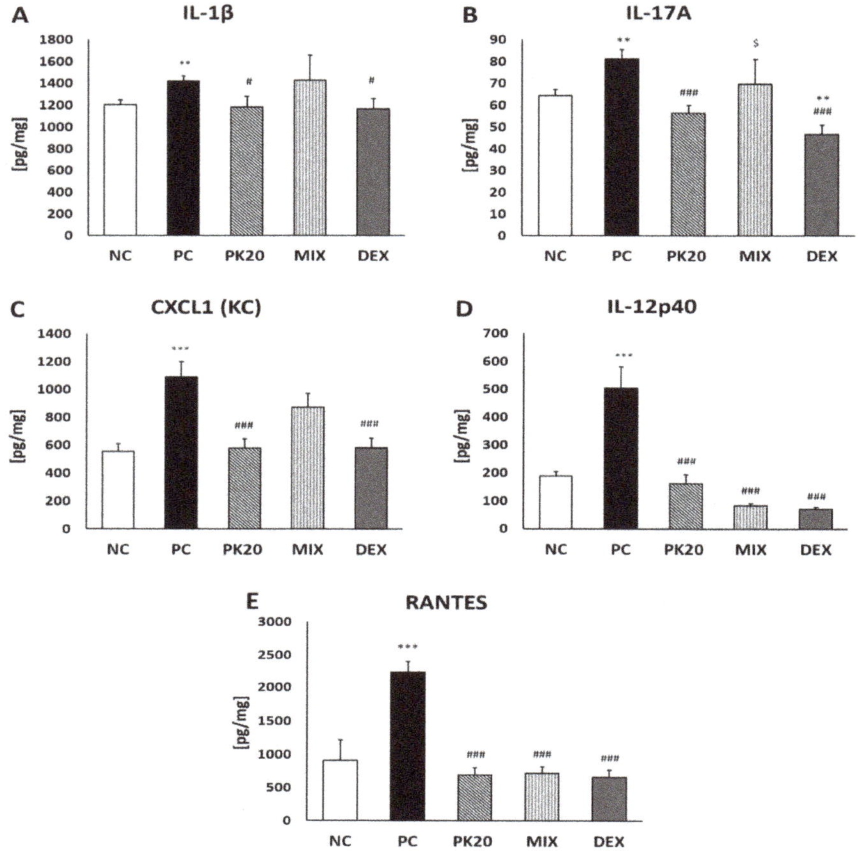

Figure 4. Concentration of pro-inflammatory cytokines in lung-tissue homogenates: IL-1β (**A**), IL-17A (**B**), IL-12p40 (**C**), KC (**D**), and RANTES (**E**) in DNFB-sensitized/DNS-challenged mice after treatment with PK20, mixture of its structural elements (MIX), and dexamethasone (DEX). Comparison to DNFB-sensitized/DNS-challenged (positive control; PC) and vehicle-sensitized/DNS-challenged group (negative control; NC) treated with NaCl. All values are the mean ± SEM ($n = 5$–7). ** $p < 0.01$, *** $p < 0.001$ compared with NC group. # $p < 0.05$, ### $p < 0.001$ compared with the PC group; $ $p < 0.05$ vs. DEX.

2.4. *Effect of PK20 Treatment on Mouse Mast Cell Protease (MCPT 1) Level in Lungs*

DNFB-sensitized/DNS-challenged mice exhibited a significantly higher level of MCPT 1 in lung tissue compared with vehicle-sensitized/DNS-challenged ones. Only PK20 and DEX treatments were effective in lowering the content of mast cell MCPT 1 (Figure 5A).

2.5. *Effect of PK20 Treatment on Malondialdehyde Level (MDA) in Lungs*

Sensitization and challenge with DNFB/DNS showed a statistically significant enhancement in MDA production compared with NC mice (Figure 5B). The content of an oxidative stress marker was significantly reduced in all treated groups (PK20, MIX, and DEX) and similar to the level present in NC mice (Figure 5B).

2.6. Effect of PK20 Treatment on Activity of Secreted Phospholipase A_2 (sPLA2) in Lungs

sPLA2 activity was significantly reduced in lungs of PK20-, MIX-, and DEX-treated groups compared to PC mice. The levels were comparable to the sPLA2 activity in vehicle-sensitized/DNS-challenged negative controls (Figure 5C).

2.7. Effect of PK20 Treatment on Albumin Concentration in BALF

Albumin concentration expressed in mg of total protein assessing pulmonary vascular leakage was significantly enhanced in PC mice compared to negative controls. The values were significantly lowered in PK20-, MIX-, and DEX-injected animals (Figure 5D).

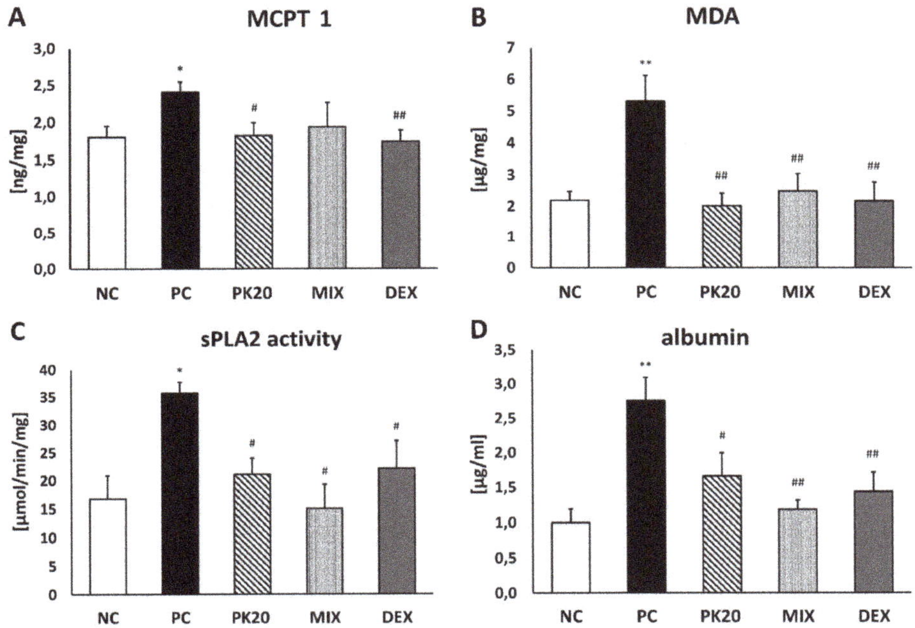

Figure 5. Concentration of (**A**) mouse mast cell protease-1 (MCPT 1), (**B**) malondialdehyde (MDA), and (**C**) activity of secreted phospholipase A_2 (sPLA2) in lung-tissue homogenates and (**D**) level of albumins in BALF of DNFB-sensitized/DNS-challenged mice after treatment with PK20, mixture of its structural elements (MIX), and dexamethasone (DEX). Comparison to DNFB-sensitized/DNS-challenged (positive control; PC) and vehicle-sensitized/DNS-challenged group (negative control; NC) treated with NaCl. All values are the mean ± SEM (n = 5–8). * $p < 0.05$, ** $p < 0.01$ compared with NC group. # $p < 0.05$, ## $p < 0.01$ compared with the PC group.

2.8. Effect of PK20 on Lung Nuclear Factor Kappa B (NF-κB) Immunohistochemistry

Challenge of DNS into airways of DNFB-sensitized mice significantly increased immunofluorescence signal for NF-κB in comparison to the vehicle-sensitized NC group (Figure 6A). In contrary to MIX, PK20 and DEX treatment apparently blunted the signal, which indicates that the NF-κB signaling pathway may be involved in the anti-inflammatory activity of both chemicals (Figure 6A–C). Note the massive enlargement of the pulmonary interalveolar septa in the mice from MIX and PC groups and more intensive lung NF-κB p50 and p65 staining in comparison to negative control, PK20-, and DEX-treated animals. This significantly increased NF-κB p50 and p65 expression most probably resulted in augmented inflammatory response present in PC and MIX animals.

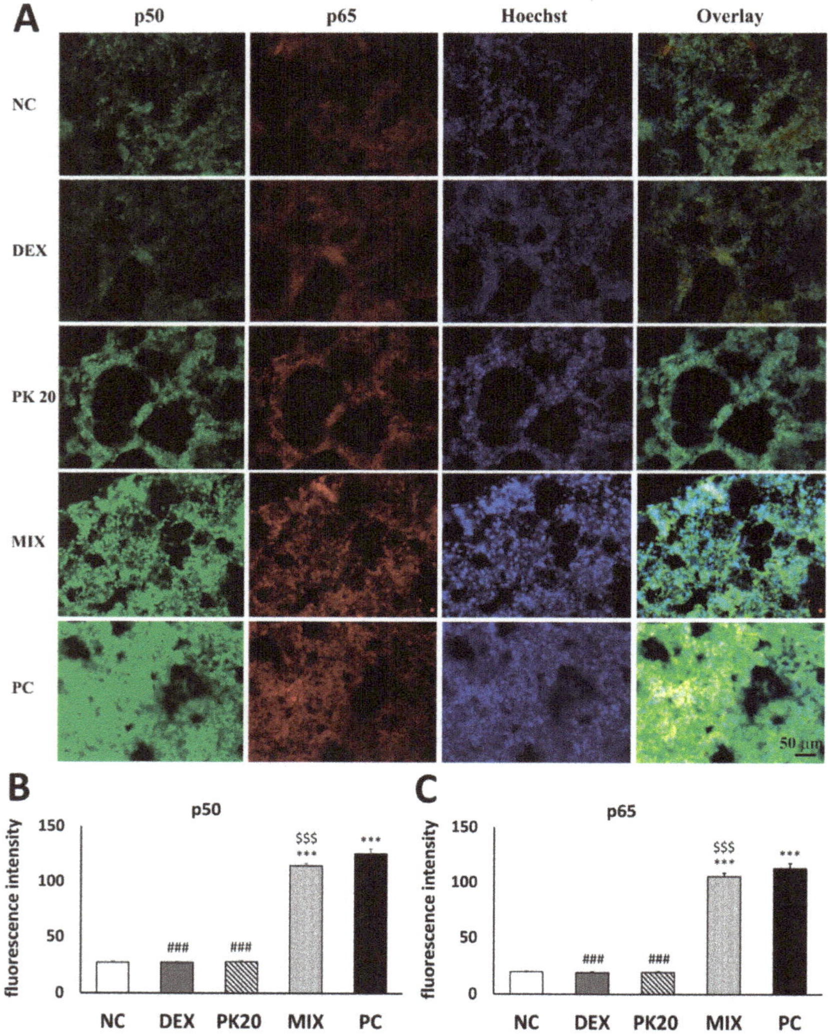

Figure 6. Effect of PK20 on NF-κB expression in DNFB-induced asthma in mice. Immunofluorescent double labeling of NF-κB p50 (green) and NF-κB p65 (red) subunits and Hoechst stain of the nuclei in lung sections from each treatment group was performed (**A**). The intensity of immunostaining was measured in grey scale and presented on panel (**B**) (p50) and (**C**) (p65). The treatment groups are: vehicle-sensitized/DNS-challenged group (negative control; NC) treated with NaCl; DNFB-sensitized/DNS-challenged group treated with NaCl (positive control; PC); and DNFB-sensitized/DNS-challenged groups treated with PK20, dexamethasone (DEX) and equimolar mixture of hybrid's structural elements (MIX). All values are mean ± SEM ($n = 4$). *** $p < 0.001$ vs. corresponding NC group, ### $p < 0.001$ vs. corresponding PC group and $$$ $p < 0.001$ vs. corresponding PK20 and DEX groups.

3. Discussion

In the murine model of non-atopic asthma, we demonstrated that the hybrid peptide engineered from pharmacophores of NT- and EM-2-based analogs may serve as a potent anti-inflammatory agent affecting a number of inflammation markers, and its potency is comparable to the effects mediated by the dexamethasone. However, glucocorticosteroid showed a stronger impact on attenuation of airway hyperreactivity. Through the comparison of the effects of chimera with simple equimolar mixture of its structural elements, we also confirmed the hypothesis that two drugs combined into one entity may be characterized by improved therapeutic activity. Indeed, PK20 treatment occurred to be more effective in substantial reduction of inflammatory cell influx, including neutrophils, typical cells for nonallergic asthma [22,24]. Reduced inflammatory cell influx was associated with diminished production of pro-inflammatory cytokines and chemokines in BALF and lung tissue. These effects were comparable to DEX, and better than MIX treatment, in which a decrease of only IL-2, IL-12p40, and RANTES was noted.

Reduction of IL-1β, TNF-α, IL-17A, and CXCL1 (KC) was parallel to PK20 evoked attenuation of macrophage and neutrophil infiltrate. Cytokines, such as TNF-α and IL-1β, augmented in sputum and BALF of asthmatics, are well-known for their role in amplifying inflammation [25] and promoting neutrophil chemotaxis [26–28]. Potency of IL-1β in eliciting neutrophil accumulation in rat BALF was shown after intratracheal interleukin administration, and this was concomitant with an increased airway responsiveness to bradykinin stimuli [29]. Importance of both IL-1 cytokines, IL-1β and IL-1α, in AHR and recruitment of inflammatory cells to the lung was evidenced after blockade with neutralizing antibodies, resulting in attenuated phenotype of murine model of asthma [30].

Contribution of TNF-α to airway hyperreactivity and airway inflammation mediated by neutrophils was demonstrated previously in normal subjects and ovalbumin and DNFB-sensitized mice [22,31,32]. Elevated levels of IL-17A displayed in sputum of asthmatics were correlated not only with neutrophilia, but also with an airway hyperresponsiveness provoked with methacholine [33]. Further reports supported the role of IL-17A in a neutrophil- and macrophage-mediated inflammation [34,35], appearing also in our asthma model. Likewise, chemokines CXCL1 and RANTES, decreased due to the hybrid administration, have been described to be potent neutrophil chemoattractant [36–38]. Cytokines such as IL-2, IL-12p40, and IL-13 increased in the present study and reduced by the hybrid are documented to be important mediators of asthma. IL-2 was elevated in BALF of nonallergic asthma patients [39]; moreover, expression of IL-2 mRNA was enhanced in infiltrated BALF cells in steroid-resistant asthmatics [40]. IL-12p40 overexpression in airway epithelial cells in asthma patients was linked with airway macrophages accumulation [41]. IL-13 is a pro-inflammatory interleukin in asthma [42], which, similarly to our study, was increased in BALF in the allergic model and reduced after treatment with dexamethasone [43]. IL-13 is thought to be the type 2 cytokine and key mediator of allergic asthma, yet it is considered to act via the pathway independent of immunoglobulin E and eosinophils [44]. Elevated mRNA encoding IL-13 has been reported in the bronchial mucosa of atopic and non-atopic asthmatics [45].

Oxidative stress and overproduction of reactive oxygen species (ROS) associated with augmented inflammatory response has been displayed in human asthma and experimental models [46]. In the present study, we showed that DNFB sensitization enhanced the level of MDA production, which was remarkably reduced by the treatment with PK20, DEX, and a mixture of pharmacophores. This was linked with a decreased number of recruited cells in BALF, which in turn are responsible for ROS production [47,48] and with the lessened levels of IL-1, RANTES and TNF-α, which are reported to enhance ROS generation [49–51]. Furthermore, augmented oxidative stress and increased release of pro-inflammatory cytokines are known to activate secreted phospholipase A_2 ($sPLA_2$), enzyme-regulating eicosanoid synthesis, airway inflammation, and AHR in asthma [52]. Secreted PLA2 group X was demonstrated to be overexpressed in airway epithelial cells and macrophages of asthmatics [53,54]. Further, it was evidenced that $sPLA_2$ increase can be provoked with allergen inhalation in humans and mice [55]. In fact, $sPLA_2$ activity was increased in our positive control group and reduced by pretreatment with all tested compounds. TNF-α, IL-1β, IL-17A, and IL-13 are cytokines, which decrease in PK20 and DEX groups is correlated with $sPLA_2$ activity reduction. All of them are implicated in asthma and has been demonstrated to upregulate $sPLA_2$ [54].

Oxidative stress is known to activate NF-κB, which is a pleiotropic transcription factor considered to be a critical signal in evoking an inflammatory response in the lung during the pathogenesis of asthma [56,57]. Pro-inflammatory stimuli enhance expression of genes encoding inflammatory mediators, such as cytokines and chemokines, important in the recruitment of inflammatory cells. Consistently in the current study, NF-κB immunoreactivity was increased in the DNFB-sensitized/DNS-challenged group, yet in DEX- and PK20- treated mice, it was significantly reduced in contrary to the MIX group. Dexamethasone's potency in inhibiting NF-κB immunoreactivity was also reported previously [43]. PK20-induced suppression of NF-κB was consistent with reduced oxidative stress and decreased expression of NF-κB-mediated pro-inflammatory cytokines and chemokines, which contribute to attenuated recruitment of inflammatory cells. All of these findings suggest that the NF-κB signaling pathway might be involved in the alleviating effect of the hybrid on hapten-induced asthma. The limitation of our study is that we measured NF-κB expression by only using immunohistochemistry; therefore, further experiments investigating NF-κB activation, that is, the levels of phosphorylated units and nuclear translocation, via Western blot are needed.

An important possibility of testing our hybrid compound and its effect on the signaling pathways involved in airway inflammation and hyperresponsiveness, such as TNF-α/IL-13, NF-κB, MAPK, c-fos, and c-jun seems to be in vitro airway organ culture targeted in the earlier studies [58,59], which can be considered in our future research.

A consistent finding of our study was improved lung vascular integrity produced with all compared substances, evidenced by decreased albumins in BALF. This might be associated with decreased migration of leukocytes, such as neutrophils and monocytes, to the lungs, evidenced in our study and reported previously to contribute to increased lung permeability [60,61]. Another marker of inflammation attenuated by PK20 and DEX in the present study was mouse mast cell protease (MCPT 1), biomarker of mast cell degranulation. The key role of mast cells in the development of non-atopic asthma induced with DNFB has been previously evidenced [62]. Their contribution to neutrophil recruitment through TNF-α was also displayed in delayed-type hypersensitivity reaction evoked with cognate hapten by Biedermann et al. [63].

In the current study, we also examined airway responsiveness to methacholine inhalation, measured by whole-body plethysmography. Validity of Penh is still under debate, as it is considered to reflect pattern of breathing rather than lung mechanics [64]. However, data published by different groups have displayed similar changes in Penh and lung resistance measured invasively [64–67]. Although Penh is affected by airway resistance only to some extent [68], we do think that this noninvasive measurement has some value in depicting airway responsiveness, at least for screening purposes. More important, substantial increases in Penh in reaction to methacholine appeared only in DNFB-sensitized/DNS-challenged mice in contrast to vehicle-sensitized/DNS-challenged ones. Dexamethasone treatment effectively reduced Penh to all doses of methacholine in comparison to positive controls. The chimeric peptide, as well as the mixture of its structural elements, achieved significant reduction solely at a dose of 20 mg/mL, and it was not as effective as DEX. One would expect that similar reduction of several markers of inflammation described above, such as BALF cell influx, cytokine, MDA, and $sPLA_2$ activity, should translate into decreased AHR. However, it is known that there is no consistent relationship between airway hyperresponsiveness to bronchoconstrictor agents and inflammation [69,70]. Furthermore, Tränkner et al. [71] showed that hyperreactive bronchoconstriction can be completely dissociated from the inflammatory immune response and ablated after silencing of the vagal sensory neurons, expressing transient receptor potential vanilloid 1 (TRPV1) ion channel. It seems that PK20's being potent in suppressing inflammation is able only partially to control airway hyperresponsiveness.

It is assumed that hybrid can exert its anti-inflammatory action via opioid receptors expressed by cells associated with inflammation, such as macrophages, mast cells, and fibroblasts [72]. Endomorphins and opioid receptors have been localized within airway smooth muscles, nerves, epithelium, and mucus glands [20]. In the respiratory system, they have been described as potent inhibitors of tachykinergic and cholinergic constriction, neurogenic mucus secretion, and goblet cells secretion [20,73]. In the review by Stein and Kuchler [18], reduction of edema, plasma extravasation, cytokine release, and tissue damage evoked by opioid-receptor activation were described in various models of inflammation. When it comes to neurotensin, its anti-inflammatory activity is far from clear. In fact, based on several studies, this neuropeptide is indicated to stimulate mast cell degranulation [74], neutrophil adherence to bronchial epithelial cells [75], and genetically modified mice deprived of production of NT had attenuated inflammatory response [76]. On the other hand, NT-treated animals in the model of inflammatory bowel disease showed a protective effect on the intestines related to anti-inflammatory, antioxidant, and anti-apoptotic actions [16]. The latest study showed that NT is a potent downregulator of inflammatory and hyperreactivity response in the nonallergic asthma model [17]. We should keep in mind that the activity exerted by the chimeric peptide PK20, used in our study, cannot be considered as a simple conjunction of activities appropriate to its parent peptides or, more precisely, pharmacophores. Hybridization of two modified structures either of EM-2 or NT (8–13) into one molecule may bring new and diverging physical and chemical properties. In the present study, we showed beneficial effect of PK20 on many markers of inflammation characteristic for asthma, such as attenuated production of MDA and cytokines, reduced activity of $sPLA_2$ and expression of NF-κB, and diminished level of MCPT 1. We also made a comparison between beneficial effects of PK20 and a mixture of its moieties, as still multicomponent drugs and/or polypharmacy play a leading role in patients' treatment, though it is associated with adverse outcomes and drug reactions. Our experiments revealed that PK20 predominates in reduction of majority of examined markers of inflammation, which may be explained by its improved stability, bioavailability, and possibly new properties, which resulted in better anti-inflammatory activity.

Overall, the present study demonstrated that chimeric peptide-PK20 attenuated DNFB induced upregulation of asthma symptoms possibly via suppression of $sPLA_2$ and NF-κB pathways. Improved anti-inflammatory potency of the hybrid shows the advantage of application of innovative chimeric compounds over the simple mixture of its moieties.

4. Materials and Methods

4.1. Drugs and Reagents

PK20 (H–Dmt–D-Lys–Phe–Phe–Lys–Lys–Pro–Phe–Tle–Leu–OH) was synthesized by using solid-phase peptide synthesis according to the method of Merrifield with Fmoc approach [77] in the Department of Neuropeptides. Chemicals and reagents were purchased from commercial companies: 1-fluoro-2, 4-dinitrobenzene (DNFB, No. 556971), dexamethasone (No. D4902), dinitrobenzene sulfonic acid (DNS, No. D1529), protease inhibitor cocktail (No. S8830), and TX-100 (No. X100) were obtained from Sigma-Aldrich (Poznań, Poland). Ketamine, xylazine, and pentobarbital sodium were purchased from Biowet (Puławy, Poland).

4.2. Animals

Animals were obtained from the Animal House of Mossakowski Medical Research Centre, Polish Academy of Sciences. Male Balb/c mice of 7–8 weeks of age were used. Animal procedures were performed in accordance with institutional guidelines accepted by the IV Warsaw Ethics Commission for the Care and Use of Laboratory Animals for Experimental Procedures (permit number: 57/2014, date of approval: 16 January 2015). Procedures were conformed to guidelines published in the European directive 2010/63/EU on the protection of animals used for scientific purposes.

4.3. DNFB-Induced Experimental Asthma

On day 1, mice were epicutaneously sensitized with either 0.5% DNFB dissolved in acetone and olive oil (4:1) or vehicle control onto shaved abdominal skin (50 μL) and paws (50 μL divided on four paws). On the next day, DNFB/vehicle was applied to the thorax alone. On day 6, mice were challenged intratracheally with 50 μL of DNS (0.6%), water soluble hapten of DNFB. The sensitization and hapten challenge were performed under intramuscular anesthesia: ketamine (70 mg/kg) and xylazine (10mg/kg). Twenty-four hours from the challenge airway hyperresponsiveness to nebulized methacholine was measured. Afterward, the mice were sacrificed with an overdose of pentobarbital sodium (150–200 mg/kg IP) and bronchoalveolar lavage fluid (BALF), and lungs were collected for histology and further analyses.

4.4. Experimental Groups

The experimental mice were randomly assigned into following groups: —treated with 0.9% NaCl, vehicle-sensitized/DNS-challenged, negative control (NC); —treated with NaCl, DNFB-sensitized/DNS-challenged, positive control (PC); —PK20 treated, DNFB-sensitized/DNS-challenged (PK20); —treated with mixture of PK20 structural pharmacophores, DNFB-sensitized/DNS-challenged (MIX); —dexamethasone treated, DNFB-sensitized/DNS-challenged (DEX). The dose of PK20 (10 mg/kg i.p.) was selected on the basis of an analgesic study performed by Kleczkowska et al. [13] and our neurotensin-based study [17]. PK20 applied via IP route of injection was previously shown to be safe, without visible adverse side effects [78]. PK20 prepared freshly from powder, (dissolved in NaCl) was administered twice (2 and 8 h after DNS challenge) in volumes of 100 μL. All groups received equivalent volumes and route of injection: NaCl—control groups (NC, PC), equimolar amount of mixture of NT and EM-2 pharmacophores—MIX group, and 4 mg/kg of dexamethasone received DEX group (all chemicals were dissolved in NaCl).

4.5. Measurement of Airway Hyperresponsiveness

Airway hyperresponsiveness to nebulized methacholine (MCh, No. A2251) was measured 24 h after DNS challenge by a whole-body barometric plethysmography (Buxco Electronics, Inc., Wilmington, NC, USA). Mice were put in the plethysmograph chamber for 10 min to adaptation. After stabilization, they were exposed to aerosolized saline as a control and every 20 min to increasing doses of nebulized methacholine for 30 s each (5, 10, 20, and 40 mg/mL). The enhanced pause (Penh) was recorded and averaged over 3 min, following each nebulization, and used as an index of airway reactivity.

4.6. Total and Differential Cell Counts in Bronchoalveolar Lavage Fluid

After sacrificing the animals, the trachea of each mouse was cannulated. Bronchoalveolar lavages were carried out by injecting and withdrawing via a cannula 4 × 1 mL of phosphate buffered saline (PBS). The first 1 mL contained protease inhibitor cocktail. Collected lavages were centrifuged (1500 rpm, 10 min) to isolate BALF cells and to remove supernatant. Supernatant from the first 1 mL was frozen and used for further analysis. Sediments of the fourth lavages were pooled and re-suspended by 150 μL of PBS. Total cells were counted by using a Bürker-Türk chamber. Remaining cell suspensions were cytocentrifuged (Thermo Shandon, Cambridge, UK) at 700 rpm for 10 min onto microscopic slides. Air-dried preparations were fixed and stained with a fast-staining kit based on eosin and azure solutions (No. AB123H, Hemastain, Analab, Warsaw, Poland), in order to differentiate morphologically the leukocyte population. Differential cell count on at least 500 cells from each sample was performed under light microscope. Results are expressed as the number of mononuclear cells (macrophages and leukocytes) and neutrophils per 1 mL of BALF.

4.7. Cytokine and Chemokine Quantification

Immediately after BALF collection, the lungs were removed and cleaned. Lungs and first 1 mL of BALF containing protease inhibitor cocktail were frozen in liquid nitrogen and stored at −80 °C. For further analysis, lungs were homogenized under liquid nitrogen and taken up in 500 μL of PBS with protease inhibitor cocktail and 1% of TX-100. Homogenates were centrifuged (14,000 rpm, 10 min, 4 °C), and the supernatants were used for cytokines measurement. Measurement of IL-1α, IL-2, IL-13, TNF-α, IL-1β, IL-12p40, IL-17A, CXCL1 (KC), and RANTES in BALF and lung homogenates was performed by using a Bio-Plex pro mouse cytokine panel for inflammatory cytokines (No. M60009 RDPD, Bio-Rad, Warsaw, Poland) according to the manufacturer's protocol. Beads were analyzed at a Bio-Plex 200 Platform (Luminex, Bio-Rad, Warsaw, Poland).

4.8. Measurement of Oxidative Stress (Malondialdehyde Level) in Lung Tissue

Malondialdehyde (MDA), aldehydic secondary product of lipid peroxidation, is a marker of oxidative stress. The determination of MDA concentration was performed in supernatants from lung tissue, using a commercially available kit (No. STA-832, OxiSelect™ MDA, Cell Biolabs, Inc., San Diego, CA, USA).

4.9. Measurement of Mouse Mast Cell Protease Level (MCPT 1) in Lung Tissue

A commercially available enzyme-linked immunosorbent assay—the Mouse MCP-1 ELISA Ready-SET kit (No. 88-7503, eBioscience, San Diego, CA, USA) was used for the quantification of MCPT 1 in lung-tissue supernatants.

4.10. Measurement of Secretory Phospholipase 2 (PLA2) Activity

The sPLA2 activity was measured in the lung supernatant by using a sPLA2 kit (No. 765001, Caymann Chemicals, Ann Arbor, MI, USA) according to the manufacturer's instructions and expressed in µmol/min/mg. The enzyme activity was read for 10 min at 414 nm, using an ELISA plate reader.

4.11. Determination of Total Protein and Albumin Concentration

Total protein assay was performed in lung supernatants and BALF with protein assay kit (No. 500-0006, Bio-Rad, Warsaw, Poland). Albumin levels were measured in collected BALF with a murine-specific albumin ELISA kit (No. E99-134, Bethyl Laboratories, Montgomery, TX, USA). All the measurements were performed according to the manufacturer's instructions. The sPLA2 activity and the content of cytokines, MDA, and MCPT 1 in lungs and albumins in BALF were expressed in units of weight per mg of total protein.

4.12. Lung Histology

The lungs were fixed for 7 days in the 4% paraformaldehyde in 0.1 M phosphate buffer at 4 °C. Afterward, the tissue was immersed in 20% (*w/v*) sucrose solutions in PBS and cut with a cryostat (CM 1850 UV, Leica, Germany) on 20 µm thick glass-mounted sections. After staining with eosin (No. 1.09844) and hematoxylin (No. 1.50174, Merck Millipore, Warsaw, Poland), they were examined under a light microscope (Olympus, Japan). To assess the severity of the immune cells' infiltration, peribronchial and perivascular cell counts were performed based on the 4-point score system described in detail by Tournoy et al. [79], where higher scores designated more severe inflammation (more inflammatory cell rings around bronchi or vessels).

4.13. NF-κB Immunohistochemistry

The studied lung sections, prepared as described in Section 4.12, were subjected to double labeling for NF-κB p50 and NF-κB p65 peptides. After preincubation with 3% normal goat serum solution in PBS with 0.2% Triton X-100 (PBST), the samples were subsequently incubated for 1 h at 37 °C with PBST containing 1% normal goat serum and primary antibodies (Abs): murine monoclonal Ab against NF-κB p65 (No. sc-8008, Santa Cruz Biotechnology, Dallas, TX, USA, dil. 1:300) and rabbit polyclonal Ab against NF-κB p50 (No. sc-114, Santa Cruz Biotechnology, Dallas, TX, USA, dil. 1:3400). Next, the sections were washed three times in PBS and incubated for 1 h at 37 °C, with a mixture of the fluorescent dye-conjugated secondary antibodies: goat anti-mouse Alexa Fluor 594 (No. A11020, Invitrogen-Molecular Probes; dil. 1:100) and goat anti-rabbit Alexa Fluor 488 (No. A11070,

Invitrogen-Molecular Probes; dil. 1:100). In the next stage, the samples were washed three times with PBS and dried.

Immunolabelled sections were washed in PBS and incubated in 1 μg/mL of PBS solution of Hoechst stain (bisbenzimide dye, No. 33258, Sigma-Aldrich, Poznań, Poland) for 2 min, at room temperature. The stain was then drained off and cover-slipped with Vectashield Mounting Medium for fluorescence microscopy (No. H1000, Vector Laboratories, Inc., Burlingame, CA, USA). Finally, the sections were analyzed with a fluorescent microscope (Nikon, Japan) and photographed with a CCD camera (Nikon, Japan). The specialist who assessed the material was blinded to the identity of the samples. Specificity of the immunolabelling was verified by performing a negative control staining procedure, in which the primary antibodies were omitted in the incubation mixture. The intensity of p65, as well as p50 immunostaining, was measured as an optical density (OD) level in grayscale values (8-bit scale), using Scion computer software (NIH, Frederick, MD, USA). Analysis was carried on within the lung sections framed by microscopic observation field. Three frames of alveolar septum per section and three sections per each mouse from all investigated groups were analyzed. Background staining was sampled on each frame individually and was detracted from the values of the cellular staining. For statistical analysis, the average gray-level value from all the measured sections of each mouse from all investigated groups was used.

4.14. Statistical Analysis

Data were presented as means ± standard error of mean (SEM). Differences between the mean values of the normally distributed data were assessed by using one-way analysis of variance (ANOVA; Newman–Keuls post hoc test) and Student's *t*-test. The *p*-values less than 0.05 were considered to be statistically significant. All data manipulations and statistical analysis were conducted by the usage of STATISTICA 12 (StatSoft, Kraków, Poland).

Author Contributions: Conceptualization, methodology, and formal analysis were performed by K.K. and E.R.; experimental investigation was performed by E.R., K.K., D.S., K.A., and P.K.; visualization was performed by E.R., K.A., D.S., and K.K.; supervision, original draft preparation, and funding acquisition were performed by K.K.; review and editing were performed by K.K., E.R., and P.K.

Funding: The study was supported by a grant from National Science Centre, Poland, no. 2014/13/B/NZ7/02247.

Conflicts of Interest: The authors declare no conflicts of interest with respect to the publication of this manuscript. The funders had no role in the design of the study; in the collection, analyses, or interpretation of data; in the writing of the manuscript, or in the decision to publish the results.

References

1. Holgate, S.T.; Arshad, H.S.; Roberts, G.C.; Howarth, P.H.; Thurner, P.; Davies, D.E. A new look at the pathogenesis of asthma. *Clin. Sci.* **2009**, *118*, 439–450. [CrossRef] [PubMed]
2. Pelaia, G.; Vatrella, A.; Busceti, M.T.; Gallelli, L.; Calabrese, C.; Terracciano, R.; Maselli, R. Cellular mechanisms underlying eosinophilic and neutrophilic airway inflammation in asthma. *Mediat. Inflamm.* **2015**, *2015*, 8. [CrossRef] [PubMed]
3. Global Asthma Report, Global Asthma Network, Auckland, New Zealand. Available online: http://www.globalasthmanetwork.org/publications/Global_Asthma_Report_2014.pdf,2014 (accessed on 9 September 2016).
4. Domínguez-Ortega, J.; Phillips-Anglés, E.; Barranco, P.; Quirce, S. Cost-effectiveness of asthma therapy: A comprehensive review. *J. Asthma* **2015**, *52*, 529–537. [CrossRef] [PubMed]
5. Behera, D.; Sehgal, I.S. Bronchial asthma-issues for the developing world. *Indian J. Med.* **2015**, *141*, 380–382. [CrossRef] [PubMed]
6. Sulaiman, I.; Lim, J.C.; Soo, H.L.; Stanslas, J. Molecularly targeted therapies for asthma: Current development, challenges and potential clinical translation. *Pulm. Pharmacol. Ther.* **2016**, *40*, 52–68. [CrossRef]
7. Abramson, M.J.; Walters, J.; Walters, E.H. Adverse effects of β-agonists. *Am. J. Respir. Med.* **2003**, *2*, 287–297. [CrossRef]

8. Chung, K.F.; Wenzel, S.E.; Brozek, J.L.; Bush, A.; Castro, M.; Sterk, P.J.; Teague, W.G. International ERS/ATS guidelines on definition, evaluation and treatment of severe asthma. *Eur. Respir. J.* **2014**, *43*, 343–373. [CrossRef]
9. Singh, J.; Shah, R.; Singh, D. Inundation of asthma target research: Untangling asthma riddles. *Pulm. Pharmacol. Ther.* **2016**, *41*, 60–85. [CrossRef]
10. Morphy, R.; Rankovic, Z. Fragments, network biology and designing multiple ligands. *Drug Discov. Today* **2007**, *12*, 156–160. [CrossRef]
11. Zimmermann, G.R.; Lehár, J.; Keith, C.T. Multi-target therapeutics: When the whole is greater than the sum of the parts. *Drug Discov. Today* **2007**, *12*, 34–42. [CrossRef]
12. Kleczkowska, P.; Kowalczyk, A.; Lesniak, A.; Bujalska-Zadrozny, M. The discovery and development of drug combinations for the treatment of various diseases from patent literature (1980-Present). *Curr. Top. Med. Chem.* **2017**, *17*, 1–20. [CrossRef] [PubMed]
13. Kleczkowska, P.; Kosson, P.; Ballet, S.; Van den Eynde, I.; Tsuda, Y.; Tourwe, D.; Lipkowski, A.W. PK20, a new opioid-neurotensin hybrid peptide that exhibits central and peripheral antinociceptive effects. *Mol. Pain* **2010**, *6*, 86. [CrossRef] [PubMed]
14. Kleczkowska, P.; Kawalec, M.; Bujalska-Zadrozny, M.; Filip, M.; Zabłocka, B.; Lipkowski, A.W. Effects of the hybridization of opioid and neurotensin pharmacophores on cell survival in rat organotypic hippocampal slice cultures. *Neurotox. Res.* **2015**, *28*, 352–360. [CrossRef] [PubMed]
15. Kaczyńska, K.; Kogut, E.; Zając, D.; Jampolska, M.; Andrzejewski, K.; Sulejczak, D.; Lipkowski, A.; Kleczkowska, P. Neurotensin-based hybrid peptide's anti-inflammatory activity in murine model of a contact sensitivity response. *Eur. J. Pharm. Sci.* **2016**, *93*, 84–89. [CrossRef] [PubMed]
16. Akcan, A.; Muhtaroglu, S.; Akgun, H.; Akyildiz, H.; Kucuk, C.; Sozuer, E.; Yurci, A.; Yilmaz, N. Ameliorative effects of bombesin and neurotensin on trinitrobenzene sulphonic acid-induced colitis, oxidative damage and apoptosis in rats. *World J. Gastroenterol.* **2008**, *14*, 1222–1230. [CrossRef]
17. Russjan, E.; Kaczyńska, K. Beneficial effects of neurotensin in murine model of hapten-induced asthma. *Int. J. Mol. Sci.* **2019**, *20*, 5025. [CrossRef]
18. Stein, C.; Kuchler, S. Targeting inflammation and wound healing by opioids. *Trends Pharmcol. Sci.* **2013**, *34*, 303–312. [CrossRef]
19. Yu, Y.; Wang, X.; Cui, Y.; Fan, Y.Z.; Liu, J.; Wang, R. Abnormal modulation of cholinergic neurotransmission by endomorphin 1 and endomorphin 2 in isolated bronchus of type 1 diabetic rats. *Peptides* **2006**, *27*, 2770–2777. [CrossRef]
20. Groneberg, D.A.; Fischer, A. Endogenous opioids as mediators of asthma. *Pulm. Pharmacol. Ther.* **2001**, *14*, 383–389. [CrossRef]
21. Börzsei, R.; Pozsgai, G.; Bagoly, T.; Elekes, K.; Pintér, E.; Szolcsányi, J.; Helyes, Z. Inhibitory action of endomorphin-1 on sensory neuropeptide release and neurogenic inflammation in rats and mice. *Neuroscience* **2008**, *152*, 82–88. [CrossRef]
22. Van der Kleij, H.P.; Kraneveld, A.D.; Van Houwelingen, A.H.; Kool, M.; Weitenberg, A.C.; Redegeld, F.A.; Nijkamp, F.P. Murine model for non-IgE-mediated asthma. *Inflammation* **2004**, *28*, 115–125. [CrossRef] [PubMed]
23. Bozkurt, T.E.; Kaya, Y.; Durlu-Kandilci, N.T.; Onder, S.; Sahin-Erdem, L.I. The effect of cannabinoids on dinitrofluorobenzene-induced experimental asthma in mice. *Respir. Physiol. Neurobiol.* **2016**, *231*, 7–13. [CrossRef] [PubMed]
24. Russjan, E.; Kaczyńska, K. Murine models of hapten-induced asthma. *Toxicology* **2018**, *410*, 41–48. [CrossRef] [PubMed]
25. Barnes, P.J. The cytokine network in asthma and chronic obstructive pulmonary disease. *J. Clin. Investig.* **2008**, *118*, 3546–3556. [CrossRef] [PubMed]
26. Smart, S.J.; Casale, T.B. Pulmonary epithelial cells facilitate TNF-alpha-induced neutrophil chemotaxis. A role for cytokine networking. *J. Immunol.* **1994**, *152*, 4087–4094.
27. Joosten, L.A.; Netea, M.G.; Dinarello, C.A. Interleukin-1β in innate inflammation, autophagy and immunity. *Semin. Immunol.* **2013**, *25*, 416–424. [CrossRef]
28. Nabe, T. Tumor necrosis factor alpha-mediated asthma? *Int. Arch. Allergy Immunol.* **2013**, *160*, 111–113. [CrossRef]

29. Tsukagoshi, H.; Sakamoto, T.; Xu, W.; Barnes, P.; Chung, F. Effect of interleukin-1 on airway hyperresponsiveness and inflammation in sensitized and nonsensitized Brown-Norway rats. *J. Allergy Clin. Immunol.* **1994**, *93*, 464–469. [CrossRef]
30. Johnson, V.J.; Yucesoy, B.; Luster, M.I. Prevention of IL-1 signaling attenuates airway hyperresponsiveness and inflammation in a murine model of toluene diisocyanate–induced asthma. *J. Allergy Clin. Immunol.* **2005**, *116*, 851–858. [CrossRef]
31. Kips, J.C.; Tavernier, J.H.; Joos, G.F.; Peleman, R.A.; Pauwels, R.A. The potential role of tumor necrosis factor α in asthma. *Clin. Exp. Allergy* **1993**, *23*, 247–250. [CrossRef]
32. Kim, H.K.; Lee, C.H.; Kim, J.M.; Ayush, O.; Im, S.Y.; Lee, H.K. Biphasic late airway hyperresponsiveness in a murine model of asthma. *Int. Arch. Allergy Immunol.* **2012**, *160*, 173–183. [CrossRef] [PubMed]
33. Barczyk, A.; Pierzchala, W.; Sozanska, E. Interleukin-17 in sputum correlates with airway hyperresponsiveness to methacholine. *Respir. Med.* **2003**, *97*, 726–733. [CrossRef] [PubMed]
34. Newcomb, D.C.; Peebles, R.S. Th17-mediated inflammation in asthma. *Curr. Opin. Immunol.* **2013**, *25*, 755–760. [CrossRef] [PubMed]
35. Alcorn, J.F.; Crowe, C.R.; Kolls, J.K. Th17 cells in asthma and COPD. *Annu. Rev. Physiol.* **2010**, *72*, 495–516. [CrossRef] [PubMed]
36. Sawant, K.V.; Poluri, K.M.; Dutta, A.K.; Sepuru, K.M.; Troshkina, A.; Garofalo, R.P.; Rajarathnam, K. Chemokine CXCL1 mediated neutrophil recruitment: Role of glycosaminoglycan interactions. *Sci. Rep.* **2016**, *6*, 33123. [CrossRef]
37. Rajarathnam, K.; Schnoor, M.; Richardson, R.M.; Rajagopal, S. How do chemokines navigate neutrophils to the target site: Dissecting the structural mechanisms and signaling pathways. *Cell. Signal.* **2019**, *54*, 69–80. [CrossRef]
38. Pan, Z.Z.; Parkyn, L.; Ray, A.; Ray, P. Inducible lung-specific expression of RANTES: Preferential recruitment of neutrophils. *Am. J. Physiol. Lung Cell. Mol. Physiol.* **2000**, *279*, 658–666. [CrossRef]
39. Walker, C.; Bode, E.; Boer, L.; Hansel, T.T.; Blaser, K.; Virchow, J. Allergic and nonallergic asthmatics have distinct patterns of T-cell activation and cytokine production in peripheral blood and bronchoalveolar lavage. *Am. Rev. Repir. Dis.* **1992**, *146*, 109–115. [CrossRef]
40. Leung, D.Y.; Martin, R.J.; Szefler, S.J.; Sher, E.R.; Ying, S.; Kay, A.B.; Hamid, Q. Dysregulation of interleukin-4, interleukin-5 and interferon-γ gene expression in steroid resistant asthma. *J. Exp. Med.* **1995**, *181*, 33–40. [CrossRef]
41. Walter, M.J.; Kajiwara, N.; Karanja, P.; Castro, M.; Holtzman, M.J. Interleukin 12p40 production by barrier epithelialc cells during airway inflammation. *J. Exp. Med.* **2001**, *193*, 339–351. [CrossRef]
42. Grünig, G.; Warnock, M.; Wakil, A.E.; Venkayya, R.; Brombacher, F.; Rennick, D.M.; Sheppard, D.; Mohrs, M.; Donaldson, D.D.; Locksley, R.M.; et al. Requirement for IL-13 independently of IL-4 in experimental asthma. *Science* **1998**, *282*, 2261–2263. [CrossRef] [PubMed]
43. Nader, M.A. Inhibition of airway inflammation and remodeling by sitagliptin in murine chronic asthma. *Int. Immunopharmacol.* **2015**, *29*, 761–769. [CrossRef] [PubMed]
44. Wills-Karp, M.; Luyimbazi, J.; Xu, X.; Schofield, B.; Neben, T.Y.; Karp, C.L.; Donaldson, D.D. Interleukin-13: Central mediator of allergic asthma. *Science* **1998**, *282*, 2258–2261. [CrossRef] [PubMed]
45. Humbert, M.; Dursham, S.R.; Kimmitt, P.; Powell, N.; Assoufi, B.; Pfister, R.; Menz, G.; Kay, A.B.; Corrigan, C.J. Elevated expression of messenger ribonucleic acid encoding IL-13 in the bronchial mucosa of atopic and nonatopic subjects with asthma. *J. Allergy Clin. Immunol.* **1997**, *99*, 657–665. [CrossRef]
46. Zuo, L.; Otenbaker, N.P.; Rose, B.A.; Salisbury, K.S. Molecular mechanisms of reactive oxygen species-related pulmonary inflammation and asthma. *Mol. Immunol.* **2013**, *56*, 57–63. [CrossRef] [PubMed]
47. Marcal, L.E.; Rehder, J.; Newburger, P.E.; Condino-Neto, A. Superoxide release and cellular gluthatione peroxidase activity in leukocytes from children with persistent asthma. *Braz. J. Med. Biol. Res.* **2004**, *37*, 1607–1613. [CrossRef] [PubMed]
48. Nadeem, A.; Siddiqui, N.; Alharbi, N.O.; Alharbi, M.M. Airway and systemic oxidant-antioxidant dysregulation in asthma: A possible scenario of oxidants spill over from lung into blood. *Pulm. Pharmacol. Ther.* **2014**, *29*, 31–40. [CrossRef]
49. Joseph, B.Z.; Routes, J.M.; Borish, L. Activities of superoxide dismutases and NADPH oxidase in neutrophils obtained from asthmatic and normal donors. *Inflammation* **1993**, *17*, 361–370. [CrossRef]

50. Chihara, J.; Yamada, H.; Yamamoto, T.; Kurachi, D.; Hayashi-Kameda, N.; Honda, K.; Kayaba, H.; Urayama, O. Priming effect of RANTES on eosinophil oxidative metabolism. *Allergy* **1998**, *53*, 1178–1182. [CrossRef]
51. Hattori, H.; Imai, H.; Furuhama, K.; Sato, O.; Nakagawa, Y. Induction of phospholipid hydroperoxide glutathione peroxidase in human polymorphonuclear neutrophils and HL60 cells stimulated with TNF-alpha. *Biochem. Biophys. Res. Commun.* **2005**, *337*, 464–473. [CrossRef]
52. Shridas, P.; Webb, N.R. Diverse Functions of Secretory Phospholipases A2. *Adv. Vasc. Med.* **2014**, *2014*, 689815. [CrossRef]
53. Hallstrand, T.S.; Chi, E.Y.; Singer, A.G.; Gelb, M.H.; Henderson, W.R. Secreted phospholipase A2 group X overexpression in asthma and bronchial hyperresponsiveness. *Am. J. Respir. Crit. Care Med.* **2007**, *176*, 1072–1078. [CrossRef] [PubMed]
54. Hallstrand, T.S.; Lai, Y.; Altemeier, W.A.; Appel, C.A.; Johnson, B.; Frevert, C.W.; Hudkins, K.L.; Bollinger, J.G.; Woodruff, P.G.; Hyde, D.M.; et al. Regulation and function of epithelial secreted phospholipase A2 group X in asthma. *Am. J. Respir. Crit. Care Med.* **2013**, *188*, 42–50. [CrossRef] [PubMed]
55. Nolin, J.D.; Lai, Y.; Ogden, H.L.; Manicone, A.M.; Murphy, R.C.; An, D.; Frevert, C.W.; Ghomashchi, F.; Naika, G.S.; Gelb, M.H.; et al. Secreted PLA2 group X orchestrates innate and adaptive immune responses to inhaled allergen. *JCI Insight* **2017**, *2*, 94929. [CrossRef]
56. Hart, L.A.; Krishnan, V.L.; Adcock, I.M.; Barnes, P.J.; Chung, K.F. Activation and localization of transcription factor, nuclear factor-kappaB, in asthma. *Am. J. Respir. Crit. Care Med.* **1998**, *158*, 1585–1592. [CrossRef]
57. Poynter, M.E.; Irvin, C.G.; Janssen-Heininger, Y.M. Rapid activation of Nuclear Factor- B in airway epithelium in a murine model of allergic airway inflammation. *Am. J. Pathol.* **2002**, *160*, 1325–1334. [CrossRef]
58. Khaddaj-Mallat, R.; Sirois, C.; Sirois, M.; Rizcallah, E.; Marouan, S.; Morin, C.; Rousseau, E. Pro-Resolving Effects of Resolvin D_2 in LTD_4 and TNF-α Pre-Treated Human Bronchi. *PLoS ONE* **2016**, *11*, e0167058. [CrossRef]
59. Khaddaj-Mallat, R.; Rousseau, E. MAG-EPA and 17,18-EpETE target cytoplasmic signalling pathways to reduce short-term airway hyperresponsiveness. *Pflugers Arch. Eur. J. Physiol.* **2015**, *467*, 1591–1605. [CrossRef]
60. Kantrow, S.P.; Shen, Z.; Jagneaux, T.; Zhang, P.; Nelson, S. Neutrophil-mediated lung permeability and host defense proteins. *Am. J. Physiol. Lung Cell. Mol. Physiol.* **2009**, *297*, 738–745. [CrossRef]
61. Maus, U.; von Grote, K.; Kuziel, W.A.; Mack, M.; Miller, E.J.; Cihak, J.; Stangassinger, M.; Maus, R.; Schlöndorff, D.; Seeger, W.; et al. The role of CC chemokine receptor 2 in alveolar monocyte and neutrophil immigration in intact mice. *Am. J. Respir. Crit. Care Med.* **2002**, *166*, 268–273. [CrossRef]
62. Kraneveld, A.D.; Van der Kleij, H.P.; Kool, M.; Van Houwelingen, A.H.; Weitenberg, A.C.; Redegeld, F.A.; Nijkamp, F.P. Key role for mast cells in nonatopic asthma. *J. Immunol.* **2002**, *169*, 2044–2053. [CrossRef] [PubMed]
63. Biedermann, T.; Kneilling, M.; Mailhammer, R.; Maier, K.; Sander, C.A.; Kollias, G.; Kunkel, S.L.; Hultner, L.; Rocken, M. Mast cells control neutrophil recruitment during T cell-mediated delayed-type hypersensitivity reactions through tumor necrosis factor and macrophage inflammatory protein 2. *J. Exp. Med.* **2000**, *192*, 1441. [CrossRef] [PubMed]
64. Verheijden, K.A.; Henricks, P.A.; Redegeld, F.A.; Garssen, J.; Folkerts, G. Measurement of airway function using invasive and non-invasive methods in mild and severe models for allergic airway inflammation in mice. *Front. Pharmacol.* **2014**, *5*, 190. [CrossRef] [PubMed]
65. Nakae, S.; Ho, L.H.; Yu, M.; Monteforte, R.; Iikura, M.; Suto, H.; Galli, S.J. Mast cell-derived TNF contributes to airway hyperreactivity, inflammation, and TH2 cytokine production in an asthma model in mice. *J. Allergy Clin. Immunol.* **2007**, *120*, 48–55. [CrossRef] [PubMed]
66. Tarkowski, M.; Vanoirbeek, J.A.; Vanhooren, H.M.; De Vooght, V.; Mercier, C.M.; Ceuppens, J.; Nemery, B.; Hoet, P.H. Immunological determinants of ventilatory changes induced in mice by dermal sensitization and respiratory challenge with toluene diisocyanate. *Am. J. Physiol. Lung Cell. Mol. Physiol.* **2007**, *292*, 207–214. [CrossRef]
67. Kanagaratham, C.; Marino, R.; Camateros, P.; Ren, J.; Houle, D.; Sladek, R.; Vidal, S.M.; Radzioch, D. Mapping of a chromosome 12 region associated with airway hyperresponsiveness in a recombinant congenic mouse strain and selection of potential candidate genes by expression and sequence variation analyses. *PLoS ONE* **2014**, *9*, e104234. [CrossRef]

68. Finkelmann, F.D. Use of unrestrained, single-chamber barometric plethysmography to evaluate sensitivity to cholinergic stimulation in mouse models of allergic airway disease. *J. Allergy Clin. Immunol.* **2008**, *121*, 334–335. [CrossRef]
69. De Vooght, V.; Vanoirbeek, J.A.; Luyts, K.; Haenen, S.; Nemery, B.; Hoet, P.H. Choice of mouse strain influences the outcome in a mouse model of chemical-induced asthma. *PLoS ONE* **2010**, *5*, e12581. [CrossRef]
70. Kumar, R.K.; Foster, P.S. Are mouse models of asthma appropriate for investigating the pathogenesis of airway hyper-responsiveness? *Front. Physiol.* **2012**, *31*, 312. [CrossRef]
71. Tränkner, D.; Hahne, N.; Suginoa, K.; Hoonb, M.A.; Zukera, C. Population of sensory neurons essential for asthmatic hyperreactivity of inflamed airways. *Proc. Natl. Acad. Sci. USA* **2014**, *111*, 11515–11520. [CrossRef]
72. Reich, A.; Szepietewski, J.C. Non-analgesic effects of opioids: Peripheral opioid receptors as promising targets for future anti-pruritic therapies. *Curr. Pharm. Des.* **2012**, *18*, 6021–6024. [CrossRef] [PubMed]
73. Kaczyńska, K.; Zając, D.; Wojciechowski, P.; Kogut, E.; Szereda-Przestaszewska, M. Neuropeptides and breathing in health and disease. *Pulm. Pharmacol. Ther.* **2018**, *48*, 217–224. [CrossRef] [PubMed]
74. Carraway, R.E.; Cochrane, D.E.; Salmonsen, R.; Muraki, K.; Boucher, W. Neurotensin elevates hematocrit and plasma levels of the leukotrienes, LTB4, LTC4, LTD4 and LTE4, in anesthetized rats. *Peptides* **1991**, *12*, 1105–1111. [CrossRef]
75. Robbins, R.A.; Nelson, K.J.; Gossman, G.L.; Rubinstein, I. Neurotensin stimulates neutrophil adherence to bronchial epithelial cells in vitro. *Life Sci.* **1995**, *56*, 1353–1359. [CrossRef]
76. Koon, H.W.; Kim, Y.S.; Xu, H.; Kumar, A.; Zhao, D.; Karagiannides, I.; Dobner, P.R.; Pothoulakis, C. Neurotensin induces IL-6 secretion in mouse preadipocytes and adipose tissues during 2,4,6,-trinitrobenzensulphonic acid-induced colitis. *Proc. Natl. Acad. Sci. USA* **2009**, *106*, 8766–8771. [CrossRef]
77. Merrifield, R.B. Solid Phase Peptide Synthesis. The synthesis of a tetrapeptide. *J. Am. Chem.* **1963**, *8*, 2149–2154. [CrossRef]
78. Kaczyńska, K.; Wojciechowski, P.; Jampolska, M.; Lipkowski, A.W.; Kleczkowska, P. Cardiovascular and respiratory activity of PK20, opioid and neurotensin hybrid peptide in anesthetized and awake rats. *Eur. J. Pharmacol.* **2017**, *797*, 20–25. [CrossRef]
79. Tournoy, K.G.; Kips, J.C.; Schou, C.; Pauwels, R.A. Airway eosinophilia is not a requirement for allergen-induced airway hyperresponsiveness. *Clin. Exp. Allergy* **2000**, *30*, 79–85. [CrossRef]

International Journal of *Molecular Sciences*

MDPI

Article

Gender Differences in the Pharmacological Actions of Pegylated Glucagon-Like Peptide-1 on Endothelial Progenitor Cells and Angiogenic Precursor Cells in a Combination of Metabolic Disorders and Lung Emphysema

Olga Victorovna Pershina [1,*], Angelina Vladimirovna Pakhomova [1], Darius Widera [2], Natalia Nicolaevna Ermakova [1], Anton Alexandrovich Epanchintsev [3], Edgar Sergeevich Pan [1], Vyacheslav Andreevich Krupin [1], Olga Evgenevna Vaizova [4], Olesia Dmitrievna Putrova [1], Lubov Alexandrovna Sandrikina [1], Irina Vitalevna Kurochkina [1], Sergey Georgievich Morozov [5], Aslan Amirkhanovich Kubatiev [5], Alexander Mikhaylovich Dygai [1,5] and Evgenii Germanovich Skurikhin [1]

1 Laboratory of Regenerative Pharmacology, Goldberg ED Research Institute of Pharmacology and Regenerative Medicine, Tomsk National Research Medical Centre of the Russian Academy of Sciences, 634028 Tomsk, Russia; angelinapakhomova2011@gmail.com (A.V.P.); nejela@mail.ru (N.N.E.); artifexpan@gmail.com (E.S.P.); vakrupin88@gmail.com (V.A.K.); olesya.putrova@mail.ru (O.D.P.); ermolaeva_la@mail.ru (L.A.S.); irinakuro4kina93@yandex.ru (I.V.K.); amdygay@gmail.com (A.M.D.); eskurihin@inbox.ru (E.G.S.)

2 Stem Cell Biology and Regenerative Medicine Group, School of Pharmacy, University of Reading, Whiteknights campus, Reading RG6 6AP, UK; d.widera@reading.ac.uk

3 Branch Federal State Unitary Enterprise "Scientific and Production Association for Immunological Preparations "Microgen" of Ministry of Health of the Russian Federation "SIC "Virion" in Tomsk, 634040 Tomsk, Russia; a.a.epanchintsev@gmail.com

4 Department of Pharmacology, Siberian State Medical University, 634050 Tomsk, Russia; vaizova@mail.ru

5 Institute of General Pathology and Pathophysiology, 125315 Moscow, Russia; biopharm@list.ru (S.G.M.); niiopp@mail.ru (A.A.K.)

* Correspondence: ovpershina@gmail.com; Tel.: +7-3822-418-375

Received: 23 September 2019; Accepted: 27 October 2019; Published: 30 October 2019

Abstract: In clinical practice, the metabolic syndrome (MetS) is often associated with chronic obstructive pulmonary disease (COPD). Although gender differences in MetS are well documented, little is known about sex-specific differences in the pathogenesis of COPD, especially when combined with MetS. Consequently, it is not clear whether the same treatment regime has comparable efficacy in men and women diagnosed with MetS and COPD. In the present study, using sodium glutamate, lipopolysaccharide, and cigarette smoke extract, we simulated lipid metabolism disorders, obesity, hyperglycemia, and pulmonary emphysema (comorbidity) in male and female C57BL/6 mice. We assessed the gender-specific impact of lipid metabolism disorders and pulmonary emphysema on angiogenic precursor cells (endothelial progenitor cells (EPC), pericytes, vascular smooth muscle cells, cells of the lumen of the nascent vessel), as well as the biological effects of pegylated glucagon-like peptide 1 (pegGLP-1) in this experimental paradigm. Simulation of MetS/COPD comorbidity caused an accumulation of EPC ($CD45^{-}CD31^{+}CD34^{+}$), pericytes, and vascular smooth muscle cells in the lungs of female mice. In contrast, the number of cells involved in the angiogenesis decreased in the lungs of male animals. PegGLP-1 had a positive effect on lipids and area under the curve (AUC), obesity, and prevented the development of pulmonary emphysema. The severity of these effects was stronger in males than in females. Furthermore, PegGLP-1 stimulated regeneration of pulmonary endothelium. At the same time, PegGLP-1 administration caused a mobilization of EPC ($CD45^{-}CD31^{+}CD34^{+}$) into the bloodstream in females and migration of precursors of angiogenesis and vascular smooth muscle cells to the lungs in male animals. Gender differences in stimulatory action of pegGLP-1 on CD31^{+} endothelial lung cells in vitro were not observed. Based on these findings, we postulated that the cellular mechanism of in vivo regeneration of lung epithelium was at least partly gender-specific. Thus, we concluded that a pegGLP-1-based treatment regime for metabolic disorder and COPD should be further developed primarily for male patients.

Keywords: gender differences; dyslipidemia; obesity; hyperglycemia; pulmonary emphysema; endothelial progenitor cells; angiogenic precursor cells; pegylated glucagon-like peptide 1; and endothelial regeneration

---

1. Introduction

In clinical practice, metabolic syndrome (MetS) is defined as the presence of at least three of the following five conditions: abdominal obesity, elevated triglyceride levels, low serum high-density lipoprotein (HDL), high blood pressure (hypertension), and elevated levels of blood glucose [1].

Gender differences in glucose homeostasis and energy balance are important factors in the development of MetS in men and women [2]. In general, women are at greater risk of metabolic disorders due to a higher fat/muscle mass ratio and more subcutaneous fat tissue. Consequently, females have higher insulin resistance than men. This is associated with hormonal differences and diet [3–6]. Gender-specific differences in the lipoprotein profile are well described [7]. In addition, gender-specific mechanisms beyond the scope of this study contribute to a higher accumulation of adipose tissue in females, whereas mobilization of fat reserves tends to be more effective in men.

In addition to MetS, gender-specific differences in major metabolic pathways have been suggested to contribute to the pathogenesis of various lung diseases [8,9].

Clinically, MetS has been found to be more frequent in patients with chronic obstructive pulmonary disease (COPD). This patient cohort has a higher body mass index compared to patients with COPD alone [10]. The clinical picture and course of COPD often differ in women and men [11]. Importantly, COPD with MetS is more frequent in female patients with a prevalence of 18.5% in men and 38.5% in women [12].

Fatty acids are an important source of energy. In this context, catabolism of fatty acids yields more Adenosine triphosphate (ATP) per mole than in the oxidation of glucose [13–15].

Recent reports indicate that changes in lipid metabolism allow lung tissue to meet the energy needs of patients with COPD [16]. On the other hand, changes in lipid metabolism (catabolism, anabolism) contribute to the pathogenesis of COPD. In particular, obesity initiates anabolic pathways involved in the synthesis of pro-inflammatory molecules [16]. This mechanism reduces lung function and increases the severity of COPD [17,18].

Understanding the gender-specific mechanisms of a disease is important for finding individual and personalized treatment strategies in patients with MetS and COPD.

In a recent review, Kim J.H. and colleagues highlighted gender-specific differences in the histology and functional activity of the lungs in patients with COPD and MetS [12]. Briefly, the authors reported that in comparison with men, women have fewer alveoli. Furthermore, their airway diameter is relatively small compared to the size of the lungs. Thus, lung function tends to decline sharply as obesity increases. Moreover, estrogens can reduce lung function. In 2018, Zore T. et al. drew attention to adipocyte progenitor cells as a potential factor affecting sex differences in adipose tissue enlargement [19]. In contrast, they postulated that hematopoietic stem cells are a factor contributing to gender differences in the inflammatory response.

Previously, we demonstrated that incretin glucagon-like peptide-1 (GLP-1) administration could stimulate the regeneration of lung endothelium in lung emphysema in obese mice [20]. The regenerative effects of GLP-1 have been at least partly mediated by effects on endothelial progenitor cells. However, in vivo, GLP-1 is rapidly metabolized by dipeptidyl peptidase-4, leading to a relatively low regenerative activity of the hormone. It is well described that pegylation can be used to preserve pharmacologically active molecules.

The aim of this research was to study gender differences in the treatment of hyperglycemia, dyslipidemia and obesity, emphysema, and alveolar endothelial injury with pegylated GLP-1 (pegGLP-1) in a C57BL/6 mice model of obesity and emphysema. We investigated the effect of pegGLP-1 on bone marrow, circulating in the blood, and pulmonary endothelial progenitor cells and other cells involved in angiogenesis in female and male animals to assess potential gender-specific differences.

2. Results

2.1. The Effect of GLP-1 and pegGLP-1 on Lee Index and Body Mass Index

To confirm obesity in the male and female C57BL/6 mice on p189—mice that received MSG (monosodium glutamate)—Lee and body mass index (BMI) indexes were assessed. Simulations of pulmonary emphysema did not affect the Lee index and BMI in animals of f3 and m3 groups (mice with lung emphysema) compared to intact controls (Figure 1). In the modeling of metabolic disorders (obesity and hyperglycemia), and in the modeling of metabolic disorders (obesity and hyperglycemia) and lungs emphysema, we observed an increase in the Lee and BMI indexes in females (groups f2 (mice with metabolic disorders) and f4 (mice with metabolic disorders and lung emphysema)) and males (groups m2 and m4), while in males, the increase in parameters was more pronounced compared to females.

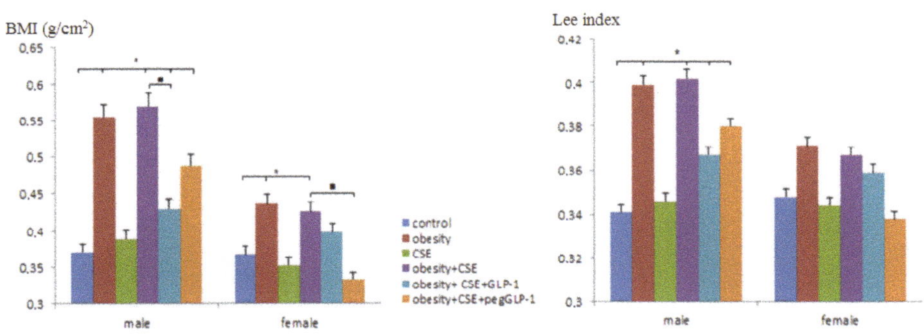

(a) (b)

Figure 1. The effect of glucagon-like peptide-1 (GLP-1) and pegylated GLP-1 (pegGLP-1) on body mass index (BMI) and Lee index of male and female C57BL/6 mice on p189: (**a**) The BMI (g/cm^2); (**b**) The Lee index. Groups: control—a control group from intact mice, obesity—mice with metabolic disorders (obesity and hyperglycemia), CSE—mice with lungs emphysema, obesity+CSE—mice with metabolic disorders (obesity and hyperglycemia) and lungs emphysema, obesity+CSE+GLP-1—mice with metabolic disorders (obesity and hyperglycemia) and lungs emphysema treated with GLP-1, obesity+CSE+pegGLP-1—mice with metabolic disorders (obesity and hyperglycemia) and lungs emphysema treated with pegGLP-1. Results are presented as the mean±SEM. *—significance of difference compared with control ($p < 0.05$); ■—significance of difference compared with the obesity+CSE group ($p < 0.05$). CSE, cigarette smoke extract.

GLP-1 or pegGLP-1 treatment had no effect on the Lee index of females and males in metabolic disorders (obesity and hyperglycemia) and emphysema compared with untreated mice of groups f4 and m4 (Figure 1b). Meanwhile, drugs significantly reduced BMI in females of groups f5 (mice with metabolic disorders and lung emphysema treated with GLP-1) and f6 (mice with metabolic disorders and lung emphysema treated with peg-GLP-1), and males of groups m5 and m6. The therapeutic effect in males m6 was more pronounced compared to females f6. This section may be divided by subheadings. It should provide a concise and precise description of the experimental results, their interpretation, as well as the experimental conclusions that can be drawn.

2.2. Changes in Serum Lipid Parameters in Emphysema, Metabolic Disorders, and the Combination of Metabolic Disorders and Emphysema

Dyslipidemia is a key component of metabolic disorders (MD) and often occurs with obesity. We studied levels of cholesterol, triglycerides (TG), high-density lipoprotein (HDL), low-density lipoprotein (LDL), and very-low-density lipoprotein (VLDL) in the serum of male and female C57BL/6 mice on p189. The m2 group showed a more pronounced increase in cholesterol, TG, HDL, and VLDL compared with the f2 group. In contrast, in group f2, there was a more marked increase in LDL than in group m2 (Figure 2c). We also observed gender-dependent differences in serum lipid levels in the development of emphysema. Thus, the levels of TG and LDL in the m3 group increased, while in the f3 group, these indicators decreased (Figure 2a,c). It should be noted that the levels of cholesterol, LDL, and HDL in males and females with emphysema of the lungs changed the same type—they increased.

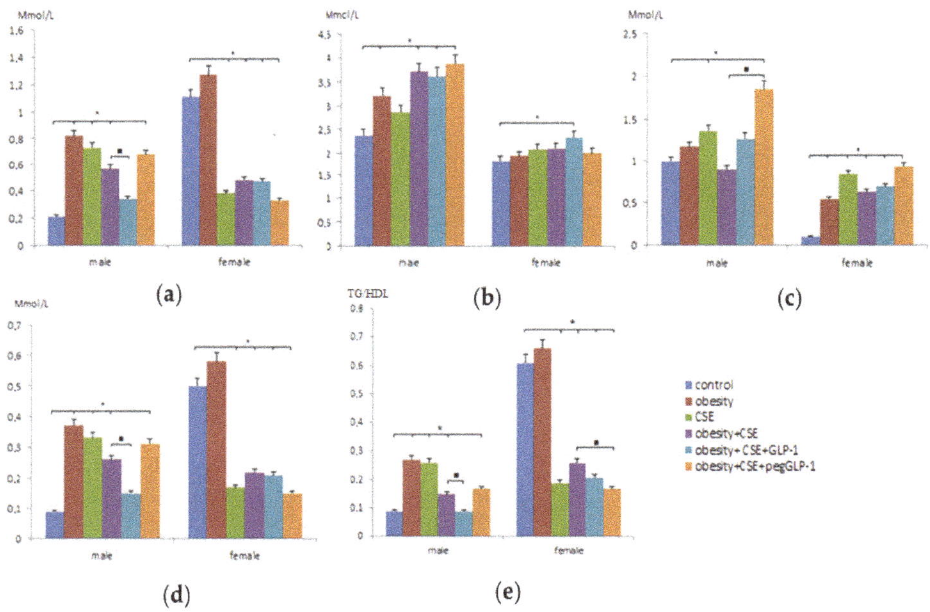

Figure 2. Lipid profile measurements in the blood of female and male C57BL/6 mice on p189: (**a**) The level of triglycerides in serum (Mmol/l); (**b**) High-density lipoprotein level (Mmol/l); (**c**) Low-density lipoprotein level (Mmol/l); (**d**) Very low-density lipoprotein level (Mmol/l); (**e**) The ratio of triglycerides to high-density lipoproteins (TG/HDL). Groups: control—a control group from intact mice, obesity—mice with metabolic disorders (obesity and hyperglycemia), CSE—mice with lungs emphysema, obesity+CSE—mice with metabolic disorders (obesity and hyperglycemia) and lungs emphysema, obesity+CSE+GLP-1—mice with metabolic disorders (obesity and hyperglycemia) and lungs emphysema treated with GLP-1, obesity+CSE+pegGLP-1—mice with metabolic disorders (obesity and hyperglycemia) and lungs emphysema treated with pegGLP-1. *—significance of difference compared with control ($p < 0.05$); ■—significance of difference compared with the obesity+CSE group ($p < 0.05$).

The combination of MD and lung emphysema revealed differences in the lipid profile of males and females. In female mice of group f4 (mice with MD and lung emphysema), serum LDL concentrations increased compared to f1 (intact control) and f2 (Figure 2c). In the m4 group, this figure did not change significantly. Decrease in triglycerides (TG) and VLDL and increase in cholesterol, VLDL in the blood serum was observed in mice m4 and f4, but the degree of severity of changes was more pronounced in males.

2.3. The Effect of GLP-1 and pegGLP-1 on Lipid Parameters of Blood Serum with a Combination of Metabolic Disorders and Emphysema

On p189, we studied the effectiveness of GLP-1 and pegGLP-1 on fat metabolism in mice of different sexes under the conditions of MSG and cigarette smoke extract (CSE) administration (a combination of MD and lung emphysema).

The introduction of GLP-1 to m5 males (with an MD and lung emphysema) reduced the concentration of TG (1.7 times) and VLDL (1.7 times) compared to m4 males. In contrast, the concentration of cholesterol and LDL increased (Figure 2c). At the same time, GLP-1 increased cholesterol and HDL cholesterol levels in females of the f5 group (with MD and lung emphysema) and did not affect other parameters.

The introduction of pegGLP-1 to m6 (with MD and lung emphysema) caused an increase in cholesterol, HDL, and LDL levels compared to the m4 group (Figure 2). On the contrary, the level of TG decreased (by 1.5 times) in group f6 females—affected by pegGLP-1—compared to group f4. The level of VLDL was reduced, while the level of LDL increased.

In addition, we studied gender differences in the ratio of triglycerides to high-density lipoproteins (TG/HDL). As can be seen from Figure 2e, the parameter value in females of the f1 group was significantly higher compared to males of the m1 group (6.8 times). Modeling of metabolic disorders, emphysema of the lungs, and a combination of metabolic disorders and emphysema of the lungs caused an increase in TG/HDL in males of group m2 (3 times), m3 (2.8 times) and m4 (1.6 times), respectively, compared to intact control. In females of group f2, the parameter increased slightly (by 8%), in groups f3 and f4, on the contrary, we observed its decrease by 69% and 57.4%, respectively, compared to f1. The GLP-1 treatment helped to reduce the ratio of TG/HDL in males and females in a combination of MD and lung emphysema, while the effect of the drug was most pronounced in the m5 group. PegGLP-1 did not affect the studied parameter in males in the m6 group and reduced it in females in the f6 group by 35%.

2.4. GLP-1 and PegGLP-1 Effect on Area Under the Curve (AUC) During the Glucose Tolerance Test

The day before the removal of animals from the experiment (p188), the glucose tolerance test (GTT) and the calculation of the area under the curve for blood glucose (AUC) were conducted. After the introduction of MSG, there was a natural increase in AUC in males of the m2 group (two times), and it was significantly less in females of the f2 group (19%) compared to the control groups (Figure 3). Interestingly, the introduction of cigarette smoke extract (CSE) also increased AUC in males of the m3 group (by 21%), while the parameter did not change in the f3 group. Simulations of the combined pathology did not affect AUC in f4 females, but in m4 males, the parameter increased 1.8 times compared to the m1 group.

The GLP-1 or pegGLP-1 treatment decreased AUC in male mice of m5 (13%) and m6 (18%), respectively (Figure 3). PegGLP-1 (11%) was more effective in females with comorbidity, but not GLP-1 (5%).

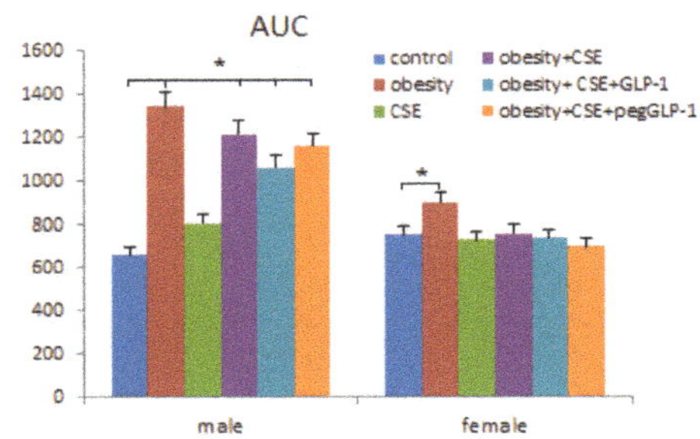

Figure 3. The area under the curve (AUC) of female and male C57BL/6 mice during the glucose tolerance test (on p188). Groups: control—a control group from intact mice, obesity—mice with metabolic disorders (obesity and hyperglycemia), CSE-mice with lungs emphysema, obesity+CSE—mice with metabolic disorders (obesity and hyperglycemia) and lungs emphysema, obesity+CSE+GLP-1—mice with metabolic disorders (obesity and hyperglycemia) and lungs emphysema treated with GLP-1, obesity+CSE+pegGLP-1—mice with metabolic disorders (obesity and hyperglycemia) and lungs emphysema treated with pegGLP-1. *—significance of difference compared with control ($p < 0.05$).

2.5. Morphological Study of Lung

Lung injuries caused by lipopolysaccharide (LPS) and CSE were similar in group f3 females and group m3 male mice. Thus, by the p148 (24 h before treatment), the animals developed moderately diffused lungs emphysema. In addition, histological preparations of the lungs revealed an increase in the size of the alveoli and alveolar passages and single ruptures of the alveolar septa due to damage to elastic membranes. In mice of groups (m3 and f3), pulmonary hyperemia and diapedesis of erythrocytes in the lumen of the alveoli were revealed; the walls of the alveoli were thickened due to inflammatory infiltration by macrophages (Figure 4b). At the same time, macrophages and single neutrophils were found in the lumen of the alveoli, and peribronchial lymph-macrophage infiltrates were observed. On p189, as well as on p148, emphysema of the lungs in group f3 and group m3 mice was diffused. Meanwhile, on p189, the thinning of the walls of the alveoli, the number of ruptures and atelectasis of the pulmonary tissue, and the area of emphysema in these mice were more significant compared to p148 (Figure 4b). The area of emphysema in the males was superior to that of the females.

When modeling metabolic disorders in the lungs of group f2 and group m2 animals, we found hyperemia of small and large vessels, a large number of macrophages, and single neutrophils on p189 (Figure 4). In males, these parameters were more pronounced than in females.

The appointment of MSG and CSE (a combination of MD and lung emphysema) caused the development of focal lung emphysema of moderate severity in males (group m4) and females (group f4) on p189 (Figure 4). The area of emphysema in females (group f4) was smaller than in males (group m4). In the lungs of mice treated with a combination of metabolic disorders and lung emphysema, we found groups of enlarged alveoli and alveolar passages, and ruptures of the alveolar walls. In the alveoli, there have been sporadic neutrophils and a greater number of macrophages compared to females f3 and males m3. It should be noted that the area of emphysema in group f4 and group m4 was inferior to that in the group f3 and group m3, respectively.

GLP-1 and pegGLP-1 treatment slightly reduced the inflammatory infiltration by macrophages of the lungs of females and males under the conditions of MSG and CSE administration on p189. In the lungs of females of groups f5 and f6, the area of emphysema-enlarged tissue significantly decreased compared to untreated females of group f4 (Figure 4). The pegGLP-1 effect was higher than that of GLP-1.

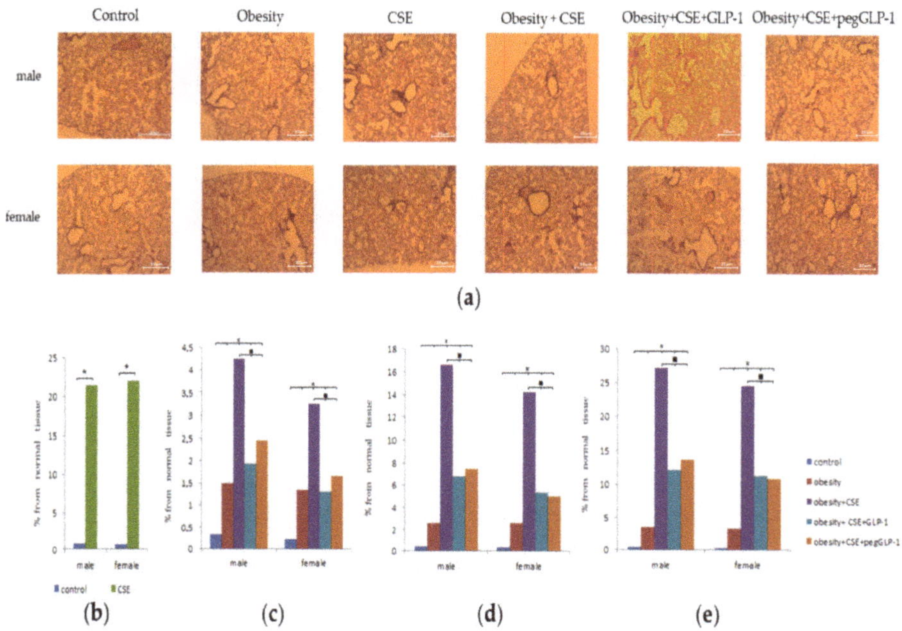

Figure 4. Morphological study of lung obtained from male and female C57BL/6 mice (n = 6): (**a**) Photomicrographs of left lung sections (lower pulmonary field) (on p189). Tissues were stained with hematoxylin-eosin; (**b**) The area of emphysema-expanded lung tissue (lower pulmonary field) of mice from all groups (on p148); (**c**) The area of emphysema-expanded lung tissue (upper pulmonary field) of mice from all groups (on p); (**d**) The area of emphysema-expanded lung tissue (middle pulmonary field) of mice from all groups (on p189); (**e**) The area of emphysema-expanded lung tissue (lower middle pulmonary field) of mice from all groups (on the p189). Groups: control—a control group from intact mice, obesity—mice with metabolic disorders (obesity and hyperglycemia), CSE—mice with lungs emphysema, obesity+CSE—mice with metabolic disorders (obesity and hyperglycemia) and lungs emphysema, obesity+CSE+GLP-1–mice with metabolic disorders (obesity and hyperglycemia) and lungs emphysema treated with GLP-1, obesity+CSE+pegGLP-1—mice with metabolic disorders (obesity and hyperglycemia) and lungs emphysema treated with pegGLP-1. * $p < 0.05$ significance of difference compared with control group, ■—significance of difference compared with the obesity+CSE group ($p < 0.05$).

2.6. Immunohistochemical Lung Study

The CSE introduction significantly reduced the number of CD31-expressing cells in the pulmonary tissue of m3 and f3 mice compared to m1 and f1 mice by the p189 (Figure 5). With the MSG and CSE introduction (a combination of MD and lung emphysema), the reduction of the number of $CD31^+$ cells in the lungs of mice (group m4 and f4) was more significant than in m3 and f3 mice. When modeling pathology, the decrease in CD31 expression in the lungs of males was more significant than in females.

GLP-1 and pegGLP-1 treatment caused a significant increase in the number of $CD31^+$ cells in mice lungs under MSG and CSE administration compared to untreated mice with MD and emphysema of the lungs (Figure 5). In this case, the therapeutic effect of males was higher than that of females.

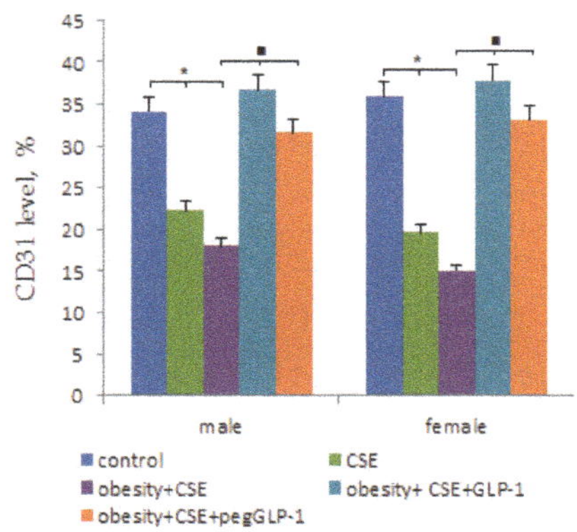

Figure 5. The relative content of cells expressing CD31 antigen in the lungs isolated from male and female C57BL/6 mice at the immunohistochemical staining for specific cellular marker: CD31 (on the p189). Groups: control—a control group from intact mice, obesity—mice with metabolic disorders (obesity and hyperglycemia), CSE—mice with lungs emphysema, obesity+CSE—mice with metabolic disorders (obesity and hyperglycemia) and lungs emphysema, obesity+CSE+GLP-1—mice with metabolic disorders (obesity and hyperglycemia) and lungs emphysema treated with GLP-1, obesity+CSE+pegGLP-1—mice with metabolic disorders (obesity and hyperglycemia) and lungs emphysema treated with pegGLP-1. *$p < 0.05$ significance of difference compared with control group, ■—significance of difference compared with the obesity+CSE group ($p < 0.05$).

2.7. Study of Stem Antigens, Epithelial and Endothelial Cells, and Other Cells Using Flow Cytometric Analysis

2.7.1. Lung

On p189, we studied the precursors and mature cells content in the lungs of healthy mice (males and females). In the lungs of the m1 male group, we found a significantly greater number of endothelial cells ($CD45^-CD31^+CD34^+$ and $CD31^+CD34^+CD146^+$) and the precursors of angiogenesis ($CD45^-CD309^+CD117^+$), cells in the lumen of the nascent vessel ($CD31^+CD34^-$), than the females of group f1 (Figure 6). At the same time, the number of vascular smooth muscle cells ($CD31^-CD34^+CD146^+$) in the m1 group was inferior to that in the f1 group, and no significant gender differences in the content of pericytes ($CD31^-CD34^-CD146^+$) were revealed.

MSG introduction caused an increase in the number of EPC and precursors of angiogenesis, vascular smooth muscle cells, pericytes in females of group f2 (mice with MD) compared to females of group f1 (Figure 6). In males of the m2 group, we found a decrease in the number of EPC and angiogenesis precursors, nascent vessel lumen cells, with vascular smooth muscle cells and pericytes accumulating in the lungs.

The LPS and CSE introduction caused a significant increase in the number of cells of the lumen of the nascent vessel and precursors of angiogenesis in group f3 compared to group f1 (Figure 6). In contrast, in the lungs of males of the m3 group, we observed a significant decrease in the number of EPC ($CD45^-CD31^+CD34^+$), vascular smooth muscle cells, and angiogenesis precursors compared to the m1 group. On the contrary, the number of pericytes increased.

Modeling of MD and lung emphysema by MSG and CSE in females of group f4 caused a significant increase in the number of EPC ($CD45^-CD31^+CD34^+$), vascular smooth muscle cells, and pericytes compared to group f1 (Figure 6). In contrast, the combination of MD and emphysema significantly reduced the number of EPC ($CD45^-CD31^+CD34^+$), precursors of angiogenesis, and pericytes in the m4 group compared to healthy males.

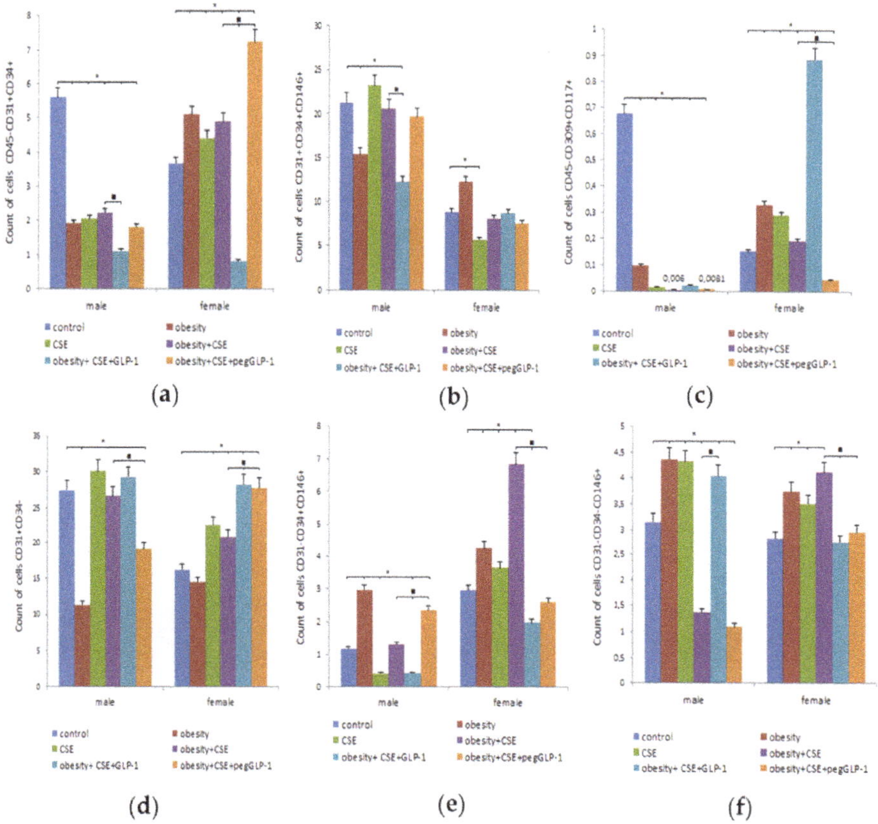

Figure 6. Characterization of cell population isolated from the lung of male and female C57BL/6 mice on the p189. (**a**) The content of endothelial progenitor cells (EPC) ($CD45^-CD31^+CD34^+$); (**b**) The content of EPC ($CD31^+CD34^+CD146^+$); (**c**) The content of angiogenesis precursors ($CD45^-CD309^+CD117^+$); (**d**) The content of cells in the lumen of the nascent vessel ($CD31^+CD34^-$); (**e**) The content of vascular smooth muscle cells ($CD31^-CD34^+CD146^+$); (**f**) The content of pericytes ($CD31^-CD34^-CD146^+$). Cells were analyzed by flow cytometry using antibodies for CD31, CD34, CD45, CD146, CD117, CD309 mice. Dot plots are representative of three independent experiments with the mean from three independent experiments. Groups: control—a control group from intact mice, obesity—mice with metabolic disorders (MD) (obesity and hyperglycemia), CSE—mice with lungs emphysema, obesity+CSE—mice with MD (obesity and hyperglycemia) and lungs emphysema, obesity+CSE+GLP-1—mice with MD (obesity and hyperglycemia) and lungs emphysema treated with GLP-1, obesity+CSE+pegGLP-1—mice with MD (obesity and hyperglycemia) and lungs emphysema treated with pegGLP-1. *—significance of difference compared with control ($p < 0.05$); ■—significance of difference compared with the obesity+CSE group ($p < 0.05$).

The GLP-1 treatment caused similar changes in mice of f5 and m5 groups, such as a decrease in the number of pulmonary EPC ($CD45^-CD31^+CD34^+$), vascular smooth muscle cells, and an increase in the number of angiogenesis precursors (Figure 6). Inter-gender differences were in the additional reduction of the population of pulmonary EPC ($CD31^+CD34^+CD146^+$) in the m5 group. In addition, GLP-1 increased the number of pericytes in the lungs of m5 mice and reduced their number in f5 mice compared to untreated animals.

The pegGLP-1 reduced the population of cells in the lumen of the nascent vessel but increased the number of vascular smooth muscle cells in group m6 (Figure 6). Changes in the cells of the lungs of group f6 females to the pegGLP-1 introduction were as follows: the number of precursors of angiogenesis, the pericytes, and vascular smooth muscle cells decreased in comparison with the female group f5, on the other hand, the number of EPC ($CD45^{-}CD31^{+}CD34^{+}$) increased.

2.7.2. Bone Marrow

On p189, we revealed gender differences in the populations of immature endothelial bone marrow cells of healthy mice. In m1 males, the number of EPC ($CD31^{+}CD34^{+}CD146^{+}$) (5.5 times) and angiogenesis precursors (4.58 times) significantly exceeded that of f1 females (Figure 7).

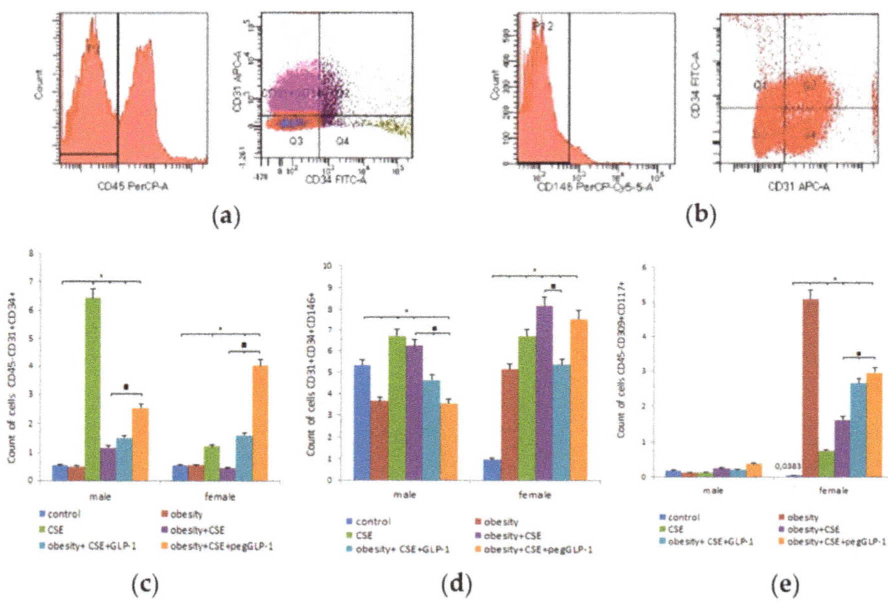

Figure 7. Characterization of cell population isolated from the bone marrow of male and female C57BL/6 mice on the p189. (**a**) Phenotype establishment and qualitative analysis of CD45 (PerCP), CD34 (FITC), and CD31 (APC) expression; (**b**) Phenotype establishment and qualitative analysis of CD34 (FITC), CD31 (APC), and CD146 (PerCP-Cy5.5) expression; (**c**) The content of EPC ($CD45^{-}CD31^{+}CD34^{+}$); (**d**) The content of EPC ($CD31^{+}CD34^{+}CD146^{+}$); (**e**) The content of angiogenesis precursors ($CD45^{-}CD309^{+}CD117^{+}$). Cells were analyzed by flow cytometry using antibodies for CD45, CD31, CD34, CD146, CD117, CD309 mice. Dot plots are representative of three independent experiments with the mean from three independent experiments. Groups: control—a control group from intact mice, obesity—mice with MD (obesity and hyperglycemia), CSE—mice with lungs emphysema, obesity+CSE—mice with MD (obesity and hyperglycemia) and lungs emphysema, obesity+CSE+GLP-1—mice with MD (obesity and hyperglycemia) and lungs emphysema treated with GLP-1, obesity+CSE+pegGLP-1—mice with MD (obesity and hyperglycemia) and lungs emphysema treated with pegGLP-1. *—significance of difference compared with control ($p < 0.05$); ■—significance of difference compared with the obesity+CSE group ($p < 0.05$).

The MSG introduction caused a decrease in the number of bone marrow EPC ($CD31^{+}CD34^{+}CD146^{+}$) and angiogenesis precursors in m2 males. On the contrary, in MSG-treated f2 females, we observed a significant increase in the number of these cells (Figure 7d).

The LPS and CSE introduction caused a significant increase in the number of all endothelial cells studied in the bone marrow of females of group f3 compared to group f1 (Figure 7c,d). In turn, in m3 males, we observed a selective accumulation of EPC ($CD45^-CD31^+CD34^+$) 11 times compared to healthy males.

Modeling of MD and lung emphysema caused an increase in the content of bone marrow EPC ($CD31^+CD34^+CD146^+$) and angiogenesis precursors in f4 females compared to healthy females (Figure 7 d,e). In males of the m4 group, a combination of MD and pulmonary emphysema led to the accumulation of EPC ($CD45^-CD31^+CD34^+$) and precursors of angiogenesis.

The GLP-1 introduction increased the number of EPC ($CD45^-CD31^+CD34^+$) and reduced the number of $CD31^+CD34^+CD146^+$-endothelial cells in the bone marrow of f5 and m5 mice compared to f4 and m4 mice, respectively (Figure 7). We found gender differences in the reaction of angiogenesis precursors to treatment: in females of group f5, the number of $CD45^-CD309^+CD117^+$-cells was 64% higher than that in group f4; males of m5 showed a tendency to decrease the number of these cells.

The pegGLP-1 treatment increased the number of bone marrow $CD45^-CD31^+CD34^+$ EPC and precursors of angiogenesis in mice of groups f6 and m6 compared to untreated mice in terms of MD and lung emphysema (Figure 7c). On the other hand, in these groups, we found a decrease in the number of $CD31^+CD34^+CD146^+$-EPC, which was more pronounced in the m6 group than in the f5 group.

2.7.3. Blood

We found gender differences in the blood cells of healthy animals. Thus, in the m1 group, the number of circulating $CD45^-CD31^+CD34^+$-EPC (by 63%), $CD31^+CD34^+CD146^+$-EPC (by 94%), and vascular smooth muscle cells (by 99.5%) was inferior to that in the f1 group (Figure 8). In contrast, the number of pericytes (21.5 times) and lumen cells of the nascent vessel (1.87 times) in healthy males was higher than in healthy females.

The MSG introduction caused a decrease in the number of $CD45^-CD31^+CD34^+$-EPCs, nascent vessel lumen cells, and pericytes in the blood of females and males compared to the corresponding healthy mice (Figure 8). Gender differences included an increase in the number of $CD31^+CD34^+CD146^+$-EPCs and vascular smooth muscle cells in females (f2) and a reduction in these cell populations in males (m2).

LPS and CSE introduction caused a significant increase in the number of EPC ($CD45^-CD31^+CD34^+$ and $CD31^+CD34^+CD146^+$), vascular smooth muscle cells, and pericytes in the blood of m3 males compared to healthy m1 males (Figure 8). Meanwhile, in the blood of f3 group, we observed a decrease in the number of endothelial cells ($CD45^-CD31^+CD34^+$ and $CD31^+CD34^+CD146^+$), vascular smooth muscle cells, and cells of the lumen of the nascent vessels; the only exception was the pericytes, the number of which sharply increased compared to group f1.

The reaction of circulating blood cells of females of the f4 group in the modeling of MD and lung emphysema coincided with that which was identified in females with emphysema (Figure 8). In males of the m4 group, a coincidence of the reaction (increase) of vascular smooth muscle cells and pericytes with that of males of the m3 group (emphysema of the lungs) was revealed. At the same time, the number of endothelial cells ($CD45^-CD31^+CD34^+$ and $CD31^+CD34^+CD146^+$) and nascent vessel cells in the blood of m4 males, on the contrary, decreased and was less than in the m1 group.

GLP-1 reduced the number of all studied cells in f5 females compared to untreated f4 females: endothelial cell populations ($CD45^-CD31^+CD34^+$ and $CD31^+CD34^+CD146^+$) and vascular smooth muscle cells were most significantly reduced (Figure 8). In contrast, GLP-1 increased the number of $CD31^+CD34^+CD146^+$-endothelial cells, vascular smooth muscle cells, and blood pericytes in m5 males compared to m4 males.

PegGLP-1 treatment caused an increase in the number of endothelial cells and vascular smooth muscle cells in the blood of females and males f6 and m6 compared to untreated animals in conditions of MD and lung emphysema (Figure 8). Gender differences were expressed in the accumulation of nascent vessel lumen cells in f6 females, and the opposite reaction of pericytes: the number of these cells in f6 females decreased, and it increased in m6 males.

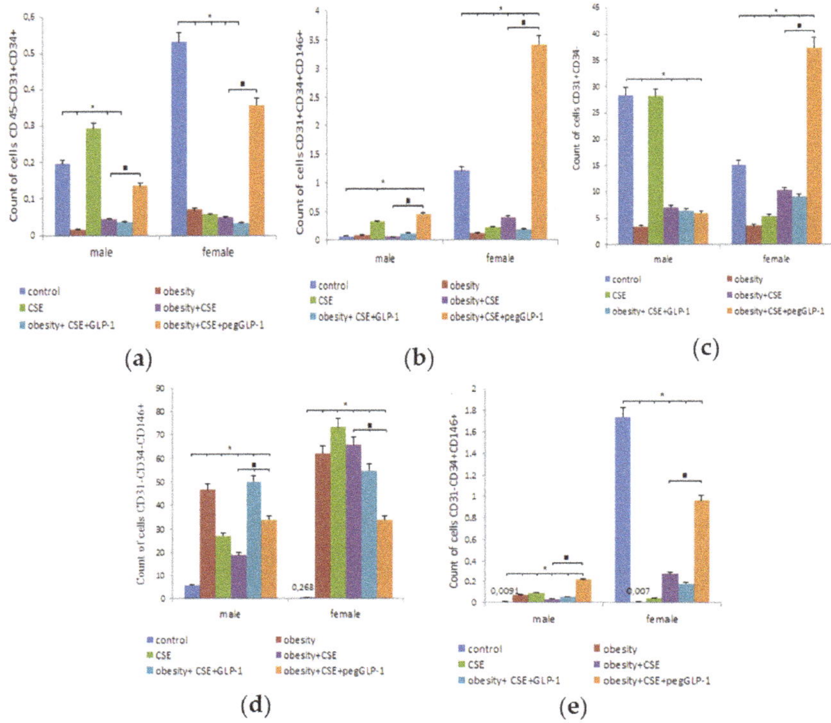

Figure 8. Characterization of cell population isolated from the blood of male and female C57BL/6 mice on the p189. (**a**) The content of EPC ($CD45^-CD31^+CD34^+$); (**b**) The content of EPC ($CD31^+CD34^+CD146^+$); (**c**) The content of cells in the lumen of the nascent vessel ($CD31^+CD34^-$); (**d**) The content of pericytes ($CD31^-CD34^-CD146^+$); (**e**) The content of vascular smooth muscle cells ($CD31^-CD34^+CD146^+$). Cells were analyzed by flow cytometry using antibodies for CD45, CD31, CD34, CD146 mice. Dot plots are representative of three independent experiments with the mean from three independent experiments. Groups: control—a control group from intact mice, obesity—mice with MD (obesity and hyperglycemia), CSE—mice with lungs emphysema, obesity+CSE—mice with MD (obesity and hyperglycemia) and lungs emphysema, obesity+CSE+GLP-1—mice with MD (obesity and hyperglycemia) and lungs emphysema treated with GLP-1, obesity+CSE+pegGLP-1—mice with MD (obesity and hyperglycemia) and lungs emphysema treated with pegGLP-1. *—significance of difference compared with control ($p < 0.05$); ■—significance of difference compared with the obesity+CSE group ($p < 0.05$).

2.8. *Study of GLP-1 and PegGLP-1 Effect on CD31⁺ Lung Cells In Vitro*

The effects of GLP-1 and pegGLP-1 on some parameters of $CD31^+$ cells obtained from the lungs of females of f1 and f4 groups and males of m1 and m4 groups were studied in vitro. GLP-1 or pegGLP-1 were introduced into the cell culture enriched with $CD31^+$ cells, and the final concentration of these drugs in the culture was 10^{-7} M.

Figure 9d shows that affected by GLP-1 and pegGLP-1, the number of apoptotic $CD31^+$ cells in the f1 and m1 groups did not change. Meanwhile, in the f4 and m4 groups, GLP-1 and pegGLP-1 significantly reduced the number of apoptotic $CD31^+$ cells compared to the solvent culture, while we did not observe gender differences in the severity of the drug effects (Figure 9g).

GLP-1 and pegGLP-1 significantly increased the expression of CD34 marker in CD31$^+$ culture of lung cells in all the studied f1 and m1 and f4 and m4 groups (Figure 9a,c,f). We did not observe gender differences in the severity of the drug effects.

GLP-1 and pegGLP-1 increased the number of CD31$^+$ lung cells with active esterases. This effect was more pronounced in the f1 and m1 groups than in the f4 and m4 groups (Figure 9b,e).

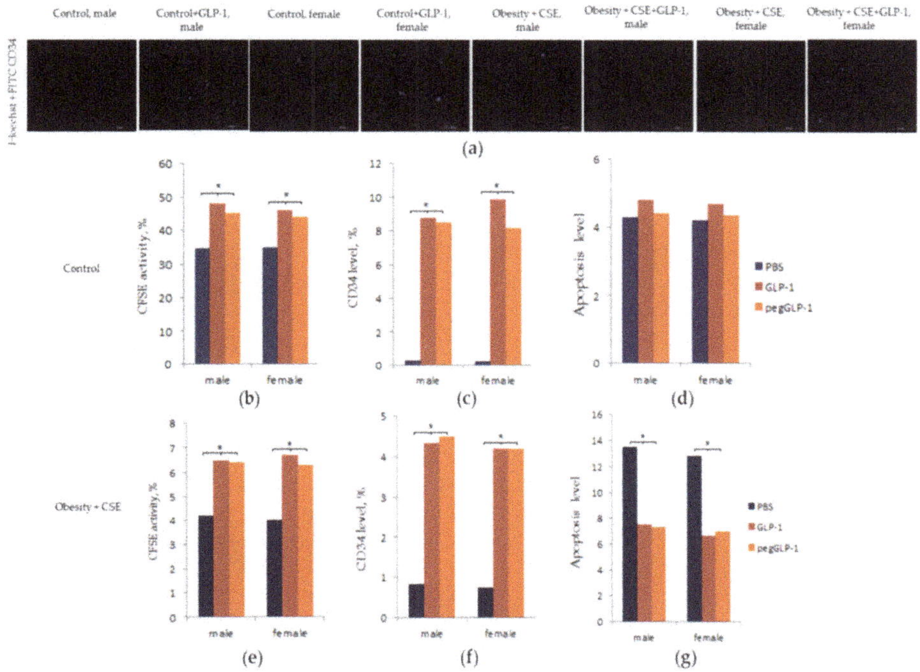

Figure 9. GLP-1 and pegGLP-1 treatment effects on CD31$^+$ endothelial cells isolated from the lungs of male and female C57BL/6 mice in vitro: (**a**) Images of CD31$^+$ cells stained with: Hoechst (blue) to identify cell nuclei; CD34 FITC (green) (Hoechst + CD34) composite image using all two colors. All scale bars are 100 µm; (**b–g**) CD31+ endothelial cells from lung were precultured for 5 days, incubated with or without GLP-1 (10^{-7} M) or pegGLP-1 (10^{-7} M) for 24 h and then labeled with Hoechst, Carboxyfluorescein succinimidyl ester (CFSE) (**b**,**e**) CD34 FITC (**c**,**f**), Annexin V and 7-Aminoactinomycin D (7-AAD) (**d**,**g**) prior to fluorescence microscopic analysis. (**b**) CFSE activity after culture of cells isolated from the lung of intact mice; (**c**) the level of CD34$^+$ cells after culture of cells isolated from the lung of intact mice; (**d**) the count of cells with apoptosis after culture of cells isolated from the lung of intact mice; (**e**) CFSE activity after culture of cells isolated from the lung of mice with MD and lung emphysema; (**f**) the level of CD34$^+$ cells after culture of cells isolated from the lung of mice with MD and lung emphysema; (**g**) the count of cells with apoptosis after culture of cells isolated from the lung of mice with MD and lung emphysema. All data are expressed as mean ± SD, *—significance of difference compared with control ($p < 0.05$).

3. Discussion

MetS is a complex clinical condition, and abdominal visceral obesity is considered to be one of its major components [21]. Gender differences in the development and progression of obesity and MetS, as well as their treatment, represent a current challenge in clinical practice. Traditionally, gender differences in metabolism (mechanisms of accumulation of adipose tissue and mobilization of fat reserves, homeostasis of glucose, secretion, and effects of insulin) and related diseases (MetS and type 1 and type 2 diabetes) are explained by the effects of sex hormones [2,7]. In our study, we focused on the combination of obesity (or MD) and COPD. Obesity contributes to increased respiratory reactivity and can lead to various respiratory pathologies, including COPD, asthma, and other lung diseases [17,22]. Moreover, obesity is common in patients diagnosed with COPD and contributes to respiratory symptoms [23]. Importantly, endothelial dysfunction and endothelial disorders are important risk factors for many complications of COPD.

While studying healthy C57BL/6 mice, we found fewer EPC ($CD45^{-}CD31^{+}CD34^{+}$; $CD31^{+}CD34^{+}CD146^{+}$) and angiogenesis precursors ($CD45^{-}CD309^{+}CD117^{+}$) in the lungs and bone marrow of females compared to males (Figures 6 and 7). In vitro, no differences between male and female mice were found regarding the numbers of apoptotic cells, expression of CD34, and the number of cells with active esterase (Figure 9).

Administration of MSG induced dyslipidemia and obesity in both male and female animals (Figure 2). The changes in lipid metabolism in females of group f2 were less pronounced compared to males in group m2. Our data is in general accordance with previously published results [24]. As in our earlier study [20], this work revealed infiltration of the parenchyma of lungs by inflammatory cells (predominantly macrophages and single neutrophils) in mice of both sexes treated with MSG (Figure 4). At the same time, hemodynamic disturbances and a decrease in the expression of CD31 in the lungs were observed. In female animals in group f2, recruitment of bone marrow EPC, angiogenic precursor cells, circulating pericytes, and vascular smooth muscle cells into the lungs was found (Figure 8). It has been reported that pericytes and vascular smooth muscle cells are involved in the restoration of the normal structure and function of the damaged endothelium through intercellular contacts [25]. In accordance with this report, the recruitment of these cell types into the lungs could be explained by ongoing regeneration of the damaged endothelium in the lungs of MSG-treated females.

We also found more pronounced hemodynamic disturbances and low CD31 expression in the lungs of male animals within the m2 group compared to females in the f2 group (Figure 5). Surprisingly, this damage to the pulmonary endothelium in male animals did not lead to the recruitment of bone marrow EPC and angiogenic precursor cells into the lung tissue. Additionally, animals in group m2 showed signs of hyperglycemia and had an increased ratio of TG/HDL (Figure 2e). High values of the TG/HDL ratio and glucose tolerance are thought to indicate a high risk of vascular complications [26]. This suggests that TG/HDL and hyperglycemia could have a prognostic significance in the simulation of emphysema in mice with obesity. LPS and CSE introduction increased the inflammatory response and caused the formation of emphysema in mice of both sexes with obesity. Meanwhile, damage to the microvascular bed in the lungs was more profound in males in the m4 group (high values of TG/HDL and hyperglycemia before emphysema modeling) than in females of the f4 group (low values of TG/HDL and the normal level of glucose before emphysema modeling) (Figures 2e and 4).

As evidenced by the findings above, the differences in the reaction of lung endothelium of females and males to external factors correlate with gender differences in fat metabolism and glucose metabolism.

LPS and CSE led to a decrease in the numbers of EPC ($CD45^{-}CD31^{+}CD34^{+}$; $CD31^{+}CD34^{+}CD146^{+}$), angiogenic precursor cells, vascular smooth muscle cells, cells of the lumen of the nascent vessels, and pericytes in the lungs of male mice in the group m4. These flow cytometric data can be explained by an impaired mobilization and migration of the cells. In f4 females, we observed recruitment of $CD45^{-}CD31^{+}CD34^{+}$ EPC, vascular smooth muscle cells, and pericytes to the lungs (Figure 6). At the same time, mobilization and migration of angiogenic precursor cells in females in the f4 group, as well as in males in group m4, were disturbed. An analysis of $CD31^{+}$ cells isolated from females in group f4 and males in group m4 in vitro revealed that there were no sex-specific differences in the rate of apoptosis, expression of CD34, and activity of esterase (Figure 9).

The main role of adult stem cells (SC) is the formation of new cells after injury [27]. In tissues of an adult organism, SCs are contained in the bone marrow and tissue-specific niches [28]. Markers of immature endothelial cells have been detected in bone marrow [28]. SC activity is regulated by internal mechanisms and external signals; the latter can come from a niche. It is believed that inflammation changes many homeostatic parameters and, thus, has a strong effect on various cells, including SC [29]. A negative impact of inflammation on the stem cell niche has been reported for intestinal stem cells, satellite cells or myogenic precursors cells, hepatic progenitor cells, epidermal stem cells, and neural stem cells [29]. Additionally, the mobilization of mesenchymal stromal/stem cells (MSC) by inflammatory factors has been demonstrated [30–32]. In our study, we found high levels of inflammatory cytokines in the lungs and disruption of EPC mobilization into the bloodstream in male mice in the m4 group and females in the group f4. In light of these findings, it is likely that mobilization of bone marrow EPC and MSC into the bloodstream and their migration to the damaged tissue is regulated by different mechanisms.

Therapy of obesity and its complications is currently limited by the lack of consideration of gender differences. Decreasing hyperglycemia and obesity could facilitate reducing the risk of vascular complications and related diseases [33]. It is well known that GLP-1 stimulates insulin production by islet β-cells, counteracts insulin resistance, improves peripheral glucose tolerance, and has anti-inflammatory properties [34,35]. In addition to the endocrine activity of GLP-1, it may also play a role in the homeostasis of the lungs. GLP-1 receptors are abundant in the alveoli, septum, airway, and smooth muscle of pulmonary vessels [36–38]. Moreover, their levels are relatively higher in the lungs than in the intestines and brain [39].

Known pegylated hormone analogs are characterized by improved pharmacokinetic characteristics without reducing the effectiveness of treatment and safety compared with native GLP-1 [40]. In our previous study, we assessed gender differences in the effects of pegGLP-1 in a streptozotocin-induced model of diabetes. Briefly, pegGLP-1 showed an anti-diabetic effect [41,42]. In the present study, pegGLP-1 showed more pronounced positive effects on the AUC, the ratio of TG/HDL, and emphysema square-extended alveolar tissue in females in group f6 compared to the f5 females treated with unpegylated GLP-1 (Figure 2). In addition, in female mice in f5 and f6 groups, both treatment regimes (pegGLP-1 and GLP-1) increased expression of CD31 in alveolar tissue. We attributed to additional recruitment of EPC ($CD45^-CD31^+CD34^+$) and angiogenic precursors to the lungs. These cellular effects were more pronounced in the case of pegGLP-1 than GLP-1. We found no differences in the effect of both agents on cultivated $CD31^+$ lung cells. Administration of pegGLP-1 or GLP-1 in cell culture resulted in a decrease of $CD31^+$ endothelial cell apoptosis and an increase in the number of CD34+ cells with active esterases (Figure 9). Thus, it is possible that similar to native GLP-1 [20], CD31 and CD34 positive endothelial progenitor cells are the target for pegGLP-1.

In the present study, we also assessed potential gender differences in the effects of GLP-1 and pegGLP-1. GLP-1 and pegGLP-1 administration in the m5 and m6 groups significantly reduced serum triglyceride levels, BMI, and AUC compared to groups f5 and f6, whereas increased serum HDL cholesterol concentration in male animals (Figure 1, Figure 2, Figure 3). Both GLP-1 and pegGLP1 administration did not affect the area of emphysema in groups m5 and m6, while the expression of CD31 in the lungs increased. However, this increase in CD31 expression was not as strong as in female groups (Figure 5). Unlike in female animals in f5 and f6 groups, angiogenic precursor cells, vascular smooth muscle cells (by pegGLP-1), and pericytes (by GLP-1) were recruited into the damaged alveolar tissue of males of groups m5 and m6 (Figure 6).

In sum, we presented evidence of gender-specific differences in lung injury, mobilization and migration of EPC, and angiogenic precursor cells in mice with MD and lung emphysema. From our point of view, genetic factors controlling fat metabolism and glucose metabolism might be involved in the gender-specific differences, and these will be a subject of future studies. In addition, our data indicated differences in the effectiveness of pegGLP-1 in COPD/MetS comorbidity. These results also suggested potentially higher therapeutic effects of pegGLP-1 for COPD treatment in obese women (or women with MD), including older women. Finally, we proposed that CD31 and CD34 positive EPCs were the cellular targets of pegGLP-1.

4. Materials and Methods

4.1. Animals

Experiments were carried out on female and male C57BL/6 mice (certified animals from the nursery of E.D. Goldberg Research Institute of Pharmacology and Regenerative Medicine) in strict adherence to the principles of European Convention for the Protection of Vertebrate Animals used for Experimental and other Scientific Purposes (Strasbourg, 1986). The study was approved by the Institutional Animal Care and Use Committee (IACUC) of the E.D. Goldberg Research Institute of Pharmacology and Regenerative Medicine (license number IACUC No. 114062016, 22.06.2016). The day of birth was considered as experimental day 0 (p0).

4.2. Induction of Obesity

Female and male C57BL/6 mice received a daily subcutaneous (sc) injection of monosodium glutamate (MSG; Sigma, St. Louis, MO, USA) diluted in buffer solution (physiological saline) at a dose of 2.2 mg/g from p0 to p10 [42]. Physiological saline was injected into control mice in equivalent volume. Obesity parameters were estimated according to the Lee index on p124th [42–44]. Briefly, Lee index was calculated as a cubic root of body weight (g) × 10/nasoanal length (mm), where an index equal to or lower than 0.300 was classified as normal. Female and male mice with Lee index values higher than 0.300 were classified as obese and included in the study [45].

4.3. Exposure to Cigarette Smoke Extract

Cigarette smoke extract (CSE) was generated from L&M RED LABEL cigarettes (2 cigarettes/mL). The composition of the cigarettes was as follows: resin 10 mg/Cigarette, nicotine 0.8 mg/ Cigarette, CO 10 mg/CIG. Before obtaining the extract, the cigarette filter was removed; the length of a cigarette with the filter was 80 mm, 55 mm with the removed filter. The extraction was carried out by stretching the smoke of a lit cigarette through the phosphate buffer at a constant speed with the help of a vacuum pump; the cigarette was burned to a length of 5 mm. The burning time of one cigarette was 180 s. To remove the particles, the extract was filtered through a bacterial filter with a pore size of 45 nm. To standardize the obtained extract, pH (pH~7) and optical density were measured at wavelengths of 405 and 540 nm (D405~237, D540~123) before and after filtration.

On p126, lung emphysema was induced by intratracheal administration of lipopolysaccharide (LPS, Sigma, St. Louis, MO, USA) and CSE [46,47]. LPS at a dose of 3 μg/mouse in 50 μL phosphate buffer and 50 μL CSE were administered intratracheally. For the introduction of LPS and CSE, general anesthesia (pentobarbital) was used. LPS was administered on p126 and p129. CSE was introduced on p127, p130, 133, p136, p139, p142, p149, p156, p163, and p170 (Figure 10).

4.4. Pharmacological Compounds

Glucagon-like peptide-1 (GLP-1) was obtained from Sigma (St. Louis, MO, USA). Pegylation on free amino groups of the peptide was carried out using succinimide pegylating agent Sunbright ME-120 TS (NOF America Corporation, San Mateo, CA, USA), for which 1 mg of lyophilizate was dissolved in 2 mL of 20 mm Na-phosphate buffer, pH 7, containing 0.01% TWEEN 20 (Sigma, St. Louis, MO, USA), then 20 mg of pegylating agent (NOF America Corporation, San Mateo, CA, USA) was added to the solution. The reaction was stopped by the application of 200 μL of 0.1 M glycine solution. The molecular weight of GLP-1 was assessed by electrophoresis in polyacrylamide gel with SDS (Sodium dodecyl sulfate, Sigma, St. Louis, MO, USA) using a standard technique. Prior to pegylation, the samples contained 100% GLP-1. After pegylation, the samples contained 7% GLP- 1, 78% of monopeg-GLP-1, and 15% double peg-GLP-1.

GLP-1 and pegGLP-1 were daily administered intraperitoneally in the region of the pancreas at a dose of 3 mmol/kg on p149, 156, 157, 173, 184, 186, and 188 (Figure 10).

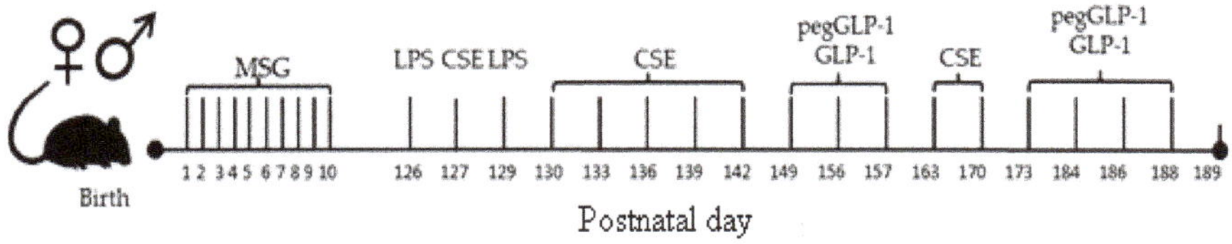

Figure 10. Schematic diagram of the experimental procedures.

4.5. *Experimental Groups*

Healthy mice treated with a saline solution formed the control groups: female control group (f1), and male control group (m1) (Table 1). Mice with metabolic disorders (MD) were divided into two groups: metabolic disorders group, females (f2), metabolic disorders group, males (m2). Mice with lung emphysema were divided into two groups: lung emphysema group, females (f3), lung emphysema group, males (m3). Mice with metabolic disorders and pulmonary emphysema were divided into two groups: metabolic disorders and pulmonary emphysema group, females (f4), and metabolic disorders and pulmonary emphysema, males (m4). GLP-1-treated mice with metabolic disorders and emphysema of the lungs were divided into two groups: GLP-1 treatment of metabolic disorders and emphysema, females (f5), and GLP-1 treatment of metabolic disorders and emphysema, males (m5). PegGLP-1-treated mice with metabolic disorders and emphysema of the lungs were divided into two groups: pegGLP-1 treatment of metabolic disorders and emphysema group, females (f6), and pegGLP-1 treatment of metabolic disorders and lung emphysema group, males (m6). All mice were culled on p189 by CO_2 asphyxia.

Table 1. Experimental groups in vivo.

	Control Groups	**Metabolic Disorders**	**Lung Emphysema**	**Metabolic Disorders + Lung Emphysema**	**Metabolic Disorders + Lung Emphysema + GLP-1**	**Metabolic Disorders + Lung Emphysema + pegGLP-1**
Females	f1 [1] (*n* = 10)	f2 (*n* = 10)	f3 (*n* = 10)	f4 (*n* = 10)	f5 (*n* = 10)	f6 (*n* = 10)
Males	m1 [2] (*n* = 10)	m2 (*n* = 10)	m3 (*n* = 10)	m4 (*n* = 10)	m5 (*n* = 10)	m6 (*n* = 10)

[1] f—Females, [2] m—males.

4.6. *Body Mass Index (BMI)*

On the day before culling (p188), BMI and Lee indexes were calculated. Briefly, BMI was calculated using the formula:

$$BMI = \text{body weight (g)}/\text{body length}^2\ \text{(cm)}$$

where body length was measured from the tip of the nose to the anus [43]. The Lee index was calculated as described above.

4.7. *Glucose Tolerance Test (GTT)*

Blood glucose levels were measured using a glucometer (Accu-Chek Performa Nano (Roche Diagnostics GmbH, Mannheim, Germany). A glucose tolerance test was performed on p188. Measurements of the initial level of glucose in the blood of animals were performed after 12 h of food deprivation. Subsequently, glucose was administered intragastrically (D-glucose, Sigma, St. Louis, MO, USA) at a dose of 2 g/kg. Blood samples to study glucose levels were taken 0, 15, 30, 60, 90 min after glucose administration [48].

4.8. Lipid Profile Determination

The lipid profile was determined on p189. Blood samples were taken from each animal in tubes without additives, kept at room temperature for 30 min, and then centrifuged at 300× *g* for 30 min. The serum was separated and used to study lipid profile parameters. The concentration of cholesterol and TG was determined by direct enzymatic methods using BioSystems reagents (Barcelona, Spain) in accordance with the manufacturer's instructions. Fractions of cholesterol, TG, high-density lipoproteins (HDL), low-density lipoproteins (LDL), and very-low-density lipoproteins (VLDL) were precipitated with phosphor-wolframate and polyvinyl sulfate, respectively, and their concentration was determined by the level of residual cholesterol. All the results were expressed as mmol/l. On p189, TG/HDL was assessed, as previously described in [49,50].

4.9. Lung Tissue Histology

The morphological examination of lungs was performed on p189. Briefly, the left lobe of the lung was fixed in a 10% solution of neutral formalin, carried out through alcohols of ascending concentrations to xylene, and poured into paraffin according to a standard procedure. Five micrometer thick dewaxed cuts were stained with hematoxylin and eosin [51]. Micro-preparations from each experimental animal were examined an Axio Lab.A1 light microscope (Carl Zeiss, MicroImaging GmbH, Göttingen, Germany) at 100× and 400× magnifications. Histoarchitecture of lung tissue and pathophysiological features of the tissue, including the presence of edema and inflammatory infiltration, venous congestion, as well as thickening of vessel walls and bronchi, were assessed [48,52,53].

4.10. Flow Cytometry

Mononuclear cells from the blood, bone marrow, and lung tissue were obtained, as previously described on p189 [20,54], followed by flow cytometric analysis of the expression of surface markers of mouse mononuclear cells. Briefly, cell suspensions were stained with the following fluorophore-conjugated monoclonal antibodies: CD45 PerCP, CD31 APC, CD34 FITC, CD146 PerCP-Cy5.5, CD309 (Flk-1) APC, and CD117 (c-kit) PeCy7 (all Becton Dickinson, San Jose, CA, USA). Appropriate isotype controls were used. Labeled cells were thoroughly washed with PBS and analyzed on a FACSCanto II flow cytometer (Becton Dickinson, San Jose, CA, USA) using FACS Diva software. At least 100,000 events were recorded for each sample.

4.11. Lung Tissue Dissociation and Magnetic Separation of $CD31^+$ Cells

The effects of GLP-1 and pegGLP-1 on $CD31^+$ lung cells in vitro were studied on p189 in cells isolated from an animal in the groups f1, m1, f4, and m4. Lungs were isolated, and the lung tissue was mechanically and enzymatically dissociated, followed by magnetic sorting for $CD31^+$, as previously described [20].

4.12. Cultivation of $CD31^+$ Cells

After 5-days of cultivation, $CD31^+$ cells from f1, m1, f4, and m4 mice were harvested using tryptic digestion, and the cells were plated in a concentration of 3×10^5 cells/1 mL medium in gelatin-coated flasks. M199 standard cultivation medium was supplemented with GLP-1 (10^{-7} M) or pegGLP-1 (10^{-7} M) (Table 2) followed by cultivation under standard conditions (3.5% CO_2, 37 °C) for 24 h. In the following, the effects of GLP-1 and pegGLP-1 on $CD31^+$ cells were evaluated by flow cytometry and imaging using the Cytation™ 3 imaging system [20].

Table 2. Experimental groups in vitro.

Drugs Introduced into the Culture CD31$^+$ Cells	CD31$^+$ Cell Culture			
	Group f1 (Control Groups)	Group m1 (Control Groups)	Group f4 (MD + Lung Emphysema)	Group m4 (MD + Lung Emphysema)
PBS	+	+	+	+
GLP-1	+	+	+	+
pegGLP-1	+	+	+	+

4.13. Cellular Imaging

Images of CD31$^+$ cells were obtained using a Cytation 3 Cell Imaging multimode reader (BioTek Instruments, Inc., Winooski, VT, USA) equipped with DAPI, GFP, and Texas Red light cubes. Cells were stained with Hoechst 33342, Annexin V-iFluor ™ 350 CFSE, and 7-AAD. Images were analyzed using Gen5 ™ data analysis software (Bad Friedrichshall, Germany), as described before [20].

4.14. Statistical Analysis

Statistical analysis was performed using SPSS statistical software (version 15.0, SPSS Inc., Chicago, IL, USA). Data were analyzed and presented as means ± standard error of the mean. Statistical significance was evaluated by Student's *t*-test (for parametric data), or Mann–Whitney test (for nonparametric data) was used according to distribution. A *p*-value of less than 0.05 (by two-tailed testing) was considered an indicator of statistical significance.

Author Contributions: Conceptualization, E.G.S.; methodology, O.V.P. and E.G.S.; software, V.A.K.; validation, E.G.S. and A.V.P.; formal analysis, E.G.S. and A.M.D.; investigation, O.V.P., A.V.P., N.N.E., E.S.P., V.A.K., O.D.P., L.A.S., and I.V.K.; resources, O.E.V. and A.A.E.; data curation, E.G.S.; writing—original draft preparation, E.G.S.; writing—review and editing, E.G.S., D.W., and O.V.P.; visualization, A.V.P., N.N.E., and V.A.K.; supervision, E.G.S.; project administration, S.G.M., A.M.D., and A.A.K.; funding acquisition, O.E.V.

Funding: This research received no external funding.

Conflicts of Interest: The authors declare no conflict of interest.

Abbreviations

COPD	Chronic obstructive pulmonary disease
EPC	Endothelial progenitor cells
GLP-1	Glucagon-like peptide 1
pegGLP-1	Pegylated Glucagon-like peptide 1
MetS	Metabolic syndrome
MD	Metabolic disorder
BMI	Body mass index
MSG	Monosodium glutamate
LPS	Lipopolysaccharide
CSE	Cigarette smoke extract
AUC	Area under the Curve
GTT	Glucose tolerance test
TG	Triglycerides
HDL	High-density lipoproteins
LDL	Low-density lipoproteins
VLDL	Very low-density lipoproteins
MSC	Mesenchymal stromal/stem cells

References

1. Lusis, A.J.; Attie, A.D.; Reue, K. Metabolic syndrome: From epidemiology to systems biology. *Nat. Rev. Genet.* **2008**, *9*, 819–830. [CrossRef] [PubMed]

2. Mauvais-Jarvis, F. Epidemiology of Gender Differences in Diabetes and Obesity. *Adv. Exp. Med. Biol.* **2017**, *1043*, 3–8. [CrossRef] [PubMed]
3. Karastergiou, K.; Smith, S.R.; Greenberg, A.S.; Fried, S.K. Sex differences in human adipose tissues e the biology of pear shape. *Biol. Sex Differ.* **2012**, *3*, 13. [CrossRef] [PubMed]
4. Mundi, M.S.; Koutsari, C.; Jensen, M.D. Effects of increased free fatty acid availability on adipose tissue fatty acid storage in men. *J. Clin. Endocrinol. Metab.* **2014**, *99*, E2635–E2642. [CrossRef]
5. Karpe, F.; Pinnick, K.E. Biology of upper-body and lower-body adipose tissue link to whole-body phenotypes. *Nat. Rev. Endocrinol.* **2015**, *11*, 90–100. [CrossRef]
6. Karastergiou, K.; Fried, S.K. Cellular mechanisms driving sex differences in adipose tissue biology and body shape in humans and mouse models. *Adv. Exp. Med. Biol.* **2017**, *1043*, 29–51. [CrossRef]
7. Link, J.C.; Reue, K. Genetic Basis for Sex Differences in Obesity and Lipid Metabolism. *Annu. Rev. Nutr.* **2017**, *37*, 225–245. [CrossRef]
8. Agarwal, A.R.; Yin, F.; Cadenas, E. Short-term cigarette smoke exposure leads to metabolic alterations in lung alveolar cells. *Am. J. Respir. Cell Mol. Biol.* **2014**, *51*, 284–293. [CrossRef]
9. Titz, B.; Boue, S.; Phillips, B.; Elamin, A.; Schneider, T.; Talikka, M.; Nury, C.; Vihervaara, T.; Guedj, E.; Peck, M.; et al. Effects of cigarette smoke, cessation, and switching to two heat-not-burn tobacco products on lung lipid metabolism in C57BL/6 and Apoe-/-mice—An integrative systems toxicology analysis. *Toxicol. Sci.* **2016**, *149*, 441–457. [CrossRef]
10. Cebron Lipovec, N.; Beijers, R.J.; van den Borst, B.; Doehner, W.; Lainscak, M.; Schols, A.M. The Prevalence of Metabolic Syndrome in Chronic Obstructive Pulmonary Disease: A Systematic Review. *COPD J. Chron. Obstr. Pulm. Dis.* **2016**, *13*, 399–406. [CrossRef]
11. Åberg, J.; Hasselgren, M.; Montgomery, S.; Lisspers, K.; Ställberg, B.; Janson, C.; Sundh, J. Sex-related differences in management of Swedish patients with a clinical diagnosis of chronic obstructive pulmonary disease. *Int. J. Chron. Obstr. Pulm. Dis.* **2019**, *14*, 961–969. [CrossRef] [PubMed]
12. Kim, J.H.; Yoo, J.Y.; Kim, H.S. Metabolic Syndrome in South Korean Patients with Chronic Obstructive Pulmonary Disease: A Focus on Gender Differences. *Asian Nurs. Res.* **2019**, *13*, 137–146. [CrossRef] [PubMed]
13. Walther, T.C.; Farese, R.V., Jr. Lipid droplets and cellular lipid metabolism. *Annu. Rev. Biochem.* **2012**, *81*, 687–714. [CrossRef] [PubMed]
14. Shimano, H.; Sato, R. SREBP-regulated lipid metabolism: Convergent physiology—Divergent pathophysiology. *Nat. Rev. Endocrinol.* **2017**, *13*, 710–730. [CrossRef]
15. Martinez-Outschoorn, U.E.; Peiris-Pages, M.; Pestell, R.G.; Sotgia, F.; Lisanti, M.P. Cancer metabolism: A therapeutic perspective. *Nat. Rev. Clin. Oncol.* **2017**, *14*, 11–31. [CrossRef]
16. Chen, H.; Li, Z.; Dong, L.; Wu, Y.; Shen, H.; Chen, Z. Lipid metabolism in chronic obstructive pulmonary disease. *Int. J. Chron. Obstr. Pulm. Dis.* **2019**, *14*, 1009–1018. [CrossRef]
17. Hanson, C.; Rutten, E.P.; Wouters, E.F.; Rennard, S. Influence of diet and obesity on COPD development and outcomes. *Int. J. Chron. Obstr. Pulm. Dis.* **2014**, *9*, 723–733. [CrossRef]
18. Lambert, A.A.; Putcha, N.; Drummond, M.B.; Boriek, A.M.; Hanania, N.A.; Kim, V.; Kinney, G.L.; McDonald, M.N.; Brigham, E.P.; Wise, R.A.; et al. Obesity is associated with increased morbidity in moderate to severe COPD. *Chest* **2017**, *151*, 68–77. [CrossRef]
19. Zore, T.; Palafox, M.; Reue, K. Sex differences in obesity, lipid metabolism, and inflammation-A role for the sex chromosomes? *Mol. Metab.* **2018**, *15*, 35–44. [CrossRef]
20. Skurikhin, E.G.; Pershina, O.V.; Pakhomova, A.V.; Pan, E.S.; Krupin, V.A.; Ermakova, N.N.; Vaizova, O.E.; Pozdeeva, A.S.; Zhukova, M.A.; Skurikhina, V.E.; et al. Endothelial Progenitor Cells as Pathogenetic and Diagnostic Factors, and Potential Targets for GLP-1 in Combination with Metabolic Syndrome and Chronic Obstructive Pulmonary Disease. *Int. J. Mol. Sci.* **2019**, *20*, 1105. [CrossRef]
21. Matsushita, K.; Dzau, V.J. Mesenchymal stem cells in obesity: Insights for translational applications. *Lab. Investig.* **2017**, *97*, 1158–1166. [CrossRef] [PubMed]
22. Lamonaca, P.; Prinzi, G.; Kisialiou, A.; Cardaci, V.; Fini, M.; Russo, P. Metabolic Disorder in Chronic Obstructive Pulmonary Disease (COPD) Patients: Towards a Personalized Approach Using Marine Drug Derivatives. *Mar. Drugs* **2017**, *15*, 81. [CrossRef] [PubMed]
23. García-Rio, F.; Soriano, J.B.; Miravitlles, M.; Muñoz, L.; Duran-Tauleria, E.; Sánchez, G.; Sobradillo, V.; Ancochea, J. Impact of obesity on the clinical profile of a population-based sample with chronic obstructive pulmonary disease. *PLoS ONE* **2014**, *9*, e105220. [CrossRef] [PubMed]

24. Sasaki, Y.; Suzuki, W.; Shimada, T.; Iizuka, S.; Nakamura, S.; Nagata, M.; Fujimoto, M.; Tsuneyama, K.; Hokao, R.; Miyamoto, K.; et al. Dose dependent development of diabetes mellitus and non-alcoholic steatohepatitis in monosodium glutamate-induced obese mice. *Life Sci.* **2009**, *85*, 490–498. [CrossRef] [PubMed]
25. Birbrair, A.; Zhang, T.; Wang, Z.M.; Messi, M.L.; Mintz, A.; Delbono, O. Pericytes at the intersection between tissue regeneration and pathology. *Clin. Sci.* **2015**, *128*, 81–93. [CrossRef] [PubMed]
26. Borrayo, G.; Basurto, L.; González-Escudero, E.; Diaz, A.; Vázquez, A.; Sánchez, L.; Hernández-González, G.O.; Barrera, S.; Degollado, J.A.; Córdova, N.; et al. TG/HDL-C Ratio as Cardio-Metabolic Biomarker Even in Normal Weight Women. *Acta Endocrinol.* **2018**, *14*, 261–267. [CrossRef]
27. He, S.; Nakada, D.; Morrison, S.J. Mechanisms of stem cell selfrenewal. *Annu. Rev. Cell Dev. Biol.* **2009**, *25*, 377–406. [CrossRef]
28. Pinho, S.; Frenette, P.S. Haematopoietic stem cell activity and interactions with the niche. *Nat. Rev. Mol. Cell Biol.* **2019**, *20*, 303–320. [CrossRef]
29. Kizil, C.; Kyritsis, N.; Brand, M. Effects of inflammation on stem cells: Together they strive? *EMBO Rep.* **2015**, *16*, 416–426. [CrossRef]
30. Keshav, S. Paneth cells: Leukocyte-like mediators of innate immunity in the intestine. *J. Leukoc. Biol.* **2006**, *80*, 500–508. [CrossRef]
31. Koning, J.J.; Kooij, G.; de Vries, H.E.; Nolte, M.A.; Mebius, R.E. Mesenchymal stem cells are mobilized from the bone marrow during inflammation. *Front. Immunol.* **2013**, *4*, 49. [CrossRef] [PubMed]
32. Mourkioti, F.; Rosenthal, N. IGF-1, inflammation and stem cells: Interactions during muscle regeneration. *Trends Immunol.* **2005**, *26*, 535–542. [CrossRef] [PubMed]
33. International Diabetes Federation. *IDF Diabetes Atlas*, 6th ed.; International Diabetes Federation: Brussels, Belgium, 2015; Available online: www.idf.org/e-library/epidemiology-research/diabetes-atlas/13-diabetes-atlas-seventh-edition.html (accessed on 10 September 2019).
34. Lee, Y.S.; Jun, H.S. Anti-Inflammatory Effects of GLP-1-Based Therapies beyond Glucose Control. *Mediat. Inflamm.* **2016**, *2016*, 3094642. [CrossRef]
35. Huang, J.; Yi, H.; Zhao, C.; Zhang, Y.; Zhu, L.; Liu, B.; He, P.; Zhou, M. Glucagon-like peptide-1 receptor (GLP-1R) signaling ameliorates dysfunctional immunity in COPD patients. *Int. J. Chron. Obstr. Pulm. Dis.* **2018**, *13*, 3191–3202. [CrossRef] [PubMed]
36. Korner, J.; Bessler, M.; Inabnet, W.; Taveras, C.; Holst, J.J. Exaggerated glucagon-like peptide-1 and blunted glucose-dependent insulinotropic peptide secretion are associated with Roux-en-Y gastric bypass but not adjustable gastric banding. *Surg. Obes. Relat. Dis.* **2007**, *3*, 597–601. [CrossRef]
37. Rogliani, P.; Ora, J.; Puxeddu, E.; Cazzola, M. Airflow obstruction: Is it asthma or is it COPD? *Int. J. Chron. Obstr. Pulm. Dis.* **2016**, *11*, 3007–3013. [CrossRef] [PubMed]
38. Nguyen, D.V.; Linderholm, A.; Haczku, A.; Kenyon, N. Glucagon-like peptide 1: A potential anti-inflammatory pathway in obesity-related asthma. *Pharmacol. Ther.* **2017**, *180*, 139–143. [CrossRef] [PubMed]
39. Viby, N.E.; Isidor, M.S.; Buggeskov, K.B.; Poulsen, S.S.; Hansen, J.B.; Kissow, H. Glucagon-like peptide-1 (GLP-1) reduces mortality and improves lung function in a model of experimental obstructive lung disease in female mice. *Endocrinology* **2013**, *154*, 4503–4511. [CrossRef]
40. Skurikhin, E.G.; Stronin, O.V.; Epanchintsev., A.A.; Pershina, O.V.; Ermakova, N.N.; Krupin, V.A.; Pakhomova, A.V.; Vaizova, O.E.; Dygai, A.M. Effects of Pegylated Glucagon-Like Peptide-1 Analogue in C57Bl/6 Mice under Optimal Conditions and During Streptozotocin-Induced Diabetes. *Bull. Exp. Biol. Med.* **2017**, *163*, 635–638. [CrossRef]
41. Skurikhin, E.G.; Pakhomova, A.V.; Epanchintsev, A.A.; Stronin, O.V.; Ermakova, N.N.; Pershina, O.V.; Ermolaeva, L.A.; Krupin, V.A.; Kudryashova, A.I.; Zhdanov, V.V.; et al. Role of Cell Precursors in the Regeneration of Insulin-Producing Pancreatic Cells under the Influence of Glucagon-Like Peptide 1. *Bull. Exp. Biol. Med.* **2018**, *165*, 644–648. [CrossRef]
42. Cameron, D.P.; Poon, T.K.-Y.; Smith, G.C. Effects of Monosodium Glutamate Administration in the Neonatal Period on the Diabetic Syndrome in KK Mice. *Diabetologia* **1976**, *12*, 621–626. [CrossRef] [PubMed]
43. Nawa, A.; Fujita-Hamabe, W.; Tokuyama, S. Altered intestinal P-glycoprotein expression levels in a monosodium glutamate-induced obese mouse model. *Life Sci.* **2011**, *89*, 834–838. [CrossRef] [PubMed]

44. Damasceno, D.C.; Sinzato, Y.K.; Bueno, A.; Dallaqua, B.; Lima, P.H.; Calderon, I.M.P.; Rudge, M.V.C.; Campos, K.E. Metabolic Profile and Genotoxicity in Obese Rats Exposed to Cigarette Smoke. *Obesity* **2013**, *21*, 1569–1601. [CrossRef] [PubMed]
45. Campos, K.E.; Sinzato, Y.K.; de Paula Pimenta, W.; Rudge, M.V.; Damasceno, D.C. Effect of maternal obesity on the diabetes developmental in adult rat offspring. *Life Sci.* **2007**, *81*, 1473–1478. [CrossRef] [PubMed]
46. Chen, Y.; Hanaoka, M.; Chen, P.; Droma, Y.; Voelkel, N.F.; Kubo, K. Protective effect of beraprost sodium, a stable prostacyclin analog, in the development of cigarette smoke extract-induced. *Am. J. Physiol. Lung Cell. Mol. Physiol.* **2009**, *296*, L648–L656. [CrossRef]
47. Hanaoka, M.; Hanaoka, P.; Chen, Y.; Agatsuma, T.; Kitaguchi, Y.; Voelkel, N.F.; Kubo, K. Carbocisteine protects against emphysema induced by cigarette smoke extract in rats. *Chest* **2011**, *139*, 1101–1108. [CrossRef]
48. Pacini, G.; Ahrén, M.; Ahrén, B. Reappraisal of the intravenous glucose tolerance index for a simple assessment of insulin sensitivity in mice. *Am. J. Physiol. Regul. Integr. Comp. Physiol.* **2009**, *296*, R1316–R1324. [CrossRef]
49. Fernández-Miranda, G.; Romero-Garcia, T.; Barrera-Lechuga, T.P.; Mercado-Morales, M.; Rueda, A. Impaired Activity of Ryanodine Receptors Contributes to Calcium Mishandling in Cardiomyocytes of Metabolic Syndrome Rats. *Front. Physiol.* **2019**, *10*, 520. [CrossRef]
50. Salazar, M.R.; Carbajal, H.A.; Espeche, W.G.; Aizpurúa, M.; Leiva Sisnieguez, C.E.; Leiva Sisnieguez, B.C. Use of the plasma triglyceride/high-density lipoprotein cholesterol ratio to identify cardiovascular disease in hypertensive subjects. *J. Am. Soc. Hypertens.* **2014**, *8*, 724–731. [CrossRef]
51. Skurikhin, E.G.; Pakhomova, A.V.; Krupin, V.A.; Pershina, O.V.; Pan, E.S.; Ermolaeva, L.A.; Vaizova, O.E.; Rybalkina, O.Y.; Dygai, A.M. Response of Inflammatory Mediators, Extracellular Matrix Proteins and Stem and Progenitor Cells to Emphysema. *Bull. Exp. Biol. Med.* **2016**, *161*, 566–570. [CrossRef]
52. Parameswaran, H.; Majumdar, A.; Ito, S.; Alencar, A.M.; Suki, B. Quantitative characterization of airspace enlargement in emphysema. *J. Appl. Physiol.* **2006**, *100*, 186–193. [CrossRef] [PubMed]
53. He, Z.H.; Chen, P.; Chen, Y.; He, S.D.; Ye, J.R.; Zhang, H.L.; Cao, J. Comparison between cigarette smoke-induced emphysema and cigarette smoke extract-induced emphysema. *Tob. Induc. Dis.* **2015**, *13*, 6. [CrossRef] [PubMed]
54. Munoz-Barrutia, A.; Ceresa, M.; Artaechevarria, X.; Montuenga, L.M.; Ortiz-de-Solorzano, C. Quantification of lung damage in an elastase-induced mouse model of emphysema. *Int. J. Biomed. Imaging* **2012**, *2012*, 734734. [CrossRef] [PubMed]

International Journal of *Molecular Sciences*

MDPI

Article

How Oxygen Availability Affects the Antimicrobial Efficacy of Host Defense Peptides: Lessons Learned from Studying the Copper-Binding Peptides Piscidins 1 and 3

Adenrele Oludiran [1], David S. Courson [1], Malia D. Stuart [2], Anwar R. Radwan [3], John C. Poutsma [3], Myriam L. Cotten [4,*] and Erin B. Purcell [1,*]

1 Department of Chemistry and Biochemistry, Old Dominion University, Norfolk, VA 23529, USA; aolud001@odu.edu (A.O.); dcourson@odu.edu (D.S.C.)
2 Biology Department, Palomar College, San Marcos, CA 92069, USA; maliastuart5@gmail.com
3 Department of Chemistry, College of William and Mary, Williamsburg, VA 23185, USA; aradwan@email.wm.edu (A.R.R.); jcpout@wm.edu (J.C.P.)
4 Department of Applied Science, College of William and Mary, Williamsburg, VA 23185, USA
* Correspondence: mcotten@wm.edu (M.L.C.); epurcell@odu.edu (E.B.P.); Tel.: +757-221-7428 (M.L.C.); +757-683-4240 (E.B.P.); Fax: +757-221-2050 (M.L.C.); +757-683-4628 (E.B.P.)

Received: 27 September 2019; Accepted: 22 October 2019; Published: 24 October 2019

Abstract: The development of new therapeutic options against *Clostridioides difficile* (*C. difficile*) infection is a critical public health concern, as the causative bacterium is highly resistant to multiple classes of antibiotics. Antimicrobial host-defense peptides (HDPs) are highly effective at simultaneously modulating the immune system function and directly killing bacteria through membrane disruption and oxidative damage. The copper-binding HDPs piscidin 1 and piscidin 3 have previously shown potent antimicrobial activity against a number of Gram-negative and Gram-positive bacterial species but have never been investigated in an anaerobic environment. Synergy between piscidins and metal ions increases bacterial killing aerobically. Here, we performed growth inhibition and time-kill assays against *C. difficile* showing that both piscidins suppress proliferation of *C. difficile* by killing bacterial cells. Microscopy experiments show that the peptides accumulate at sites of membrane curvature. We find that both piscidins are effective against epidemic *C. difficile* strains that are highly resistant to other stresses. Notably, copper does not enhance piscidin activity against *C. difficile*. Thus, while antimicrobial activity of piscidin peptides is conserved in aerobic and anaerobic settings, the peptide–copper interaction depends on environmental oxygen to achieve its maximum potency. The development of pharmaceuticals from HDPs such as piscidin will necessitate consideration of oxygen levels in the targeted tissue.

Keywords: host defense peptides; membrane activity; copper; piscidins; *Clostridioides difficile*

1. Introduction

Clostridioides (formerly *Clostridium*) *difficile* infection (CDI), whose symptoms can include inflammation, profuse diarrhea, and pseudomembranous colitis, has been recognized as an urgent public health threat in the United States and other industrialized nations [1,2]. CDI is primarily a hospital-acquired disease, as disruption of the native gut microbiota by prior antibiotic usage is the major risk factor for *C. difficile* colonization, although the number of community-acquired infections has increased in recent years [3,4]. The severity of CDI has also increased during the 21st century with the emergence of so-called "hypervirulent" epidemic ribotypes of the bacterium, most notably ribotype 027, that are associated with higher levels of disease recurrence and death in infected patients [3,5–7].

C. difficile is resistant to several families of antibiotics, including penicillin-family beta lactams and fluoroquinolones, and is increasingly resistant to next-generation therapeutics including fidaxomicin and vancomycin [6,8,9]. Currently, the most clinically effective treatment for CDI is replenishment of the protective gut microbiota through fecal transplants. As these procedures have a high inherent risk of introducing uncharacterized pathogens and are not recommended for immunocompromised patients, there is great interest in the development of new strategies for prevention and treatment [10–13].

C. difficile persists in the environment in the form of metabolically dormant spores, which are highly resilient to chemical and physical stresses and remain viable for months [14,15]. If mammals ingest these spores, amino acids and bile salts in the digestive system trigger their germination into metabolically active vegetative cells [16–18]. Vegetative *C. difficile* often cannot integrate well into the diverse, metabolically efficient microbial ecosystem of a healthy intestinal microbiome but can take advantage of the loss of bacterial species diversity and rise in nutrient availability induced by antibiotic exposure to establish colonization [19–22]. *C. difficile* colonization triggers the innate immune response, including the release of reactive oxygen species (ROS) and cationic host defense peptides (HDPs) [23–25]. These antimicrobial peptides can kill bacterial cells directly through a number of mechanisms, attacking the cell membrane and/or intracellular targets, and indirectly by activating the host innate immune response [26–29]. As these peptides have multiple cellular targets, bacteria cannot quickly develop or transmit genetically encoded resistance to them, and they are a promising precursor for the development of stand-alone antibiotics or adjuvants designed to work synergistically with existing antibiotics [30,31].

Piscidins are a family of HDPs found in teleost (bony) fish species with demonstrated efficacy against a wide range of bacteria and viruses [32–35]. The piscidins p1 (FFHHIFRGIVHVGKTIHRLVTG) and p3 (FIHHIFRGIVHAGRSIGRFLTG), which are derived from the mast cells of hybrid striped sea bass, exhibit broad spectrum antibacterial activity although their mechanisms of action differ [36,37]. Both peptides localize to bacterial cell membranes and are internalized at sub-lethal concentrations. While p1 is more damaging to membrane integrity than p3, the latter is more disruptive to DNA [36]. Furthermore, studies done on live bacteria and model membranes indicate that the peptides, especially p1, take advantage of lipid heterogeneity to deploy their mechanism of membrane disruption [36,38]. Recently, we demonstrated that under aerobic conditions both peptides use their amino-terminal copper- and nickel-binding (ATCUN) motifs to coordinate Cu^{2+} with picomolar affinity [37,39]. Piscidin-copper complexes form ROS and exhibit nuclease activity against double stranded DNA, resulting in increased lethality against multiple bacterial species [37]. Such copper-ATCUN complexes can serve as sources of oxidative stress, increasing peptide lethality against bacteria in an aerobic environment [40–42]. Oxidative stress can be harmful or lethal to organisms, depending on their oxidative stress tolerance. Obligate anaerobes such as *Clostridia* are considered completely intolerant to oxygen, although they can employ scavenger and reductase enzymes to survive transient exposure to environmental oxygen or immune-mediated oxidative bursts [15,23,43–45]. Application of antimicrobial peptides sensitizes *C. difficile* to antibiotics, although epidemic strains from ribotype 027 are less sensitive than other strains [31,46].

Importantly, as HDPs, piscidins have immunomodulatory effects. In particular, our investigations have demonstrated that both p1 and p3 induce chemotaxis in neutrophils [39]. These effects are exclusively mediated by formyl peptide receptors 1 and 2 (FPR1 and FPR2), both of which are G-protein coupled receptors (GPCRs) that play important functions in the immune system [47–51]. Interestingly, Cu^{2+}-coordination decreases the chemotactic effects of p1 and p3, suggesting a regulatory effect of copper between the direct and indirect antimicrobial effects of the peptides [39]. Given the role of FPR2 for the resolution of inflammation, it has become an important drug target [47–49]. P1 has also been shown to decrease the inflammatory response through a process that may involve binding lipopolysaccharides and decreasing toll-like receptor (TLR)-mediated inflammatory pathways [52,53]. The immunomodulatory properties of HPDs such as piscidin have emerged as an important topic of research given that these effects are indirect, and thus unlikely to activate mechanisms of drug resistance observed with traditional antibiotics that directly attack bacteria [54–59]. In addition, HDP modulation of the inflammatory immune response can mitigate infection symptoms and has been shown to reduce toxin-dependent inflammation in mouse models of *C. difficile* infection [60].

As indicated above, the antimicrobial effects of piscidin have previously been measured in aerobic environments. However, there is a ten-fold range of partial oxygen pressure among the tissues of the human body, with organs such as the large intestine providing a habitat for anaerobic microbes, both commensal and pathogenic [61]. Here, we report that the antimicrobial activity of p1 and p3 differ in aerobic and anaerobic environments. In an anaerobic environment, both p1 and p3 are incorporated into *C. difficile* cells, inhibit bacterial proliferation, and are highly toxic against actively dividing *C. difficile*. Both peptides associate extensively with bacterial cell membranes, exhibiting preferential localization at sites of high curvature such as cell poles and septa. In contrast to previously observed aerobic data, anaerobic piscidin antibacterial activity does not appear to be enhanced by metal complex formation. Our findings suggest that the mechanism by which these peptides induce bacterial cell death is influenced by the availability of environmental oxygen. It is clear that future mechanistic investigations of HDPs focused on potential medical applications must account for oxygen levels at the desired site of action in order to accurately model antimicrobial activity.

2. Results

2.1. Piscidins Are Incorporated into C. difficile and Appear to Localize to Sites of Membrane Curvature

Confocal microscopy of fixed bacterial cells exposed to fluorescently labeled p1 and p3 has previously shown that they enter both Gram-negative and Gram-positive bacterial cells and appear to be concentrated at bacterial nucleoids and cell septa [36,39]. We exposed live *C. difficile* R20291 cells to 0.75 μM 5-carboxytetramethylrhodamine (TAMRA)-labeled p1 and p3 and observed peptide uptake and localization in unfixed live cells. As *C. difficile* exhibits green autoflouresence, the red TAMRA labeling was distinct from any intrinsic signal produced by the cells [62]. Exponential-phase cells and peptides were mixed and sealed within microscopy chambers in an anaerobic chamber and then transported to the microscope, resulting in a 6-min delay between the onset of peptide exposure and the first image [63]. Mean fluorescence intensity within cells was stable over the course of 1 h of monitoring, indicating that peptide incorporation into cells occurs within the first few min of exposure (Figure 1A,B). Peptide integration appeared to be complete within 6 min even at lower peptide concentrations of 0.25 and 0.075 μM (data not shown). The addition of additional unlabeled peptide or unlabeled peptide complexed with Cu^{2+} did not increase fluorescence intensity, and thus there was no evidence of potential cooperativity in peptide uptake. There were distinct fluorescent puncta at the septa of predivisional cells (Figure 1C,E,F), consistent with prior observations in *Escherichia coli* (*E. coli*) and *Bacillus megaterium* (*B. megaterium*) [36]. In addition, there were fluorescent puncta at cell poles, suggesting that piscidins generally localize to sites of high curvature (Figure 1D). While unlabeled cells were motile and maintained rod-like shapes, many of the fluorescently labeled cells exhibited curvature or surface irregularities suggestive of cell envelope damage (Figure 1E). Performing these experiments on live cells allowed real-time observation of cellular response to peptide intoxication. We observed a motile chain of predivisional rod-shaped cells over the course of 10 min (Figure 1G). During this time, the chain of cells took up labeled p1 at one pole and subsequently developed progressively severe curvature at cell septa and separated into smaller fragments (Figure 1G). The resulting pieces were asymmetrically curved and non-motile, indicating that lysis rather than healthy cell division had occurred.

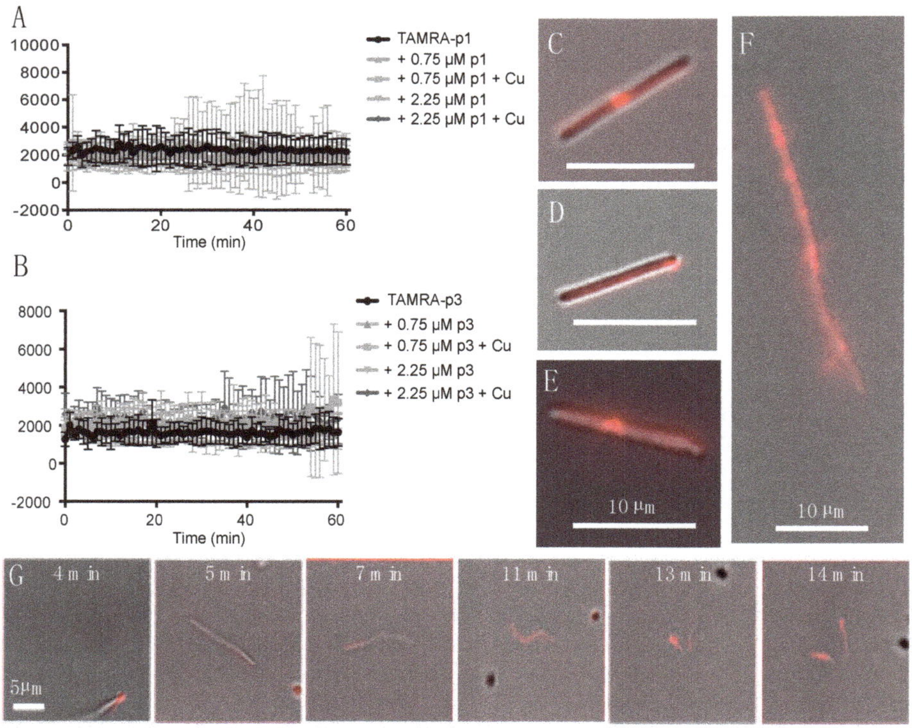

Figure 1. Incorporation of TAMRA-piscidin into live *C. difficile*. (**A**,**B**) Fluorescent signal per cell of 0.75 μM TAMRA-labeled p1 (**A**) and p3 (**B**) mixed with live *C. difficile* R20291. Cells had already reached maximum peptide incorporation by the time recording began, roughly 6 min after peptides and cells were mixed. Addition of unlabeled peptide, in the presence or absence of equimolar amounts of copper sulfate, did not cooperatively increase peptide incorporation. (**C**–**F**) Representative images of *C. difficile* labeled with: (**C**) 0.75 μM TAMRA-labeled p1 plus 0.75 μM unlabeled p1; (**D**) 0.75 μM TAMRA-labeled p3; (**E**) 0.75 μM TAMRA-labeled p1 plus 2.25 μM unlabeled p1; (**F**) 0.75 μM TAMRA-labeled p1 plus 0.75 μM unlabeled p1. (**G**) Timecourse showing the rupture of a pre-divisional cell labeled with 0.75 μM TAMRA-labeled p1 plus 0.75 μM unlabeled p1. Scale bars in panels (**C**–**F**) represent 10 μm. Scale bar in panel (**G**) represents 5 μm.

2.2. *Piscidins Prevent C. difficile Proliferation*

In order to measure the inhibitory effects of piscidin peptides on *C. difficile* growth we inoculated *C. difficile* strains 630Δ*erm* and R20291 into a medium containing the peptides. *C. difficile* 630Δ*erm* is an erythromycin-sensitive derivative of the reference strain *C. difficile* 630, while R20291 is an epidemic strain isolated from a 2003–2005 hospital outbreak of *C. difficile* infection in the United Kingdom [64,65]. R20291 is a so-called "hypervirulent" strain of ribotype 027 and is more resistant than 630 to multiple classes of antibiotics including clindamycin and ciprofloxacin [66–69]. The presence of piscidin peptides prevented *C. difficile* proliferation. Notably, R20291 was as susceptible as 630Δ*erm* to growth inhibition by piscidin. p1 inhibited proliferation of both strains at concentrations at or above 4.00 μM and p3 inhibited growth at or above 8.00 μM (Figure 2).

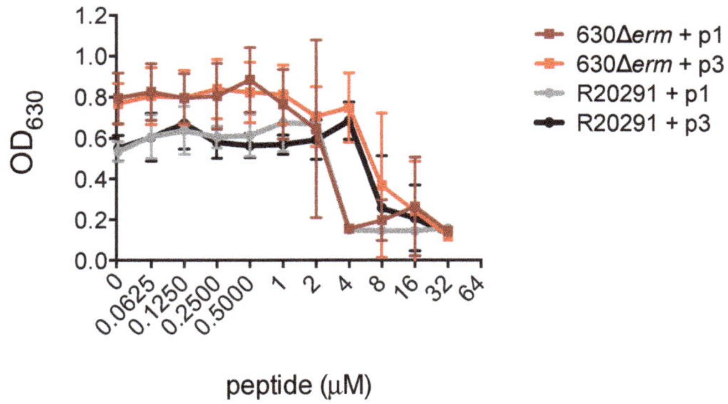

Figure 2. Piscidins inhibit *C. difficile* growth. Optical densities of overnight *C. difficile* 630Δ*erm* and R20291 cultures grown in the presence of the indicated concentrations of piscidins. Data shown are the means and standard deviations of four biologically independent samples.

2.3. Piscidins Reduce Established *C. difficile* Populations

Growth inhibition assays do not distinguish between substances that kill cells and bacteriostatic substances that inhibit growth only if compounds are present in sufficient quantities prior to bacterial proliferation. To confirm that piscidins are capable of reducing the number of viable cells in established bacterial populations, we performed time-kill assays to confirm that the number of viable *C. difficile* R20291 cells in exponentially growing culture decreases with exposure to p1 and p3 at sub-inhibitory concentrations. As shown in Figure 3, both p1 and p3 reduce *C. difficile* viability at half of the concentration needed to inhibit bacterial growth. The addition of 2.00 μM p1 significantly reduces the number of viable cells in the culture within 30 min, with continued loss of colony forming units over the course of 4 h (Figure 3A). Similarly, incubation with 4.00 μM p3 significantly reduced the number of viable cells within 30 min (Figure 3B). The bacterial killing by both p1 and p3 against *E. coli* in aerobic environments is exacerbated by the addition of equimolar copper, which complexes with the peptides, resulting in covalent damage to lipids and DNA [37]. We investigated the effect of adding copper sulfate to the anaerobic *C. difficile* killing assays at the same molar concentration as the piscidins. The addition of copper ions had no observable effect on p1 lethality at any of the timepoints examined (Figure 3A). Copper did appear to accelerate killing by p3 at 30 min and 2 h post-treatment, but the differences between peptide alone and peptide with copper had disappeared by 4 h post-treatment (Figure 3B). To investigate the possibility that copper ions were being chelated by other factors present in the tryptone-yeast (TY) medium and not actually forming complexes with piscidins, we repeated the assays with pre-formed piscidin-copper complexes. Copper allowed to form complexes with p1 prior to addition to the bacteria cultures still had no additional impact on p1 killing of *C. difficile* (Figure 3A) compared to apo p1. Pre-formed p3-copper complexes behaved like p3 in the absence of copper, killing *C. difficile* more slowly than p3 with copper added separately, but differences between the two conditions disappeared within 4 h (Figure 3B). This confirms that copper does not enhance the antimicrobial effects of piscidins under anaerobic conditions.

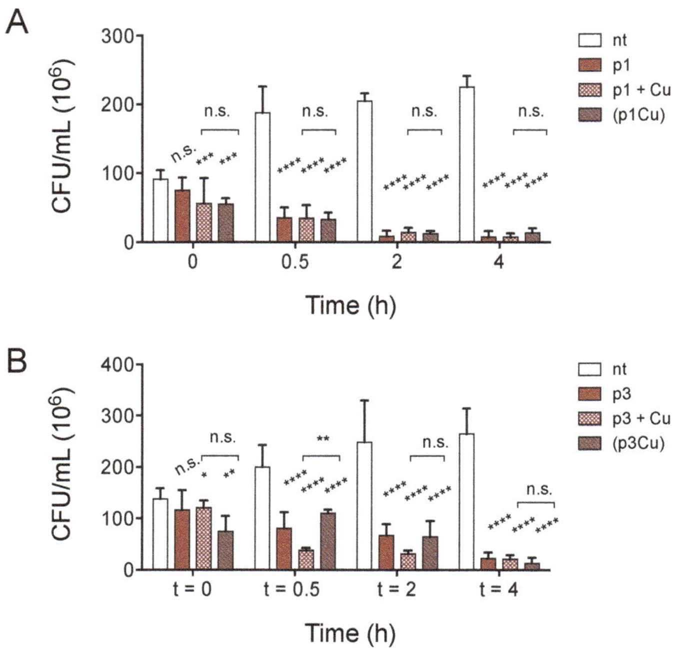

Figure 3. Copper does not accelerate anaerobic *C. difficile* killing by piscidins. Time-kill assays comparing viable colony forming units per milliliter (CFU/mL) of bacterial culture before exposure to p1 (**A**) and p3 (**B**) with the CFU/mL 30 min, 2, and 4 h post-exposure. Cells were exposed to peptides (p1 and p3), peptides and equimolar copper sulfate added simultaneously (p1 + Cu and p3 + Cu), and peptides allowed to form piscidin-copper complexes in an aerobic environment prior to addition to the anaerobic bacterial cultures ((p1Cu) and (p3Cu)). CFU/mL in treated samples were compared to those in untreated samples and to each other using two-way ANOVA with Tukey's post-test comparison. nt, not treated; n.s., not significant; * $p < 0.05$; ** $p < 0.01$; *** $p < 0.001$; **** $p < 0.0001$.

2.4. Copper Is Toxic to Anaerobically Growing C. difficile

It has been previously suggested the antimicrobial synergy of ATCUN-containing HDPs depends upon environmental dioxygen, which could account for the inability of copper ions to enhance piscidin activity against *C. difficile* [40,70,71]. To define the effects of copper alone against *C. difficile*, we determined that the copper concentration of the TY growth medium used to culture *C. difficile* under anaerobic conditions (Figure 4A) is 1.46 µM, which is lower than that of the Mueller-Hinton broth previously used to culture *E. coli* and *Pseudomonas aeruginosa* under aerobic conditions [37,72]. We found that copper salts are capable of inhibiting *C. difficile* growth at the same concentrations used in the time-kill assays with piscidins. The epidemic strain R20291 is less susceptible to copper inhibition that the historical strain 630Δ*erm* (Figure 4B). Actively dividing R20291 cells in the exponential phase are also killed by exposure to micromolar concentrations of copper sulfate (Figure 4C).

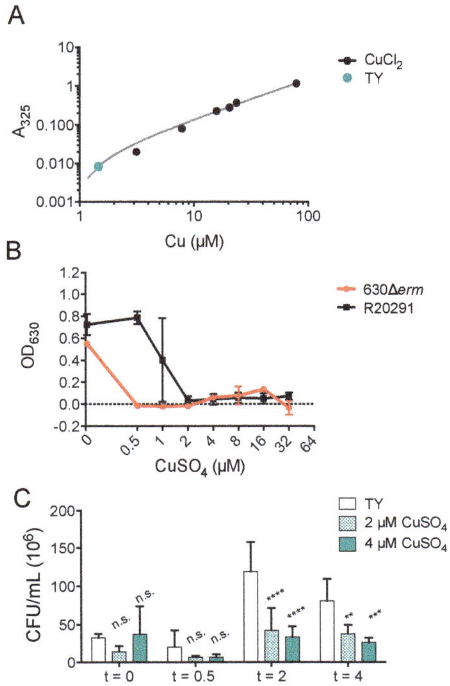

Figure 4. Copper is still antimicrobial in anaerobic environments. (**A**) Copper concentration of TY medium. (**B**) Optical densities of overnight *C. difficile* 630Δ*erm* and R20291 cultures grown in the presence of the indicated concentrations of copper sulfate. Data shown are the means and standard deviations of four biologically independent samples. (**C**) Time-kill assays comparing viable colony forming units per milliliter (CFU/mL) of bacterial culture before exposure to the indicated concentrations of copper sulfate with the CFU/mL 30 min, 2, and 4 h post-exposure. CFU/mL in treated samples were compared to those in untreated samples and to each other using two-way ANOVA with Tukey's post-test comparison. nt, not treated; n.s., not significant; ** $p < 0.01$; *** $p < 0.001$; **** $p < 0.0001$.

3. Discussion

The development of new therapies to combat antibiotic-resistant infections, including *C. difficile* infection, is an urgent public health priority. Antimicrobial peptides capable of simultaneously killing bacterial pathogens and stimulating the host innate system are a promising avenue for the development of new antimicrobial therapies. Piscidins previously showed efficacy against Gram-negative and -positive bacteria in aerobic environments [37]. Here, we investigated their effect on anaerobic bacteria. While piscidins are still highly lethal against *C. difficile* growing anaerobically, we found that piscidin growth inhibition and killing against *C. difficile* were not enhanced by the addition of Cu^{2+}. This is true despite the fact that copper alone is capable of inhibiting *C. difficile* growth and killing actively growing *C. difficile* cells. While in aerobic environments, the bacterial response to piscidins with Cu^{2+} is greater than that to either piscidins or copper alone, it appears that the anaerobic response to piscidin with copper is less than the sum of the parts, and thus does not feature the synergistic effects observed aerobically.

Bacteria employ general stress response pathways that can be activated by multiple diverse extracellular stresses. Bacterial cells that have previously been exposed to an extracellular stress, such as starvation or oxidative stress, demonstrate increased resilience against unrelated threats, such as antibiotic exposure [73,74]. The fact that the *C. difficile* strains 630Δ*erm* and R20291 exhibit identical inhibition in response to p1 and p3, but differential inhibition to copper alone, suggests that piscidins and copper inhibit bacterial growth and viability by different mechanisms. This makes sense because Cu^{2+} has no specificity while piscidin does. It should be noted that free Cu^{2+} in vivo is highly toxic to mammalian as well as bacterial cells, and complexation by HDPs such as p1 and p3 can provide the critical specificity of targeting the metal ions to bacterial rather than host cells. HDPs are a viable treatment option against both aerobic and anaerobic bacteria, and in the case of *C. difficile*, it is extremely encouraging that the epidemic strain R20291 is just as susceptible to HDP inhibition and killing as the less robust 630Δ*erm* strain. As the symptoms of *C. difficile* infection are largely inflammatory, and treatments based on piscidins could potentially reduce inflammation while killing the causative pathogen, this is a very promising strategy to pursue. However, it is clear that while clinical antibiotics derived from HDPs would benefit from the inclusion of copper in tissues with high levels of oxygen, such as the lungs, antibiotics targeted to less aerobic tissues, such as the kidneys or large intestine, may not. Future work to investigate the interaction of piscidins with *C. difficile* in animal models of infection will be necessary to determine whether copper could or should be included with the peptides. More broadly, it appears that clinical treatments developed from HDPs should be designed in a tissue-specific manner, as metal ion adjuvants may be beneficial or necessary in some organs and unneeded in others, based on the oxygen levels at the site of activity.

4. Materials and Methods

4.1. Materials, Chemicals, Bacterial Strains and Growth Conditions

Materials and chemicals were purchased from Fisher Scientific (Hampton, NH, USA) unless otherwise indicated. The bacterial strains used in this study are listed in Table S1. *C. difficile* 630Δ*erm* and R20291 were maintained on brain-heart infusion supplemented with 5% yeast extract (BHIS) agar plates and liquid cultures were grown in TY medium [16,75,76]. All anaerobic bacterial culture took place at 37 °C in a Coy anaerobic chamber (Coy Laboratory Products, Grass Lake, MI) with an atmosphere of 85% N_2, 10% CO_2, 5% H_2. All plastic consumables were allowed to equilibrate in the anaerobic chamber for a minimum of 72 h.

4.2. Peptide Synthesis

Caboxyamidated p1 (MW 2571) and p3 (MW 2492) were synthesized using Fmoc chemistry at the University of Texas Southwestern Medical Center (Dallas, TX, USA). For the TAMRA-labeled forms of the peptides, the fluorescent label was attached to the amino-end of the peptides before cleavage from the resin. The peptides were purified at William and Mary on a Waters HPLC system using a C18 X-Bridge Waters column (Milford, MA, USA) and acetonitrile/water gradient acidified with 0.1% trifluoroacetic acid, as previously described [77,78]. After removal of the organic phase, the peptides were lyophilized. Next, they were dissolved in dilute HCl and dialyzed to remove residual trifluoroacetic acid. The purification steps yielded 98% pure peptides based on HPLC chromatograms and mass spectrometry data. HPLC chromatograms and mass spectra collected at William and Mary on the purified peptides are included in the Supplementary Information (Figures S1 and S2). The purified peptides were dissolved in nanopure water. The concentrations of p1 and p3 were determined by amino acid analysis at the Texas A&M Protein Chemistry Center (College Station, TX, USA). Metallation of each peptide was achieved when an equimolar of $CuSO_4$ was added to the medium (see below). For the TAMRA-labeled peptides, the absorbance at 547 nm was used to quantify their concentrations. To avoid photobleaching of the fluorescent probe, the TAMRA-labeled peptides were protected from light by wrapping containers with foil.

4.3. Microscopy

Live-cell, time-lapse, wide-field fluorescence, and differential interference contrast (DC) microscopy of the interaction between TAMRA-labeled piscidin peptides and *C. difficile* R20291 bacteria was performed on a Nikon Ti-E inverted microscope equipped with apochromatic TIRF 60X oil immersion objective lens

(N.A. 1.49), pco.edge 4.2 LT sCMOS camera, and SOLA SE II 365 Light Engine as well as complementary DIC components (Nikon Instruments Inc, Melville, NY, USA). Mid-logarithmic phase cells and peptides at the indicated concentration were mixed inside the anaerobic chamber and injected into home-built anaerobic rose-type imaging chambers as previously described [63]. Imaging chambers were removed from the anaerobic chamber and placed on the microscope. The microscope was maintained at 37 °C using a home-built enclosure and a Nevtek Air Stream microscope stage warmer (Nevtek, Williamsville, VA, USA). Nikon Perfect Focus system (Nikon Instruments Inc, Melville, NY, USA) was employed to eliminate focal drift during recordings. Movies consisting of a fluorescence and DIC image each minute for 60 min were then recorded for each condition. Movies started 6–8 min after the bacteria and peptide were mixed. Data analysis was performed using the Nikon Elements imaging suite. During the recordings the amount of fluorescence background increased with time, presumably as peptide was deposited on the coverslip surface. During analysis this background change was corrected using background-leveling tools, then a second rolling-ball type background correction was used to remove imaging artifacts. A thresholded binary mask was then applied to the fluorescence images to isolate and count each fluorescent object (peptide-labeled bacteria) in the movie. Fluorescence levels of objects were monitored as a function of time.

4.4. Growth Inhibition Assays

Two-fold dilution series of TY medium containing the indicated concentrations of peptide and/or copper salts were prepared in sterile 96-well plates as detailed in Wiegand et al. [79]. Wells containing 200 µL of medium were inoculated with 20 µL of saturated overnight culture of *C. difficile* 630Δ*erm* or R20291 containing approximately 10^8 CFU/mL and incubated anaerobically for 16 h at 37 °C. Closed microplates were removed from the anaerobic chamber and the outsides of the plates were disinfected with 10% bleach before examination to determine the minimum concentration of each peptide and/or metal ion sufficient to complete inhibit visible growth. Culture density at 630 nm was measured in a BioTek (Winooski, VT, USA) microplate reader. As removal from the anaerobic chamber killed the anaerobic *C. difficile* bacteria, we were not able to plate samples to determine CFU/mL after spectroscopic measurements. Inhibitory concentrations were reported as the peptide concentration necessary to reduce the overnight OD_{630} by at least 50% from that of untreated samples. Data reported are the means and standard deviations of four biologically independent samples.

4.5. Time-Kill Assays

3 mL of TY media were inoculated with single colonies of *C. difficile* R20291 and allowed to grow at 37 °C to an optical density at 600 nm (OD600) of 0.5–0.7. At the onset of the experiment 20 µL aliquots were removed from the exponentially growing culture and inoculated into fresh TY medium containing the indicated concentration of peptide and/or copper sulfate ($CuSO_4$). The final volume was adjusted to 1 mL with fresh TY medium. After 0, 0.5, 2, and 4 h of incubation at 37 °C, 10 µL aliquots were removed for serial 10-fold dilution in TY. 10^6 dilutions were plated in duplicate on BHIS agar plates for colony enumeration. Colony forming units (CFU) were counted after 24 h. Data reported are the averages of three biologically independent samples measured in duplicate. Treated samples were compared to untreated samples and to each other by two-way ANOVA using Tukey's multiple comparison test with Prism (GraphPad Software, San Diego, CA, USA).

4.6. Atomic Absorption Spectroscopy

The copper concentration in TY medium was measured using an AA-7000 atomic absorption spectrophotometer (Shimadzu Scientific Instruments, Columbia, MD, USA) with a hollow cathode lamp using an acetylene flame [80]. Copper from TY medium was detected at 324.8 nm and quantified using a standard curve of copper chloride ($CuCl_2$) diluted in water.

Supplementary Materials: Supplementary materials can be found at http://www.mdpi.com/1422-0067/20/21/5289/s1.

Author Contributions: Conceptualization, M.L.C. and E.B.P.; Methodology, M.L.C. and E.B.P.; Validation, D.S.C., E.B.P., and M.L.C.; Formal Analysis, E.B.P. and M.L.C.; Investigation, A.O., A.R.R., J.C.P., M.D.S., M.L.C., and D.S.C.; Resources, E.B.P. and M.L.C.; Data Curation, E.B.P., M.L.C., and J.C.P.; Writing—Original Draft Preparation, A.O. and E.B.P.; Writing—Review and Editing, M.L.C. and E.B.P.; Visualization, D.S.C., M.L.C., and E.B.P.; Supervision, M.L.C. and E.B.P.; Project Administration, M.L.C. and E.B.P.; Funding Acquisition, M.L.C. and E.B.P.

Funding: This research was funded by the National Institutes of Health (NIGMS 1R15GM126527-01A1 and NIAID 1K22AI118929-01).

Acknowledgments: This work was funded by NIGMS 1R15GM126527-01A1 to Myriam L. Cotten and NIAID 1K22AI118929-01 to Erin B. Purcell, Malia D. Stuart was supported by NSF REU CHE-1659476. The authors thank Kory Castro and Watson Stahl for assistance with the atomic absorption spectrophotometer and members of the Cotten lab for assisting with the purification of the peptides.

Conflicts of Interest: The authors declare no conflict of interest.

References

1. Miller, M. Fidaxomicin (OPT-80) for the treatment of *Clostridium difficile* infection. *Expert Opin. Pharm.* **2010**, *11*, 1569–1578. [CrossRef] [PubMed]
2. Centers for Disease Control and Prevention. Nearly Half a Million Americans Suffered from *Clostridium difficile* Infections in a Single Year. Available online: https://www.cdc.gov/media/releases/2015/p0225-clostridium-difficile.html (accessed on 10 October 2019).
3. Denève, C.; Janoir, C.; Poilane, I.; Fantinato, C.; Collignon, A. New trends in *Clostridium difficile* virulence and pathogenesis. *Int. J. Antimicrob. Agents* **2009**, *33*, S24–S28. [CrossRef] [PubMed]
4. Gupta, A.; Khanna, S. Community-acquired *Clostridium difficile* infection: An increasing public health threat. *Infect Drug Resist.* **2014**, *7*, 63–72. [CrossRef] [PubMed]
5. Loo, V.G.; Poirier, L.; Miller, M.A.; Oughton, M.; Libman, M.D.; Michaud, S.; Bourgault, A.M.; Nguyen, T.; Frenette, C.; Kelly, M.; et al. A predominantly clonal multi-institutional outbreak of *Clostridium difficile*-associated diarrhea with high morbidity and mortality. *N. Engl. J. Med.* **2005**, *353*, 2442–2449. [CrossRef]
6. Bartlett, J.G. *Clostridium difficile*: Progress and challenges. *Ann. N. Y. Acad. Sci.* **2010**, *1213*, 62–69. [CrossRef]
7. Smits, W.K. Hype or hypervirulence: A reflection on problematic *C. difficile* strains. *Virulence* **2013**, *4*, 592–596. [CrossRef]
8. Louie, T.J.; Miller, M.A.; Mullane, K.M.; Weiss, K.; Lentnek, A.; Golan, Y.; Gorbach, S.; Sears, P.; Shue, Y.K. Fidaxomicin versus vancomycin for *Clostridium difficile* infection. *N. Engl. J. Med.* **2011**, *364*, 422–431. [CrossRef]
9. Drekonja, D.M.; Butler, M.; MacDonald, R.; Bliss, D.; Filice, G.A.; Rector, T.S.; Wilt, T.J. Comparative effectiveness of *Clostridium difficile* treatments: A systematic review. *Ann. Intern. Med.* **2011**, *155*, 839–847. [CrossRef]
10. Boyle, M.L.; Ruth-Sahd, L.A.; Zhou, Z. Fecal microbiota transplant to treat recurrent *Clostridium difficile* infections. *Crit. Care Nurse* **2015**, *35*, 51–64. [CrossRef]
11. McDonald, L.C.; Gerding, D.N.; Johnson, S.; Bakken, J.S.; Carroll, K.C.; Coffin, S.E.; Dubberke, E.R.; Garey, K.W.; Gould, C.V.; Kelly, C.; et al. Clinical practice guidelines for *Clostridium difficile* infection in adults and children: 2017 Update by the Infectious Diseases Society of America (IDSA) and Society for Healthcare Epidemiology of America (SHEA). *Clin. Infect. Dis.* **2018**, *66*, 987–994. [CrossRef]
12. Van Nood, E.; Vrieze, A.; Nieuwdorp, M.; Fuentes, S.; Zoetendal, E.G.; de Vos, W.M.; Visser, C.E.; Kuijper, E.J.; Bartelsman, J.F.; Tijssen, J.G.; et al. Duodenal infusion of donor feces for recurrent *Clostridium difficile*. *N. Engl. J. Med.* **2013**, *368*, 407–415. [CrossRef] [PubMed]
13. Cammarota, G.; Ianiro, G.; Gasbarrini, A. Fecal microbiota transplantation for the treatment of *Clostridium difficile* infection: A systematic review. *J. Clin. Gastroenterol.* **2014**, *48*, 693–702. [CrossRef] [PubMed]
14. Rodriguez-Palacios, A.; Lejeune, J.T. Moist-heat resistance, spore aging, and superdormancy in *Clostridium difficile*. *Appl. Env. Microbiol.* **2011**, *77*, 3085–3091. [CrossRef]

15. Edwards, A.N.; Karim, S.T.; Pascual, R.A.; Jowhar, L.M.; Anderson, S.E.; McBride, S.M. Chemical and stress resistances of *Clostridium difficile* spores and vegetative cells. *Front. Microbiol.* **2016**, *7*, 1698. [CrossRef] [PubMed]
16. Sorg, J.A.; Sonenshein, A.L. Bile salts and glycine as cogerminants for *Clostridium difficile* spores. *J. Bacteriol.* **2008**, *190*, 2505–2512. [CrossRef]
17. Howerton, A.; Ramirez, N.; Abel-Santos, E. Mapping interactions between germinants and *Clostridium difficile* spores. *J. Bacteriol.* **2011**, *193*, 274–282. [CrossRef]
18. Sarker, M.R.; Paredes-Sabja, D. Molecular basis of early stages of *Clostridium difficile* infection: Germination and colonization. *Future Microbiol.* **2012**, *7*, 933–943. [CrossRef]
19. Cochetière, M.-F.; Montassier, E.; Hardouin, J.-B.; Carton, T.; Vacon, F.; Durand, T.; Lalande, V.; Petit, J.; Potel, G.; Beaugerie, L. Human Intestinal Microbiota Gene Risk Factors for Antibiotic-Associated Diarrhea: Perspectives for Prevention. *Microb. Ecol.* **2010**, *59*, 830–837. [CrossRef]
20. Manges, A.R.; Labbe, A.; Loo, V.G.; Atherton, J.K.; Behr, M.A.; Masson, L.; Tellis, P.A.; Brousseau, R. Comparative metagenomic study of alterations to the intestinal microbiota and risk of nosocomial *Clostridum difficile*-associated disease. *J. Infect. Dis.* **2010**, *202*, 1877–1884. [CrossRef]
21. Ng, K.M.; Ferreyra, J.A.; Higginbottom, S.K.; Lynch, J.B.; Kashyap, P.C.; Gopinath, S.; Naidu, N.; Choudhury, B.; Weimer, B.C.; Monack, D.M.; et al. Microbiota-liberated host sugars facilitate post-antibiotic expansion of enteric pathogens. *Nature* **2013**, *502*, 96–99. [CrossRef]
22. Theriot, C.M.; Koenigsknecht, M.J.; Carlson, P.E.; Hatton, G.E.; Nelson, A.M.; Li, B.; Huffnagle, G.B.; Li, J.Z.; Young, V.B. Antibiotic-induced shifts in the mouse gut microbiome and metabolome increase susceptibility to *Clostridium difficile* infection. *Nat. Commun.* **2014**, *5*, 3114. [CrossRef] [PubMed]
23. Frädrich, C.; Beer, L.-A.; Gerhard, R. Reactive oxygen species as additional determinants for cytotoxicity of *Clostridium difficile* Toxins A and B. *Toxins* **2016**, *8*, 25. [CrossRef] [PubMed]
24. Gudmundsson, G.H.; Agerberth, B. Neutrophil antibacterial peptides, multifunctional effector molecules in the mammalian immune system. *J. Immunol. Methods* **1999**, *232*, 45–54. [CrossRef]
25. McBride, S.M.; Sonenshein, A.L. The dlt operon confers resistance to cationic antimicrobial peptides in *Clostridium difficile*. *Microbiology* **2011**, *157*, 1457–1465. [CrossRef] [PubMed]
26. Hancock, R.E.; Sahl, H.G. Antimicrobial and host-defense peptides as new anti-infective therapeutic strategies. *Nat. Biotechnol.* **2006**, *24*, 1551–1557. [CrossRef]
27. Nicolas, P.; Rosenstein, Y. Multifunctional host defense peptides. *Febs. J.* **2009**, *276*, 6464. [CrossRef]
28. Bahar, A.A.; Ren, D. Antimicrobial peptides. *Pharmaceuticals (Basel)* **2013**, *6*, 1543–1575. [CrossRef]
29. Yount, N.Y.; Yeaman, M.R. Immunocontinuum: Perspectives in antimicrobial peptide mechanisms of action and resistance. *Protein Pept. Lett.* **2005**, *12*, 49–67. [CrossRef]
30. Wong, C.C.; Zhang, L.; Ren, S.X.; Shen, J.; Chan, R.L.; Cho, C.H. Antibacterial peptides and gastrointestinal diseases. *Curr. Pharm. Des.* **2011**, *17*, 1583–1586. [CrossRef]
31. Nuding, S.; Frasch, T.; Schaller, M.; Stange, E.F.; Zabel, L.T. Synergistic effects of antimicrobial peptides and antibiotics against *Clostridium difficile*. *Antimicrob. Agents Chemother.* **2014**, *58*, 5719–5725. [CrossRef]
32. Lauth, X.; Shike, H.; Burns, J.C.; Westerman, M.E.; Ostland, V.E.; Carlberg, J.M.; Van Olst, J.C.; Nizet, V.; Taylor, S.W.; Shimizu, C.; et al. Discovery and characterization of two isoforms of moronecidin, a novel antimicrobial peptide from hybrid striped bass. *J. Biol. Chem.* **2002**, *277*, 5030–5039. [CrossRef]
33. Lee, S.A.; Kim, Y.K.; Lim, S.S.; Zhu, W.L.; Ko, H.; Shin, S.Y.; Hahm, K.S.; Kim, Y. Solution structure and cell selectivity of piscidin 1 and its analogues. *Biochemistry* **2007**, *46*, 3653–3663. [CrossRef]
34. Wang, G.; Watson, K.M.; Peterkofsky, A.; Buckheit, R.W. Identification of novel human immunodeficiency virus type 1-inhibitory peptides based on the antimicrobial peptide database. *Antimicrob. Agents Chemother.* **2010**, *54*, 1343–1346. [CrossRef]
35. Chen, W.; Cotten, M.L. Expression, purification, and micelle reconstitution of antimicrobial piscidin 1 and piscidin 3 for NMR studies. *Protein Expr. Purif.* **2014**, *102*, 63–68. [CrossRef]
36. Hayden, R.M.; Goldberg, G.K.; Ferguson, B.M.; Schoeneck, M.W.; Libardo, M.D.; Mayeux, S.E.; Shrestha, A.; Bogardus, K.A.; Hammer, J.; Pryshchep, S.; et al. Complementary effects of host defense peptides Piscidin 1 and Piscidin 3 on DNA and lipid membranes: Biophysical insights into contrasting biological activities. *J. Phys. Chem. B* **2015**, *119*, 15235–15246. [CrossRef]

37. Libardo, M.D.J.; Bahar, A.A.; Ma, B.; Fu, R.; McCormick, L.E.; Zhao, J.; McCallum, S.A.; Nussinov, R.; Ren, D.; Angeles-Boza, A.M.; et al. Nuclease activity gives an edge to host-defense peptide piscidin 3 over piscidin 1, rendering it more effective against persisters and biofilms. *Febs. J.* **2017**, *284*, 3662–3683. [CrossRef]
38. Comert, F.; Greenwood, A.; Maramba, J.; Acevedo, R.; Lucas, L.; Kulasinghe, T.; Cairns, L.S.; Wen, Y.; Fu, R.; Hammer, J.; et al. The host-defense peptide piscidin P1 reorganizes lipid domains in membranes and alters activation energies in mechanosensitive ion channels. *J. Biol. Chem.* **2019**, in press. [CrossRef]
39. Kim, S.Y.; Zhang, F.; Gong, W.; Chen, K.; Xia, K.; Liu, F.; Gross, R.; Wang, J.M.; Linhardt, R.J.; Cotten, M.L. Copper regulates the interactions of antimicrobial piscidin peptides from fish mast cells with formyl peptide receptors and heparin. *J. Biol. Chem.* **2018**, *293*, 15381–15396. [CrossRef]
40. Joyner, J.C.; Reichfield, J.; Cowan, J.A. Factors influencing the DNA nuclease activity of iron, cobalt, nickel, and copper chelates. *J. Am. Chem. Soc.* **2011**, *133*, 15613–15626. [CrossRef]
41. Joyner, J.C.; Hodnick, W.F.; Cowan, A.S.; Tamuly, D.; Boyd, R.; Cowan, J.A. Antimicrobial metallopeptides with broad nuclease and ribonuclease Activity. *Chem. Commun.* **2013**, *49*, 2118–2120. [CrossRef]
42. Libardo, M.D.; Nagella, S.; Lugo, A.; Pierce, S.; Angeles-Boza, A.M. Copper-binding tripeptide motif increases potency of the antimicrobial peptide Anoplin via Reactive Oxygen Species generation. *Biochem. Biophys. Res. Commun.* **2015**, *456*, 446–451. [CrossRef]
43. Hillmann, F.; Fischer, R.J.; Saint-Prix, F.; Girbal, L.; Bahl, H. PerR acts as a switch for oxygen tolerance in the strict anaerobe *Clostridium acetobutylicum*. *Mol. Microbiol.* **2008**, *68*, 848–860. [CrossRef]
44. Riebe, O.; Fischer, R.J.; Wampler, D.A.; Kurtz, D.M.; Bahl, H. Pathway for H2O2 and O2 detoxification in *Clostridium acetobutylicum*. *Microbiology* **2009**, *155*, 16–24. [CrossRef]
45. Zhang, L.; Nie, X.; Ravcheev, D.A.; Rodionov, D.A.; Sheng, J.; Gu, Y.; Yang, S.; Jiang, W.; Yang, C. Redox-responsive repressor Rex modulates alcohol production and oxidative stress tolerance in *Clostridium acetobutylicum*. *J. Bacteriol.* **2014**, *196*, 3949–3963. [CrossRef]
46. McQuade, R.; Roxas, B.; Viswanathan, V.K.; Vedantam, G. *Clostridium difficile* clinical isolates exhibit variable susceptibility and proteome alterations upon exposure to mammalian cationic antimicrobial peptides. *Anaerobe* **2012**, *18*, 614–620. [CrossRef]
47. Corminboeuf, O.; Leroy, X. FPR2/ALXR agonists and the resolution of inflammation. *J. Med. Chem.* **2015**, *58*, 537–559. [CrossRef]
48. Le, Y.; Murphy, P.M.; Wang, J.M. Formyl-peptide receptors revisited. *Trends Immunol.* **2002**, *23*, 541–548. [CrossRef]
49. Migeotte, I.; Communi, D.; Parmentier, M. Formyl peptide receptors: A promiscuous subfamily of G protein-coupled receptors controlling immune responses. *Cytokine Growth Factor Rev.* **2006**, *17*, 501–519. [CrossRef]
50. Pundir, P.; Catalli, A.; Leggiadro, C.; Douglas, S.E.; Kulka, M. Pleurocidin, a novel antimicrobial peptide, induces human mast cell activation through the FPRL1 receptor. *Mucosal. Immunol.* **2014**, *7*, 177–187. [CrossRef]
51. Park, Y.J.; Lee, S.K.; Jung, Y.S.; Lee, M.; Lee, H.Y.; Kim, S.D.; Park, J.S.; Koo, J.; Hwang, J.S.; Bae, Y.S. Promotion of formyl peptide receptor 1-mediated neutrophil chemotactic migration by antimicrobial peptides isolated from the centipede *Scolopendra subspinipes mutilans*. *BMB Rep.* **2016**, *49*, 520–525. [CrossRef]
52. Chen, W.F.; Huang, S.Y.; Liao, C.Y.; Sung, C.S.; Chen, J.Y.; Wen, Z.H. The use of the antimicrobial peptide piscidin (PCD)-1 as a novel anti-nociceptive agent. *Biomaterials* **2015**, *53*, 1–11. [CrossRef]
53. Lee, E.; Shin, A.; Jeong, K.W.; Jin, B.; Jnawali, H.N.; Shin, S.; Shin, S.Y.; Kim, Y. Role of phenylalanine and valine10 residues in the antimicrobial activity and cytotoxicity of piscidin-1. *PLoS ONE* **2014**, *9*, e114453. [CrossRef]
54. Mansour, S.C.; de la Fuente-Nunez, C.; Hancock, R.E. Peptide IDR-1018: Modulating the immune system and targeting bacterial biofilms to treat antibiotic-resistant bacterial infections. *J. Pept. Sci.* **2015**, *21*, 323–329. [CrossRef]
55. Hilchie, A.L.; Wuerth, K.; Hancock, R.E. Immune modulation by multifaceted cationic host defense (antimicrobial) peptides. *Nat. Chem. Biol.* **2013**, *9*, 761–768. [CrossRef]
56. Fjell, C.D.; Hiss, J.A.; Hancock, R.E.; Schneider, G. Designing antimicrobial peptides: Form follows function. *Nat. Rev. Drug Discov.* **2011**, *11*, 37–51. [CrossRef]
57. Yeaman, M.R.; Yount, N.Y. Unifying themes in host defence effector polypeptides. *Nat. Rev. Microbiol.* **2007**, *5*, 727–740. [CrossRef]

58. Fox, J.L. Antimicrobial peptides stage a comeback. *Nat. Biotechnol.* **2013**, *31*, 379–382. [CrossRef]
59. Hale, J.D.; Hancock, R.E. Alternative mechanisms of action of cationic antimicrobial peptides on bacteria. *Expert Rev. Anti. Infect* **2007**, *5*, 951–959. [CrossRef]
60. Hing, T.C.; Ho, S.; Shih, D.Q.; Ichikawa, R.; Cheng, M.; Chen, J.; Chen, X.; Law, I.; Najarian, R.; Kelly, C.P.; et al. The antimicrobial peptide cathelicidin modulates *Clostridium difficile*-associated colitis and toxin A-mediated enteritis in mice. *Gut* **2013**, *62*, 1295–1305. [CrossRef]
61. Carreau, A.; El Hafny-Rahbi, B.; Matejuk, A.; Grillon, C.; Kieda, C. Why is the partial oxygen pressure of human tissues a crucial parameter? Small molecules and hypoxia. *J. Cell Mol. Med.* **2011**, *15*, 1239–1253. [CrossRef]
62. Buckley, A.M.; Jukes, C.; Candlish, D.; Irvine, J.J.; Spencer, J.; Fagan, R.P.; Roe, A.J.; Christie, J.M.; Fairweather, N.F.; Douce, G.R. Lighting up *Clostridium difficile*: Reporting gene expression using fluorescent LOV domains. *Sci. Rep.* **2016**, *6*, 23463. [CrossRef]
63. Courson, D.S.; Pokhrel, A.; Scott, C.; Madrill, M.; Rinehold, A.J.; Tamayo, R.; Cheney, R.E.; Purcell, E.B. Single cell analysis of nutrient regulation of *Clostridioides (Clostridium) difficile* motility. *Anaerobe* **2019**, *59*, 205–211. [CrossRef]
64. Hussain, H.A.; Roberts, A.P.; Mullany, P. Generation of an erythromycin-sensitive derivative of *Clostridium difficile* strain 630 (630Deltaerm) and demonstration that the conjugative transposon Tn916DeltaE enters the genome of this strain at multiple sites. *J. Med. Microbiol.* **2005**, *54*, 137–141. [CrossRef]
65. Stabler, R.A.; He, M.; Dawson, L.; Martin, M.; Valiente, E.; Corton, C.; Lawley, T.D.; Sebaihia, M.; Quail, M.A.; Rose, G.; et al. Comparative genome and phenotypic analysis of *Clostridium difficile* 027 strains provides insight into the evolution of a hypervirulent bacterium. *Genome Biol.* **2009**, *10*, R102. [CrossRef]
66. Lachowicz, D.; Pituch, H.; Obuch-Woszczatyński, P. Antimicrobial susceptibility patterns of *Clostridium difficile* strains belonging to different polymerase chain reaction ribotypes isolated in Poland in 2012. *Anaerobe* **2015**, *31*, 37–41. [CrossRef]
67. Kuijper, E.J.; Coignard, B.; Tull, P. Emergence of *Clostridium difficile*-associated disease in North America and Europe. *Clin. Microbiol. Infect.* **2006**, *12*, 2–18. [CrossRef]
68. Drudy, D.; Kyne, L.; O'Mahony, R.; Fanning, S. gyrA Mutations in Fluoroquinolone-resistant *Clostridium difficile* PCR-027. *Emerg. Infect. Dis.* **2007**, *13*, 504–505. [CrossRef]
69. Banawas, S.S. *Clostridium difficile* infections: A global overview of drug sensitivity and resistance mechanisms. *Biomed. Res. Int.* **2018**, *2018*, 9. [CrossRef]
70. Libardo, M.D.; Cervantes, J.L.; Salazar, J.C.; Angeles-Boza, A.M. Improved bioactivity of antimicrobial peptides by addition of amino-terminal copper and nickel (ATCUN) binding motifs. *Chem. Med. Chem.* **2014**, *9*, 1892–1901. [CrossRef]
71. Jin, Y.; Lewis, M.A.; Gokhale, N.H.; Long, E.C.; Cowan, J.A. Influence of stereochemistry and redox potentials on the single- and double-strand DNA cleavage efficiency of Cu(II) and Ni(II) Lys-Gly-His-derived ATCUN metallopeptides. *J. Am. Chem. Soc.* **2007**, *129*, 8353–8361. [CrossRef]
72. Fernandez-Mazarrasa, C.; Mazarrasa, O.; Calvo, J.; del Arco, A.; Martinez-Martinez, L. High concentrations of manganese in Mueller-Hinton agar increase MICs of tigecycline determined by Etest. *J. Clin. Microbiol.* **2009**, *47*, 827–829. [CrossRef]
73. Manteca, A.; Alvarez, R.; Salazar, N.; Yague, P.; Sanchez, J. Mycelium differentiation and antibiotic production in submerged cultures of *Streptomyces coelicolor*. *Appl. Env. Microbiol.* **2008**, *74*, 3877–3886. [CrossRef]
74. Poole, K. Bacterial stress responses as determinants of antimicrobial resistance. *J. Antimicrob. Chemother.* **2012**, *67*, 2069–2089. [CrossRef]
75. Sorg, J.A.; Dineen, S.S. Laboratory maintenance of *Clostridium difficile*. *Curr. Protoc. Microbiol.* **2009**, *12*, 1–10. [CrossRef]
76. Purcell, E.B.; McKee, R.W.; McBride, S.M.; Waters, C.M.; Tamayo, R. Cyclic diguanylate inversely regulates motility and aggregation in *Clostridium difficile*. *J. Bact.* **2012**, *194*, 3307–3316. [CrossRef]
77. Chekmenev, E.Y.; Vollmar, B.S.; Forseth, K.T.; Manion, M.N.; Jones, S.M.; Wagner, T.J.; Endicott, R.M.; Kyriss, B.P.; Homem, L.M.; Pate, M.; et al. Investigating molecular recognition and biological function at interfaces using piscidins, antimicrobial peptides from fish. *Biochim. Biophys. Acta* **2006**, *1758*, 1359–1372. [CrossRef]
78. Perrin, B.S.; Tian, Y.; Fu, R.; Grant, C.V.; Chekmenev, E.Y.; Wieczorek, W.E.; Dao, A.E.; Hayden, R.M.; Burzynski, C.M.; Venable, R.M.; et al. High-resolution structures and orientations of antimicrobial peptides

piscidin 1 and piscidin 3 in fluid bilayers reveal tilting, kinking, and bilayer immersion. *J. Am. Chem. Soc.* **2014**, *136*, 3491–3504. [CrossRef]
79. Wiegand, I.; Hilpert, K.; Hancock, R.E. Agar and broth dilution methods to determine the minimal inhibitory concentration (MIC) of antimicrobial substances. *Nat. Protoc.* **2008**, *3*, 163–175. [CrossRef]
80. Ghaedi, M.; Ahmadi, F.; Shokrollahi, A. Simultaneous preconcentration and determination of copper, nickel, cobalt and lead ions content by flame atomic absorption spectrometry. *J. Hazard. Mater.* **2007**, *142*, 272–278. [CrossRef]

International Journal of *Molecular Sciences*

Article

The Synthetic Dipeptide Pidotimod Shows a Chemokine-Like Activity through CXC Chemokine Receptor 3 (CXCR3)

Francesca Caccuri [1], Antonella Bugatti [1], Silvia Corbellini [2], Sara Roversi [1], Alberto Zani [1], Pietro Mazzuca [1], Stefania Marsico [3], Arnaldo Caruso [1] and Cinzia Giagulli [1,*]

1 Section of Microbiology, Department of Molecular and Translational Medicine, University of Brescia, 25123 Brescia, Italy; francesca.caccuri@unibs.it (F.C.); antonella.bugatti@unibs.it (A.B.); s.roversi003@unibs.it (S.R.); a.zani033@unibs.it (A.Z.); p.mazzuca@unibs.it (P.M.); arnaldo.caruso@unibs.it (A.C.)

2 Laboratory of Microbiology and Virology, Azienda Socio Sanitaria Territoriale Spedali Civili, 25123 Brescia, Italy; silvia.corbellini@gmail.com

3 Department of Pharmacy, Health and Nutritional Sciences, University of Calabria, Arcavacata di Rende, 87036 Cosenza, Italy; stefania.marsico@unical.it

* Correspondence: cinzia.giagulli@unibs.it; Tel.: +39-030-399-6698; Fax: +39-030-395-258

Received: 9 October 2019; Accepted: 22 October 2019; Published: 24 October 2019

Abstract: In recent years immunomodulators have gained a strong interest and represent nowadays an active expanding area of research for the control of microbial diseases and for their therapeutic potential in preventing, treating and reducing the morbidity and mortality of different diseases. Pidotimod (3-L-pyroglutamyl-L-thiaziolidine-4carboxylic acid, PDT) is a synthetic dipeptide, which possesses immunomodulatory properties and exerts a well-defined pharmacological activity against infections, but its real mechanism of action is still undefined. Here, we show that PDT is capable of activating tyrosine phosphorylation-based cell signaling in human primary monocytes and triggering rapid adhesion and chemotaxis. PDT-induced monocyte migration requires the activation of the PI3K/Akt signaling pathway and chemokine receptor CXCR3. Indeed, a mAb to CXCR3 and a specific receptor inhibitor suppressed significantly PDT-dependent chemotaxis, and CXCR3-silenced primary monocytes lost responsiveness to PDT chemoattraction. Moreover, our results highlighted that the PDT-induced migratory activity is sustained by the CXCR3A isoform, since CXCR3-transfected L1.2 cells acquired responsiveness to PDT stimulation. Finally, we show that PDT, as CXCR3 ligands, is also able to direct the migration of IL-2 activated T cells, which express the highest levels of CXCR3 among CXCR3-expressing cells. In conclusion, our study defines a chemokine-like activity for PDT through CXCR3A and points on the possible role that this synthetic dipeptide may play in leukocyte trafficking and function. Since recent studies have highlighted diverse therapeutic roles for molecules which activates CXCR3, our findings call for an exploration of using this dipeptide in different pathological processes.

Keywords: pidotimod; CXCR3; monocyte; migration; PI3K/Akt pathway; T cell; immunomodulant

1. Introduction

Viruses are the main agents responsible for Acute Respiratory Tract Infections (ARTIs) during the pediatric age [1]. The introduction of new antibiotics and vaccines has surely contributed to control the most life-threatening ARTIs but has not had a strong impact on viral ARTIs. Therefore, one efficient approach in preventing and treating ARTIs is to increase the immune response by enhancing the child's innate defense mechanisms. The last decade has seen the emergence of different kinds of natural and synthetic molecules with different mechanisms of action, called immunomodulators, which have been introduced for prophylaxis and treatment of various infectious diseases and inflammation [2–6]. Some of these substances are granulocyte colony-stimulating factor (G-CSF), interferons, imiquimod and bacterial-derived preparations, which are already licensed for use in patients. Others including IL-12, various chemokines, synthetic cytosine phosphate-guanosine (CpG) oligodeoxynucleotides and glucans, have been investigated extensively in clinical and preclinical studies [7]. Another compound, which exerts a well-defined immunomodulatory and antimicrobial activity against infections is Pidotimod (3-L-pyroglutamyl-L-thiaziolidine-4carboxylic acid, PDT) [8,9], a synthetic dipeptide on which research focused particular attention for its properties in prevention and treatment of ARTIs in childhood. Indeed, different studies have shown that PDT exerts a beneficial effect in children reducing the number of ARTIs, the severity of signs and symptoms of acute episodes [8,10–14].

In vitro studies, performed using either murine or human cells, have shown that PDT is able to modulate both innate and adaptive immune responses [15]. In particular, this dipeptide upregulates the expression of HLA-DR and of the co-stimulatory molecules CD83 and CD86 on dendritic cells (DCs) inducing their maturation. It also stimulates DCs to release pro-inflammatory molecules, stimulates natural killer cell activity, inhibits thymocyte apoptosis and potentiates phagocytosis [16–19]. This dipeptide has also been described to enhance the proliferation of mitogen-activated peripheral blood mononuclear cells (PBMCs) and increase the production of cytokines, as IFN-α, IFN-γ and IL-12, crucial to drive T cell proliferation and differentiation towards a Th1 phenotype [16,19,20]. More recently, Fogli et al. (2014) showed that the treatment of PBMCs and monocytes with PDT led to a significant attenuation of the inflammatory response to TLR agonists. Since TLRs are key components in pathogen recognition and critical mediators in the early response to foreign microorganisms, attenuation of the inflammatory response to TLR agonists by PDT represents an important immunomodulating effect [21].

Currently, the research focused on this molecule has attempted to better elucidate its mechanism of action. The ability of PDT to modulate different aspects of both innate and adaptive immune response leaves to speculate that this dipeptide might exert its function through the involvement of one or more cytokine/chemokine receptors. Indeed, these receptors play an essential role in the immune response and are responsible for activation and recruitment of immune cells at infection and inflammation sites, contributing to the induction and exacerbation of chronic inflammatory reactions. However, despite the numerous efforts, there are limited information about the cellular receptor(s) engaged on immune cells and intracellular signaling pathways triggered by this molecule. Currently, a new research input could be now essential on the role of PDT as immunomodulant, since immunomodulation represents an adjunct modality which looks promising for control of microbial diseases and in the future could play a key role in treating and reducing the morbidity and mortality of different diseases.

In this study we investigated the PDT mechanism of action, thus, to better understand how this dipeptide exerts its biological activities and the why of its efficacy and safety.

Here, we demonstrate that PDT shows a chemokine-like activity through the activation of CXCR3 receptor, in particular CXCRA isoform, and of the PI3K/Akt signaling pathway.

2. Results

PDT activates rapid intracellular tyrosine phosphorylation-based protein signaling on human primary monocytes.

Rapid phosphorylation of signaling proteins is triggered by immune cell surface receptors and plays a pivotal role in the regulation of innate and adaptive immune functions. In particular, tyrosine phosphorylation (pTyr) of signaling proteins is indispensable in the regulatory pathways and represents a key activation signal, promoted by cytokine/chemokine receptors [22–24]. Therefore, in order to understand if PDT is able to trigger a cytokine/chemokine receptor, we investigated its capability to activate in a few min tyrosine phosphorylation-based cell signaling on human primary monocytes.

To this aim monocytes were stimulated at 37 °C for 5 min with the dipeptide PDT at different concentrations (1, 5, 25, 50, 100 µg/mL) and the tripeptide fMLP (10 nM), used as positive control. Data obtained from western blot analysis, using a protein specific anti-phosphotyrosine antibody, show that extracts from untreated cells have a low basal level of Tyr phosphorylated proteins. At the same time, PDT dose-response analysis shows that the immunomodulant is able to trigger a rapid increase of Tyr phosphorylated proteins at concentrations ranging from 5 to 100 μg/mL (Figure 1). As expected, fMLP induced a significant level of Tyr phosphorylated proteins.

These data show the PDT ability to induce protein tyrosine phosphorylation in monocytes and suggest the capability of the dipeptide to act probably through a cytokine/chemokine receptor activation.

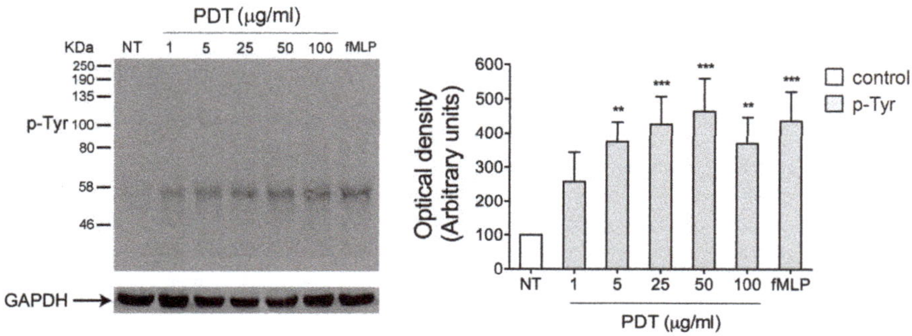

Figure 1. Effect of PDT stimulation on protein tyrosine phosphorylation in monocytes. Monocytes were treated for 5 min with 1, 5, 25, 50, 100 μg/mL of PDT and 10 nM of fMLP. Untreated cells were used as control (NT). Western blot analysis of cells lysates shows that PDT is able to induce an increase of Tyr phosphorylated proteins at different concentrations tested, as shown by densitometry analysis and plotting of the pTyr/GAPDH. In the left panel blots from one representative experiment of three with similar results are shown. In the right panels, values reported for protein Tyr phosphorylation are the mean ± SD of three independent experiments. Statistical analysis was performed by one-way ANOVA and the Bonferroni's post-test was used to compare data, ** $p < 0.01$, *** $p < 0.001$.

2.1. PDT Induces Monocyte Adhesion and Migration

Chemokines, through chemokine receptor activation, trigger intracellular signaling events, which control leukocyte recruitment, a key multi-step process in regulation of immune responses involving rapid integrin-dependent adhesion and migration of leukocytes [25]. In order to assess the ability of PDT to functionally activate a chemokine receptor on monocytes, we performed static adhesion and migration assays. Static adhesion assays were performed on immobilized ligands, as ICAM-1 and VCAM-1, in response to different concentration of the synthetic dipeptide (1, 5, 10, 50, 100 μg/mL). Figure 2 shows that PDT triggered a rapid (2 min) concentration-dependent adhesion of primary human monocytes to ICAM-1 (Figure 2A) and V-CAM (Figure 2B). In particular, PDT significantly stimulated monocyte adhesion on ICAM-1 at a concentration ranging from 10–50 μg/mL with a peak at 10 μg/mL (Figure 2A). On the other hand, PDT-induced adhesion on VCAM-1 occurred at a lower concentration, ranging from 5 to 10 μg/mL and reaching a peak at 5 μg/mL (Figure 2B).

Then, we performed monocyte migration in Transwell chemotaxis assays in response to different concentrations of the dipeptide (0.05, 0.1, 0.5, 1, 5, 25 μg/mL). In Figure 2C we show that PDT stimulates chemoattraction of monocytes at a concentration ranging between 0.1 and 5 μg/mL.

These data show that monocyte adhesion requires a higher PDT concentration than that required for chemotaxis. This phenomenon is common to chemokines and can be elucidated by the findings of Campbell et al. (1996), who demonstrated that adhesion requires a high agonist concentration with the simultaneous occupancy of many receptors, whereas chemotaxis occurs at low agonist concentration. These different requirements for triggering adhesion and chemotaxis are necessary for their independent regulation [26].

Overall, these results show the capability of PDT to stimulate rapid adhesion and migration of human primary monocytes, suggesting a chemokine-like role for the dipeptide and its ability to transduce, through a chemokine receptor, intracellular signals involved in regulation of cell motility.

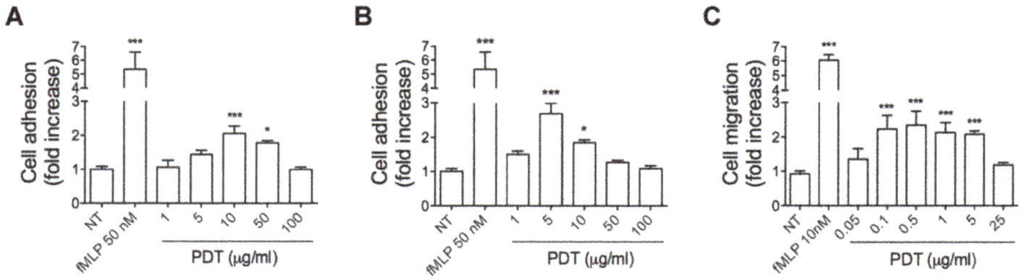

Figure 2. Effect of PDT on monocytes adhesion and migration. (**A**,**B**) Static adhesion assay on ICAM-1 (**A**) and VCAM-1 (**B**). Monocytes were stimulated or not (NT) for 2 min at 37 °C with PDT at the indicated concentrations. Bars represent the means ± SD of 3 independent experiments performed in triplicate. Statistical analysis was performed by one-way ANOVA and the Bonferroni's post-test was used to compare data, *** $p < 0.001$, * $p < 0.05$. (**C**) Transwell migration assays of monocytes in response to the indicated treatments. Bars represent the means ± SD of 3 independent experiments performed in triplicate. Statistical analysis was performed by one-way ANOVA and the Bonferroni's post-test was used to compare data, *** $p < 0.001$. NT = not treated.

2.2. PTx Treatment Inhibits PDT-Induced Chemokine Activity and Tyrosine Phosphorylation-based Protein Signaling in Monocytes

Chemokines bind and signal through seven-transmembrane receptors coupled with the Gi class of heterotrimeric G proteins. Pertussis toxin (PTx) is known to prevent the Gi proteins interaction with G protein–coupled receptors, thus blocking intracellular signaling cascade. In order to determine if PDT receptor is coupled to Gi proteins, monocytes were pretreated with 500 ng/mL of PTx for 2 h at 37 °C, then stimulated with the dipeptide and tested for their capability to adhere, migrate and trigger phosphorylation-based cell signaling.

PDT-triggered monocyte adhesion on ICAM-1 and migration were completely inhibited by PTx pretreatment (Figure 3A and 3C, respectively). In the same assays, as expected, the strong adhesion and chemotaxis induced by fMLP (100 nM and 10 nM, respectively), a reference chemoattractant able to transduce intracellular signals through Gi proteins, was inhibited by PTx (Figure 3A,C) [26–28]. The inhibitory effect of PTx was not attributable to a generic toxic effect of PTx pretreatment, because adhesion of monocytes was retained after 10 min of PMA (phorbol 12-myristate 13-acetate) stimulation (100 ng/mL) (Figure 3B), and cell migration occurred in response to LPC (L-α-lysophosphatidylcholine, palmitoyl C16:0, Sigma-Aldrich) (10 μM) (Figure 3C), in line with previous observations [27,29,30].

In addition, to determine if rapid tyrosine phosphorylation-based protein signaling triggered by the dipeptide was dependent from a receptor coupled with Gi proteins, monocytes were pretreated with PTx, then stimulated with PDT (5 μg/mL) and tested for their capability to induce protein tyrosine phosphorylation. Data obtained from western blot analysis, using a protein specific anti-phosphotyrosine antibody, showed that PDT-triggered protein tyrosine phosphorylation was completely inhibited by PTx pretreatment (Figure 3D).

Overall, these data showed that PDT induces monocyte adhesion, migration and intracellular protein tyrosine phosphorylation through a cell surface receptor coupled with the Gi class of heterodimeric G proteins.

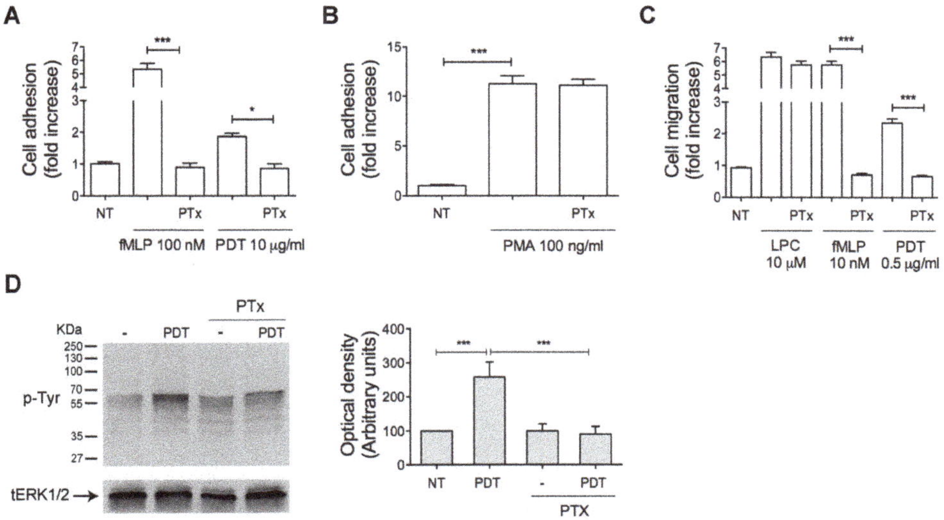

Figure 3. Effect of PTx on PDT-induced monocytes adhesion, migration and protein tyrosine phosphorylation. (A, B) Static adhesion assay of monocytes on ICAM-1. Cells pretreated with 500 ng/mL of PTx for 2 h at 37 °C were stimulated for 2 min at 37 °C with PBS (NT), fMLP (100 nM), PDT (10 μg/mL) (**A**) or for 10 min with PMA (100 ng/mL) (**B**). Bars represent the means ± SD of 3 independent experiments performed in triplicate. Statistical analysis was performed by paired 2-tail Student *t* test, *** $p < 0.001$, * $p < 0.05$. (**C**) Transwell migration assay of monocytes in response to the indicated treatments. Monocytes pretreated with PTx were stimulated for 90 min at 37 °C with PBS (NT), LPC (10 μM), fMLP (10 nM) or PDT (0.5 μg/mL). Bars represent the means ± SD of 3 independent experiments performed in triplicate. Statistical analysis was performed by paired 2-tail Student *t* test, *** $p < 0.001$. (**D**) Monocytes pretreated or not with PTx were stimulated or not for 5 min at 37 °C with PDT (5 μg/mL). Western blot analysis of cell lysates shows that PDT-triggered protein tyrosine phosphorylation was inhibited by PTx, as shown by densitometry analysis and plotting of the pTyr/tERK1/2. In the left panel blots from one representative experiment of three with similar results are shown. In the right panels, values reported for protein Tyr phosphorylation are the mean ± SD of three independent experiments. Statistical analysis was performed by paired 2-tail Student *t* test and the Bonferroni's post-test was used to compare data, *** $p < 0.001$. NT = not treated.

2.3. Monocyte Migration Triggered by PDT Requires the Activation of PI3K/Akt Signaling Pathway

Cell migration is governed by a complex network of signal transduction pathways, which involve lipid second messengers, small GTPases, kinases, cytoskeleton-modifying proteins, and culminates in the cytoskeletal remodeling and chemoattraction response to ligand-induced receptor activation [31]. In order to investigate the possible signaling pathways responsible for monocyte recruitment triggered by PDT stimulation, primary human monocytes were pretreated for 60 min at 37 °C with inhibitors of different key signaling molecules and then stimulated with the dipeptide (0.5 μg/mL) to migrate in a Transwell chemotaxis system. As shown in the Figure 4A, AG490 (10 μM), a specific and potent inhibitor of the Janus kinase 2 protein (JAK2), as well as PD98059 (10 μM), a MAP kinase/extracellular signal-regulated kinase (ERK) inhibitor, and staurosporine, a strong inhibitor of Protein Kinase C (PKC) and other protein kinases, did not impact on monocyte migration induced by PDT (Figure 4A). On the other hand, PDT-triggered monocyte migration was completely inhibited by two specific inhibitors of PI3K (wortmannin 100 nM and LY294002 10 μM) and by the specific Akt inhibitor VIII (1 μM) (Figure 4A). Therefore, in order to confirm the capability of PDT to activate PI3K/Akt signaling pathways, we assessed by western blot analysis the Akt phosphorylation status at Ser473 in cytosolic extracts of cells, treated for 5, 30 and 60 min with the dipeptide (5 μg/mL). As shown in Figure 4B, monocytes treated with PDT (0.5 g/mL) for 30 min showed a significant Akt phosphorylation, which remained sustained up to 60 min.

Overall, these data show that PDT-induced monocyte chemotaxis is due to the activation of the PI3K/Akt signaling pathway, which is known to be implicated in cell migration and in downstream signaling of cytokine/chemokine receptors [32,33].

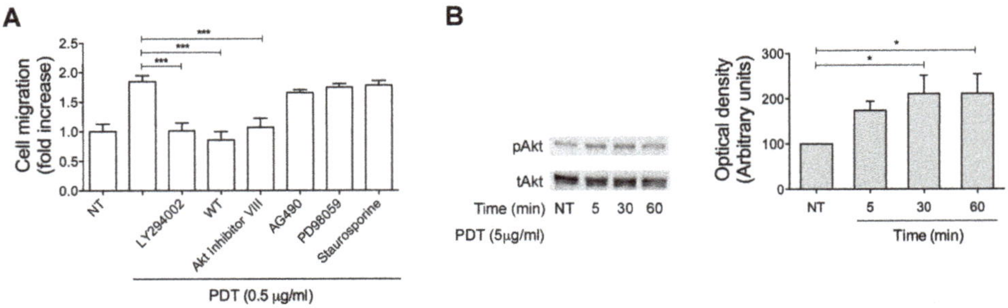

Figure 4. Signaling molecules involved in PDT-triggered monocyte migration. (**A**) Transwell migration assay of monocytes in response to the indicated treatments. Monocytes pretreated for 1 h at 37 °C with the inhibitors LY294002 (25 μM), wortmannin (100 nM), Akt inhibitor VIII (1 μM), AG490 (100 μM), PD98059 (10 μM) and staurosporine (1 nM) were stimulated for 90 min at 37 °C with PBS (NT) or PDT (0.5 μg/mL). Bars represent the means ± SD of 3 independent experiments performed in triplicate. Statistical analysis was performed by one-way ANOVA and the Bonferroni's post-test was used to compare data, *** $p < 0.001$. (**B**) Monocytes were stimulated with 5 μg/mL of PDT at 37 °C for the indicated times. Not treated cells (NT) were used as control (lane 1). Western blot analysis of monocyte lysates shows that PDT activates Akt, as shown by the respective phosphorylation state, verified by densitometric analysis and plotting of the phospho-Akt/total Akt (pAkt/tAkt). In the left panel blots from one representative experiment of three with similar results are shown. In the right panel, values reported for Akt phosphorylation are the mean ± SD of three independent experiments. Statistical analysis was performed by one-way ANOVA and the Bonferroni's post-test was used to compare data, * $p < 0.05$.

2.4. PDT-Triggered Monocyte Migration Is Mediated by CXCR3

Purinergic receptors are G protein-coupled receptors (GPCR) involved in several cellular functions, including cell migration and its function is mediated by Gi proteins [34]. So, in order to exclude the activation of purinergic receptors from the nucleotide release due to a PDT cytotoxic effect, we evaluate the viability of monocytes treated for 2h with different concentrations of PDT (5, 25, 100 μg/mL) by CellTiter-Glo®Luminescent Cell Viability Assay, which is based on determination of intracellular ATP concentration, as indicator of metabolically active cells, and propidium iodide, which can stain nucleic acids inside of dead or damaged cells. As shown in Figure 5A,B, the cells were metabolically active at any tested concentration. Therefore, we can exclude a purinergic receptor activation by the dipeptide.

Human monocytes express numerous chemokine receptors, among which CXCR1, CXCR2, CXCR3, CXCR4 might play a role in PDT chemokine-like activity. Indeed, these receptors are coupled with the Gi class of G proteins and involved in promoting monocyte migration [35–37]. Therefore, first we tested whether PDT-triggered monocyte migration could be ascribed to the activation by the dipeptide of one of these receptors.

To this aim we performed Transwell chemotaxis assays with monocytes pre-treated for 1 h at 37 °C with neutralizing mAbs to CXCR1, CXCR2, CXCR3, CXCR4 or with a control mAb. The specificity of the antibodies is reported in Supplementary Material (Figure S1). As shown in Figure 5C, mAb to CXCR3 significantly inhibited PDT-dependent chemotaxis. Moreover, no inhibition of PDT activity was observed in monocytes pre-treated with the control mAb or with mAbs to CXCR1, CXCR2 and CXCR4.

In order to confirm the involvement of CXCR3 in monocyte migration induced by PDT, we also performed chemotaxis assays with cells pre-treated 36 h at 37 °C with the CXCR3 antagonist AMG487 (0.5 μM). As expected, the antagonist was able to significantly inhibit monocyte migration triggered by the dipeptide (Figure 5D). The specific effect of AMG487 was confirmed by the fact that fMLP-induced monocyte migration (10 nM) was not influenced by the presence of this antagonist, and on the other side, CXCL11-stimulated chemotaxis (10 nM), known to be mediated by CXCR3, was completely inhibited (Figure 5D).

To further assess the involvement of CXCR3 in PDT chemokine activity, the expression of this receptor on monocytes was suppressed by silencing, using the AMAXA nucleofection technology to deliver specific siRNAs. Efficiency of nucleofection was evaluated after 16 h post-nucleofection by flow cytometric analysis and showed an average of 88% ± 8% nucleoporated monocytes. The effects of siRNA on CXCR3 expression were evaluated by real-time PCR analysis. As shown in Figure 5E, approximately 52% inhibition of CXCR3 transcripts was observed in monocytes at 16 h after nucleofection with CXCR3 siRNAs as compared to monocytes nucleofected with scr siRNA. The CXCR3 silencing induced a significant inhibition of PDT-triggered migration of monocytes as compared to cells nucleofected with an unrelated scrambled (scr) siRNAs (Figure 5F). Moreover, as expected, a significant reduction of monocyte migratory activity to CXCL11 (10 nM) (Figure 5F), one of the physiological ligands of CXCR3, confirmed the silencing of the receptor. Finally, the specificity of CXCR3 silencing was confirmed by the finding that fMPL-induced monocyte migration (10 nM) was not influenced by inhibition of transcripts (Figure 5F).

In conclusion, all these data suggest that PDT-triggered migratory activity of PDT is mediated by CXCR3.

2.5. The PDT-Induced Migratory Activity Is Sustained by the CXCR3A Isoform

CXCR3 gene is alternatively spliced, generating at least three functional isoforms differing in either their N or C-terminus. The more studied CXCR3 isoforms are CXCR3A and CXCR3B. It's known that G protein recruitment by CXCR3 isoforms is different and leads to different signaling cascades. Indeed, CXCR3A isoform is linked to Gαi or Gαq proteins, which promote cell migration by activating the PI3K/Akt pathway [38], while CXCR3B variant has been reported to be pertussis toxin-insensitive and possibly Gs coupled [39]. Since we have established that PDT acts as a chemokine in a pertussis toxin-sensitive manner, through activation of PI3K/Akt signaling pathway, we explored if a de novo expression of CXCR3A was responsible for the dipeptide activity. To this aim, L1.2 cells were used to express CXCR3A. Cells were nucleofected with fluorescent positive control vector (pmaxGFPTM Control vector) or pTarget empty vector or expressing human CXCR3A. The efficiency of nucleofection, evaluated by flow cytometry, was found to be of 86% ± 6% in 3 different experiments. Flow cytometric analysis showed that a high percentage of cells nucleofected with pTarget-CXCR3A expressed the receptor on their surface (88% ± 8%), when compared to cells nucleofected with the control empty vector. The migratory behavior of L1.2 cells expressing the empty vector and CXCR3A receptor following CXCL11 (10 nM) or PDT stimulation (0.5, 5, 50, 500 ng/mL) was assessed by Transwell chemotaxis assays. Figure 5G shows that empty vector expression did not induce cells to migrate in response to PDT or CXCL11. On the other hand, expression of CXCR3A was found to significantly promote migration in response to CXCL11, the physiological ligand of this receptor, and PDT. The migratory activity triggered by PDT in CXCR3A expressing L1.2 cells was dose dependent with a peak of activity at the concentrations of 5 and 50 ng/mL.

Therefore, our results show that a de novo expression of CXCR3A in L1.2 cells is sufficient to mediate cell migration in response to PDT stimulation.

2.6. Pidotimod Induces Migration of IL-2-Activated T Lymphocytes

Chemokine receptors are specifically and differently expressed on different subsets of leukocytes to control a selective tissue recruitment of functional immune cells [40]. CXCR3 is rapidly induced on resting T cells following activation and plays an important role in T cell trafficking and function. Therefore, in order to confirm a chemokine-like activity for PDT through the CXCR3 receptor, we performed migration assays with a 5-µm Transwell chemotaxis system on IL-2-activated T cells, which express high levels of this receptor on their surface [40], in response to different PDT concentrations (0.01, 0.05, 0.1, 0.5, 1, 5, 25, 50 µg/mL). As expected, PDT stimulated a statistically significant chemoattraction of IL-2-activated T lymphocytes and its activity occurred at a concentration ranging between 0.05 and 5 µg/mL with a bell-shaped chemotactic dose-response curve and a peak activity at 0.1 µg/mL (2.5-fold increase) (Figure 5H). Moreover, as expected, a strong induction of migration was obtained with IL-2-activated T lymphocytes using CXCL11 (10 nM), a reference chemoattractant for CXCR3 receptor (Figure 5H).

Overall, these data show that PDT is able also to stimulate migration of IL-2-activated T lymphocytes.

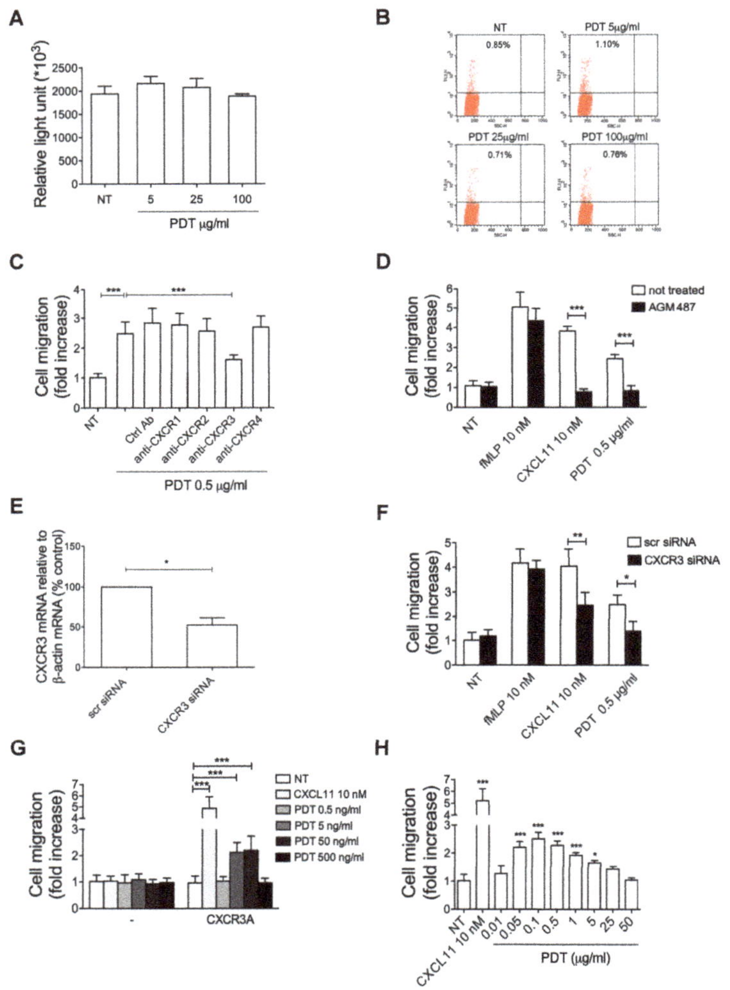

Figure 5. PDT–induced chemotactic activity is mediated by CXCR3. (**A**) Cells were treated or not (NT) with PDT (5, 25, 100 µg/mL) for 2 h. Cell viability was assessed on 10^3 cells by CellTiter-Glo® Luminescent Cell Viability Assay, based on determination of intracellular ATP concentration, according

to manufacturer's instructions. Bars represent the means ± SD of 3 independent experiments. (**B**) Cells were treated or not (NT) with PDT (5, 25, 100 μg/mL) for 2 h, stained with propidium iodide and analyzed by flow cytometer. Data were analyzed by CellQuest software. Figures are representative of one experiment of three with similar results. (**C**) Transwell migration assay of monocytes in response to the indicated treatments. Monocytes pretreated for 1 h at 37 °C with 50 μg/mL of Ctrl mAb or mAb to anti-CXCR1 or anti-CXCR2 or anti-CXCR3 or anti-CXCR4 were stimulated for 90 min at 37 °C with PBS (NT) or PDT 0.5 μg/mL. Bars represent the mean ± SD of three independent experiments performed in triplicate. Statistical analysis was performed by one-way ANOVA and the Bonferroni's post-test was used to compare data, *** $p < 0.001$. (**D**) Transwell migration assay of monocytes pretreated for 1 h at 37°C with the inhibitor AGM487 (0.5 μM) and stimulated for 90 min at 37 °C with PBS (NT), fMLP (10 nM) or CXCL11 (10 nM) or PDT (0.5 μg/mL). Bars represent the means ± SD of 3 independent experiments performed in triplicate. Statistical analysis was performed by paired 2-tail Student *t* test, *** $p < 0.001$. (**E**) Analysis of CXCR3 gene expression performed using quantitative real time PCR. Monocytes were nucleoporated with scrambled siRNAs, used as negative control or with a pool of four distinct siRNAs specific for four distinct regions of CXCR3. Analysis of real time PCR data was performed with the $2^{-\Delta\Delta Ct}$ method using Relative Quantitation Study software. Quantification of CXCR3 mRNA was normalized in each reaction according to the internal β-actin control. Bars represent the mean ± SD of three independent experiments performed in triplicate. (**F**) Transwell migration assay of monocytes nucleoporated with specific siRNAs for CXCR3 or with scrambled fluorescein-labeled siRNAs (scr siRNA), used as negative control, in response to the indicated treatments. Bars represent the mean ± SD of three independent experiments performed in triplicate. Statistical analysis was performed by paired two-tail Student's *t*-test, ** $p < 0.01$, * $p < 0.05$. NT = not treated (**G**) Transwell migration assay of L1.2 cells transfected with pTarget empty vector (-) or expressing CXCR3A and stimulated for 90 min at 37°C with PBS (NT), CXCL11 (10 nM) and PDT (0.5, 5, 50, 500 ng/mL). Bars represent the mean ± SD of three independent experiments performed in triplicate. Statistical analysis was performed by two-way ANOVA. Bonferroni's post-test was used to compare data, *** $p < 0.001$. (**H**) Transwell migration assay of IL-2-activated T lymphocytes in response to the indicated treatments. Bars represent the means ± SD of 3 independent experiments performed in triplicate. Statistical analysis was performed by one-way ANOVA and the Bonferroni's post-test was used to compare data, * $p < 0.05$, *** $p < 0.001$.

2.7. CXCL11, as PDT, Triggers Monocyte Migration through Akt Kinase Activation

Here, we have shown that PDT exerts its chemotactic activity through CXCR3 activation triggering PI3K/Akt signaling pathway. In order to confirm that PDT induces monocyte migration with the same mechanism used by the physiological ligand of CXCR3, CXCL11, we performed monocyte migration assays and western blot analysis. First, monocytes were stimulated or not at 37 °C for 5 min with CXCL11 (10 nM) and data obtained from western blot analysis show that, as expected, CXCL11 is able to trigger a rapid increase of Tyr phosphorylated proteins (Figure 6A). Then, in order to investigate if the PI3K/Akt signaling pathway, responsible for PDT-induced monocyte recruitment, is also triggered by CXCL11, monocytes were pretreated at 37 °C with PTx, Akt inhibitor VIII (1 μM) and MAPK/ERK inhibitor PD98059 (10 μM) and then stimulated with CXCL11 (10 nM) to migrate in a Transwell chemotaxis system. As shown in the Figure 6B, PD98059 did not influence monocyte migration induced by CXCL11, while PTx and the specific Akt inhibitor VIII (1 μM) completely inhibited CXCL11-triggered monocyte migration (Figure 4A). Finally, in order to confirm the capability of CXCL11 to activate Akt kinase, we assessed by western blot analysis the Akt phosphorylation status at Ser473 in cytosolic extracts of monocytes, treated for 5, 30 and 60 min with CXCL11 (10 nM). As shown in Figure 6C, monocytes stimulated with CXCL11 for 30 min showed a significant Akt phosphorylation.

Overall, these data show that CXCL11-induced monocyte chemotaxis, like PDT one, is due to the activation of Akt kinase.

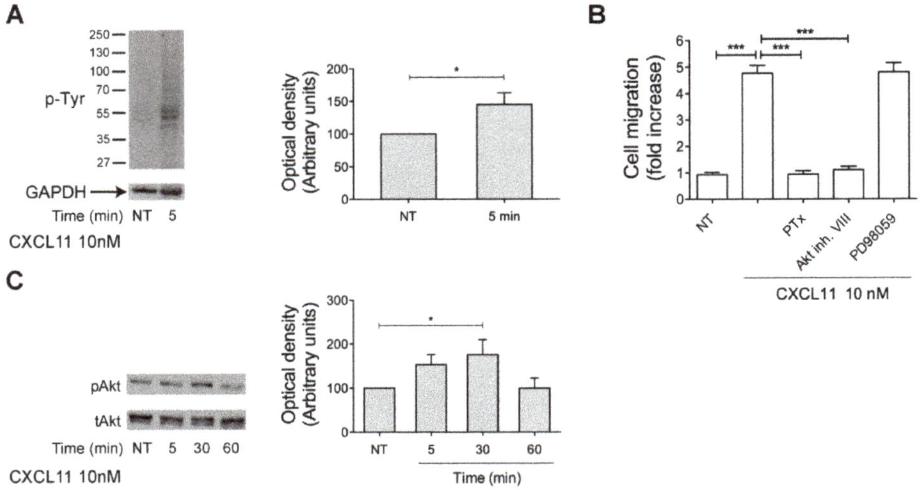

Figure 6. CXCL11–induced chemotactic activity, as PDT one, is mediated by Akt kinase. (**A**) Monocytes were treated or not for 5 min with 10nM of CXCL11. Western blot analysis of cells lysates shows that CXCL11 is able to induce an increase of Tyr phosphorylated proteins, as shown by densitometry analysis and plotting of the pTyr/GAPDH. In the left panel blots from one representative experiment of three with similar results are shown. In the right panels, values reported for protein Tyr phosphorylation are the mean ± SD of three independent experiments. Statistical analysis was performed by Student's *t*-test, * $p < 0.05$. NT = not treated. (**B**) Transwell migration assay of monocytes in response to the indicated treatments. Monocytes pretreated with PTx (500 ng/mL) or Akt inhibitor VIII (1 μM) or PD98059 (10 μM) were stimulated for 90 min at 37 °C with PBS (NT) or CXCL11 (10 nM). Bars represent the means ± SD of 3 independent experiments performed in triplicate. Statistical analysis was performed by one-way ANOVA and the Bonferroni's post-test was used to compare data. *** $p < 0.001$. (**C**) Monocytes were stimulated with 10 nM of CXCL11 at 37 °C for the indicated times. Not treated cells (NT) were used as control (lane 1). Western blot analysis of monocyte lysates shows that CXCL11 activates Akt, as shown by the respective phosphorylation state, verified by densitometric analysis and plotting of the phospho-Akt/total Akt (pAkt/tAkt). In the left panel blots from one representative experiment of three with similar results are shown. In the right panel, values reported for Akt phosphorylation are the mean ± SD of three independent experiments. Statistical analysis was performed by one-way ANOVA and the Bonferroni's post-test was used to compare data, * $p < 0.05$.

3. Discussion

In the present study for the first time we demonstrate that PDT exerts a chemokine-like activity triggering monocyte adhesion and migration. In this regard, PDT can be considered the shortest synthetic peptide known up to date with a chemokine activity.

Our experimental evidences highlight the ability of the dipeptide to functionally activate the G protein-coupled receptor CXCR3, in particular the CXCR3A isoform, which is linked to heterotrimeric GTP-binding proteins of the Gi family. Indeed, PDT-induced chemokine activity was PTx-sensitive and neutralized by an anti-human CXCR3 mAb, a specific CXCR3 inhibitor and CXCR3 siRNA. Moreover, the specificity of CXCR3-mediated effects was confirmed by the capability of dipeptide to chemoattract L1.2 cells stably expressing the human CXCR3A isoform. The study of intracellular signaling downstream of CXCR3 showed that PDT triggers the PI3K/Akt pathway. The activation of such signaling cascade has also been previously observed in response to a wide number of CXCR3 agonists in different cell types and it is in line with the activity previously reported for this canonical receptor [41–43].

GPCRs represent one of the therapeutically most relevant protein families and around one-third of the available small-molecule drugs target GPCRs [44]. Surprisingly, PDT, which is simply a dipeptide and doesn't show any amino acid sequence identity with CXCR3 ligands, showed to activate this chemokine receptor. The 3D structure of CXCR3 is currently not known, but a general "two-step" model has been proposed to describe the chemokine receptor activation [45]. The first step is governed by binding of the ligand to the N-terminus and extracellular loops of the GPCR [46], subsequently, the N-terminus of the chemokine is able to interact with the transmembrane domains (TM), leading to activation of the receptor [45,46]. In contrast, it is likely that this two-step binding model cannot not be applied in the case of small molecule modulators, like PDT. Indeed, GPCRs possess multiple binding sites and different ligands can selectively stabilize different "active" conformations [47,48]. Small molecules, which don't mimic the N-terminal regions of chemokines, rather point out other important interactions and generally bind to the TM part [49,50], as recently shown for example for a small molecule CXCR4 antagonist using X-ray crystallography [51]. In addition, Xanthou et al. (2003) did not validate the two-step model for CXCR3 interactions with its ligands, but instead support a multi-site model, in which multiple and distinct extracellular domains contribute to receptor activation [52]. Consequently, small molecule modulators are considered to allosterically modulate GPCR function from binding sites that are distinct or only partially overlapped with the chemokine interaction points [53–56]. Future studies on PDT structure–activity relationship will be necessary to reveal the modality of dipeptide to trigger CXCR3 and could be of both fundamental and practical relevance.

CXCR3 is known to play a key role in mediating leukocyte recruitment to mucosal tissues and inflammatory sites [57–61]. In particular, different studies demonstrated a key role for CXCR3 in recruitment, trafficking and function of Th1 CD4+ and effector CD8 T cells to infection sites and the establishment on Th1 amplification loop mediated by IFNγ and IFNγ-inducible CXCR3 ligands 61. Moreover, CXCR3 is known to play a role in the migration of T cells within peripheral tissue and lymphoid compartment, facilitating their interaction with antigen presenting cells and leading to the generation of effector, regulatory and memory T cells [61]. On the other side, it has been demonstrated that PDT is capable to enhance the proliferation of PBMCs and their production of crucial cytokines, as IFN-γ and IL-12 (20), to induce T cell migration (Figure 5E), drive T cell proliferation and differentiation towards a Th1 phenotype [16,19], and induce maturation of dendritic cells (DCs) [18]. In this regard, we can hypothesize that the dipeptide could attract T cells through the CXCR3 receptor, co-stimulate their proliferation, differentiation and activation, priming T cell responses with consequent therapeutic implications following local delivery. Moreover, like many chemokines, PDT might direct the migration of circulating leukocytes in the lymphoid tissues regulating and ensure continued recirculation and surveillance of lymphoid tissues. In particular, PDT administration in defined anatomically restricted locations (in vivo) might act as homing molecule in the control of T-cell trafficking over the course of immune response in mucosa-associated lymphoid tissues (MALTs). However, how PDT could be capable of coordinating T cell responses in the inflamed periphery through CXCR3 remains to be determined and future studies should be aimed at studying the role of this interaction in the control of T cell function.

Finally, since diverse studies suggested that CXCR3 activation could be beneficial in skin wound healing [62–64] and CXCR3 agonists possess antitumor activity [65–67], it might be interesting to evaluate the use of PDT in different pathological processes. Therefore, from a therapeutic point of view, our demonstration of a chemokine-like activity of PDT through CXCR3, will call for a future exploration of therapeutic potential of using this synthetic dipeptide in different diseases.

4. Materials and Methods

4.1. Pidotimod Preparation

PDT (purity > 99.6%) was a kind gift of Polichem SA (Lugano, CH). For all experiments the same PDT stock solution was used (10 mg/mL in ultrapure endotoxin-free water, stored in aliquots at −20 °C).

Biological effects were evaluated using PDT at the final concentrations of 0.01, 0.05, 0.1, 0.5, 1, 5, 25, 50 and 100 μg/mL. The absence of endotoxin contamination (<0.25 unit/500 μg) in the PDT stock solution was assessed by Limulus amoebocyte assay (Associates of Cape Cod, East Falmouth, MA, USA) and all experiments were performed in endotoxin free-condition.

4.2. Cell Culture

The murine pre-B L1.2 cells were obtained from the American Type culture collection (Manassas, VA) and cultured in RPMI 1640 containing 10% fetal bovine serum (FBS, Lonza, Basel, CH), 2 mM L-glutamine, 1 mM sodium pyruvate and 50 μM 2-mercaptoethanol (complete medium).

4.3. Isolation of Human Primary Cells

Blood was collected from healthy donors who gave informed consent for this research according to the Declaration of Helsinki. PBMCs were isolated by using Ficoll-Paque (Sigma-Aldrich, St. Louis, MO, USA) density gradient centrifugation and then human primary monocytes were isolated from PBMCs by negative selection using the monocyte isolation Kit (Miltenyi Biotec, Bergish Gladbach, NRW, DA) according to the manufacturer's protocol. Monocytes were cultured (2×10^6/mL) in RPMI 1640 containing 10% fetal bovine serum (low endotoxin FBS, Lonza) and 2 mM L-glutamine (complete medium). In each experiment, cell purity was evaluated by flow cytometry and was always greater than 97%.

Human primary lymphocytes were isolated from PBMC by Percoll gradient sedimentation, as previously described [68]. T lymphocytes were isolated by negative selection (Miltenyi Biotec) according to manufacturer's instructions. Purity of T lymphocyte preparation was evaluated by flow cytometry after staining with fluorochrome-conjugated anti-CD3 antibodies and was more than 95%. Isolated T lymphocytes were seeded (2×10^6/mL) in 6 well plates and kept in culture for 6 days in complete medium supplemented with IL-2 (20 ng/mL) (R&D System, Minneapolis, MN, USA) to obtain activated T cells. After 3 days the medium was changed with complete medium containing the cytokine. After 6 days of culture, the cells were harvested, pooled together and counted for the assays.

Monocytes were pretreated for 2 h at 37 °C, when indicated, with PTx (500 ng/mL) or 30 min with LY294002 (25 μM) or wortmannin (100 nM) or Akt inhibitor VIII (1 μM) (Enzo Life Sciences, Farmingdale, NY, USA) or Tyrphostin AG490 (10 μM) (Vinci-Biochem, Firenze, IT) or PD98059 (10 μM) or staurosporine (1 nM) (Sigma-Aldrich) or AMG 487 (0.5 μM, a CXCR3 antagonist) (Tocris Bioscience, Bristol, GB) or with a control IgG antibody and a neutralizing mAb to CXCR1 (MAb330) or CXCR2 (MAb331) or CXCR3 (MAb160) or CXCR4 (MAb171) (50 μg/mL) (R&D System).

4.4. Chemotaxis Assays

Migration of monocytes, T lymphocytes and L1.2 cells was assessed using 5-μm pore-size Transwells (BD Biosciences, Franklin Lakes, NJ, USA). Chemotaxis assays were performed as previously described [27]. Briefly, cells, pre-treated or not with PTx or different inhibitors, were suspended at 2×10^6/mL in adhesion buffer. Cell suspension (100 μL) was added to the top well and adhesion buffer (600 μL) containing Pidotimod (0.01, 0.05, 0.1, 0.5, 1, 5, 25 and 50 μg/mL) to the bottom well. LPC (L-α-lysophosphatidylcholine, palmitoyl C16:0, Sigma-Aldrich) (10 μM) and fMLP (10 nM) were also used as reference chemoattractants to evaluate the selective and not toxic effect of PTx; CXCL8, CXCL11 and CXCL12 (R&D System) (10 nM) were used as chemokine ligands to evaluate the specificity of antibodies. Chemotaxis was performed for 90 min at 37 °C and then filters were removed. After fixation with 1.5% paraformaldehyde, migrated cells were counted in 5 high-power fields by light microscopy at a 10× magnification. Results are expressed as the fold increase compared with control.

4.5. Static Adhesion Assay

Adhesion assays were performed as previously described [27]. Monocytes, pre-treated or not with PTx, were suspended at 5×10^6/mL in adhesion buffer (PBS, 1 mM $CaCl_2$, 1 mM $MgCl_2$, 10%

FCS, pH 7.2) and 20 μL of cell suspension was added to 18-well glass slides (Thermo Fisher Scientific, Waltham, MA, USA) coated overnight at 4 °C with ICAM-1 and V-CAM (1 μg/mL, R&D Systems). Cells were then stimulated for 2 min at 37 °C with 5 μL of PDT at different concentrations (1, 5, 10, 50, 100 μg/mL). PMA (phorbol 12-myristate 13-acetate) stimulation (100 ng/mL) at 37 °C for 10 min was also evaluated to exclude a toxic effect of PTx. After washing, adherent cells were fixed in 1.5% glutaraldehyde. Computer-assisted enumeration of cells in 4 high-power fields by light microscopy was performed. Results are expressed as the -fold increase compared with control.

4.6. Western Blot Analysis

Monocytes (10×10^6/mL) were stimulated with fMLP (10 nM) and PDT at different concentrations 1, 5, 25, 50, 100 μg/mL and then lysed in 200 μL of buffer (pH 7.2) containing 20 mM MOPS (pH 7.0), 2 mM EGTA, 5 mM EDTA, 30 mM sodium fluoride, 60 mM glycerophosphate (pH 7.2), 20 mM sodium pyrophosphate, 1 mM sodium orthovanadate, 1 mM Dithiothreitol (DTT), 1% Triton X-100 and a mixture of protease inhibitors (Complete Mini Roche, Hoffmann-La Roche, Basel, CH). The total concentration of proteins was detected by QuantiProTM BCA Assay (Sigma-Aldrich). Equal amounts of total protein were resolved on a 12% SDS-polyacrylamide gel and then electroblotted using PVDF (polyvinylidene difluoride) membrane. The blots were incubated overnight at 4 °C with 1) mouse monoclonal p-Tyr antibody (PY99) (Santa Cruz Biotechnology, Dallas, TX, USA), 2) rabbit polyclonal phospho-Akt (Ser473) antibody (Santa Cruz Biotecnology), 3) mouse monoclonal Akt antibody (Cell Signaling Technology, Danvers, MA, USA), 4) rabbit polyclonal antibody to ERK1/2 (Santa Cruz Biotechnology), 5) mouse monoclonal GADPH antibody (Santa Cruz Biotechnology). Antigen-antibody complexes were revealed by incubating the membranes at room temperature for 1h with peroxidase-conjugated anti-mouse or anti-rabbit antibodies (Thermo Fisher Scientific) and using the ECL (Enhanced Chemiluminescence) System (Santa Cruz Biotechnology). The images were captured by ChemiDoc-It Imaging System and the integrated optical density (IOD) was determined using the Gel-Pro Analyzer 6.0 software (Houston, TX, USA.

4.7. siRNA Technique

Monocytes were nucleoporated (program Y-001) with specific siRNAs for CXCR3 receptor (Origene, Rockville, MD, USA) using the Amaxa Nucleofector System (Amaxa Biosystems, Cologne, NW, DE). In particular, siRNAs (100 nM) were added to 3×10^6 cells resuspended in 100 mL of nucleofection buffer. Fluorescein-labeled irrelevant (scr) siRNAs (Invitrogen, Carisbad, CA, USA) were used as negative control and to assess the efficiency of siRNA nucleoporation by flow cytometry. The efficacy of CXCR3 siRNA was evaluated by real-time PCR analysis.

4.8. Real-Time PCR for Gene Expression Analysis

Total RNA was isolated from monocytes (1×10^6 cells) using RNeasy Plus Mini Kit (Qiagen, Valencia, CA, USA). Following retrotranscription, 50 ng of cDNA mixed with sterile water and SYBR Green qPCR Master Mix (Promega, Madison, WI, USA) were amplified using the PrimeTime qPCR primer Assay for CXCR3 and the following PCR primers (0.2 μM each): human β actin, 5′- GGCACCCAGCACAATGAAG -3′ (forward), and 5′- GCTGATCCACATCTGCTGG -3′ (reverse) (Integrated DNA technologies, Coralville, IA, USA). Quantification of CXCR3 cDNA was normalized in each reaction according to the internal β-actin control. Results are expressed as percentage of control.

4.9. Cell Viability Assay

Cell viability was evaluated according to the manufacturer's instructions by CellTiter-Glo Luminescent Cell Viability Assay (Promega), which is based on the quantification of ATP. The same number of treated (PDT 5, 25, 100 μg/mL) and not treated cells for 120 min were analyzed.

4.10. Flow Cytometry Analysis

Cell integrity was analyzed by flow cytometry. In brief, treated (PDT 5, 25, 100 µg/mL) or not treated cells with PDT were harvested, washed with phosphate-buffered saline (PBS) and pelleted by centrifugation. Then, the cells were stained for nuclear DNA content with 50 µg/mL propidium iodide (PI) (Sigma-Aldrich). Determination of PI-stained cells was performed in FL-2A channel on FACSCalibur flow cytometer and data were acquired and analyzed by using CellQuest Software (BD Bioscience).

4.11. Transfections

Murine pre-B cells L1.2 were cultured in RPMI 1640 containing 1 mM L-glutamine, 1 mM sodium pyruvate, 10% FBS and 50 µM 2-mercaptoethanol. Transfections of L1.2 cells were performed using Amaxa nucleofector (Lonza) (program U-015), according to the manufacturer's instructions. Cells were plated at 1×10^6/mL and, after 24 h, 4×10^6 cells were nucleofected with 3 mg endotoxin-free plasmids pTarget or pTarget-CXCR3A. The vector pTarget was from Promega and was used as negative control. In pTarget-CXCR3A construct (a kind gift from Alan Wells, Department of Pathology, University of Pittsburgh, Pittsburgh, Pennsylvania, USA) CXCR3A (NM_001504.1) was cloned into the EcoRI and SalI sites of pTarget. After 4 h from transfection, in order to enhance CXCR3A cell surface expression, transient transfectants were incubated overnight at 37 °C in medium supplemented with 10 mM sodium butyrate (Sigma-Aldrich). The transfection efficiency was evaluated as GFP expression by flow cytometry using control vector pmaxGFP included in Amaxa nucleofector kit. Cell surface expression of CXCR3A was evaluated by flow cytometry using an anti-human CXCR3 PE-conjugated mAb (Thermo Fisher Scientific).

4.12. Statistical Analysis

Data obtained from multiple independent experiments are expressed as the means ± the standard deviations (SD). The data were analyzed for statistical significance using a paired 2-tail Student t test or one-way ANOVA. Bonferroni's post-test was used to compare data. Differences were considered significant at $p < 0.05$. Statistical tests were performed using Prism 5 software (GraphPad Software, La Jolla, CA, USA).

Supplementary Materials: Supplementary materials can be found at http://www.mdpi.com/1422-0067/20/21/5287/s1.

Author Contributions: Conceptualization, F.C. and C.G.; formal analysis, F.C., A.B, S.C., S.R., A.Z., P.M., S.M., C.G; investigation, F.C., A.B., S.C., S.R., P.M. and A.Z.; methodology, F.C., A.B., and C.G.; project administration, A.C.; supervision, F.C. and C.G.; writing—original draft, C.G.; writing—review and editing, A.C.

Funding: This research received no external funding.

Acknowledgments: We thank Alan Wells, (Department of Pathology, University of Pittsburgh, Pittsburgh, Pennsylvania, USA) for supplying pTarget-CXCR3A construct. This study was kindly supported by Polichem SA, Lugano, Switzerland.

Conflicts of Interest: The authors declare no conflict of interest.

References

1. Griffin, M.R.; Walker, F.J.; Iwane, M.K.; Weinberg, G.A.; Staat, M.A.; Erdman, D.D. Epidemiology of respiratory infections in young children: Insights from the new vaccine surveillance network. *Pediatr. Infect. Dis. J.* **2004**, *23*, 188–192. [CrossRef] [PubMed]
2. Schaad, U.B. OM-85 BV, an immunostimulant in pediatric recurrent respiratory tract infections: A systematic review. *World J. Pediatr.* **2010**, *6*, 5–12. [CrossRef] [PubMed]
3. Rozy, A.; Chorostowska-Wynimko, J. Bacterial immunostimulants–mechanism of action and clinical application in respiratory diseases. *Pneumonol. Alergol. Pol.* **2008**, *76*, 353–359. [PubMed]

4. Ounis, I. Determination of the antiinfectious activity of RU 41740 (Biostim) as an example of an immunomodulator. *Adv. Exp. Med. Biol.* **1992**, *319*, 165–174. [PubMed]
5. Fiocchi, A.; Terracciano, L.; Martelli, A.; Bernardo, L.; Calcinai, E.; Marcassa, S. Ribosome-component immune modulation of respiratory tract infections in children. *Allergy Asthma Proc.* **2009**, *30*, 21–31. [CrossRef] [PubMed]
6. Lavigne, P.; LeeS, E. Immunomodulators Chronic Rhinosinusitis. *World J. Otorhinolaryngol. Head Neck Surg.* **2018**, *4*, 186–192. [CrossRef]
7. Masihi, K.N. Fighting infection using immunomodulatory agents. *Expert Opin. Biol. Ther.* **2001**, *1*, 641–653. [CrossRef]
8. Burgio, G.R.; Marseglia, G.L.; Severi, F.; de Benedetti, F.; Masarone, M.; Ottolenghi, A.; Pagliano, L.; Serra, U.; Nespoli, L. Immunoactivation by pidotimod in children with recurrent respiratory infections. *Arzneimittelforschung* **1994**, *44*, 1525–1529.
9. Esposito, S.; Musio, A. Immunostimulants and prevention of recurrent respiratory tract infections. *J. Biol. Regul. Homeost. Agents* **2013**, *27*, 627–636.
10. Motta, G.; De Campora, E.; de Vita, C.; Esposito, S.; Galletti, C.; Incutti, V.; Mallardi, V.; Motta, S.; Pucci, V.; Salonna, F. Immunoactivity of Pidotimod against episodes of recurrent tonsillitis in childhood. *Arzneimittelforschung* **1994**, *44*, 1521–1524.
11. Careddu, P.; Mei, V.; Venturoli, V.; Corsini, A. Pidotimod in the treatment of recurrent respiratory infections in paediatric patients. *Arzneimittelforschung* **1994**, *44*, 1485–1489. [PubMed]
12. Caramia, G.; Clemente, E.; Solli, R.; Mei, V.; Cera, R.; Carnelli, V.; Venturoli, V.; Corsini, A. Efficacy and safety of Pidotimod in the treatment of recurrent respiratory infections in children. *Arzneimittelforschung* **1994**, *44*, 1480–1484. [PubMed]
13. Passali, D.; Calearo, C.; Conticello, S. Pidotimod in the management of recurrent pharyngotonsillar infections in childhood. *Arzneimittelforschung* **1994**, *44*, 1511–1516. [PubMed]
14. La Mantia, I.; Grillo, C.; Mattina, T.; Zaccone, P.; Xiang, M.; Di Mauro, M.; Meroni, P.L.; Nicoletti, F. Prophylaxis with the novel immunomodulator Pidotimod reduces the frequency and severity of upper respiratory tract infections in children with Down's syndrome. *J. Chemother.* **1999**, *11*, 126–130. [CrossRef]
15. Riboldi, P.; Gerosa, M.; Meroni, P.L. Pidotimod: A reappraisal. *Int. J. Immunopathol. Pharmacol.* **2009**, *22*, 255–362. [CrossRef]
16. Auteri, A.; Pasqui, A.L.; Bruni, F.; Saletti, M.; Di Renzo, M.; Bova, G. Effect of Pidotimod, a new immunostimulating agent, on some aspects of immune response. *Vitr. Study. Pharmacol. Res.* **1992**, *26*, 196–197. [CrossRef]
17. Migliorati, G.; Nicoletti, I.; Riccardi, C. Immunomodulating activity of Pidotimod. *Arzneimittelforschung* **1994**, *44*, 1421–1424.
18. Giagulli, C.; Noerder, M.; Avolio, M.; Becker, P.; Fiorentini, S.; Guzman, C.; Caruso, A. Pidotimod promotes functional maturation of dendritic cells and displays adjuvant properties at the nasal mucosa level. *Int. Immunopharmacol.* **2009**, *9*, 1366–1373. [CrossRef]
19. Zuccotti, G.V.; Mameli, C. Pidotimod: The past and the present. *J. Pediatrics* **2013**, *39*, 75–78. [CrossRef]
20. Di Renzo, M.; Pasqui, A.L.; Bruni, F.; Saletti, M.; Bova, G.; Chiaron, C.; Girardello, R.; Ferri, P.; Auteri, A. The in vitro effect of Pidotimod on some immune functions in cancer patients. *Immunopharmacol. Immunotoxicol.* **1997**, *19*, 37–51. [CrossRef]
21. Fogli, M.; Caccuri, F.; Iaria, M.L.; Giagulli, C.; Corbellini, S.; Campilongo, F.; Caruso, A.; Fiorentini, S. The immunomodulatory molecule Pidotimod induces the expression pf the nod-like receptor NLRP12 and attenuates TLR-induced inflammation. *J. Biol. Regul. Homeost. Agents* **2014**, *28*, 753–766. [PubMed]
22. Mertens, I.; Vandingenen, A.; Meeusen, T.; de Loof, A.; Schoofs, L. Postgenomic characterization of G-protein-coupled receptors. *Pharmacogenomics* **2004**, *5*, 657–672. [CrossRef] [PubMed]
23. Weiss, A.; Littman, D.R. Signal transduction by lymphocyte antigen receptors. *Cell* **1994**, *76*, 263–274. [CrossRef]
24. Seet, B.T.; Dikic, L.; Zhou, M.M.; Pawson, T. Reading protein modifications with interaction domains. *Nat. Rev. Mol. Cell Biol.* **2006**, *7*, 473–483. [CrossRef]
25. Laudanna, C.; Kim, J.Y.; Constantin, G.; Butcher, E. Rapid leukocyte integrin activation by chemokines. *Immunol. Rev.* **2002**, *186*, 37–46. [CrossRef]

26. Campbell, J.J.; Qin, S.; Bacon, K.B.; Mackay, C.R.; Butcher, E.C. Biology of chemokine and classical chemoattractant receptors: Differential requirements for adhesion-triggering versus chemotactic responses in lymphoid cells. *J. Cell Biol.* **1996**, *134*, 255–266. [CrossRef]
27. Giagulli, C.; Magiera, A.; Bugatti, A.; Caccuri, F.; Marsico, S.; Rusnati, M.; Vermi, W.; Fiorentini, S.; Caruso, A. HIV-1 matrix protein p17 binds to the IL-8 receptor CXCR1 and shows IL-8–like chemokine activity on monocytes through Rho/ROCK activation. *Blood* **2012**, *119*, 2274–2283. [CrossRef]
28. Sozzani, S.; Luini, W.; Molino, M.; Jílek, P.; Bottazzi, B.; Cerletti, C.; Matsushima, K.; Mantovani, A. The signal transduction pathway involved in the migration induced by a monocyte chemotactic cytokine. *J. Immunol.* **1991**, *147*, 2215–2221.
29. Fontana, L.; Giagulli, C.; Minuz, P.; Lechi, A.; Laudanna, C. 8-Iso-PGF2 alpha induces beta 2-integrin-mediated rapid adhesion of human polymorphonuclear neutrophils: A link between oxidative stress and ischemia/reperfusion injury. *Arterioscler. Thromb. Vasc. Biol.* **2001**, *21*, 55–60. [CrossRef]
30. Schilling, T.; Eder, C. Lysophosphatidylcholine- and MCP-1-induced chemotaxis of monocytes requires potassium channel activity. *Pflug. Arch.* **2009**, *459*, 71–77. [CrossRef]
31. Welf, E.S.; Haugh, J.M. Signaling pathways that control cell migration: Models and analysis. *Wiley Interdiscip. Rev. Syst. Biol. Med.* **2011**, *3*, 231–240. [CrossRef] [PubMed]
32. Finlay, D.; Cantrell, D. Phosphoinositide 3-kinase and the mammalian target of rapamycin pathways control T cell migration. *Ann. N. Y. Acad. Sci.* **2010**, *1183*, 149–157. [CrossRef] [PubMed]
33. Xue, G.; Zippelius, A.; Wicki, A.; Mandalà, M.; Tang, F.; Massi, D.; Hemmings, B.A. Integrated Akt7PKB signaling in immunomodulation and its potential role in cancer immunotherapy. *J. Natl. Cancer Inst.* **2015**, *11*, 107.
34. Xue, G.; Hemmings, B.A. PKB/Akt-dependent regulation of cell motility. *J. Natl. Cancer Inst.* **2013**, *105*, 393–404. [CrossRef] [PubMed]
35. Le Duc, D.; Schulz, A.; Lede, V.; Schulze, A.; Thor, D.; Brüser, A.; Schöneberg, T. P2Y Receptors in Immune Response and Inflammation. *Adv. Immunol.* **2017**, *136*, 85–121. [PubMed]
36. Gerszten, R.E.; Garcia-Zepeda, E.A.; Lim, Y.C.; Yoshida, M.; Ding, H.A.; Gimbrone, M.A., Jr.; Luster, A.D.; Luscinskas, F.W.; Rosenzweig, A. MCP-1 and IL-8 trigger firm adhesion of monocytes to vascular endothelium under flow conditions. *Nature* **1999**, *398*, 718–723. [CrossRef]
37. Struckhoff, A.P.; Vitko, J.R.; Rana, M.K.; Davis, C.T.; Foderingham, K.E.; Liu, C.H.; Vanhoy-Rhodes, L.; Elliot, S.; Zhu, Y.; Burow, M.; et al. Dynamic regulation of ROCK in tumor cells controls CXCR4-driven adhesion events. *J. Cell Sci.* **2010**, *123*, 401–412. [CrossRef]
38. Costa, C.; Traves, S.L.; Tudhope, S.J.; Fenwick, P.S.; Belchamber, K.B.; Russell, R.E.; Barne, P.J.; Donnelly, L.E. Enhanced monocyte migration to CXCR3 and CCR5 chemokines in COPD. *Eur. Respir. J.* **2016**, *47*, 1093–1102. [CrossRef]
39. Billottet, C.; Quemener, C.; Bikfalvi, A. CXCR3, a double-edged sword in tumor progression and angiogenesis. *Biochim. Biophys. Acta* **2013**, *1836*, 287–295. [CrossRef]
40. Lasagni, L.; Francalanci, M.; Annunziato, F.; Lazzeri, E.; Giannini, S.; Cosmi, L.; Sagrinati, C.; Mazzinghi, B.; Orlando, C.; Maggi, E.; et al. An alternatively spliced variant of CXCR3 mediates the inhibition of endothelial cell growth induced by IP-10, Mig, and I-TAC, and acts as functional receptor for platelet factor 4. *J. Exp. Med.* **2003**, *197*, 1537–1549. [CrossRef]
41. Sallusto, F.; Mackay, C.R. Chemoattractants and their receptors in homeostasis and inflammation. *Curr. Opin. Immunol.* **2004**, *16*, 724–731. [CrossRef] [PubMed]
42. Bonacchi, A.; Romagnani, P.; Romanelli, R.G.; Efsen, E.; Annunziato, F.; Lasagni, L.; Francalanci, M.; Serio, M.; Laffi, G.; Pinzani, M.; et al. Signal transduction by the chemokine receptor CXCR3: Activation of Ras/ERK, Src, and phosphatidylinositol 3-kinase/Akt controls cell migration and proliferation in human vascular pericytes. *J. Biol. Chem.* **2001**, *276*, 9945–9954. [CrossRef] [PubMed]
43. Korniejewska, A.; McKnight, A.J.; Johnson, Z.; Watson, M.L.; Ward, S.G. Expression and agonist responsiveness of CXCR3 variants in human T lymphocytes. *Immunology* **2011**, *132*, 503–515. [CrossRef] [PubMed]
44. Scholten, D.J.; Canals, M.; Wijtmans, M.; de Munnik, S.; Nguyen, P.; Verzijl, D.; de Esch, I.J.; Vischer, H.F.; Smit, M.J.; Leurs, R. Pharmacological characterization of a small-molecule agonist for the chemokine receptor CXCR3. *Br. J. Pharmacol.* **2012**, *166*, 898–911. [CrossRef]

45. Overington, J.P.; Al-Lazikani, B.; Hopkins, A.L. How many drug targets are there? *Nat. Rev. Drug Discov.* **2006**, *5*, 993–996. [CrossRef]
46. Kufareva, I.; Salanga, C.L.; Handel, T.M. Chemokine and chemokine receptor structure and interactions: Implications for therapeutic strategies. *Immunol. Cell Biol.* **2015**, *93*, 372–383. [CrossRef]
47. Allen, J.A.; Halverson-Tamboli, R.A.; Rasenick, M.M. Lipid raft microdomains and neurotransmitter signaling. *Nat. Rev. Neurosci.* **2007**, *8*, 128–140. [CrossRef]
48. Kenakin, T. Collateral efficacy in drug discovery: Taking advantage of the good (allosteric) nature of 7TM receptors. *Trends Pharmacol. Sci.* **2007**, *28*, 407–415. [CrossRef]
49. Baker, J.G.; Hill, S.J. Multiple GPCR conformations and signaling pathways: Implications for antagonist affinity estimates. *Trends Pharmacol. Sci.* **2007**, *28*, 374–381. [CrossRef]
50. Blanpain, C.; Doranz, B.J.; Bondue, A.; Govaerts, C.; De Leener, A.; Vassart, G.; Doms, R.W.; Proudfoot, A.; Parmentier, M. The core domain of chemokines binds CCR5 extracellular domains while their amino terminus interacts with the transmembrane helix bundle. *J. Biol. Chem.* **2003**, *101*, 746–754. [CrossRef]
51. Kofuku, Y.; Yoshiura, C.; Ueda, T.; Terasawa, H.; Hirai, T.; Tominaga, S.; Hirose, M.; Maeda, Y.; Takahashi, H.; Terashima, Y.; et al. Structural basis of the interaction between chemokine stromal cell-derived factor-1/CXCL12 and its G-protein-coupled receptor CXCR4. *J. Biol. Chem.* **2009**, *284*, 35240–35250. [CrossRef] [PubMed]
52. Wu, B.; Chien, E.Y.; Mol, C.D.; Fenalti, G.; Liu, W.; Katritch, V.; Abagyan, R.; Brooun, A.; Wells, P.; Bi, F.C.; et al. Structures of the CXCR4 chemokine GPCR with small-molecule and cyclic peptide antagonists. *Science* **2010**, *330*, 1066–1071. [CrossRef] [PubMed]
53. Xanthou, G.; Williams, T.J.; Pease, J.E. Molecular characterization of the chemokine receptor CXCR3: Evidence for the involvement of distinct extracellular domains in a multi-step model of ligand binding and receptor activation. *Eur. J. Immunol.* **2003**, *33*, 2927–2936. [CrossRef] [PubMed]
54. Kondru, R.; Zhang, J.; Ji, C.; Mirzadegan, T.; Rotstein, D.; Sankuratri, S.; Dioszegi, M. Molecular interactions of CCR5 with major classes of small-molecule anti-HIV CCR5 antagonists. *Mol. Pharmacol.* **2008**, *73*, 789–800. [CrossRef] [PubMed]
55. Verzijl, D.; Storelli, S.; Scholten, D.J.; Bosch, L.; Reinhart, T.A.; Streblow, D.N.; Tensen, C.P.; Fitzsimons, C.P.; Zaman, G.J.; Pease, J.E.; et al. Noncompetitive antagonism and inverse agonism as mechanism of action of nonpeptidergic antagonists at primate and rodent CXCR3 chemokine receptors. *J. Pharmacol. Exp. Ther.* **2008**, *325*, 544–555. [CrossRef]
56. De Kruijf, P.; van Heteren, J.; Lim, H.; Conti, P.; van der Lee, M.; Bosch, L.; Ho, K.K.; Auld, D.; Ohlmeyer, M.; Smit, M.J.; et al. Nonpeptidergic allosteric antagonists differentially bind to the CXCR2 chemokine receptor. *J. Pharmacol. Exp. Ther.* **2009**, *329*, 783–790. [CrossRef]
57. Garcia-Perez, J.; Rueda, P.; Staropoli, I.; Kellenberger, E.; Alcami, J.; Arenzana-Seisdedos, F.; Lagane, B. New insights into the mechanisms whereby low molecular weight CCR5 ligands inhibit HIV-1 infection. *J. Biol. Chem.* **2011**, *286*, 4978–4990. [CrossRef]
58. Qin, S.; Rottman, J.B.; Myers, P.; Kassam, N.; Weinblatt, M.; Loetscher, M.; Koch, A.E.; Moser, B.; Mackay, C.R. The chemokine receptors CXCR3 and CCR5 mark subsets of T cells associated with certain inflammatory reactions. *J. Clin. Invest.* **1998**, *101*, 746–754. [CrossRef]
59. Cella, M.; Jarrossay, D.; Facchetti, F.; Alebardi, O.; Nakajima, H.; Lanzavecchia, A.; Colonna, M. Plasmacytoid monocytes migrate to inflamed lymph nodes and produce large amounts of type I interferon. *Nat. Med.* **1999**, *5*, 919–923. [CrossRef]
60. Thomas, S.Y.; Hou, R.; Boyson, J.E.; Means, T.K.; Hess, C.; Olson, D.P.; Strominger, J.L.; Brenner, M.B.; Gumperz, J.E.; Wilson, S.B.; et al. CD1d-restricted NKT cells express a chemokine receptor profile indicative of Th1-type inflammatory homing cells. *J. Immunol.* **2003**, *171*, 2571–2580. [CrossRef]
61. Nanki, T.; Takada, K.; Komano, Y.; Morio, T.; Kanegane, H.; Nakajima, A.; Lipsky, P.E.; Miyasaka, N. Chemokine receptor expression and functional effects of chemokines on B cells: Implication in the pathogenesis of rheumatoid arthritis. *Arthritis Res. Ther.* **2009**, *11*, R149. [CrossRef] [PubMed]
62. Groom, J.R.; Luster, A.D. CXCR3 in T cell function. *Exp. Cell Res.* **2011**, *317*, 620–631. [CrossRef] [PubMed]
63. Yates, C.C.; Whaley, D.; Kulasekeran, P.; Hancock, W.W.; Lu, B.; Bodnar, R.; Newsome, J.; Hebda, P.A.; Wells, A. Delayed and deficient dermal maturation in mice lacking the CXCR3 ELR-negative CXC chemokine receptor. *Am. J. Pathol.* **2007**, *171*, 484–495. [CrossRef] [PubMed]

64. Yates, C.C.; Whaley, D.; Y-Chen, A.; Kulasekeran, P.; Hebda, P.A.; Wells, A. ELR-negative CXC chemokine CXCL11 (IP-9/I-TAC) facilitates dermal and epidermal maturation during wound repair. *Am. J. Pathol.* **2008**, *173*, 643–652. [CrossRef] [PubMed]
65. Yates, C.C.; Whaley, D.; Hooda, S.; Hebda, P.A.; Bodnar, R.J.; Wells, A. Delayed reepithelialization and basement membrane regeneration after wounding in mice lacking CXCR3. *Wound Repair Regen.* **2009**, *17*, 34–41. [CrossRef] [PubMed]
66. Luster, A.D.; Leder, P. IP-10, a -C-X-C- chemokine, elicits a potent thymus-dependent antitumor response in vivo. *J. Exp. Med.* **1993**, *178*, 1057–1065. [CrossRef] [PubMed]
67. Sgadari, C.; Farber, J.M.; Angiolillo, A.L.; Liao, F.; Teruya-Feldstein, J.; Burd, P.R.; Yao, L.; Gupta, G.; Kanegane, C.; Tosato, G. Mig, the monokine induced by interferon-gamma, promotes tumor necrosis in vivo. *Blood* **1997**, *89*, 2635–2643. [CrossRef]
68. Hensbergen, P.J.; Wijnands, P.G.; Schreurs, M.W.; Scheper, R.J.; Willemze, R.; Tensen, C.P. The CXCR3 targeting chemokine CXCL11 has potent antitumor activity in vivo involving attraction of CD8+ T lymphocytes but not inhibition of angiogenesis. *J. Immunother.* **2005**, *28*, 343–351. [CrossRef]

International Journal of
Molecular Sciences

MDPI

Article

Antiseptic Effect of Ps-K18: Mechanism of Its Antibacterial and Anti-Inflammatory Activities

Mihee Jang [1], Jieun Kim [1], Yujin Choi [2], JeongKyu Bang [3] and Yangmee Kim [1,*]

1 Department of Bioscience and Biotechnology, Research Institute for Bioactive-Metabolome Network, Konkuk University, Seoul 05029, Korea; smileday1229@konkuk.ac.kr (M.J.); za3524@konkuk.ac.kr (J.K.)

2 Chuncheon Center, Korea Basic Science Institute, Chuncheon 24341, Korea; cyj4854@gmail.com

3 Protein Structure Group, Korea Basic Science Institute, Ochang, Cheongju, Chung-Buk 28199, Korea; bangjk@kbsi.re.kr

* Correspondence: ymkim@konkuk.ac.kr; Tel.: +822-450-3421; Fax: +822-447-5987

Received: 4 September 2019; Accepted: 30 September 2019; Published: 2 October 2019

Abstract: Recently, bioactive peptides have attracted attention for their therapeutic applications in the pharmaceutical industry. Among them, antimicrobial peptides are candidates for new antibiotic drugs. Since pseudin-2 (Ps), isolated from the skin of the paradoxical frog *Pseudis paradoxa*, shows broad-spectrum antibacterial activity with high cytotoxicity, we previously designed Ps-K18 with a Lys substitution for Leu^{18} in Ps, which showed high antibacterial activity and low toxicity. Here, we examined the potency of Ps-K18, aiming to develop antibiotics derived from bioactive peptides for the treatment of Gram-negative sepsis. We first investigated the antibacterial mechanism of Ps-K18 based on confocal micrographs and field emission scanning electron microscopy, confirming that Ps-K18 targets the bacterial membrane. Anti-inflammatory mechanism of Ps-K18 was investigated by secreted alkaline phosphatase reporter gene assays and RT-PCR, which revealed that Ps-K18 activates innate defense via Toll-like receptor 4-mediated nuclear factor-kappa B signaling pathways. Moreover, we investigated the antiseptic effect of Ps-K18 using a lipopolysaccharide or *Escherichia coli* K1-induced septic shock mouse model. Ps-K18 significantly reduced bacterial growth and inflammatory responses in the septic shock model. Ps-K18 showed low renal and liver toxicity and attenuated lung damage effectively. This study suggests that Ps-K18 is a potent peptide antibiotic that could be applied therapeutically to Gram-negative sepsis.

Keywords: pseudin-2; antimicrobial peptide; antisepsis, peptide antibiotics

1. Introduction

Naturally occurring bioactive peptides in various organisms are selective and effective cellular signaling molecules that play an important role either directly or indirectly in physiological processes. The peptides released through systems such as food processing or microbial fermentation play physicochemical roles to regulate important processes and exert beneficial effects on body functions [1]. Most peptides bind certain cell surface receptors such as G protein-coupled receptors (GPCRs) or ion channels, causing intracellular effects [2]. They can act as hormones, neurotransmitters, growth factors, ion channel ligands, or anti-infectious agents, and accordingly, these characteristics have become valuable to treat diseases that could not be cured previously [2,3]. Recently, bioactive peptides derived from nature have received attention due to their pharmacological effects and clinical potential [2,4]. For example, Lyxumia® (Lixisenatide), an analogue of glucagon-like peptide 1 (GLP-1) isolated from Gila monster venom, was launched by Sanofi and is a GLP-1 receptor agonist for the treatment of type II diabetes that received approval from the FDA [5,6]. Further, Ile-Pro-Pro (IPP) and Val-Pro-Pro (VPP) are found in milk fermented with *Lactobacillus helveticus*, and are known as angiotensin-converting enzyme inhibitors that lower blood pressure [7]. Currently, many peptide drugs approved by the FDA have been supplied to the market and the number of peptide drugs under clinical development is steadily increasing. [2,6].

Various antibiotics have been identified since the discovery of penicillin, and proven effective in treating bacterial infections [8]. Imipenem, the first clinically available carbapenem antibiotic, and gentamicin, one of aminoglycoside antibiotics, are highly efficacious antibiotics against Gram-negative bacteria [9,10]. However, emergence of their multidrug-resistant (MDR) Gram-negative bacteria is a growing threat and the development of new antibiotic drugs are necessary [11,12]. Antimicrobial peptides (AMPs), molecules that have come under the spotlight as new antibiotics, are important components of the immune system found in all living organisms [13]. AMPs not only effectively inhibit bacterial infections but also exert an immunosuppressive effect and have potential as future therapeutics. However, these peptides have some limitations in that they are difficult to use directly as a treatment due to their poor physicochemical stability and cytotoxicity [2]. Polymyxin E, also known as colistin, shows potent antimicrobial activities against Gram-negative bacteria but it is used as a last resort due to kidney toxicity [4,14]. The decrease in toxicity mediated by chemical and sequence modification suggests an important strategy for the development of bioactive peptide-derived antibiotics [4]. Therefore, the design of highly functional antimicrobial peptides based on structure–activity relationships is expected to provide new possibilities for the development of antibiotics, since structural features such as hydrophobic and amphipathic properties affect their activities [15,16].

With the emergence of MDR Gram-negative bacteria and the failure of most new antibiotics in clinical trials, the development of new antibiotics has great importance in Gram-negative sepsis treatment [17]. Sepsis, a disease caused by bacterial infections, still has a high mortality rate of 35%–45% despite advances in antibiotic development and early targeted treatments [18,19]. Sepsis is caused mainly by the host's reaction to lipopolysaccharide (LPS), an endotoxin [20]. LPS is located in the outer membrane of Gram-negative bacteria and is released in the infected host. This molecule is recognized by Toll-like receptors (TLRs) as a pathogen-associated molecular pattern (PAMP). TLR4 is activated by lipid A of LPS to induce the release of pro-inflammatory cytokines, which are important to activate the host immune response. An inflammatory reaction activated by the bacterial infection can lead to sepsis, a systemic inflammatory response syndrome [21–23].

Pseudin-2 (Ps; GLNALKKVFQGIHEAIKLINNHVQ) is a naturally occurring AMP derived from the skin of the paradoxical frog *Pseudis paradoxa*. Ps has a linear amphipathic α-helix structure from Leu^2 to Glu^{24} residues and has shown efficacy against various bacteria, but it exerts severe cytotoxicity against mammalian cells [13]. In previous studies, we designed a Ps-K18 analogue by substituting Lys for Leu^{18} of Ps, making it less toxic, while maintaining the antibacterial activity of the parent peptide [24]. However, the detailed antibacterial and anti-inflammatory mechanism of Ps-K18 in relation to TLR4 signaling needs to be elucidated further.

The objective of the current study was to elucidate the antibacterial and anti-inflammatory mechanisms and evaluate the benefits of Ps-K18 as a bioactive peptide against sepsis induced by LPS or bacterial infection. With these aims, we examined the ability of Ps-K18 to inhibit bacterial growth and inflammation in vitro and in vivo. We elucidated the mechanism underlying the high antibacterial potency of Ps-K18, as well as its anti-inflammatory activity through TLR4 signaling. We also demonstrated the beneficial effects of Ps-K18 in vivo by monitoring the recovery of damaged organs in mouse models of endotoxemia or *E. coli* K1-induced septic shock. Our research results might provide insight into development of potent bioactive peptides for the treatment of sepsis caused by bacterial infection.

2. Results

2.1. Properties of Ps-K18

Most AMPs target bacteria by permeabilizing the cell membrane, resulting in its loss of function [25]. Structural features such as net charge, helicity, hydrophobicity, and amphipathicity are important factors affecting their antibacterial activity including their interactions with bacterial membranes [26].

Thus, design based on the structure–activity of peptides is one of the most basic methods for developing new antibiotics [15,16]. Net charge is an important factor for the early interaction with negatively-charged bacterial membranes [27]. The solution structure of Ps has been determined as a linear α-helix from Leu^2 to Glu^{24} residues and an amphipathic structure wherein one side is hydrophobic and the other side is hydrophilic with only a +2 net charge [24]. These specific structural features of Ps result in high cytotoxicity against mammalian cells, as well as bacteria. In our previous study, we designed Ps-K18, increasing the cationicity to +3 by replacing Leu^{18} between hydrophilic and hydrophobic sides with a Lys residue in the amphipathic α-helix (Table 1, Figure 1B); this resulted in lower toxicity but the maintenance of Ps antibacterial activity [24]. Peptide properties of Ps and Ps-K18 are listed in Table 1. The helical-wheel diagrams of Ps and Ps-K18 in Figure 1 show that replacing the Leu^{18} of Ps with Lys resulted in increased amphipathicity of the α-helical structure and cationicity, as well as decreased hydrophobicity. Decreased hydrophobicity can also reduce the antimicrobial activity of AMPs, whereas increased hydrophobicity can kill not only bacteria but also eukaryotic cells [27].

Table 1. Amino acid sequences and peptide properties.

Peptide	Sequence	Molecular Weight[a]	Net Charge	Hydrophobicity[b] <H>
Ps	GLNALKKVFQGIHEAIKLINNHVQ	2685	+2	0.407
Ps-K18	GLNALKKVFQGIHEAIKKINNHVQ	2702	+3	0.295

[a] The molecular weight (MW) was measured by mass spectroscopy. [b] Hydrophobicity <H> was calculated online at: http://heliquest.ipmc.cnrs.fr/cgi-bin/ComputParams.py.

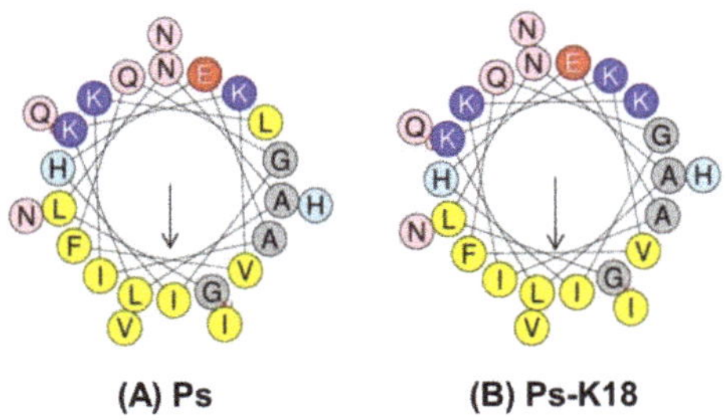

Figure 1. Helical-wheel diagrams of Ps and Ps-K18 using HeliQuest (http://heliquest.ipmc.cnrs.fr/). Helical-wheel diagrams of (**A**) Ps and (**B**) Ps-K18. Positively charged residues are shown in blue, negatively charged residues in red, and hydrophobic residues in yellow at the bottom of the wheel. In addition, Gly and Ala are shown in gray, Asn and Gln in pink, and His in sky blue. The arrows are presented based on the helical hydrophobic moment.

2.2. Antibacterial Activity

Previous studies showed high antibacterial activity of Ps and Ps-K18 against broad-spectrum Gram-negative bacteria [24]. We confirmed the antibacterial activities of the peptide only against Gram-negative bacteria and multidrug-resistant (MDR) Gram-negative bacteria to conduct further experiments on Ps-K18. Antibacterial activities were examined for four Gram-negative bacteria (*Escherichia coli* KCTC1682, *E. coli* K1, *Pseudomonas aeruginosa* KCCM11328, and *Acinetobacter baumannii*

KCCM 40203) and three MDR Gram-negative bacteria (*E. coli* CCARM 1229, *P. aeruginosa* CCARM 2003, *A. baumannii* CCARM 12010). The antimicrobial activities of Ps and Ps-K18 are summarized in Table 2. Melittin, with high antibacterial activity and toxicity, was used as a control to compare antibacterial activity. Ps-K18 had similar antibacterial activity to Ps against standard and MDR Gram-negative bacteria. These results indicated that Ps-K18 is as effective as Ps against Gram-negative bacteria, and the results revealed that Ps-K18 is a potent antibiotic against MDR strains as well as standard strains.

Table 2. Antimicrobial activities of the Ps and Ps-K18 against standard bacterial strains and MDR bacterial strains.

MIC (μM)	Ps	Ps-K18	Melittin
Standard Gram-negative bacteria			
E. coli KCTC1682	4	4	4
E. coli K1	2	2	2
P. aeruginosa KCCM11328	4	4	8
A. baumannii KCCM40203	2	2	2
MDR Gram-negative bacteria			
E. coli CCARM 1229	2	2	1
P. aeruginosa CCARM 2003	2	2	2
A. baumannii CCARM 12010	2	2	1
GM[a]	2.57	2.57	2.86

[a] The geometric means (GM) are the average values of minimum inhibitory concentrations (MICs).

2.3. Visualization of the Interaction Between E. coli and Ps Peptides based on Confocal Micrographs and Field Emission Scanning Electron Microscopic (FE-SEM) Micrographs

Ps and Ps-K18 showed high depolarization activity against *E. coli* as the concentration increased, implying that Ps and Ps-K18 target the bacterial membrane, in our previous study [24]. We further investigated the interaction between Ps peptides and bacteria. We first investigated the interaction between *E. coli* and FITC-labeled Ps or Ps-K18 by confocal microscopy (Figure 2A). For this, we treated *E. coli* with each peptide at the 1× MIC for 1 h. The Ps peptides were mainly localized on the surface of the *E. coli* membrane and no fluorescence was observed inside the cell.

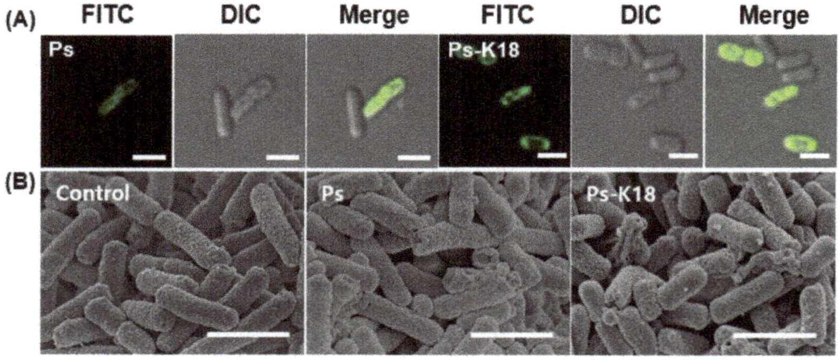

Figure 2. Confocal and scanning electron micrographs of *E. coli* treated with Ps and Ps-K18. (**A**) Confocal micrographs of Ps and Ps-K18 at a 1× MIC after 1 h incubation. The scale bar is 2 μm. (**B**) FE-SEM micrographs of control, no peptide treatment, and Ps and Ps-K18 (1× MIC) after 4 h incubation. The scale bar is 2 μm.

The results suggested that Ps peptides attached to the surface of the bacterial membrane and caused disruption of the *E. coli* membrane. These results agree with the previous results obtained from dye leakage and depolarization experiments, indicating that Ps and Ps-K18 target the bacterial membrane [24]. We next visualized membrane damage induced by Ps or Ps-K18 using FE-SEM (Figure 2B). *E. coli* cells treated with a 1× MIC of Ps for 4 h were wrinkled and contracted, whereas control cells had a normal oval shape (Figure 2B). *E. coli* cells treated with a 1× MIC of Ps-K18 were crumpled and contracted, similar to that with Ps treatment. This proved that Ps-K18, which retained high antibacterial activity similar to that of Ps, effectively interacts with the membrane to cause fatal damage to the bacterium.

2.4. Cytotoxicity of Ps and Ps-K18 In Vitro and In Vivo

Since toxicity of highly active peptides has always been a limit for the therapeutic application of peptide drugs, we examined the toxicity of Ps and Ps-K18 against mammalian cells [2]. The survival rates of mouse macrophage RAW264.7 cells and human embryonic kidney HEK-293 cells are shown in Figure 3A,B. Melittin showed extremely high cytotoxicity at lower concentrations. At 50 μM, survival rates were 18% and 31% for Ps and 82% and 84% for Ps-K18 against RAW264.7 cells and HEK-293 cells, respectively.

We then evaluated the in vivo cytotoxicity of Ps peptides by measuring levels of alanine aminotransferase (ALT), aspartate aminotransferase (AST), and blood urea nitrogen (BUN) in the serum of mice. ALT and AST are used as indicators of liver damage, whereas BUN is used to assess kidney damage [28]. When Ps was injected into mice, levels of AST, ALT, and BUN increased with higher concentrations (Figure 3C–E). However, Ps-K18 did not increase the level of AST and BUN as concentrations increased, but increased ALT only at very high dose, 20 mg/kg, although much lower than Ps. These results support that Ps-K18 has improved potential as a therapeutic agent because it is less toxic than the Ps while maintaining high antibacterial activity.

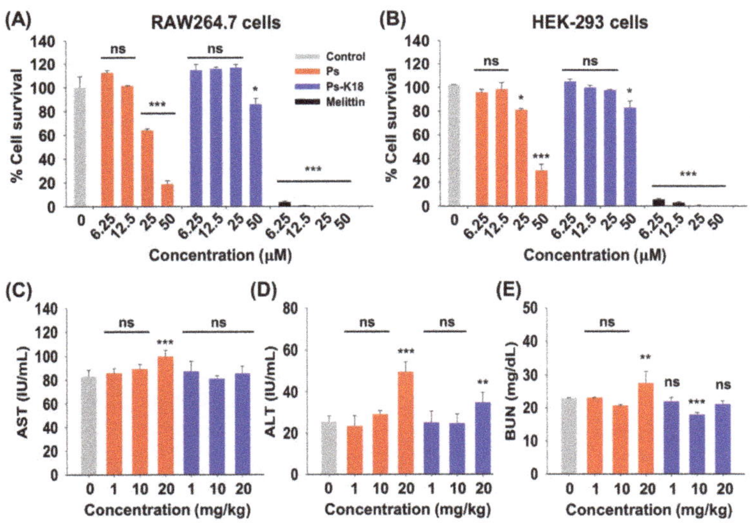

Figure 3. Cytotoxicity of Ps and Ps-K18 in vitro and in vivo. Concentration-dependent toxicities of pseudin-2 (Ps) and Ps-K18 against (**A**) RAW264.7 cells and (**B**) HEK-293 cells. (**C**) Aminotransferase (AST); (**D**) alanine aminotransferase (ALT); and (**E**) blood urea nitrogen (BUN) levels in mouse serum. The error bars represent means ± standard error of the mean. (*$p < 0.05$; **$p < 0.01$; ***$p < 0.001$; ns. represents no significance).

2.5. Specificity of Ps-K18 towards Various TLRs

TLRs, which play an important role in the innate immune system, are activated by the specific agonist molecules, causing an inflammatory reaction [29]. Pam_2CSK_4 activates TLR2/6 signaling and

Pam_3CSK_4 activates the TLR2/1 inflammation response. Polyinosinic:polycytidylic acid (Poly(I:C)) activates TLR3, and LPS activates TLR4 inflammatory signaling. TLR7 and TLR9 can be activated by imiquimod and ODN1826, respectively [30].

We investigated the specificity of Ps-K18 towards various TLRs. For this, we treated RAW264.7 cells with Ps-K18 and known TLR-specific agonists (Figure 4A). Ps-K18 significantly suppressed TLR4-mediated signaling in RAW264.7 cells by 15% and 43% based on LPS-induced nitrite production at 10 μM and 20 μM, respectively. Ps-K18 did not effectively inhibit nitrite production in the presence of other agonists (Pam_2CSK_4, Pam_3CSK_4, polyinosinic:polycytidylic acid (Poly(I:C)), imiquimod, ODN1826) and TLRs, implying that Ps-K18 selectively modulates LPS-induced TLR4 signaling.

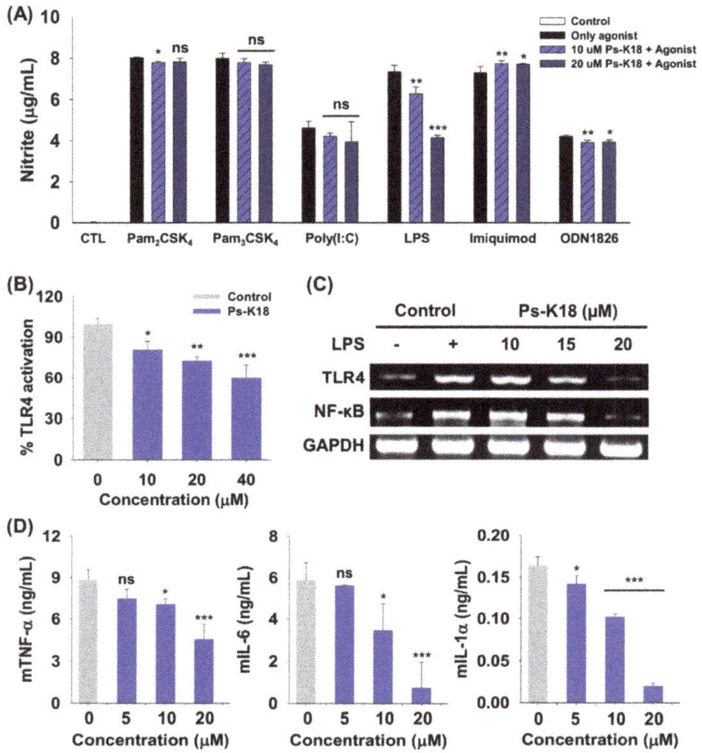

Figure 4. Specificity of Ps-K18 toward various Toll-like receptors (TLRs) and TLR-specific agonists, and anti-inflammatory activity of Ps-K18. (**A**) Specificity of Ps-K18 for TLRs and TLR-specific agonists that selectively activate TLRs, as determined by measuring nitrite production in RAW264.7 cells. Different TLRs were selectively activated by Pam_2CSK_4 (200 ng/mL), Pam_3CSK_4 (200 ng/mL), Poly(I:C) (2 μg/mL), LPS (20 ng/mL), imiquimod (1 μg/mL), and ODN1826 (20 μg/mL). Control is the data without agonist treatment. (**B**) Dose-dependent reduction in secreted alkaline phosphatase reporter gene (SEAP) activity by Ps-K18 in LPS (20 ng/mL)-stimulated HEK-Blue™ hTLR4 cells. (**C**) Inhibitory effect of Ps-K18 on inflammation-related gene expression in LPS (50 ng/mL)-stimulated RAW264.7 cells. The cells were pre-treated with Ps analogues for 1 h and then treated with LPS for 12 h. (**D**) Inhibitory effects on the production of inflammatory cytokines tumor necrosis factor-α (TNF-α), interleukin (IL)-6, and IL-1α by Ps-K18 (5 μM, 10 μM, and 20 μM) in LPS (20 ng/mL)-stimulated RAW264.7 cells. The error bars represent means ± standard error of the mean (*$p < 0.05$; **$p < 0.01$; ***$p < 0.001$; ns. represents no significance).

2.6. Ps-K18 Suppresses Inflammation Response through TLR4-Mediated Signaling in RAW264.7 Cells and HEK-Blue^{TM} hTLR4 Cells

To investigate the mechanism underlying the anti-inflammatory activity of Ps-K18, we performed a secreted alkaline phosphatase reporter (SEAP) gene assay (Figure 4B). LPS, released from Gram-negative bacteria, stimulates cells by activating TLR4-mediated signaling to induce inflammatory cytokine and nitric oxide release, leading to sepsis [31]. HEK-Blue^{TM} hTLR4 cells that express TLR4 contain an SEAP gene located downstream from the nuclear factor-kappa B (NF-κB) promoter, which results in TLR4-mediated reporter activation when the cells are stimulated with LPS [32]. Ps-K18 inhibited 19%, 27%, and 40% of SEAP activity efficiently at 10, 20, and 40 μM, respectively.

TLR4 activated by LPS induces NF-κB phosphorylation. Further, phosphorylated NF-κB is translocated to the nucleus and activates inflammatory cytokine gene transcription [30]. We also confirmed that the gene expression of TLR4 and NF-κB can be inhibited by Ps-K18 in LPS-stimulated RAW264.7 cells through Reverse transcription polymerase chain reaction (RT-PCR). By increasing the concentration of Ps-K18, it effectively suppressed the gene expression of TLR4 and NF-κB (Figure 4C). When we quantified relative gene expression of TLR4 and NF-κB using ImageJ software, they were decreased by approximately 83% and 85%, respectively, at 20 μM. These results showed that Ps-K18 exerts anti-inflammatory activity by inhibiting LPS-induced TLR4-mediated signaling.

We further investigated the anti-inflammatory activity of Ps-K18 to confirm inhibitory effects on TLR4-mediated signaling (Figure 4D). Ps-K18 effectively inhibited the production of the inflammatory cytokines mTNF-α, mIL-6 and mIL-1α depending on concentration. Ps-K18 suppressed cytokines mTNF-α by 49%, mIL-6 by 87% and mIL-1α by 88%, at 20 μM. These results showed that Ps-K18 has potentially high anti-inflammatory activity that can effectively inhibit inflammatory responses with low cytotoxicity.

2.7. Antisepsis Effect of Ps-K18 in LPS-Induced Endotoxemia Mouse Model

Since Ps-K18 is a potent peptide with high antibacterial and anti-inflammatory activity in vitro, we next confirmed its potential as a therapeutic agent to treat sepsis in vivo. For this, we first investigated the anti-inflammatory activity of Ps-K18 using an endotoxemia mouse model induced by LPS. LPS is an endotoxin located in the outer membrane of Gram-negative bacteria, and triggers a severe inflammatory response, causing sepsis in the body [20]. Cytokine storms that can lead to organ damage and death aggravate the inflammatory response during endotoxemia [33]. The group injected only with LPS showed high levels of inflammatory cytokine expression (Figure 5A,B). Only the Ps-K18 treatment group showed similar results to the control group (Figure 5). Ps-K18 resulted in a marked reduction in inflammatory cytokines mTNF-α and mIL-6. In the serum, Ps-K18 effectively reduced inflammatory cytokines mTNF-α by 65% and mIL-6 by 57% when Ps-K18 was injected with LPS in mice. (Figure 5A). Ps-K18 also inhibited inflammatory cytokines mTNF-α by 64% and mIL-6 by 56% in the lung (Figure 5B).

We also tested whether Ps-K18 can remove LPS endotoxin in vivo by performing a limulus amebocyte lysate (LAL) assay that measures the amount of endotoxin in the sample (Figure 5C). Ps-K18 reduced the endotoxin levels by 53% when injected with LPS. These results revealed that Ps-K18 can effectively decrease endotoxin levels in vivo. In addition, the levels of AST, ALT, and BUN were increased by LPS but were effectively reduced by Ps-K18 treatment (Figure 5D). Levels of AST, ALT, and BUN were reduced by 18%, 19%, and 49%, respectively, when Ps-K18 was injected with LPS into the mice. These results showed that Ps-K18 treatment can improve the functions of organs damaged by the endotoxin LPS in vivo.

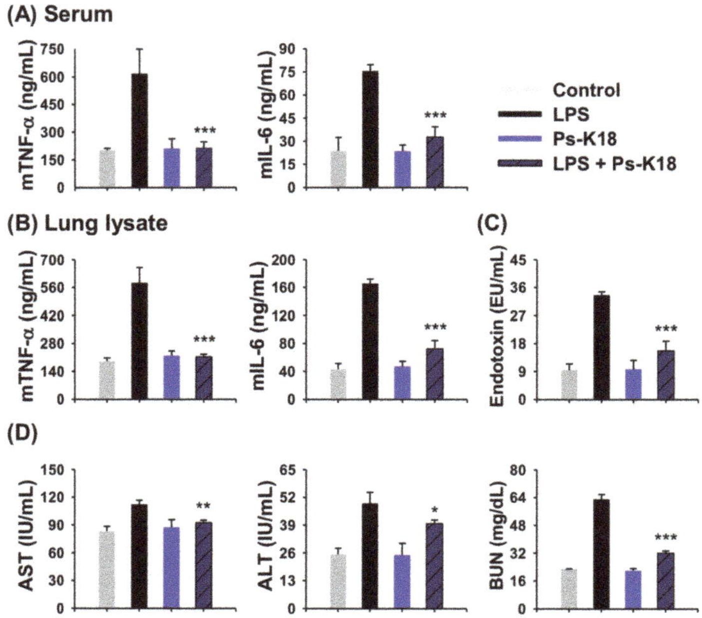

Figure 5. Effects of Ps-K18 on LPS-induced endotoxemia mouse model. (**A**) Inhibition of cytokine production (mTNF-α and mIL-6) in the serum. (**B**) Inhibition of cytokine production (mTNF-α and mIL-6) in lung lysates. (**C**) Reduction of LPS endotoxin by Ps-K18 in serum of the LPS-induced sepsis model. (**D**) AST, ALT, and BUN levels in the mouse septic shock model induced by LPS. BALB/c mice were pre-treated with 1 mg/kg of peptide, followed by intraperitoneal injection with 15 mg/kg of LPS. Data are presented as means ± standard errors of the mean (*$p < 0.05$; **$p < 0.01$; ***$p < 0.001$; ns. represents no significance) compared to LPS-injected group. All measurements were obtained five times.

2.8. Antisepsis Effect of Ps-K18 Based on an E. coli K1-Induced Mouse Model of Septic Shock

Since sepsis is one of diseases caused by bacterial infection and there are urgent needs to develop new antiseptic agents [18,19], we examined the potency of Ps-K18 as a candidate antiseptic peptide drug. We used an intact *E. coli* K1-induced septic shock mouse model to confirm its potential as a therapeutic agent for sepsis. Ps-K18 reduced the number of bacteria in the mouse lung, liver, and kidney by approximately 73%, 87%, and 71%, respectively. This indicated that Ps-K18 can effectively inhibit the growth of bacteria, *E. coli* K1 in vitro and in vivo (Figure 6A). Because Ps-K18 had high anti-inflammatory activity in vitro, we then measured the in vivo anti-inflammatory activity using serum and lung lysates of mice with *E. coli* K1-induced septic shock. We detected high levels of mTNF-α and mIL-6 in the experimental group only treated *E. coli* K1. However, Ps-K18 reduced TNF-α levels by approximately 45% and mIL-6 by 43% in the serum when injected with *E. coli* K1 (Figure 6B). Ps-K18 also reduced mTNF-α levels by 51% and mIL-6 by 41% in the lungs of septic mice (Figure 6C).

We also performed LAL assays to confirm the potency of Ps-K18 using the *E. coli* K1-induced sepsis model. Ps-K18 reduced levels of endotoxin by 61% when administered with *E. coli* K1 (Figure 6D). The levels of AST, ALT, and BUN were also measured using the blood of mice with sepsis induced by *E. coli* K1. Ps-K18 reduced the levels of AST by 35%, ALT by 69%, and BUN by 45%, when injected with *E. coli* K1 (Figure 6E). This showed that Ps-K18 is effective for the treatment sepsis by inhibiting the growth of bacteria and restoring liver and kidney function after damage mediated by bacterial infection in vivo.

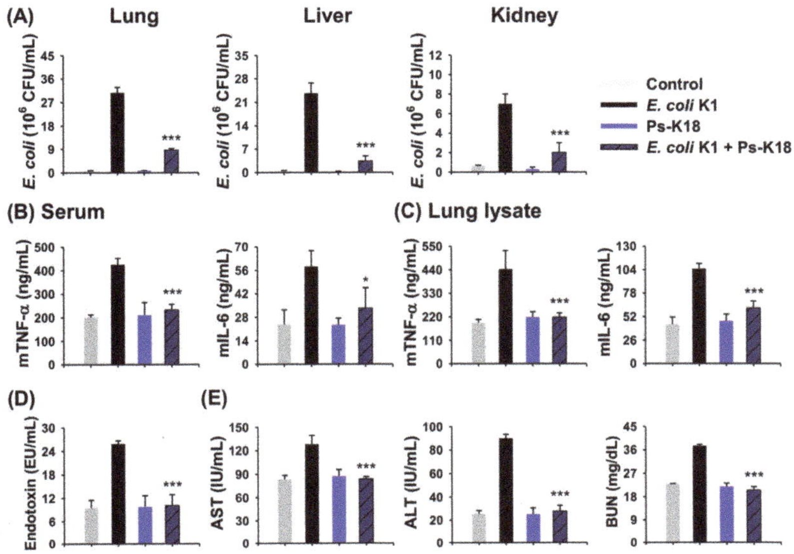

Figure 6. Effects of Ps-K18 on septic shock in mice induced by *E. coli* K1. (**A**) Inhibition of bacterial growth in the lung, liver, and kidney of the septic shock model. BALB/c mice were pre-treated with 1 mg/kg of peptide, followed by intraperitoneal injection with *E. coli* K1 (1×10^6 CFU/mouse). (**B**) Inhibition of cytokine production (mTNF-α and mIL-6) in the serum. (**C**) Inhibition of cytokine production (mTNF-α and mIL-6) in lung lysates. (**D**) Reduction of LPS endotoxin by Ps-K18 in serum of the *E. coli* K1-induced sepsis model. (**E**) AST, ALT, and BUN levels in the mouse septic shock model induced by *E. coli* K1. Data are presented as means ± standard errors of the mean (*$p < 0.05$; **$p < 0.01$; ***$p < 0.001$; ns. represents no significance) compared to *E. coli* K1-injected group. All measurements were obtained five times.

2.9. Ps-K18 Treatment Effectively Suppresses Polymorphonuclear Lymphocyte Infiltration in LPS-Induced Endotoxemia Mouse Model and E. coli K1-Induced Mouse Model

We also pathologically investigated the lung tissues obtained from LPS-induced endotoxemia and *E. coli* K1-induced mouse models. Capillary-alveolar barrier dysfunction occurs during the infiltration of polymorphonuclear lymphocytes (PMNs) into the lungs, which is followed by pulmonary edema with impaired pulmonary gas exchange [34,35]. PMNs are associated with lung damage and are an important cause of acute lung injury, which results in high mortality and morbidity during sepsis. Therefore, reducing the number of PMNs is of great importance for the treatment of sepsis [36]. Figure 7 shows the pathological characteristics of lung damage induced by LPS or *E. coli* K1 infection, which were improved by Ps-K18. Control mice, not treated with LPS or *E. coli* K1, and Ps-K18 mice, only injected with Ps-K18, showed normal lung morphology, whereas LPS or *E. coli* K1 treatment induced morphological changes and PMN infiltration (Figure 7A–D). However, Ps-K18 treatment effectively suppressed PMN infiltration and significantly improved lung morphology, similar to that observed in the control group (Figure 7E,F). All results obtained from the septic shock model revealed that Ps-K18 can be used to treat sepsis in vivo.

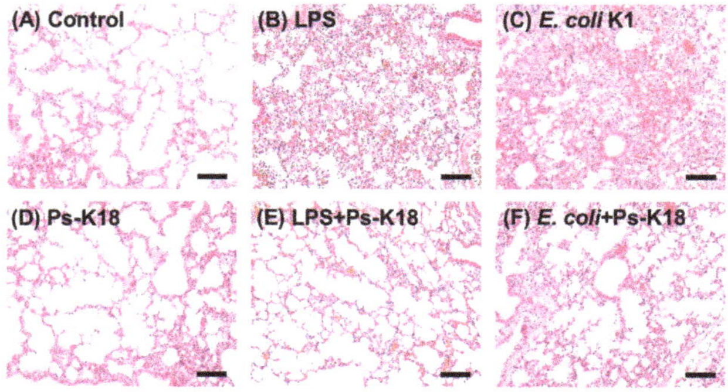

Figure 7. Effect of Ps-K18 on neutrophil infiltration in the lungs of septic shock mouse induced by LPS or *E. coli* K1. Lung morphology of (**A**) control, no peptide treatment, (**B**) LPS (15 mg/kg), (**C**) *E. coli* K1 (1×10^6 CFU/mouse), (**D**) Ps-K18 (1 mg/kg), (**E**) LPS + Ps-K18, injected with both LPS and Ps-K18, and (**F**) *E. coli* K1 + Ps-K18, injected with both *E. coli* K1 and Ps-K18. LPS or *E. coli* K1 was pre-injected in mice for 1h, and then Ps-K18 injected in mice. The scale bar is 100 μm.

3. Discussion

Sepsis is a highly lethal disease that induces severe inflammatory reactions throughout the body [37,38]. In particular, sepsis caused by MDR Gram-negative bacteria is more difficult to treat as it is associated with reductions in the efficacy of existing antibiotics [39]. To meet these social needs, new drug development is urgently needed. Among the antibiotics that have been developed, peptide antibiotics have received attention as new therapeutic agents due to their outstanding potency with strong antibacterial activity against drug-resistant strains and immunosuppressive properties, and act by controlling host defense systems [4,15,24,40]. Despite these properties, low cell selectivity and high cytotoxicity of AMPs are obstacles that impede their use in disease treatment. Because the activity of AMPs is affected by physiochemical properties such as net charge helicity, hydrophobicity, and amphipathicity, the optimization of these properties is one of the most common means to develop highly-functional AMPs [41,42]. Colistin, also called polymyxin E, is a peptide antibiotic that represents the last resort against MDR Gram-negative sepsis due to its high toxicity [43,44]. Therefore, there are rapid needs for the development of new peptide antibiotics that can protect against bacterial infection with high safety; this offers the potential for new solutions to eliminate rapidly emerging multidrug-resistant bacterial strains.

AMPs have a variety of net charge ranges from +2 to +9 [45]. The cationicity of peptides is known to be essential for their insertion into and destruction of cell membranes because it leads to structural instability in bacteria due to interactions with their negatively charged membrane, thereby improving the antibacterial activity [42,46,47]. It was reported that a truncated Ps analogue, with increased net charge, has excellent inhibitory effects on bacterial growth [48]. In a previous study, we designed Ps-K18 (+3) with a Lys substitution for Leu^{18} in Ps and showed that it has high antibacterial activity and low toxicity [24]. Here, we investigated the detailed antibacterial mechanism of Ps-K18 as a potent peptide antibiotic, using confocal micrographs and SEM analysis (Figure 2). Ps peptides were located at the surface of the membranes of *E. coli* (Figure 2A) and the morphology of *E. coli* changed into the wrinkled and disrupted shape after Ps or Ps-K18 treatment (Figure 2B). These results revealed the antibacterial mechanism of Ps or Ps-K18, which effectively inhibited the growth of bacteria by targeting bacterial membranes, confirming our previous results obtained from dye leakage and depolarization assays against various membranes [24].

We checked the toxicity of Ps peptides for its safe therapeutic application. Ps-K18 had very low toxicity compared to Ps in vivo and in vitro (Figure 3). LPS released from Gram-negative bacteria activates TLR4 to induce the release of pro-inflammatory cytokines, which are important to activate the host immune response [20]. Because serious inflammatory responses can induce sepsis, there is extensive interest in peptide antibiotics that suppress inflammatory responses and inhibit bacterial growth. Furthermore, Ps-K18 exerted significant effects in specifically suppressing LPS-induced TLR4 signaling, among various TLRs (Figure 4A). We then investigated the anti-inflammatory activity of Ps-K18 in LPS-stimulated RAW264.7 cells (Figure 4B–D). Ps-K18 effectively suppressed the production of inflammatory TNF-α, IL-6, and IL-α (Figure 4D). Our previous study showed that Ps and Ps-K18 bind TLR4 tightly with 0.8 and 1.5 μM affinity, respectively [24]. The detailed mechanism underlying the anti-inflammatory activity of Ps-K18 was demonstrated by SEAP assays and RT-PCR, implying that Ps-K18 has remarkable anti-inflammatory activity by suppressing TLR4-mediated NF-κB signaling pathways.

We demonstrated that Ps-K18 is a potent peptide antibiotic in vitro, with antibacterial activity that effectively relies on interactions with bacterial membranes to induce the death of bacteria, as well as anti-inflammatory effects that inhibit TLR4-mediated NF-κB signaling pathways. To prove that Ps-K18 is an effective antibiotic not only in vitro but also in vivo, we conducted in vivo experiments using LPS-induced endotoxemia or *E. coli* K1-induced septic shock mouse model. Sepsis, mainly induced by LPS released from bacteria, activates all components of the immune system, including various blood cells, endothelial cells, and bone marrow, to remove infecting bacteria from the body. During this process, major organs such as the lungs, abdomen, and urinary tract appear to be damaged by the release of different mediators [49]. The in vivo efficacy of Ps-K18 was examined by monitoring organ injury in LPS-induced endotoxemia and *E. coli* K1-induced septic shock models (Figures 5 and 6). Importantly, Ps-K18 effectively inhibited the growth of Gram-negative bacteria and the associated production of inflammatory cytokines mTNF-α and mIL-6, ameliorated liver and kidney damage induced by infection, and reduced the amount of endotoxin LPS in serum. From LAL assay in septic shock model, we showed that Ps-K18 can effectively decrease endotoxin LPS levels in vivo. Antimicrobial peptides such as polymyxin B and LL-37 are well known LPS neutralizing agents which can prevent the initiation of the inflammatory responses [50]. LPS in the blood can stimulate the expression of inflammatory cytokines, resulting in septic shock [51]. In case of Gram-negative infection, where the innate immunity mediator TLR4 is already significantly enhanced, antiseptic agent should be able to suppress the LPS or bacteria-induced TLR4 activation which eventually causes septic shock. As shown in this study, Ps-K18 can be a potent antiseptic agent, having both LPS neutralizing ability as well as inhibiting ability of systemic TLR4-mediated inflammatory signaling by blocking TLR4/NF-κB activation in septic shock mouse model. Our results confirmed that Ps-K18 could be an excellent therapeutic agent for treatment of sepsis.

4. Materials and Methods

4.1. Peptide Synthesis

All peptides listed in Table 1, synthesized by solid-phase synthesis using N-(9-fluorenyl) methoxycarbonyl solid-phase synthesis technique, were purified by reversed-phase preparative high-performance liquid chromatography [52]. Purities of the peptides (>95%) were measured using an analytical C_{18} column. Molecular masses and retention times of peptides were determined by matrix-assisted laser-desorption ionization-time-of-flight (MALDI-TOF) mass spectrometry at Korea Basic Science Institute (KBSI, Ochang, Korea).

4.2. Antibacterial Activity

We purchased *E. coli* KCTC 1682, *P. aeruginosa* KCCM 11328, *A. baumannii* KCCM 40203 from the Korean Collection for Type Cultures (KCTC) (Taejon, Korea), and Korean Culture Center of Microorganisms (KCCM) (Seoul, Korea). MDR Gram-negative bacteria (*E. coli* CCARM 1229, *P. aeruginosa* CCARM 2003, *A. baumannii* CCARM 12010) were obtained from the Culture Collection of Antibiotic-Resistant Microbes (CCARM) at Seoul Women's University in Korea. We were kindly provided *E. coli* K1 strain RS218 (O18:K1:H7) by Jang-Won Yoon of Kangwon National University

(Gangwon-do, Korea). MIC of peptides was determined by the broth microdilution assay using 1% peptone media, as reported previously [24]. MIC was determined by the average value of three independent experiments. Peptides were diluted using 1% peptone media treated with 2×10^5 CFU/mL of bacterial suspensions in 1% peptone media for 16 h at 37 °C.

4.3. Confocal Microscope Analysis

When *E. coli* KCTC 1682 became a mid-log phage, it was washed and diluted in 10 mM PBS buffer. Bacterial suspensions (1×10^7 CFU/mL) were incubated with 1× MIC of peptides for 1 h. After 1 h, the cells were washed three times using 10 mM PBS buffer and visualized with a confocal laser scanning microscope (LSM 800; Carl Zeiss, Oberkochen, Germany).

4.4. FE-SEM Analysis

We visualized the membrane damage of *E. coli* KCTC 1682 by FE-SEM to confirm the antibacterial mechanisms. The method was the same as a previously reported method [28]. The 1 × MIC of peptide treated to *E. coli* KCTC 1682 was diluted in 10 mM PBS buffer to an OD_{600} of 0.2 and incubated for 4 h at 37 °C. After the fixation and dehydration step, the cells were visualized with a FE-SEM (SU8020; Hitachi, Tokyo, Japan).

4.5. Cytotoxicity In Vitro

RAW264.7 cells and HEK-293 cells were purchased from Korean cell line bank (Seoul, Korea). The cytotoxicity of peptides towards RAW264.7 cells and HEK-293 cells were evaluated by Cell Counting Kit-8 purchased from Abbkine Scientific (Wuhan, China). The assay was the same as previously reported method [28,53].

4.6. Specificity Against TLRs Selectively Activated by Agonists

The specificity of peptides was assessed by measuring nitrate production in RAW264.7 cells after activating different TLRs. Agonists including Pam_2CSK_4 (200 ng/mL), Pam_3CSK_4 (200 ng/mL), Poly(I:C) (2 µg/mL), LPS (20 ng/mL), and ODN1826 (20 µg/mL) were purchased from Invivogen (San Diego, CA, USA). Imiquimod (1 µg/mL) was purchased from Sigma-Aldrich (St. Louis, MO, USA). The method was similar to that previously reported [30]. We pretreated cells with Ps peptide for 1 h. Next, we added the aforementioned agonists at the indicated concentrations. To confirm the specificity of peptides, nitrate production was detected at an absorbance of 540 nm, as previously reported [28].

4.7. Quantification of Inflammatory Cytokine Production in LPS-Stimulated RAW264.7 Cells

ELISA assay was used to check the level of inflammatory cytokine. Raw264.7 cells plated at a density of 1×10^5 cells/well were stimulated with 20 ng/mL LPS of *E. coli* O111:B4 (Sigma-Aldrich, St. Louis, MO, USA) with or without peptide for 16 h. The inhibition effect of peptide on mTNF-α, mIL-6 and mIL-1α expression, which belong to inflammatory cytokines, was measured by enzyme-linked immunosorbent assays (ELISA; R&D Systems, Minneapolis, MN, USA), as previously reported [54]. Glyceraldehyde 3-phosphate dehydrogenase (GAPDH) was used control in the RT-PCR. Relative intensity of the DNA bands was quantified by ImageJ software (Version 1.52q, National Institutes of Health, Bethesda, MD, USA).

4.8. Inhibition Effect of Ps-K18 on TLR4-Mediated Inflammatory Response in LPS-Stimulated HEK-Blue^{TM} hTLR4 and RAW264.7 Cells

HEK-Blue^{TM} hTLR4 cells (InvivoGen, San Diego, CA, USA) were plated at a density of 2.5×10^4 cells/well with Ps-K18 in HEK-Blue detection media (InvivoGen, San Diego, CA, USA). After 1 h, the cells were stimulated by 20 ng/mL LPS and incubated for 8 h at 37 °C in a humidified 5% CO_2 atmosphere. SEAP activity was determined by measuring absorbance at 630 nm.

RT-PCR was performed to measure inhibition activity of Ps-K18 on expression of TLR4 mediating inflammatory response signaling as previously reported [55]. The results visualized by UV illumination using Alpha Innotech gel documentation system (Alphaimager ® HP; Alpha Innotech Corporation, San Leandro, CA, USA).

4.9. Sepsis Mouse Model

Female BALB/c mice (6-week-old) were purchased from Orient (Daejeon, Korea). All mice were housed under specific pathogen-free (SPF) conditions and allowed free access to food and water in a temperature- and humidity-controlled environment. All mice were used for LPS or *E. coli* K1-induced sepsis model. *E. coli* K1 possesses K1 capsular polysaccharide antigen which is an essential virulence determinant that protects the bacteria from immune attack [56,57]. All procedures were approved by the Institutional Animal Care and Use Committee (IACUC) of Konkuk University, South Korea (IACUC number: KU18163-1, approved on 17 April 2019).

4.10. Cytotoxicity In Vivo

Cytotoxicity was determined by detecting AST, ALT, and BUN levels in mice serum. They were detected five times independently using a standard kit available for purchase from Asan pharmaceutical (Seoul, Korea), as described previously [58]. The control group was only injected with PBS and the Ps or Ps-K18 groups were only administered peptide (1, 10, 20 mg/kg). After 16 h, the mice were sacrificed for analysis.

4.11. Measurement of Antiseptic Activity of Peptides in LPS-Induced Endotoxemia and E. coli K1 Septic Shock Mouse Model

To determine the antiseptic activity of Ps-K18, experiments were performed based on four different groups. The control group was only injected with PBS and the peptide group was only administered Ps-K18 (1 mg/kg). The LPS group was only injected with LPS (15 mg/kg) and the LPS + Ps-K18 group received an injection of LPS (15 mg/kg) 1 h after the administration of Ps-K18 (1 mg/kg). The *E. coli* group was only injected with *E. coli* K1 (1×10^6 CFU/mouse), whereas the *E. coli* + Ps-K18 group received an injection of *E. coli* K1 (1×10^6 CFU/mouse) 1 h after the administration of Ps-K18 (1 mg/kg). Mice were inoculated via intraperitoneal (i.p.) injection. After 16 h, the mice were sacrificed for analysis.

To measure cytokine levels in the serum or lung lysates of the sepsis model, we used ELISA kits (R&D Systems, Minneapolis, MN, USA), as reported previously [54]. The experiments were performed five times independently. We additionally confirmed the inhibitory effect of the peptide on endotoxin LPS using a LAL kit purchased from Lonza Group Ltd. (Allendale, NJ, USA), as previously reported [28]. We counted the number of bacteria in organ tissues, as previously reported, to determine whether Ps-K18 could inhibit the growth of bacteria in vivo [28]. Serum AST, ALT, and BUN levels were detected five times independently using a standard kit to confirm recovery ability of peptide for damaged organ by LPS or *E. coli* K1, as described previously [58]. To determine the suppressive effect of the peptide on neutrophil infiltration in vivo, tissue slides were produced, as previously described [59]. Briefly, lung tissues extracted from mice were fixed in 4% paraformaldehyde solution and dehydrated to produce paraffin blocks. The paraffin blocks were cut to an appropriate thickness to prepare tissue slides and the remaining paraffin was removed with xylene. The slides were then stained with hematoxylin and eosin and visualized with a microscope (Eclipse Ni; Nikon, Tokyo, Japan).

4.12. Statistical Analysis

All measurements were obtained at least three times and statistical analysis was calculated using Graphpad Prism (GraphPad Software Inc., La Jolla, CA, USA). One-way ANOVA followed by post-hoc Bonferroni tests (Prism 8.0) were used to compare with the experimental groups.

The error bars reveal standard error of measurement (±SEM). Values indicate statistically significance at *$p < 0.05$; **$p < 0.01$; ***$p < 0.001$; n.s. represents no significance.

5. Conclusions

Our study showed that Ps-K18 has low toxicity upon Lys substitution at Leu^{18} in Ps and functions through antibacterial mechanisms to restrict bacterial growth by targeting bacterial membranes and anti-inflammatory mechanism-inhibiting inflammatory reactions by suppressing LPS-induced TLR4 signaling with high specificity in vivo and in vitro. These properties led to excellent therapeutic effects in treating sepsis caused by endotoxin or *E. coli* K1. In our future studies, we must overcome obstacles such as the stability of the peptide in vivo, reduced activity mediated by proteases, cost, and the scaling of production of this peptide [60]. Developing methods to overcome these limitations will enable the application of Ps-K18 as a peptide drug derived from bioactive peptides. All results in this study clearly showed that Ps-K18 is a potential antimicrobial candidate that functions through dual mechanisms to suppress bacterial growth and inhibit inflammatory reactions in vitro and in vivo. Accordingly, Ps-K18 could be a potent peptide antibiotic for the effective treatment of sepsis caused by Gram-negative bacterial infections. This study may provide insight into design and development of a potent peptide drug which could be used as an effective treatment for sepsis and into a pharmaceutical application of peptide antibiotics.

Author Contributions: Conceptualization, Y.K.; methodology, Y.K.; investigation, M.J., J.K.; resources, Y.K., J.B; data curation, M.J., J.K., Y.C.; writing—original draft preparation, Y.K., M.J.; writing—review and editing, Y.K., M.J.; visualization, M.J.; supervision, Y.K; funding acquisition, Y.K.

Funding: This work was supported by the National Research Foundation of Korea (NRF) grant funded by the Korea government (MSIT) (No. 2019R1H1A2079889).

Conflicts of Interest: The funders had no role in the design of the study; in the collection, analyses, or interpretation of data; in the writing of the manuscript, or in the decision to publish the results.

References

1. Chakrabarti, S.; Guha, S.; Majumder, K. Food-derived bioactive peptides in human health: Challenges and opportunities. *Nutrients* **2018**, *10*, 1738. [CrossRef] [PubMed]
2. Fosgerau, K.; Hoffmann, T. Peptide therapeutics: Current status and future directions. *Drug Discov. Today* **2015**, *20*, 122–128. [CrossRef]
3. Hayashi, M.A.; Ducancel, F.; Konno, K. Natural peptides with potential applications in drug development, diagnosis, and/or biotechnology. *Int. J. Pept.* **2012**, *2012*, 757838. [CrossRef] [PubMed]
4. Hancock, R.E.; Chapple, D.S. Peptide antibiotics. *Antimicrob. Agents Chemother.* **1999**, *43*, 1317–1323. [CrossRef] [PubMed]
5. Koppkubel, S. International nonproprietary names (inn) for pharmaceutical substances. *Bull. World Health Organ.* **1995**, *73*, 275–279.
6. Kaspar, A.A.; Reichert, J.M. Future directions for peptide therapeutics development. *Drug Discov. Today* **2013**, *18*, 807–817. [CrossRef] [PubMed]
7. Iwaniak, A.; Darewicz, M.; Minkiewicz, P. Peptides derived from foods as supportive diet components in the prevention of metabolic syndrome. *Compr. Rev. Food Sci. Food Saf.* **2018**, *17*, 63–81. [CrossRef]
8. Williams, K.J. The introduction of 'chemotherapy' using arsphenamine-the first magic bullet. *J. R. Soc. Med.* **2009**, *102*, 343–348. [CrossRef]
9. Drusano, G.L. An overview of the pharmacology of imipenem/cilastatin. *J. Antimicrob. Chemother.* **1986**, *18* (Suppl. E), 79–92. [CrossRef]
10. Lin, L.; Wagner, M.C.; Cocklin, R.; Kuzma, A.; Harrington, M.; Molitoris, B.A.; Goebl, M.G. The antibiotic gentamicin inhibits specific protein trafficking functions of the arf1/2 family of gtpases. *Antimicrob. Agents Chemother.* **2011**, *55*, 246–254. [CrossRef]
11. Codjoe, F.S.; Donkor, E.S. Carbapenem resistance: A review. *Med. Sci. (Basel)* **2017**, *6*, 1. [CrossRef] [PubMed]

12. Moellering, R.C., Jr.; Wennersten, C.; Kunz, L.J. Emergence of gentamicin-resistant bacteria: Experience with tobramycin therapy of infections due to gentamicin-resistant organisms. *J. Infect. Dis.* **1976**, *134*, S40–S49. [CrossRef] [PubMed]
13. Reddy, K.V.; Yedery, R.D.; Aranha, C. Antimicrobial peptides: Premises and promises. *Int. J. Antimicrob. Agents* **2004**, *24*, 536–547. [CrossRef] [PubMed]
14. Falagas, M.E.; Grammatikos, A.P.; Michalopoulos, A. Potential of old-generation antibiotics to address current need for new antibiotics. *Expert Rev. Anti Infect. Ther.* **2008**, *6*, 593–600. [CrossRef] [PubMed]
15. Mahlapuu, M.; Hakansson, J.; Ringstad, L.; Bjorn, C. Antimicrobial peptides: An emerging category of therapeutic agents. *Front. Cell. Infect. Microbiol.* **2016**, *6*, 194. [CrossRef] [PubMed]
16. Kang, H.K.; Kim, C.; Seo, C.H.; Park, Y. The therapeutic applications of antimicrobial peptides (amps): A patent review. *J. Microbiol.* **2017**, *55*, 1–12. [CrossRef] [PubMed]
17. Akin, A.; Alp, E.; Altindiş, M.; Azak, E.; Batirel, A.; Çağ, Y.; Durmuş, G.; Kepenek Kurt, E.; Sağiroğlu, P.; Türe, Z.; et al. Current diagnosis and treatment approach to sepsis. *Mediterr. J. Infect. Microbes Antimicrob.* **2018**, *7*, 17. [CrossRef]
18. Fleischmann, C.; Scherag, A.; Adhikari, N.K.; Hartog, C.S.; Tsaganos, T.; Schlattmann, P.; Angus, D.C.; Reinhart, K. International Forum of Acute Care Trialists. Assessment of global incidence and mortality of hospital-treated sepsis. Current estimates and limitations. *Am. J. Respir. Crit. Care Med.* **2016**, *193*, 259–272. [CrossRef]
19. Xu, C.; Guo, Z.; Zhao, C.; Zhang, X.; Wang, Z. Potential mechanism and drug candidates for sepsis-induced acute lung injury. *Exp. Ther. Med.* **2018**, *15*, 4689–4696. [CrossRef] [PubMed]
20. Park, B.S.; Lee, J.O. Recognition of lipopolysaccharide pattern by tlr4 complexes. *Exp. Mol. Med.* **2013**, *45*, e66. [CrossRef]
21. Akira, S.; Uematsu, S.; Takeuchi, O. Pathogen recognition and innate immunity. *Cell* **2006**, *124*, 783–801. [CrossRef] [PubMed]
22. Hotchkiss, R.S.; Moldawer, L.L.; Opal, S.M.; Reinhart, K.; Turnbull, I.R.; Vincent, J.L. Sepsis and septic shock. *Nat. Rev. Dis. Prim.* **2016**, *2*, 16045. [CrossRef] [PubMed]
23. Cohen, J.; Vincent, J.L.; Adhikari, N.K.; Machado, F.R.; Angus, D.C.; Calandra, T.; Jaton, K.; Giulieri, S.; Delaloye, J.; Opal, S.; et al. Sepsis: A roadmap for future research. *Lancet Infect. Dis.* **2015**, *15*, 581–614. [CrossRef]
24. Jeon, D.; Jeong, M.C.; Jacob, B.; Bang, J.K.; Kim, E.H.; Cheong, C.; Jung, I.D.; Park, Y.; Kim, Y. Investigation of cationicity and structure of pseudin-2 analogues for enhanced bacterial selectivity and anti-inflammatory activity. *Sci. Rep.* **2017**, *7*, 1455. [CrossRef] [PubMed]
25. Jiang, Z.; Vasil, A.I.; Hale, J.; Hancock, R.E.; Vasil, M.L.; Hodges, R.S. Effects of net charge and the number of positively charged residues on the biological activity of amphipathic alpha-helical cationic antimicrobial peptides. *Adv. Exp. Med. Biol.* **2009**, *611*, 561–562. [PubMed]
26. Lee, E.; Shin, A.; Jeong, K.W.; Jin, B.; Jnawali, H.N.; Shin, S.; Shin, S.Y.; Kim, Y. Role of phenylalanine and valine10 residues in the antimicrobial activity and cytotoxicity of piscidin-1. *PLoS ONE* **2014**, *9*, e114453. [CrossRef]
27. Bahar, A.A.; Ren, D. Antimicrobial peptides. *Pharmaceuticals (Basel)* **2013**, *6*, 1543–1575. [CrossRef] [PubMed]
28. Kim, J.; Jacob, B.; Jang, M.; Kwak, C.; Lee, Y.; Son, K.; Lee, S.; Jung, I.D.; Jeong, M.S.; Kwon, S.H.; et al. Development of a novel short 12-meric papiliocin-derived peptide that is effective against gram-negative sepsis. *Sci. Rep.* **2019**, *9*, 3817. [CrossRef]
29. Kawasaki, T.; Kawai, T. Toll-like receptor signaling pathways. *Front. Immunol.* **2014**, *5*, 461. [CrossRef]
30. Kim, J.; Durai, P.; Jeon, D.; Jung, I.D.; Lee, S.J.; Park, Y.M.; Kim, Y. Phloretin as a potent natural tlr2/1 inhibitor suppresses tlr2-induced inflammation. *Nutrients* **2018**, *10*, 868. [CrossRef]
31. Lee, E.; Kim, J.K.; Jeon, D.; Jeong, K.W.; Shin, A.; Kim, Y. Functional roles of aromatic residues and helices of papiliocin in its antimicrobial and anti-inflammatory activities. *Sci. Rep.* **2015**, *5*, 12048. [CrossRef]
32. Bronstein, I.; Fortin, J.J.; Voyta, J.C.; Juo, R.R.; Edwards, B.; Olesen, C.E.; Lijam, N.; Kricka, L.J. Chemiluminescent reporter gene assays: Sensitive detection of the gus and seap gene products. *Biotechniques* **1994**, *17*, 172–174, 176–177. [PubMed]
33. Munford, R.S. Endotoxemia-menace, marker, or mistake? *J. Leukoc. Biol.* **2016**, *100*, 687–698. [CrossRef] [PubMed]

34. Bian, Z.; Guo, Y.; Ha, B.; Zen, K.; Liu, Y. Regulation of the inflammatory response: Enhancing neutrophil infiltration under chronic inflammatory conditions. *J. Immunol.* **2012**, *188*, 844–853. [CrossRef] [PubMed]
35. Bijli, K.M.; Kanter, B.G.; Minhajuddin, M.; Leonard, A.; Xu, L.; Fazal, F.; Rahman, A. Regulation of endothelial cell inflammation and lung polymorphonuclear lymphocyte infiltration by transglutaminase 2. *Shock* **2014**, *42*, 562–569. [CrossRef] [PubMed]
36. Martin, T.R. Neutrophils and lung injury: Getting it right. *J. Clin. Investig.* **2002**, *110*, 1603–1605. [CrossRef] [PubMed]
37. Van der Poll, T.; van de Veerdonk, F.L.; Scicluna, B.P.; Netea, M.G. The immunopathology of sepsis and potential therapeutic targets. *Nat. Rev. Immunol.* **2017**, *17*, 407–420. [CrossRef] [PubMed]
38. Wang, J.; Gong, S.; Wang, F.; Niu, M.; Wei, G.; He, Z.; Gu, T.; Jiang, Y.; Liu, A.; Chen, P. Granisetron protects polymicrobial sepsis-induced acute lung injury in mice. *Biochem. Biophys. Res. Commun.* **2019**, *508*, 1004–1010. [CrossRef] [PubMed]
39. Ventola, C.L. The antibiotic resistance crisis: Part 1: Causes and threats. *P T* **2015**, *40*, 277–283.
40. Diamond, G.; Beckloff, N.; Weinberg, A.; Kisich, K.O. The roles of antimicrobial peptides in innate host defense. *Curr. Pharm. Des.* **2009**, *15*, 2377–2392. [CrossRef]
41. Matsuzaki, K. Control of cell selectivity of antimicrobial peptides. *Biochim. Biophys. Acta* **2009**, *1788*, 1687–1692. [CrossRef] [PubMed]
42. Zelezetsky, I.; Tossi, A. Alpha-helical antimicrobial peptides–using a sequence template to guide structure-activity relationship studies. *Biochim. Biophys. Acta* **2006**, *1758*, 1436–1449. [CrossRef] [PubMed]
43. Al-Lawama, M.; Aljbour, H.; Tanash, A.; Badran, E. Intravenous colistin in the treatment of multidrug-resistant acinetobacter in neonates. *Ann. Clin. Microbiol. Antimicrob.* **2016**, *15*, 8. [CrossRef] [PubMed]
44. Lin, Z.; Zhao, X.; Huang, J.; Liu, W.; Zheng, Y.; Yang, X.; Zhang, Y.; Lamy de la Chapelle, M.; Fu, W. Rapid screening of colistin-resistant escherichia coli, acinetobacter baumannii and pseudomonas aeruginosa by the use of raman spectroscopy and hierarchical cluster analysis. *Analyst* **2019**, *144*, 2803–2810. [CrossRef] [PubMed]
45. Jenssen, H.; Hamill, P.; Hancock, R.E. Peptide antimicrobial agents. *Clin. Microbiol. Rev.* **2006**, *19*, 491–511. [CrossRef] [PubMed]
46. Pal, T.; Sonnevend, A.; Galadari, S.; Conlon, J.M. Design of potent, non-toxic antimicrobial agents based upon the structure of the frog skin peptide, pseudin-2. *Regul. Pept.* **2005**, *129*, 85–91. [CrossRef]
47. Tossi, A.; Sandri, L.; Giangaspero, A. Amphipathic, alpha-helical antimicrobial peptides. *Biopolymers* **2000**, *55*, 4–30. [CrossRef]
48. Kang, H.K.; Seo, C.H.; Luchian, T.; Park, Y. Pse-t2, an antimicrobial peptide with high-level, broad-spectrum antimicrobial potency and skin biocompatibility against multidrug-resistant pseudomonas aeruginosa infection. *Antimicrob. Agents Chemother.* **2018**, *62*, e01493-18. [CrossRef]
49. Angus, D.C.; van der Poll, T. Severe sepsis and septic shock. *N. Engl. J. Med.* **2013**, *369*, 840–851. [CrossRef]
50. Martin, L.; van Meegern, A.; Doemming, S.; Schuerholz, T. Antimicrobial peptides in human sepsis. *Front. Immunol.* **2015**, *6*, 404. [CrossRef]
51. Lu, Y.C.; Yeh, W.C.; Ohashi, P.S. Lps/tlr4 signal transduction pathway. *Cytokine* **2008**, *42*, 145–151. [CrossRef] [PubMed]
52. Lee, E.; Shin, A.; Kim, Y. Anti-inflammatory activities of cecropin a and its mechanism of action. *Arch. Insect Biochem. Physiol.* **2015**, *88*, 31–44. [CrossRef] [PubMed]
53. Cheon, D.; Kim, J.; Jeon, D.; Shin, H.C.; Kim, Y. Target proteins of phloretin for its anti-inflammatory and antibacterial activities against propionibacterium acnes-induced skin infection. *Molecules* **2019**, *24*, 1319. [CrossRef] [PubMed]
54. Jnawali, H.N.; Lee, E.; Jeong, K.W.; Shin, A.; Heo, Y.S.; Kim, Y. Anti-inflammatory activity of rhamnetin and a model of its binding to c-jun nh2-terminal kinase 1 and p38 mapk. *J. Nat. Prod.* **2014**, *77*, 258–263. [CrossRef] [PubMed]
55. Xu, W.; Huang, M.; Zhang, Y.; Li, H.; Zheng, H.; Yu, L.; Chu, K.; Lin, Y.; Chen, L. Extracts of bauhinia championii (benth.) benth. Attenuate the in fl ammatory response in a rat model of collagen-induced arthritis. *Mol. Med. Rep.* **2016**, *13*, 4167–4174. [CrossRef]
56. Mushtaq, N.; Redpath, M.B.; Luzio, J.P.; Taylor, P.W. Prevention and cure of systemic escherichia coli k1 infection by modification of the bacterial phenotype. *Antimicrob. Agents Chemother.* **2004**, *48*, 1503–1508. [CrossRef]

57. Kaczmarek, A.; Budzynska, A.; Gospodarek, E. Detection of k1 antigen of escherichia coli rods isolated from pregnant women and neonates. *Folia Microbiol. (Praha)* **2014**, *59*, 419–422. [CrossRef]
58. Jnawali, H.N.; Jeon, D.; Jeong, M.C.; Lee, E.; Jin, B.; Ryoo, S.; Yoo, J.; Jung, I.D.; Lee, S.J.; Park, Y.M.; et al. Antituberculosis activity of a naturally occurring flavonoid, isorhamnetin. *J. Nat. Prod.* **2016**, *79*, 961–969. [CrossRef]
59. Park, H.J.; Lee, S.J.; Cho, J.; Gharbi, A.; Han, H.D.; Kang, T.H.; Kim, Y.; Lee, Y.; Park, W.S.; Jung, I.D.; et al. Tamarixetin exhibits anti-inflammatory activity and prevents bacterial sepsis by increasing il-10 production. *J. Nat. Prod.* **2018**, *81*, 1435–1443. [CrossRef]
60. Otvos, L., Jr.; Wade, J.D. Current challenges in peptide-based drug discovery. *Front. Chem.* **2014**, *2*, 62. [CrossRef]

International Journal of
Molecular Sciences

MDPI

Article

The Bactericidal Activity of Temporin Analogues Against Methicillin Resistant *Staphylococcus aureus*

Anna Golda [1], Paulina Kosikowska-Adamus [2], Aleksandra Kret [1], Olena Babyak [1], Kinga Wójcik [1], Ewelina Dobosz [1], Jan Potempa [1,3], Adam Lesner [2] and Joanna Koziel [1,*]

1 Department of Microbiology, Faculty of Biochemistry, Biophysics and Biotechnology, Jagiellonian University, 30-387 Krakow, Poland; anna.b.golda@uj.edu.pl (A.G.); ola.kret57@gmail.com (A.K.); olenka.babyak@gmail.com (O.B.); kinga.wojcik@uj.edu.pl (K.W.); ewelina.blazusiak@uj.edu.pl (E.D.); jspote01@louisville.edu (J.P.)

2 Faculty of Chemistry, University of Gdansk, 80-309 Gdansk, Poland; paulina.kosikowska-adamus@ug.edu.pl (P.K.-A.); adam.lesner@ug.edu.pl (A.L.)

3 Department of Oral Immunity and Infectious Diseases, University of Louisville School of Dentistry, Louisville, KY 40202, USA

* Correspondence: Joanna.koziel@uj.edu.pl; Tel.: +48-126646377

Received: 4 September 2019; Accepted: 23 September 2019; Published: 25 September 2019

Abstract: *Staphylococcus aureus* is a major infectious agent responsible for a plethora of superficial skin infections and systemic diseases, including endocarditis and septic arthritis. Recent epidemiological data revealed the emergence of resistance to commonly used antibiotics, including increased numbers of both hospital- and community-acquired methicillin-resistant *S. aureus* (MRSA). Due to their potent antimicrobial functions, low potential to develop resistance, and immunogenicity, antimicrobial peptides (AMPs) are a promising alternative treatment for multidrug-resistant strains. Here, we examined the activity of a lysine-rich derivative of amphibian temporin-1CEb (DK5) conjugated to peptides that exert pro-proliferative and/or cytoprotective activity. Analysis of a library of synthetic peptides to identify those with antibacterial potential revealed that the most potent agent against multidrug-resistant *S. aureus* was a conjugate of a temporin analogue with the synthetic Leu-enkephalin analogue dalargin (DAL). DAL-PEG-DK5 exerted direct bactericidal effects via bacterial membrane disruption, leading to eradication of both planktonic and biofilm-associated staphylococci. Finally, we showed that accumulation of the peptide in the cytoplasm of human keratinocytes led to a marked clearance of intracellular MRSA, resulting in cytoprotection against invading bacteria. Collectively, the data showed that DAL-PEG-DK5 might be a potent antimicrobial agent for treatment of staphylococcal skin infections.

Keywords: temporin; MRSA; antimicrobial peptide; human keratinocytes

1. Introduction

Recent epidemiological and clinical data indicate that infectious diseases are a fast-growing health concern. Overuse of antibiotics in humans and animals contributes to the selection of a variety of pathogens that are resistant to conventional drugs. *Staphylococcus aureus* has the ability to become resistant to nearly all clinically useful antibiotics through the development of resistance mechanisms acquired by horizontal transfer and by chromosomal mutation [1]. Methicillin-resistant *S. aureus* (MRSA) was reported for the first time among nosocomial isolates in 1961 [2]. Up until the 1990s, MRSA was restricted to hospitals and intensive care units, after which novel strains of MRSA emerged in the community [3]. At present, MRSA is one of the most common antibiotic-resistant *S. aureus* strains that cause serious, often detrimental, infections in many parts of the world, including Europe, the Americas, North Africa, the Middle East, and East Asia [4]. Indeed, most nosocomial staphylococcal infections in intensive care units (40–70%) are caused by MRSA strains [3,5–7]. Skin and soft tissue infections (SSTIs), severe necrotizing pneumonia, and bacteremia are the most common clinical presentations strongly associated with MRSA [3,7,8]. The rise of MRSA and other multidrug-resistant (MDR) staphylococci is a serious problem with respect to treatment and control. Therefore, there is an urgent need to develop new therapeutics. Among candidates are host defense peptides, also called antimicrobial peptides (AMPs), which are part of the natural defenses of all phyla of multicellular organisms, including plants, insects, amphibians, and mammals [9,10]. Most AMPs are small cationic molecules that act as potent antimicrobial compounds against a broad spectrum of microorganisms [11].

Amphibian skin granular glands are one of the richest sources of AMPs, especially those of the *Rana* genus, a widely distributed group with over 250 species worldwide [12]. Among the AMP families identified in frog skin are temporins, which are short (10–14 amino acids), hydrophobic, C-terminally amidated AMPs [13]. Temporins act as potent AMPs against staphylococci [14–16]. Here, we focused on temporin-1CEb, a naturally occurring 12-residue peptide derived from frog skin secretions that display antimicrobial activity against a broad spectrum of gram-positive and (albeit to a lesser extent) gram-negative bacteria [17,18]. Moreover, temporins, including temporin-1CEb, play a significant anti-inflammatory role both in vitro and in vivo [19]. The emergence of MRSA has resulted in an epidemic of SSTIs; thus, we created a library of compounds based on an analogue of temporin-1CEb (DK5) conjugated to peptides showing cytoprotective, pro-proliferative, and pro-migratory effects against human keratinocytes (cells that facilitate wound-healing) [20–24]. All selected peptides, including carnosine (CAR), dalargin (DAL), and COMB1, were linked to DK5 by a short biologically inert and non-immunogenic PEG linker (8-amino-3,6-dioxaoctanoic acid) [25].

We investigated the antibacterial activity of the designed peptides against different clinical isolates of MDR *S. aureus*. The data revealed the unique properties of DAL-PEG-DK5, which inhibited the growth of all tested MRSA strains, including planktonic and biofilm bacteria. We also assessed the mechanism by which DAL-PEG-DK5 disrupts the cytoplasmic membrane of staphylococci and evaluated the antibacterial activity of the conjugate against MRSA-infected human keratinocytes. Importantly, the data strongly support the potential of DAL-PEG-DK5 as a topical agent for treatment of MRSA.

2. Results

2.1. Antibacterial Activity of Temporin Conjugates Against Staphylococci

The antimicrobial activity of peptide conjugates (Table S1) was tested against staphylococci—*S. aureus* (Newman, ATCC 25923, USA300) and *S. epidermidis* (ATCC 12228). A broth microdilution method was used to determine the MIC (minimal inhibitory concentration) as a standard protocol. According to the obtained results (Table 1), the strongest bacteriostatic activity against all tested strains was observed for DAL-PEG-DK5 (MIC of 40 µg/mL against *S. aureus* USA300 and ATCC 25923, and 60 µg/mL against *S. epidermidis*). The only exception was the Newman strain, for which the MIC was 140 µg/mL. Moreover, CAR3-PEG-DK5, CAR-PEG-DK5, and COMB-1-PEG-DK5 did not inhibit growth of tested strains at concentrations up to 190 µg/mL.

Table 1. MIC of tested peptides conjugates against *S. aureus* strains and *S. epidermidis*.

Staphylococcal Strain	Strain ID	MIC [µg/mL]			
		DAL-PEG-DK5	CAR-PEG-DK5	CAR3-PEG-DK5	COMB1-PEG-DK5
S. aureus	USA300	40	>190	>190	>190
	Newman	140	>190	>190	>190
	ATCC 25923	40	>190	>190	>190
S. epidermidis	ATCC 12228	60	40	110	>190

Based on the above results, we selected DAL-PEG-DK5 for further studies to examine how the composition of this conjugate can influence its bacteriostatic effects. Thus, we measured the MIC of native peptides (DAL and DK5), two peptide conjugates (DAL-PEG-DK5 and DK5-PEG-DAL), and a scramble peptide (SCR). We tested these against the MRSA strain *S. aureus* USA300. MIC analysis revealed that neither the native peptides DK5 and DAL, nor the scramble peptide SRC, inhibited bacterial growth (Table 2).

Table 2. MIC of native peptides, their peptides conjugates, and SCR against methicillin resistant *S. aureus* (MRSA) isolate.

USA300	MIC [μg/mL]
DK5	>190
DAL	>190
DK5-PEG-DAL	160
DAL-PEG-DK5	40
SCR	>190

Notably, DK5-PEG-DAL inhibited USA300 growth less effectively (four times less) than its sister analogue with the reversed connection between three components of DAL-PEG-DK5. The MIC values for DAL-PEG-DK5 and DK5-PEG-DAL against USA300 were 40 and 160 μg/mL, respectively. These results indicate that the DK5 conjugate shows potential antibacterial activity when DAL is localized at the C-terminus of the molecule.

Because USA300 was highly sensitive to the test conjugate, we decided to introduce other MRSA *S. aureus* strains. The majority of these strains (11/13) were isolated from the blood of infected patients, whereas two were obtained from skin. All strains were characterized (Table S2) before their sensitivity to DAL-PEG-DK5 was tested. The MIC assay revealed that all were highly sensitive to DAL-PEG-DK5. For 84% of the tested strains (11/13), the MIC was $\leq$ 40 μg/mL (Table 3). Taken together, the results show that DAL-PEG-DK5 inhibits the growth of different MRSA strains.

Table 3. MIC of DAL-PEG-DK5 against methicillin resistant *S. aureus* (MRSA) clinical isolates.

MRSA strain ID	MIC [μg/mL]		
	DAL-PEG-DK5	Vancomycin	Linezolid
56A1	40	-	-
52B	40	0.5	2
1694	70	-	-
2492	40	-	-
2706	40	-	-
2872cv	40	-	-
3417	40	-	-
4187	40	-	-
6674	40	-	-
7219	30	-	-
7501	40	-	-
7569	40	-	-
7718	50	-	-
USA300	40	1	1

- not determined.

Biofilm formation is one of strategies used by *S. aureus* to compromise host responses and protect the bacteria from antibiotic and antiseptics agents. Currently, there are no antimicrobials that treat biofilm infections specifically [26]; thus, we aimed to examine whether DAL-PEG-DK5 disrupts mature biofilms (formed after 24 h) of USA300. As shown in Figure 1, the peptide conjugate (at 4 × MIC, equivalent to 160 μg/mL) disrupted the mature biofilm of MRSA, reducing biofilm mass by 50%.

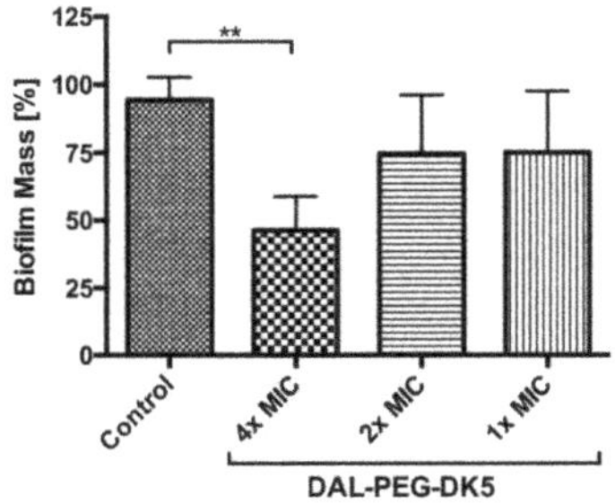

Figure 1. The effect of DAL-PEG-DK5 on the biofilm formed by *S. aureus*. USA300 strain was grown in 96-well plates at 37 °C for 24 h to form biofilm. Peptide was added at desired concentrations and plates were incubated for additional 24 h. After incubation with DAL-PEG-DK5, the biofilm was stained with crystal violet, then the dye was extracted with ethanol, measured OD and presented as percentage of biofilm reduction compared to untreated wells (Control). Mean ± SD $n = 3$. ** $p < 0.0151$; one-way ANOVA.

Because the conjugate of DK5 with dalargin exhibited strong antibacterial activity against MRSA and efficiently reduced biofilm formation by staphylococci, we used this compound in all further experiments.

2.2. Mechanism Underlying the Bactericidal Activity of DAL-PEG-DK5

The ability of DAL-PEG-DK5 to inhibit the growth of staphylococci and disrupt the mature biofilms formed by *S. aureus* suggests that the peptide also exerts bactericidal activity. To verify this hypothesis, we used LIVE/DEAD staining with DNA-binding dyes SYTO9 and propidium iodide (PI). SYTO9 is a membrane permeable (green) DNA stain that labels all bacteria in a population (those with intact membranes and those with damaged membranes). By contrast, PI (red) penetrates only bacteria with damaged membranes, causing a reduction in SYTO9 fluorescence when both dyes are present. Using this approach, we counted the total number of bacteria (green) and the number of non-viable bacteria (red). *S. aureus* USA300 was incubated with different concentrations of DAL-PEG-DK5 (0.5–50 µg/mL) for 3 h, and then stained with SYTO9 and PI (Figure 2A). We observed 20% and 5% live bacteria after exposure to 5 and 10 µg/mL, respectively, DAL-PEG-DK5.

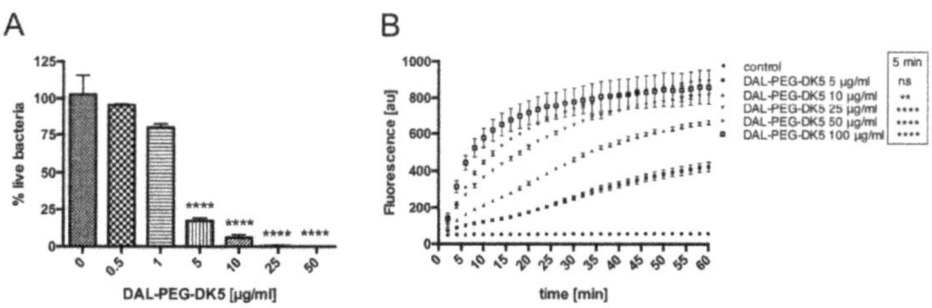

Figure 2. Bactericidal effect of DAL-PEG-DK5 on *S. aureus*. (**A**) SYTO 9 and PI staining. Staphylococci were treated with DAL-PEG-DK5 (0.5–50 µg/mL) for 3 h and stained with SYTO 9 (all bacteria in the population) and PI (bacteria with damaged membrane). The fluorescence was measured (excitation at 485 nm and emission at 530 nm (SYTO9) and 630 nm (PI)). Mean ± SD $n = 2$. **** $p < 0.0001$; one-way ANOVA. (**B**) Time dependent influx of Sytox green into USA300. Staphylococci were incubated with 1 µM Sytox green for 15 min and then DAL-PEG-DK5 was added (5–100 µg/mL). Probes were incubated 60 min and fluorescence was measured at indicated time points (excitation at 485 nm and emission at 520 nm). The data shown is representative of three separate experiments performed in triplicate. Mean ± SD. ns – non-significant, ** $p < 0.005$; **** $p < 0.0001$; one-way ANOVA.

Next, to explore the mechanism underlying DAL-PEG-DK5-mediated killing of MRSA, we examined the disruption of the cytoplasmic membrane of USA300 cells loaded with 1 μM Sytox green. The fluorescent dye does not cross intact membranes; thus, there is no change in intracellular Sytox green fluorescence without membrane disruption. When the peptide conjugate was added to USA300, we observed a dose-dependent increase in fluorescence intensity (reflecting bacterial membrane disruption) after only 5 min (Figure 2B). Moreover, at a concentration of 25 μg/mL peptide, permeabilization of the bacterial membrane reached a plateau. Taken together, these data suggest that DAL-PEG-DK5 exerts bactericidal effects against *S. aureus* MRSA strains by disrupting the bacterial cell membrane.

2.3. *Activity of DAL-PEG-DK5 Against Intracellular S. aureus*

For many years, *S. aureus* was considered an extracellular pathogen; however, recent data indicate that the bacteria can invade and persist within a variety of mammalian cells, including epithelial cells and even professional phagocytes [27,28]. One type of host cell that is efficiently colonized by *S. aureus* is the keratinocyte [29]. This strategy of staphylococci seems to be important for chronic and recurrent skin infections [30,31]. Therefore, it is important to evaluate the intracellular activity of DAL-PEG-DK5 against *S. aureus* that colonize the cytoplasm of human keratinocytes.

First, we evaluated the dose-dependent toxicity of DAL-PEG-DK5 against human keratinocytes using MTT and LDH assays. We found that the peptide conjugate did not affect the metabolism of HaCaT cells at a concentration ranging from 5 to 75 μg/mL at exposure times of 3, 6, and 18 h (Figure 3A). However, higher concentrations (100 or 200 μg/mL) of DAL-PEG-DK5 reduced the number of metabolically active cells in a concentration-dependent manner (Figure 3A). By contrast, the LDH assay showed that DAL-PEG-DK5 triggered a low, but still significant, release of dehydrogenase at a concentration of 75 μg/mL (Figure 3B). Note that when human keratinocytes were incubated with 200 μg/mL of the compound, it was cytotoxic to 40% (3, 6 h) and 100% (18 h) of cells (Figure 3B). Based on the obtained data, we examined the effect of the peptide conjugate at a non-toxic dose (≤50 μg/mL) on the morphology of HaCaT cells. The data revealed that exposure of cells to the peptide did not affect their morphology (Figure 3C; bright field). Moreover, we used the CFS-labeled peptide conjugate to examine the cellular localization of DAL-PEG-DK5 to determine whether DAL-PEG-DK5 is located on the surface of HaCaT cells or internalized into the cytoplasm or nucleus. The HaCaT cells were treated with CFS-conjugated DAL-PEG-DK5 (25 μg/mL) for 30 min at 37 °C. Next, the keratinocytes were stained with DAPI and phalloidin (nuclear and actin cytoskeleton stains, respectively). As shown by confocal microscopy (Figure 3C; fluorescence), CFS-labeled DAL-PEG-DK5 appeared to localize to the cell membrane and of the cells within 30 min post-addition, without entering the nucleus.

As we demonstrated that DAL-PEG-DK5 exerted no cytotoxic effects on keratinocytes at a dose that is strongly bactericidal against *S. aureus*, we examined the ability of DAL-PEG-DK5 to kill intracellular bacteria. For this, human keratinocytes were infected with USA300. All extracellular bacteria were removed (data not shown) by washing and treatment with lysostaphin (10 μg/mL) before adding DAL-PEG-DK5. We examined the eradication of intracellular bacteria from keratinocytes at different multiplicities of infection (MOI 1:5, 1:25, 1:50, and 1:100) and at different incubation times (1, 3, and 6 h). As depicted in Figure 4A, treatment with DAL-PEG-DK5 (50 μg/mL) reduced the intracellular USA300 load by more than 60% at an MOI 1:50 and 1:100.

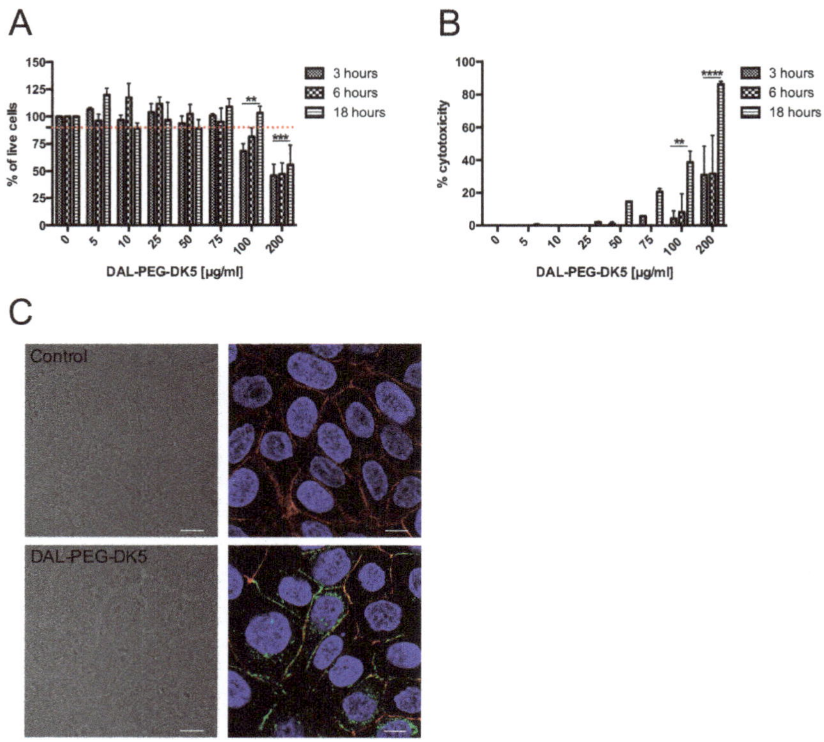

Figure 3. The influence of DAL-PEG-DK5 on physiology of human keratinocytes. The potentially toxic effect of DAL-PEG-DK5 on HaCaT cells was evaluated using (**A**) MTT and (**B**) LDH assay. Cells were plated on 96-well plates and incubated overnight. Next, keratinocytes were treated with the peptide at different concentrations (5–200 µg/mL) for 3, 6, 18 h. Mean ± SD $n = 2$. ** $p < 0.005$ *** $p < 0.001$ **** $p < 0.0001$; 2way ANOVA. (**C**) Morphology of HaCaT cells was examined by confocal laser scanning microscopy. HaCaT cells were treated with CFS-conjugated DAL-PEG-DK5 (25 µg/mL) for 30 min at 37 °C and were stained with: DAPI and phalloidin for nuclear detection and actin cytoskeleton detection, respectively. Blue—DNA; red—f-actin; green-peptide conjugate; scale bar: 10 µm. Images present single slice of XY stacks.

Efficient elimination of the most intracellular *S. aureus* was observed at 1 h post-exposure of keratinocytes to DAL-PEG-DK5 (Figure 4B). The marked decreased in the load of intercellular *S. aureus* was also observed during prolonged (6 h) cultivation with infected cells in the presence of the peptide. To verify the vitality of intracellular bacteria, we stained cells with a LIVE/DEAD Kit (Figure 4B; insert). In concordance with the CFU (colony forming units) count, the staining results showed both live and dead intracellular staphylococci in human keratinocytes after incubation of infected cells with DAL-PEG-DK5 (50 µg/mL for 3 h). Finally, we compared the antibacterial activity of DAL-PEG-DK5 against intercellular *S. aureus* and planktonic *S. aureus*. Interestingly, we found that the ability of the peptide conjugate to kill MRSA inside HaCaT cells and in suspension was comparable (Figure 4C). The MRSA load in keratinocytes was reduced by 71.5% (compared with a 74% reduction in the number of live bacteria exposed to the peptide in suspension). Taken together, the results show that the selected temporin analogue conjugated to enkephalin exhibits potent intracellular anti-staphylococcal efficacy at a dose that is not cytotoxic to host keratinocytes.

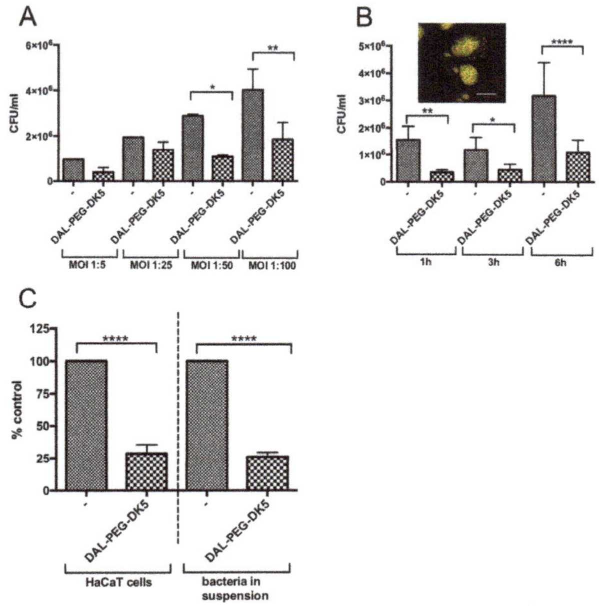

Figure 4. Bactericidal activity of DAL-PEG-DK5 against intracellular *S. aureus*. (**A**) USA300 survival within infected HaCaT cells. Keratinocytes were infected with USA300 (MOI 1:5, 1:25, 1:50, 1:100) for 2.5 h and treated with DAL-PEG-DK5 (50 µg/mL) for 3 h. Afterwards, keratinocytes were lysed and plated on agar plates for counting of bacteria. The number of viable bacterial cells is expressed as CFU/mL with respect to the number of intracellular bacteria in the corresponding control samples. The data shown is representative of two separate experiments performed in triplicate. Mean ± SD $n = 2$. * $p < 0.01$; ** $p < 0.005$; one-way ANOVA. (**B**) Keratinocytes were infected with USA300 (MOI 1:50) for 2.5 h and treated with DAL-PEG-DK5 (50 µg/mL) for indicated time points (1, 3, 6 h). Afterwards, keratinocytes were lysed and plated on agar plates for counting of bacteria. The number of viable bacterial cells is expressed as a CFU/mL with respect to the number of intracellular bacteria in the corresponding control samples. Mean ± SD $n = 2$. * $p < 0.01$; ** $p < 0.005$; **** $p < 0.0001$; one-way ANOVA. For confocal laser scanning microscopy HaCaT cells were infected with USA300 (MOI 1:50) for 2.5 h, then treated with DAL-PEG-DK5 (50 µg/mL) for 3 h. Afterwards cells were stained with SYTO 9 and PI. Viable *S. aureus* cells are stained green while red signals represent dead bacteria and the host cell's nuclear DNA stained with SYTO 9 and PI. Scale bar: 10 µm. Image presents single slice of XY stacks. (**C**) Keratinocytes were infected with USA300 (MOI 1:50) for 2.5 h and treated with DAL-PEG-DK5 (50 µg/mL) for 1 h. Afterwards, keratinocytes were lysed and plated on agar plates for counting of bacteria. In parallel, MRSA (2×10^6 CFU/mL) were incubated with DAL-PEG-DK5 (50 µg/mL) for 1 h and then plated on agar plates. The number of viable bacterial cells is expressed as a % of control with respect to the number of intracellular bacteria/bacteria in suspension in the corresponding control samples. Mean ± SD n=2. **** $p < 0.0001$; one-way ANOVA.

3. Discussion

AMPs have gained increasing attention as novel agents for treating antibiotic-resistant *S. aureus* infections [32]. Among naturally existing host defense peptides, temporins are active against methicillin- and vancomycin-resistant staphylococci [33,34]. Here, we analyzed a set of compounds based on the sequence of an antibacterial peptide, an analogue of temporin-1CEb (DK5). This compound shows activity against MDR *S. aureus*, although it is barely described [18]. This is in contrast to other temporins such as A, B, 1Tb, CPa, CPb, 1Ga, 1Oc, 1Ola, 1Spa, and PTa [16,33,35–37]. To fill this gap, we performed a comparative analysis of the properties of DK5 conjugated to compounds that resemble the sequence of human peptides with known cytoprotective activity (dalargin, carnosine, and COMB1). The conjugates were first evaluated for bacteriostatic and bactericidal activity. The data revealed that the most efficient peptide conjugate was a combination of an amidated form of DK5C linked to dalargin at the C-terminus by means of PEG linker (DAL-PEG-DK5); this compound displayed marked bacteriostatic activity against all tested staphylococci.

To examine the clinical relevance of our observations, we expanded the study by examining the susceptibility of 13 isolates of MDR *S. aureus* strains to DAL-PEG-DK5. The data revealed that DAL-PEG-DK5 showed strong antibacterial activity against all tested MRSA strains. We estimated that the MIC against the majority of tested MRSA strains (90%) was 40 µg/mL (17.56 µM). This value corroborates findings by Capparelli et al., who showed that temporin A and temporin B analogues (TB-YK) exhibit antibacterial activity against clinical isolates of *S. aureus* (strains A170, A172, and 007) with MIC values ranging from 10 to 25 µg/mL [36]. Consistent with this, Ciandrini and co-workers tested the antibacterial effects of temporin A against 15 MRSA clinical isolates and found that the MIC values ranged from 4 to 16 µg/mL [16]. Moreover, temporin-PTa and its analogues display antibacterial activity against MRSA, with MIC of 0.78–3.12 µM [37]. A similar trend is seen for temporins CPa, CPb, 1Ga, 1Oc, 1Ola, and 1Spa, with MIC values ranging from 1.6 to 12.5 µM against the USA300 strain [33]. Interestingly, we noted that the Newman strain was less sensitive than MRSA strains to the tested conjugates. This can be explained by the observation published by Boyle-Vavra and co-workers, who showed that the majority of USA300 isolates were CP (capsular polysaccharide)-negative. Moreover, whole-genome sequence analysis of USA300 isolates revealed that they all carry a cap5 locus with four conserved mutations [38]. By contrast, the Newman strain shows wild-type CP5 expression [39,40]. Therefore, we suggest that the presence of CP hinders access by the peptide to the cell membrane.

We found that DAL-PEG-DK5 rapidly disrupts the staphylococcal cell membrane; at 5 min post-treatment with peptide (10 µg), there was marked permeabilization of the cell membrane, which corroborates data showing the bactericidal effects of the peptide. The efficient antibacterial activity of the DAL-PEG-DK5 conjugate may be due to the fact that it has a higher cationic charge than DK5 (+7 vs. +6, respectively) [25]. However, a α-helicity must also play an important role in this case. According to a common view about the mechanism underlying the antimicrobial activity of AMPs, formation of helices by AMPs upon binding to the membranes of target cells is a driving force behind further incorporation into the deeper layers of the membrane structure; this ultimately leads to disruption and cell leakage. For example, our previously reported circular dichroism studies show that the conjugate (DK5-PEG-DAL) was less able to form helices (in the presence of 30 mM SDS in water) than its DAL-PEG-DK5 counterpart (20% vs. 35% α-helical content, respectively) [25]. The latter (partially) explains the difference in antimicrobial activity between these two compounds. However, it should be also kept in mind that the unique amino acid sequences of these peptides define amphipathicity, charge distribution, and hydrophobicity; these factors may also play a role in the results of the present study comparing the antimicrobial activity of active DAL-PEG-DK5 with that of moderately active DK5-PEG-DAL and its inactive scrambled analogue.

Biofilms play a central role in the pathogenesis of severe staphylococcal infections (e.g., chronic wound infections or medical device-related diseases) [41–43]. To broaden our study, we investigated the effect of DAL-PEG-DK5 on mature biofilms formed by USA300 and showed that the conjugate is bactericidal not only against planktonic bacteria, but also against bacterial cells encrusted in the dense structure of a biofilm. Thus, we conclude that the tested peptide conjugate might be a potential new therapeutic agent for biofilm-associated infections, as revealed for analogues of temporin 1Tb [35]. However, as DAL-PEG-DK5 is cytotoxic against human keratinocytes in a dose, which is active against biofilm, therefore its topical application should be carefully considered.

The success of therapy using AMPs depends mainly on their bactericidal effects, which may be limited against intercellular pathogens. This limitation is mitigated by the amphipathic nature and high cationic charge of DAL-PEG-DK5, which enables the conjugate to enter host cells. Such a feature is deemed important for the design of therapies against chronic and relapsing staphylococcal infections, which arise as a consequence of the intracellular persistence of *S. aureus* [27,44]. Here, we show that DAL-PEG-DK5 can enter keratinocytes easily without affecting cell morphology and viability. Di Grazia and co-workers observed similar effects when studying the interaction between temporin A and B and HaCaT cells. After 30 min of incubation with keratinocytes, temporins were distributed evenly throughout the cytoplasm but did not enter the nucleus [15]. We also observed that DAL-PEG-DK5 penetrates the cell membrane and accumulates in the cytoplasm, mainly in the perinuclear area. A marked amount of the peptide was also associated with the cell membrane. Importantly, our data indicate that the peptide conjugate did not exert any cytotoxic effects against HaCaT cells at a concentration that was effective against staphylococci.

Cumulatively, these findings encouraged us to test the ability of the peptide conjugate to kill invasive intracellular MRSA. We demonstrated that after 1 h, DAL-PEG-DK5 (MIC 40 μg/mL, equal to 17.56 μM) reduced the number (by 71.5%) of viable MRSA cells inside human keratinocytes when used at 1.25 × MIC. These results agree with those of other studies on peptides with intracellular activity against *S. aureus*, which show that temporin A (at 2 × MIC (MIC 8 μM)) and temporin B (at 4 × MIC (MIC 16 μM)) killed 20% and 40% of the intracellular MRSA clinical isolate, respectively [15]. In addition, a recent report shows that another antimicrobial peptide, CPPTat-JDlys (MIC 40 μg/mL), reduced the USA300 load in keratinocytes to 20% of that of the untreated control when used at 2 × MIC [45]. Remarkably, DAL-PEG-DK5 was as effective at eradicating MRSA inside HaCaT cells as it was at killing *S. aureus* in suspension (reduction in the CFU of 71.5% vs. 74%, respectively). This appears to be due to the peptide's ability to permeate and accumulate inside mammalian eukaryotic cells and then exert direct harmful effects against the bacterial membrane. Intracellular activity of DAL-PEG-DK5 against *S. aureus* was confirmed using infected macrophages (data not shown). We showed that temporin-1CEb is an efficient antibacterial peptide against staphylococci engulfed by human and murine macrophages, but only when combined with dalargin. Although we documented the direct bactericidal activity of DAL-PEG-DK5 and its efficient permeation of eukaryotic cells, we cannot rule out the possibility that an indirect mechanism contributes to the eradication of intracellular staphylococci from macrophages. Such a scenario is highly possible in light of the finding that DAL-PEG-DK5 exerts significant immunomodulatory effects on human macrophages [19].

In conclusion, we demonstrate that modification of temporin-1CEb to generate a peptide conjugate, DAL-PEG-DK5, makes the peptide an attractive candidate lead compound for the generation of a new agent to treat MRSA-related skin infections. Detailed in vivo studies are needed to confirm this hypothesis; however, it should be mentioned that we previously documented the stability and activity of this peptide in a murine model of sepsis [19]. The peptide inhibits planktonic growth of different clinical MRSA strains and kills them in the biofilm. DAL-PEG-DK5 acts by disrupting the integrity of the bacterial cell membrane without damaging keratinocytes. Such a unique mechanism is probably the consequence of combining temporin with dalargin, which is known for its cytoprotective properties towards human epithelial cells, including keratinocytes [25,46]. Taken together, the data presented herein indicate that the DAL-PEG-DK5 conjugate is a candidate treatment for skin infections caused by MRSA.

4. Materials and Methods

4.1. Reagents

Gentamicin, vancomycin, linezolid, and lysostaphin were from Sigma-Aldrich (St. Louis, MO, USA). FBS, DMEM, RPMI 1640, calcium and magnesium free phosphate-buffered saline (PBS without Ca^{2+} and Mg^{2+}), and penicillin-streptomycin (PEST) were obtained from Gibco (Life Technologies, Paisley, UK). CytoTox96 non-radioactive cytotoxicity assay kit was obtained from Promega (Promega, Madison, WI, USA). Sytox green and LIVE/DEAD BacLight Bacterial Viability kit were purchased from ThermoFisher Scientific (Invitrogen, ThermoFisher Scientific, Eugene, OR, USA).

4.2. Peptides Synthesis and Purification

All compounds were synthesized manually by means of the solid phase method applying Fmoc (fluorenyl-9methoxycarbonyl) chemistry under the standard conditions. S RAM (substitution 0.25 meq/g, RAPP Polymere, Tubingen, Germany) resins were used as solid support. The peptide chain was elongated by means of Fmoc-protected amino acids (3 equiv) using HOBt (N-hydroxybenzotriazole)/HBTU (N,N,N′,N′-Tetramethyl-O-(1H-benzotriazol-1-yl) uronium hexafluorophosphate) (3 equiv) as coupling reagents in the presence of DIPEA (N,N-Diisopropylethylamine) (6 equiv). The completeness of each coupling was monitored by the Kaiser test. The Fmoc-protection after each step of coupling was removed with 20% piperidine in dimethylformamide (DMF). For fluorescent labeling, N-terminal fluorescein moiety was introduced to the sequence via its succinimide derivative. Cleavage of the peptides from the resin was achieved using a TFA/phenol/triisopropylsilane/H2O mixture (88:5:2:5, *v*/*v*).

The synthesized compounds were purified by reverse phase high performance chromatography (RP-HPLC) on Waters system (Phenomenex Jupiter 4 μ Proteo 90 Å column, 250 × 10 mm). The linear gradient from 10% to 80% B within 60 min (A: 0.1% TFA in water; B: 80% acetonitrile in A) with a flow rate 5 mL/min was employed. The homogeneity of the final fractions of peptides were analyzed on Shimadzu HPLC System (Shimadzu Europe GmbH, Duisburg, Germany) equipped with Phenomenex Jupiter 4 μ Proteo 90 Å column, 250 × 4.60 mm column. Mass spectra of the synthesized peptides were recorded using a Biflex III MALDI TOF mass spectrometry (Bruker, Mannheim, Germany) with α-cyano-4-hydroxy-cinnamic acid (CCA) or 2,5-dihydroxybenzoic acid (DHB) used as the matrix.

4.3. Cell Culture

A well-established line of human immortalized keratinocytes (HaCaT cell line) were obtained from American Type Culture Collection (ATCC, Manassas, VA, USA) and cultured in DMEM (Gibco, Life Technologies, Paisley, UK) supplemented with 10% heat-inactivated fetal bovine serum (FBS) and PEST (100 U/mL penicillin and 100 U/mL streptomycin) at 37 °C in humidified 5% CO_2 atmosphere. Cells were passaged every 4–5 days.

4.4. Microorganisms

The staphylococcus strains used in this study, listed in Table 4, were stored in tryptic soy broth (TSB, Sigma Aldrich, St. Louis, MO, USA) containing glycerol (50% *v*/*v*) at −80 °C. Cultures were inoculated from stocks into 10 mL media. Strains were grown overnight under constant rotation (180 rpm) to mid-logarithmic growth phase at 37 °C, centrifuged at 5000× *g* for 5 min, washed in PBS, and resuspended in PBS to the desired OD (600 nm).

Table 4. Staphylococcus strains used in this study.

Staphylococcus Strains	Relevant Properties	Source
S. aureus		
USA300	Wilde type strain	L.N. Shaw [47]
ATCC 25923	Clinical isolate	ATCC
Newman	Wilde type laboratory strain	T.J. Foster
56A1	Clinical isolate	*
52B	Clinical isolate	*
1694	Clinical isolate	*
2492	Clinical isolate	*
2706	Clinical isolate	*

Table 4. *Cont.*

Staphylococcus Strains	Relevant Properties	Source
2872cv	Clinical isolate	*
3417	Clinical isolate	*
4187	Clinical isolate	*
6674	Clinical isolate	*
7219	Clinical isolate	*
7501	Clinical isolate	*
7569	Clinical isolate	*
7718	Clinical isolate	*
S. epidermidis		
ATCC 12228	Wilde type strain	ATCC

* The clinical strains were collected from nonrelated patients admitted to the Stefan Zeromski Specialist Muncipal Hospital in Krakow, Poland.

4.5. Cell Viability Test

The viability of HaCaT cells was examined by a 3-(4,5-dimethylthiazol-2-yl)-2,5-diphenyltetrazolium bromide (MTT, Sigma-Aldrich) reduction assay. In brief, cells were incubated with 10% (*v*/*v*) of a 5 mg/mL MTT solution for 1 to 2 h at 37 °C until purple precipitate is visible and acidified isopropanol was added followed by measuring of the absorbance at a wavelength of 570 nm.

To assess the effect of peptides on the integrity of the plasma membrane, the LDH release assay was performed using the CytoTox96 nonradioactive cytotoxicity assay kit (Promega, Madison, WI, USA) according to the manufacturer's instructions. Cytotoxicity was calculated with the formula: % cytotoxicity = 100 × (experimental LDH release/ maximum LDH release), where maximum LDH release is after lysis solution addition (Triton X-100). Relative amounts of LDH release were measured (absorbance at 490 nm) using plate reader SpectraMax (Molecular Device, Wokingham, UK). All assays were performed in triplicate.

4.6. Antimicrobial Activity

Antimicrobial activity of peptides was analyzed through determination of MIC parameter according to the standard microdilution technique performed on 96-well plates. Mueller-Hinton broth (MHB, Sigma Aldrich, St. Louis, MO, USA) was used as the working medium for all bacterial strains. Briefly, inoculum was prepared from freshly grown cultures of bacteria being at their exponential phase of growth. Each well of 96-well plates containing 100 µL of serially diluted peptides in PBS at desired concentration was inoculated with 100 µL of 10^5 CFU/mL of bacterial suspension. Then plates were incubated overnight 37 °C and absorbance was read at 600 nm after 24 h. Wells without peptide were treated as the positive control, while uninoculated MHB was defined as the negative control. All measurements were run in triplicates. MIC was defined as the minimal concentration that completely inhibits growth of microorganisms was performed according to Clinical and Laboratory Standards Institute (CLSI).

4.7. Efficacy of Peptides on S. aureus Biofilms

The efficacy of peptides to disrupt mature biofilms was followed as described before [48]. Briefly, USA300 strain grown overnight was diluted 1:100 in TSB + 1% glucose and incubated in 96-well plates at 37 °C for 24 h. After removing media, wells were rinsed with PBS to remove planktonic bacteria before re-filling wells with fresh MHB. Peptide conjugates were added at desired concentrations and plates were incubated at 37 °C for 24 h. After incubation, wells were washed and biofilms were stained with 0.5% (*w/v*) crystal violet for 30 min. The dye was solubilized with ethanol (95%) and the OD of biofilms was measured.

4.8. Sytox Green Uptake Analysis

The USA300 strain was grown to mid-logarithmic phase at 37 °C, washed, and resuspended in 10 mM sodium phosphate buffer (pH 7.2) Bacterial cells were then incubated with 1 μM Sytox green for 15 min in the dark. After the addition of desired concentration of peptide conjugate (5–100 μg/mL), the time-dependent increases in fluorescence caused by the binding of the cationic dye to intracellular DNA were monitored (excitation at 485 nm and emission at 520 nm) [49].

4.9. Assessment of Bacterial Viability by Using the LIVE/ DEAD BacLight KIT

USA300 cells were collected from overnight cultures, at the end of the exponential growth phase and the beginning of the stationary phase, washed and suspended at 2×10^7 bacteria/mL and staphylococci were treated with DAL-PEG-DK5 (0.5–50 μg/mL) for 3 h. In order to obtain a standard curve, five different proportions of live and dead bacteria were mixed prior to staining. After the incubation, the 100 μL of bacterial suspensions were transferred to flat bottom black 96-well microtitration plate. Staining solution containing SYTO9 and PI (100 μL) prepared according to manufacturer's instructions was then mixed with bacterial suspensions. Samples were incubated at room temperature in the dark for 15 min and fluorescence intensity was measured with FlexStation3 Multimode Microplate Reader using a 485 nm excitation filter (for both SYTO9 and PI) and a 530 nm (SYTO9 emission wavelength) and 630 nm (PI emission wavelength) emission filter. The data were analyzed by dividing the fluorescence intensity of the stained bacterial suspensions at green emission by the cell fluorescence intensity of red fluorescence. All samples were prepared in triplicates.

4.10. Antibacterial Efficacy of Peptides Against Intracellular S. aureus

Infection of HaCaT keratinocytes was conducted as previously described [50]. Human keratinocytes were grown in standard medium in 12 (5×10^5 cells) or 24-well (3×10^5 cells) tissue culture plates. Following incubation, the cells were infected with USA300 at different multiplicity of infection (MOI; 1:5, 1:25, 1:50, 1:100) in DMEM supplemented with 10% FBS, without PEST for 2.5 h. After infection, the cells were washed with ice cold PBS and further incubated for 30 min with lysostaphin (10 μg/mL) to kill extracellular bacteria. DAL-PEG-DK5 was diluted in culture media to 50 μg/mL and added to the cells for an additional 1, 3, and 6 h. The medium alone was used as a negative control. After incubation, media were aspirated and human keratinocytes were washed twice with PBS to remove any residual peptide. Then HaCaT cells were lysed with ice cold H_2O. The cell lysates were serially diluted and plated on TSA plates.

4.11. Confocal Microscopy

HaCaT keratinocytes were seeded on coverslips for 24 h in DMEM supplemented with 10% FBS at 37 °C in humidified 5% CO_2 atmosphere. After 24 h, cells were washed with PBS and treated with FITC conjugated DAL-PEG-DK5 (25 μg/mL) for 30 min at 37 °C. Then cells were washed with cold PBS and fixed with 3.7% formaldehyde for 15 min at RT. Afterwards, HaCaT keratinocytes were stained with DAPI and phalloidin for nuclear and actin cytoskeleton detection, respectively. The coverslips were placed on glass slide with mounting media and visualized using Zeiss LSM 880 confocal system.

To distinguish vivid bacteria with the intact cell membrane from dead bacteria with the compromised membrane, the LIVE/DEAD BacLight Bacterial Viability kit was used. SYTO 9 and PI were mixed in proper ratios, and gave fluorescence signals indicative of alive or dead bacteria. Before staining HaCaT cells were permeabilized with 0.2% Triton X-100 to allow PI to bind to the dead bacteria and keratinocyte DNA. Confocal laser scanning microscopy (CSLM) was carried out using Zeiss LSM 880 confocal system equipped with 100× oil immersion objectives and acquired images analyzed in Zeiss ZEN Microscope Software.

4.12. Statistical Analysis

Statistical comparisons were performed with Prism 6.0 software (GraphPad Software, Inc., San Diego, CA, USA), using one- or two-way ANOVA test for multiple comparisons. p value < 0.05 was considered to be significant.

Supplementary Materials: Supplementary materials can be found at http://www.mdpi.com/1422-0067/20/19/4761/s1.

Author Contributions: A.G and J.K. conceived and designed the experiments; A.G., A.K., O.B., E.D., and J.K. performed the experiments; P.K.-A. synthesized peptides; K.W. provided and characterized MRSA clinical strains; A.G. and J.K. wrote the manuscript A.G., P.K.-A., A.L., J.P., and J.K. edited the manuscript; J.K. supervised all research.

Funding: This work was supported by National Science Center, Poland, Sonata Bis 6 2016/22/E/NZ6/00336, 2016/23/D/ST5/02234 and Miniatura 1 2017/01/X/NZ6/01803 (to J.K., P.K.-A., and A.G. respectively).

Conflicts of Interest: The authors declare no conflict of interest.

Abbreviations

AMPs	Antimicrobial peptides
DMEM	Dulbecco's Modified Eagle's Medium
CFU	Colony forming units
FBS	Fetal bovine serum
MIC	Minimal inhibitory concentration
MOI	Multiplicity of infection
OD	Optical density
PBS	Phosphate buffered saline
PI	Propidium iodide
SCR	Scramble peptide

References

1. Peterson, E.; Kaur, P. Antibiotic Resistance Mechanisms in Bacteria: Relationships Between Resistance Determinants of Antibiotic Producers, Environmental Bacteria, and Clinical Pathogens. *Front. Microbiol.* **2018**, *9*, 2928. [CrossRef] [PubMed]
2. Jevons, M.P. "Calbenin"-resistant staphylococci. *BMJ* **1961**, *1*, 124–125. [CrossRef]
3. Zetola, N.; Francis, J.S.; Nuermberger, E.L.; Bishai, W.R. Community-acquired meticillin-resistant Staphylococcus aureus: An emerging threat. *Lancet Infect. Dis.* **2005**, *5*, 275–286. [CrossRef]
4. Grundmann, H.; Aires-de-Sousa, M.; Boyce, J.; Tiemersma, E. Emergence and resurgence of meticillin-resistant Staphylococcus aureus as a public-health threat. *Lancet* **2006**, *368*, 874–885. [CrossRef]
5. Diekema, D.J.; Pfaller, M.A.; Schmitz, F.J.; Smayevsky, J.; Bell, J.; Jones, R.N.; Beach, M.; SENTRY Partcipants Group. Survey of infections due to Staphylococcus species: Frequency of occurrence and antimicrobial susceptibility of isolates collected in the United States, Canada, Latin America, Europe, and the Western Pacific region for the SENTRY Antimicrobial Surveillance Program, 1997-1999. *Clin. Infect Dis.* **2001**, *32*, 114–132. [CrossRef]
6. McCaig, L.F.; McDonald, L.C.; Mandal, S.; Jernigan, D.B. Staphylococcus aureus-associated skin and soft tissue infections in ambulatory care. *Emerg. Infect. Dis.* **2006**, *12*, 1715–1723. [CrossRef] [PubMed]
7. Stryjewski, M.E.; Chambers, H.F. Skin and soft-tissue infections caused by community-acquired methicillin-resistant Staphylococcus aureus. *Clin. Infect. Dis.* **2008**, *46*, 368–377. [CrossRef] [PubMed]
8. Gorak, E.J.; Yamada, S.M.; Brown, J.D. Community-acquired methicillin-resistant Staphylococcus aureus in hospitalized adults and children without known risk factors. *Clin. Infect. Dis.* **1999**, *29*, 797–800. [CrossRef] [PubMed]
9. Hancock, R.E.; Haney, E.F.; Gill, E.E. The immunology of host defence peptides: Beyond antimicrobial activity. *Nat. Rev. Immunol.* **2016**, *16*, 321–334. [CrossRef] [PubMed]
10. Choi, K.Y.; Chow, L.N.; Mookherjee, N. Cationic host defence peptides: Multifaceted role in immune modulation and inflammation. *J. Innate Immun.* **2012**, *4*, 361–370. [CrossRef] [PubMed]

11. Yeung, A.T.; Gellatly, S.L.; Hancock, R.E. Multifunctional cationic host defence peptides and their clinical applications. *Cell. Mol. Life Sci.* **2011**, *68*, 2161–2176. [CrossRef] [PubMed]
12. Conlon, J.M.; Kolodziejek, J.; Nowotny, N. Antimicrobial peptides from ranid frogs: Taxonomic and phylogenetic markers and a potential source of new therapeutic agents. *Biochim. Biophys. Acta* **2004**, *1696*, 1–14. [CrossRef] [PubMed]
13. Mangoni, M.L.; Grazia, A.D.; Cappiello, F.; Casciaro, B.; Luca, V. Naturally Occurring Peptides from Rana temporaria: Antimicrobial Properties and More. *Curr. Top. Med. Chem.* **2016**, *16*, 54–64. [CrossRef] [PubMed]
14. Mangoni, M.L. Temporins, anti-infective peptides with expanding properties. *Cell. Mol. Life Sci.* **2006**, *63*, 1060–1069. [CrossRef] [PubMed]
15. Di Grazia, A.; Luca, V.; Segev-Zarko, L.A.; Shai, Y.; Mangoni, M.L. Temporins A and B stimulate migration of HaCaT keratinocytes and kill intracellular Staphylococcus aureus. *Antimicrob. Agents Chemother.* **2014**, *58*, 2520–2527. [CrossRef]
16. Ciandrini, E.; Morroni, G.; Arzeni, D.; Kamysz, W.; Neubauer, D.; Kamysz, E.; Cirioni, O.; Brescini, L.; Baffone, W.; Campana, R. Antimicrobial Activity of Different Antimicrobial Peptides (AMPs) Against Clinical Methicillin-resistant Staphylococcus aureus (MRSA). *Curr. Top. Med. Chem.* **2018**, *18*, 2116–2126. [CrossRef]
17. Shang, D.; Yu, F.; Li, J.; Zheng, J.; Zhang, L.; Li, Y. Molecular cloning of cDNAs encoding antimicrobial peptide precursors from the skin of the Chinese brown frog, Rana chensinensis. *Zool. Sci.* **2009**, *26*, 220–226. [CrossRef]
18. Shang, D.; Li, X.; Sun, Y.; Wang, C.; Sun, L.; Wei, S.; Gou, M. Design of potent, non-toxic antimicrobial agents based upon the structure of the frog skin peptide, temporin-1CEb from Chinese brown frog, Rana chensinensis. *Chem. Biol. Drug Des.* **2012**, *79*, 653–662. [CrossRef]
19. Golda, A.; Kosikowska-Adamus, P.; Babyak, O.; Lech, M.; Wysocka, M.; Lesner, A.; Potempa, J.; Koziel, J. Conjugate of Enkephalin and Temporin Peptides as a Novel Therapeutic Agent for Sepsis. *Bioconjug. Chem.* **2018**, *29*, 4127–4139. [CrossRef]
20. Shekhter, A.B.; Solov'eva, A.I.; Spevak, S.E.; Titov, M.I. Effects of opioid peptide dalargin on reparative processes in wound healing. *Biull. Eksp. Biol. Med.* **1988**, *106*, 487–490. [CrossRef]
21. Legeza, V.P.; Koshcheev, A.G.; Konovalova, L.N. Effect of dalargin on healing of a bullet wound of the soft tissues in rabbits. *Patol. Fiziol. Eksp. Ter.* **1995**, *4*, 45–48.
22. Hipkiss, A.R. Aging, Proteotoxicity, Mitochondria, Glycation, NAD and Carnosine: Possible Inter-Relationships and Resolution of the Oxygen Paradox. *Front. Aging Neurosci.* **2010**, *2*, 10. [CrossRef] [PubMed]
23. Cripps, M.J.; Hanna, K.; Lavilla, C., Jr.; Sayers, S.R.; Caton, P.W.; Sims, C.; De Girolamo, L.; Sale, C.; Turner, M.D. Carnosine scavenging of glucolipotoxic free radicals enhances insulin secretion and glucose uptake. *Sci. Rep.* **2017**, *7*, 13313. [CrossRef] [PubMed]
24. Demidova-Rice, T.N.; Geevarghese, A.; Herman, I.M. Bioactive peptides derived from vascular endothelial cell extracellular matrices promote microvascular morphogenesis and wound healing in vitro. *Wound Repair Regen.* **2011**, *19*, 59–70. [CrossRef] [PubMed]
25. Kosikowska, P.; Pikula, M.; Langa, P.; Trzonkowski, P.; Obuchowski, M.; Lesner, A. Synthesis and Evaluation of Biological Activity of Antimicrobial–Pro-Proliferative Peptide Conjugates. *PLoS ONE* **2015**, *10*, e0140377. [CrossRef] [PubMed]
26. Pletzer, D.; Hancock, R.E. Antibiofilm Peptides: Potential as Broad-Spectrum Agents. *J. Bacteriol.* **2016**, *198*, 2572–2578. [CrossRef] [PubMed]
27. Bayles, K.W.; Wesson, C.A.; Liou, L.E.; Fox, L.K.; Bohach, G.A.; Trumble, W.R. Intracellular Staphylococcus aureus escapes the endosome and induces apoptosis in epithelial cells. *Infect. Immun.* **1998**, *66*, 336–342.
28. Kubica, M.; Guzik, K.; Koziel, J.; Zarebski, M.; Richter, W.; Gajkowska, B.; Golda, A.; Maciag-Gudowska, A.; Brix, K.; Shaw, L.; et al. A potential new pathway for Staphylococcus aureus dissemination: The silent survival of S. aureus phagocytosed by human monocyte-derived macrophages. *PLoS ONE* **2008**, *3*, e1409. [CrossRef]
29. Sayedyahossein, S.; Xu, S.X.; Rudkouskaya, A.; McGavin, M.J.; McCormick, J.K.; Dagnino, L. Staphylococcus aureus keratinocyte invasion is mediated by integrin linked kinase and Rac1. *FASEB J.* **2015**, *29*, 711–723. [CrossRef]
30. Bitschar, K.; Wolz, C.; Krismer, B.; Peschel, A.; Schittek, B. Keratinocytes as sensors and central players in the immune defense against Staphylococcus aureus in the skin. *J. Dermatol. Sci.* **2017**, *87*, 215–220. [CrossRef]

31. Travers, J.B.; Norris, D.A.; Leung, D.Y. The keratinocyte as a target for staphylococcal bacterial toxins. *J. Investig. Derm. Symp. Proc.* **2001**, *6*, 225–230. [CrossRef] [PubMed]
32. Ulvatne, H. Antimicrobial peptides: Potential use in skin infections. *Am. J. Clin. Derm.* **2003**, *4*, 591–595. [CrossRef] [PubMed]
33. Mishra, B.; Wang, X.; Lushnikova, T.; Zhang, Y.; Golla, R.M.; Narayana, J.L.; Wang, C.; McGuire, T.R.; Wang, G. Antibacterial, antifungal, anticancer activities and structural bioinformatics analysis of six naturally occurring temporins. *Peptides* **2018**, *106*, 9–20. [CrossRef] [PubMed]
34. Mangoni, M.L.; Maisetta, G.; Di Luca, M.; Gaddi, L.M.; Esin, S.; Florio, W.; Brancatisano, F.L.; Barra, D.; Campa, M.; Batoni, G. Comparative analysis of the bactericidal activities of amphibian peptide analogues against multidrug-resistant nosocomial bacterial strains. *Antimicrob. Agents Chemother.* **2008**, *52*, 85–91. [CrossRef] [PubMed]
35. Grassi, L.; Maisetta, G.; Maccari, G.; Esin, S.; Batoni, G. Analogs of the Frog-skin Antimicrobial Peptide Temporin 1Tb Exhibit a Wider Spectrum of Activity and a Stronger Antibiofilm Potential as Compared to the Parental Peptide. *Front. Chem.* **2017**, *5*, 24. [CrossRef]
36. Capparelli, R.; Romanelli, A.; Iannaccone, M.; Nocerino, N.; Ripa, R.; Pensato, S.; Pedone, C.; Iannelli, D. Synergistic antibacterial and anti-inflammatory activity of temporin A and modified temporin B in vivo. *PLoS ONE* **2009**, *4*, e7191. [CrossRef] [PubMed]
37. Mishra, B.; Wang, G. Ab initio design of potent anti-MRSA peptides based on database filtering technology. *J. Am. Chem. Soc.* **2012**, *134*, 12426–12429. [CrossRef]
38. Boyle-Vavra, S.; Li, X.; Alam, M.T.; Read, T.D.; Sieth, J.; Cywes-Bentley, C.; Dobbins, G.; David, M.Z.; Kumar, N.; Eells, S.J.; et al. USA300 and USA500 clonal lineages of Staphylococcus aureus do not produce a capsular polysaccharide due to conserved mutations in the cap5 locus. *MBio* **2015**, *6*, e02585-14. [CrossRef]
39. Kuipers, A.; Stapels, D.A.; Weerwind, L.T.; Ko, Y.P.; Ruyken, M.; Lee, J.C.; van Kessel, K.P.; Rooijakkers, S.H. The Staphylococcus aureus polysaccharide capsule and Efb-dependent fibrinogen shield act in concert to protect against phagocytosis. *Microbiology* **2016**, *162*, 1185–1194. [CrossRef]
40. Herbert, S.; Newell, S.W.; Lee, C.; Wieland, K.P.; Dassy, B.; Fournier, J.M.; Wolz, C.; Döring, G. Regulation of Staphylococcus aureus type 5 and type 8 capsular polysaccharides by CO(2). *J. Bacteriol.* **2001**, *183*, 4609–4613. [CrossRef]
41. Davis, S.C.; Ricotti, C.; Cazzaniga, A.; Welsh, E.; Eaglstein, W.H.; Mertz, P.M. Microscopic and physiologic evidence for biofilm-associated wound colonization in vivo. *Wound Repair Regen.* **2008**, *16*, 23–29. [CrossRef] [PubMed]
42. Schulin, T.; Voss, A. Coagulase-negative staphylococci as a cause of infections related to intravascular prosthetic devices: Limitations of present therapy. *Clin. Microbiol. Infect.* **2001**, *7*, 1–7. [CrossRef] [PubMed]
43. Giormezis, N.; Kolonitsiou, F.; Foka, A.; Drougka, E.; Liakopoulos, A.; Makri, A.; Papanastasiou, A.D.; Vogiatzi, A.; Dimitriou, G.; Marangos, M.; et al. Coagulase-negative staphylococcal bloodstream and prosthetic-device-associated infections: The role of biofilm formation and distribution of adhesin and toxin genes. *J. Med. Microbiol.* **2014**, *63*, 1500–1508. [CrossRef] [PubMed]
44. Nuzzo, I.; Sanges, M.R.; Folgore, A.; Carratelli, C.R. Apoptosis of human keratinocytes after bacterial invasion. *FEMS Immunol. Med. Microbiol.* **2000**, *27*, 235–240. [CrossRef] [PubMed]
45. Wang, Z.; Kong, L.; Liu, Y.; Fu, Q.; Cui, Z.; Wang, J.; Ma, J.; Wang, H.; Yan, Y.; Sun, J. A Phage Lysin Fused to a Cell-Penetrating Peptide Kills Intracellular Methicillin-Resistant Staphylococcus aureus in Keratinocytes and Has Potential as a Treatment for Skin Infections in Mice. *Appl. Environ. Microbiol.* **2018**, *84*. [CrossRef] [PubMed]
46. Pencheva, N.; Pospisek, J.; Hauzerova, L.; Barth, T.; Milanov, P. Activity profiles of dalargin and its analogues in mu-, delta- and kappa-opioid receptor selective bioassays. *Br. J. Pharm.* **1999**, *128*, 569–576. [CrossRef] [PubMed]
47. Koziel, J.; Chmiest, D.; Bryzek, D.; Kmiecik, K.; Mizgalska, D.; Maciag-Gudowska, A.; Shaw, L.N.; Potempa, J. The Janus face of α-toxin: A potent mediator of cytoprotection in staphylococci-infected macrophages. *J. Innate Immun.* **2015**, *7*, 187–198. [CrossRef]
48. Mohamed, M.F.; Abdelkhalek, A.; Seleem, M.N. Evaluation of short synthetic antimicrobial peptides for treatment of drug-resistant and intracellular Staphylococcus aureus. *Sci. Rep.* **2016**, *6*, 29707. [CrossRef]

49. Lee, J.K.; Seo, C.H.; Luchian, T.; Park, Y. Antimicrobial Peptide CMA3 Derived from the CA-MA Hybrid Peptide: Antibacterial and Anti-inflammatory Activities with Low Cytotoxicity and Mechanism of Action in Escherichia coli. *Antimicrob. Agents Chemother.* **2015**, *60*, 495–506. [CrossRef]
50. Von Eiff, C.; Becker, K.; Metze, D.; Lubritz, G.; Hockmann, J.; Schwarz, T.; Peters, G. Intracellular persistence of Staphylococcus aureus small-colony variants within keratinocytes: A cause for antibiotic treatment failure in a patient with darier's disease. *Clin. Infect. Dis.* **2001**, *32*, 1643–1647. [CrossRef]

International Journal of
Molecular Sciences

MDPI

Review

Antimicrobial Peptides as Anti-Infective Agents in Pre-Post-Antibiotic Era?

Tomislav Rončević [1,2,*], Jasna Puizina [1] and Alessandro Tossi [3]

1 Department of Biology, Faculty of Science, University of Split, 21000 Split, Croatia; puizina@pmfst.hr
2 Laboratory for Aquaculture, Institute of Oceanography and Fisheries, 21000 Split, Croatia
3 Department of Life Sciences, University of Trieste, 34127 Trieste, Italy; atossi@units.it
* Correspondence: troncevic@pmfst.hr; Tel.: +385-21-619-274

Received: 16 October 2019; Accepted: 11 November 2019; Published: 14 November 2019

Abstract: Resistance to antibiotics is one of the main current threats to human health and every year multi-drug resistant bacteria are infecting millions of people worldwide, with many dying as a result. Ever since their discovery, some 40 years ago, the antimicrobial peptides (AMPs) of innate defense have been hailed as a potential alternative to conventional antibiotics due to their relatively low potential to elicit resistance. Despite continued effort by both academia and start-ups, currently there are still no antibiotics based on AMPs in use. In this study, we discuss what we know and what we do not know about these agents, and what we need to know to successfully translate discovery to application. Understanding the complex mechanics of action of these peptides is the main prerequisite for identifying and/or designing or redesigning novel molecules with potent biological activity. However, other aspects also need to be well elucidated, i.e., the (bio)synthetic processes, physiological and pathological contexts of their activity, and a quantitative understanding of how physico-chemical properties affect activity. Research groups worldwide are using biological, biophysical, and algorithmic techniques to develop models aimed at designing molecules with the necessary blend of antimicrobial potency and low toxicity. Shedding light on some open questions may contribute toward improving this process.

Keywords: antimicrobial peptides; antimicrobial resistance; AMP identification and design; biosynthesis; mode of action; physico-chemical properties; therapeutic potential

---

1. Antibiotics and Antimicrobial Resistance—History is Important

Despite a common misconception, exposure to antibiotics is not confined to the modern antibiotic era starting from the early 20th century. Ancient civilizations used antibiotics to treat bacterial infections, with topical applications of moldy bread to wounds being well documented in ancient Egypt, China, Greece, and the Roman empire [1]. Traces of molecules related to modern day antibiotics, the tetracyclines, have been found in skeletal remains from ancient Sudanese Nubia (350–550 A.D.) and from the Roman period in Egypt [2–4]. In both cases, it is presumed that they were ingested with grains contaminated by Streptomycetes [3,4] and most likely had a preventive role rather than a "systemic" effect on infection. Remedies used in traditional Chinese medicine have been the source of potent anti-infective agents for millennia, with artemisinin or "qinghaosu" which is a potent anti-malarial drug, being one of the best known examples [5].

The modern antibiotic era begins with Paul Erlich's concept of the "magic bullet" at the beginning of the 20th century. Together with the chemist Alfred Bertheim and bacteriologist Sahachiro Hata, he discovered the arsenic dye arsphenamine, which was later called Salvarsan or 606 (it was the 606th compound tested) as a potent anti-syphilis drug [2,6]. This was known as antimicrobial chemotherapy, and the first widely available antibiotic, introduced in 1935 by Gerhard Domagk, was sulfonamidochrysoidine or Prontosil, with antibacterial activity occurring in a number of infectious diseases. It was soon determined that Prontosil is a precursor to *p*-aminophenylsulfonamide, already discovered in 1908 and not patentable, but which led to easily modified derivatives and the era of sulfonamide antibiotics [7]. Penicillin, which is one of the best known antibiotics, was discovered in 1928 by Alexander Fleming, even though mass production started more than 15 years later during World War II, following synthesis and purification work carried out by Howard Florey and Ernest Chain [2,8]. In 1944, Selman Waksman (considered "the father of antibiotics" and who first coined the term) discovered an aminoglycoside antibiotic from *Streptomyces griseus* and named it streptomycin [9,10]. This marked the beginning of a golden age, which led to the discovery of more than 20 different groups of antibiotics in the following decades, of several different types: *sulfonamides*, ***β****-lactams* (with penicillin, cephalosporin, monobactam and carbapenem subclasses), *aminoglycosides*, *quinolones*, *cyclic peptides* (including gramicidins, polymyxins, glycopeptides, lipopeptides and lipoglycopeptides), *tetracyclines* (including glcyclines), *macrolides* (including ketolides), *amphenicols*, *nitroimidazoles*, *dihydrofolate reductase (DHFR) inhibitors*, *lincosamides*, *oxazolidinones*, and *ansamycins* (see Figure 1).

There is no doubt that antibiotics have changed the course of medicine and saved untold millions of lives worldwide since they were first introduced. Infectious diseases that could not be previously treated, and their catastrophic effects, could now be easily controlled. However, resistance started to emerge rapidly. Resistance of *Staphylococcus* to penicillin had already emerged by 1940, before mass production had even begun, while resistance toward methicillin was first reported only two years after its introduction (see Figure 1). Fortunately, during the golden age, novel antibiotics kept being discovered and developed for clinical use. The glycopeptide vancomycin, in use from 1958, was widely believed to be resistance-proof, but, by 1986, a resistant *Enterococcus* had appeared, and resistant staphylococci emerged before the end of the century (see Figure 1). Multi-drug-resistant (MDR) bacteria have, by now, become a major concern, with the emergence of pan-drug resistant (PDR) or extensively-drug resistant strains (XDR, see Figure 1) such as *Mycobacterium tuberculosis* resistant to fluoroquinolones and all second-line injectable drugs (capreomycin, kanamycin, or amikacin) [11]. It is conservatively estimated that, in the US and Europe, 2.5 million people are affected by such infections each year, and approximately 50,000 people die as a result of the infection [11].

A related major concern is the drying up of the pipeline over the last 30 years, and novel classes of antibiotics entering into it (see Figure 1). Among the limited number of promising compounds at various stages of clinical investigations and development [12], few, if any, represent a truly new class, with a completely different mode-of-action to previously approved drugs. Such an example are oxazolidinones, with linezolid being introduced in 2000, but, unfortunately, staphylococcal resistance emerged shortly after (see Figure 1) [13]. In this period, ketolides, glycylcyclines, and lipopeptides have also been introduced (see Figure 1). However, telithromycin (ketolide) and tigecycline (glycylcycline) showed adverse side effects, which prompted the Food and Drug Administration (FDA) to label the products with their strongest form of warning known as the "black box warning" [14,15]. As a consequence, telithromycin was discontinued from further use. Daptomycin, which is a lipopeptide, was introduced in 2003 and showed success, but was removed from the World Health Organization (WHO) List of Essential Medicines in 2019 [16].

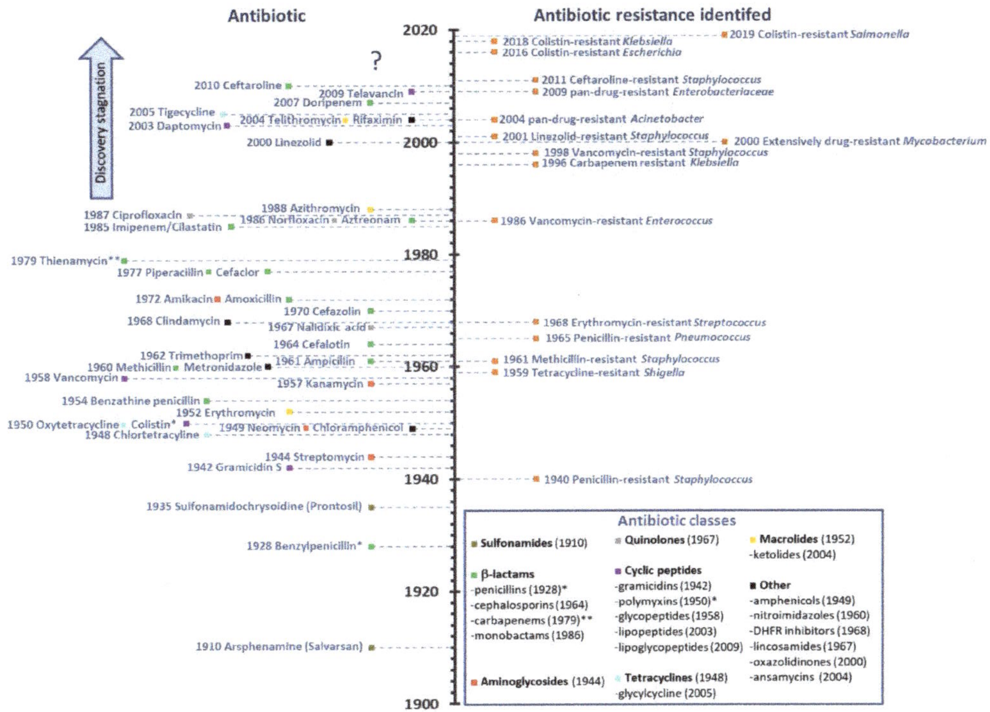

Figure 1. Timeline of antibiotic development as released and in parallel with the timeline for emergence of drug-resistant bacteria [2,6,11,13,17–34]. All antibiotic classes are represented by at least one antibiotic. For figure clarity, not all antibiotics nor all antibiotic-resistant bacteria are represented in the timeline. (*) Date of discovery not of release, (**) date of discovery, never introduced due to instability in an aqueous solution [35].

The drying pipeline has resulted in some shelved antibiotics returning to clinical practice despite inadequate pharmacological properties, such as colistin (polymyxin E). Discovered in 1950 by Japanese researchers [21], it was abandoned shortly afterward due to its nephrotoxicity [36] and the abundance of other equally potent antibiotics with less pronounced side effects. It is now being used as a "drug of last resort" against Gram-negative bacterial infections [37]. However, cases of resistance to it have recently been reported [38], with plasmid-mediated dissemination of the *mcr-1* gene was reported in *Escherichia coli* in 2016 [24], of *mcr-8* in *Klebsiella pneumoniae* [25], and *mcr-9* in *Salmonella enterica* [26] soon after (see Figure 1).

It is, therefore, evident that novel anti-infective agents with alternative modes of action are urgently required to battle the ever evolving, multi-drug resistant bacteria. In 2017, the WHO posted a shortlist of the critical strains to combat: carbapenem- and 3rd generation cephalosporin-resistant *Acinetobacter baumannii, Pseudomonas aeruginosa*, and *Enterobacteriaceae* [39]. Many research groups worldwide are now devoted to solving this crisis and, among other molecules being considered [40,41], antimicrobial peptides (AMPs) are hailed as possible alternatives to conventional antibiotics for some therapeutic uses, which have a relatively low potential to elicit resistance [42]. Few believe that completely resistance-free antimicrobials can be developed any more.

2. Antimicrobial Peptides—What Are They?

"I Like the Dreams of the Future Better than the History of the Past."

-Thomas Jefferson (1743–1826)

There is no univocal answer to this question, but we can find a consensus of what the antimicrobial peptides research community says that they are: multifunctional effector molecules, often gene encoded, produced by almost all organisms, and having a direct antimicrobial activity and/or immunomodulatory properties [43,44]. In light of the burgeoning resistance problem, peptides with a direct antimicrobial activity (AMPs) and often immune-regulatory capacities are easy to fit to the "dream (molecules) of the future" definition, to paraphrase Thomas Jefferson. Many AMPs have been discovered to date (see below), and they are reported to be active against Gram-negative and Gram-positive bacteria, against infective fungi and sometimes with antiviral, antiparasitic, or antiprotozoal properties [45]. Many of these have also been shown to modulate host immunity by activating immunocytes, and modulating inflammation, alternatively suppressing or promoting it [46]. To emphasize their pleiotropic nature in higher organisms, natural AMPs are often referred to as '*host defense peptides*' (HDP), or, more specifically, as '*innate defense regulatory* (IDR) peptides,' since reports on their immunomodulatory activities have mostly been confirmed at the level of innate immunity [46,47]. Their impact and importance to innate immunity is supported by their abundance in all eukaryotic organisms (fungi, algae, plants, invertebrate, and vertebrate animals), and distribution in cells and tissues at the front line of host defense against infection (mainly circulating immunocytes and epithelia) (see Figure 2). The dedicated Collection of Anti-Microbial Peptides (CAMP^{R3}) database currently contains 8164 entries for peptides, most of which (~74%) come from animals [48]. A particularly abundant source are anuran species, with almost two thousand peptide sequences reported in the dedicated Database of Anuran Defense Peptides (DADP) [49]. AMPs are, however, also well represented in prokaryotes, produced by both Gram-negative and Gram-positive bacteria, with one abundant class being the bacteriocins [50]. In this case, however, their role is somewhat different to that in eukaryotes, as they are principally used to clear the immediate environment of producer bacteria from competition by closely related, bacterial strains [50].

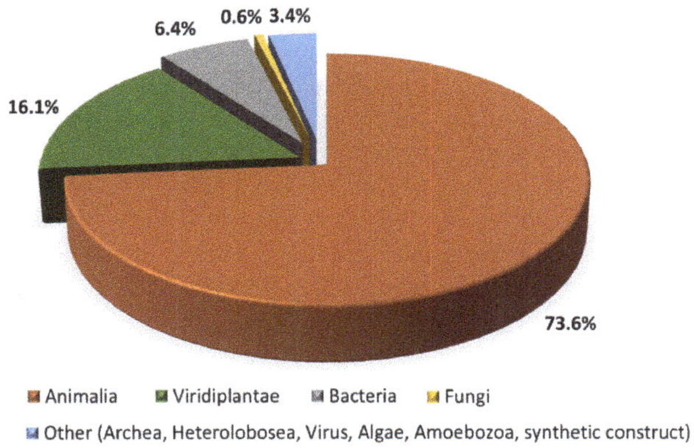

Figure 2. Distribution of AMPs across kingdoms based on sequences in the CAMP^{R3} database. (http://www.camp.bicnirrh.res.in/dbStat.php) [48].

3. Ribosomal vs. Non-ribosomal Synthesis and Antimicrobial Peptide Precursors

AMPs are often gene encoded and ribosomally synthesized (this is, by far, the most common case in eukaryotes), or can be assembled by large multi-functional enzymes known as non-ribosomal peptide synthetases (NRPSs) [51,52]. The latter process is used by bacteria and fungi [52] and allows incorporation of non-proteinogenic amino acids into the peptides (often the D-enantiomers of natural residues) and to further modify the peptides with ring formation, glycosylation, hydroxylation, or acylation [53,54]. There are ~500 non-proteinogenic amino acids known today, which possess added structural and functional features that can contribute significantly to a peptide's potency. In fact, the cyclic peptide antibiotics polymyxin B, gramicidin S, and vancomycin are synthesized in this manner [54] and all contain some non-proteinogenic amino acids in their sequences [53,55].

Gene encoded, ribosomally synthesized peptides are produced by almost all forms of life, including bacteria [42,48,56]. Quite often, multiple AMP genes are clustered at a single chromosomal locus, which is the case with α- and β-defensins [57] and can be co-expressed. They are, furthermore, frequently expressed as inactive precursors, containing a signal peptide region and a pro-piece that can serve to keep the mature peptide inactive until it is conveyed to the site of infection, where it is proteolytically released (see Figure 3). For this reason, the pro-piece is often anionic to complement the mature peptide, which is generally cationic. In most cases, the pro-region precedes (it is N-terminal to) the AMP sequence, but cases are known where it is the C-terminal (such as for some fish and plant peptides) [58,59].

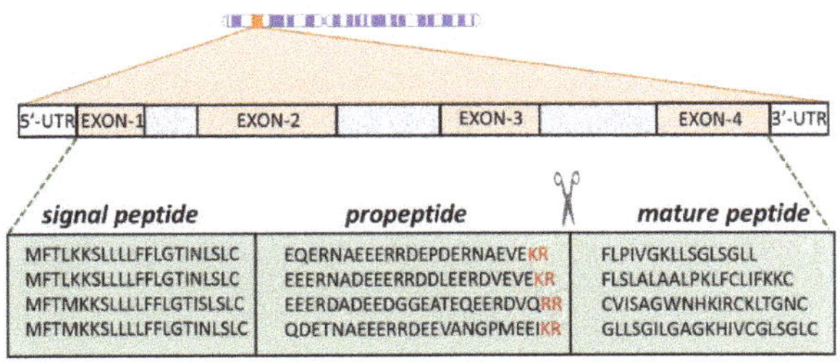

Figure 3. Representation of AMP expression. Peptides are cleaved at dibasic cleavage sites (-KR, -RR, in red). The proregion has a distinctly anionic nature. Depicted peptide sequences belong to anuran AMPs from the Ranidae family [60–63].

The activity of AMPS is, therefore, regulated not only by the expression level but also by the presence and abundance of appropriate proteases at the right place and the right time to cleave the peptide, which is generally at dibasic cleavage sites (see Figure 3) [57,64]. The signal peptide is a common feature of prokaryotic and eukaryotic proteins that need to enter secretory pathways [65]. A very useful aspect of antimicrobial pro-peptides is that signal regions for a given class can be much more evolutionarily conserved than the mature peptides themselves (see Figure 3) [58,66–68]. This provides a very useful handle for sequence mining in databases. The diversity of the mature AMP sequence most likely occurs as species' adaptation to specific microbial communities in a particular environment. While there is still no solid explanation for the phenomenon of signal sequence conservation, it gives valuable insights into the evolution of some AMP families, as in the case with the anuran ones [69].

Lastly, it is worth noting that a majority of gene encoded AMPs undergo post-translational modifications, currently classified into more than 15 types, including disulfide-bridge formation, N-terminal or C-terminal capping (acetylation, pyroglutamic acid formation, amidation), halogenation, hydroxylation, phosphorylation, glycosylation, etc. Peptides can be modified to a greater or lesser extent, which contributes to peptide potency and/or stability [70].

4. Physico-Chemical Properties

Certain physico-chemical properties of AMPs are undeniably related to their interaction with lipid molecules that make up the bilayer system in the membrane of the target cells and correlate directly with the peptides' biological activity and specificity. In fact, the same consideration could be made for other components of the microbial cell wall, but the data is less consistent. However, even limiting one's focus to the lipid bilayer, there is still an imperfect understanding of the complex relationship between the AMP and membrane, since even peptides with very similar structures can have remarkably different reported mechanisms of action (e.g., buforin and magainin 2, or LL-37 and the rhesus orthologue RL-37) [71–73]. A better understanding of the relationship between physico-chemical properties and biological activity is, therefore, required to identify features that are responsible for potency and specificity of AMPs.

4.1. Molecular Type, Size, and Structure

AMPs have been divided into several categories, according to particular features of their secondary structure. The simplest and most widely used classification divides them into extended structures, linear α-helical peptides and peptides with β-sheet or hairpin-like structures, generally braced by disulfide bridges [74]. Guha et al. [75] have, however, proposed a more elaborate taxonomy based on secondary motifs, with six different groups including *i*) α-helix, *ii*) 3/10 helix, *iii*) pi-helix, *iv*) β-strand, *v*) β-turn, and *vi*) disordered coil, which may concern the entire peptide or only parts of it, where some scaffolds combine to make different structural motifs. Essentially, this classification still depicts three major classes of AMPs: helical, β strand, and extended. In this review, the latest Wang terminology will be used to classify peptides in four classes [76]: α helical (e.g., human LL-37), β sheet (e.g., the human α defensin HD-4), αβ peptides (e.g., the human β defensin hBD-2), and non αβ peptides (e.g., the small, extended Trp-rich peptide indolicidin) (see Figure 4).

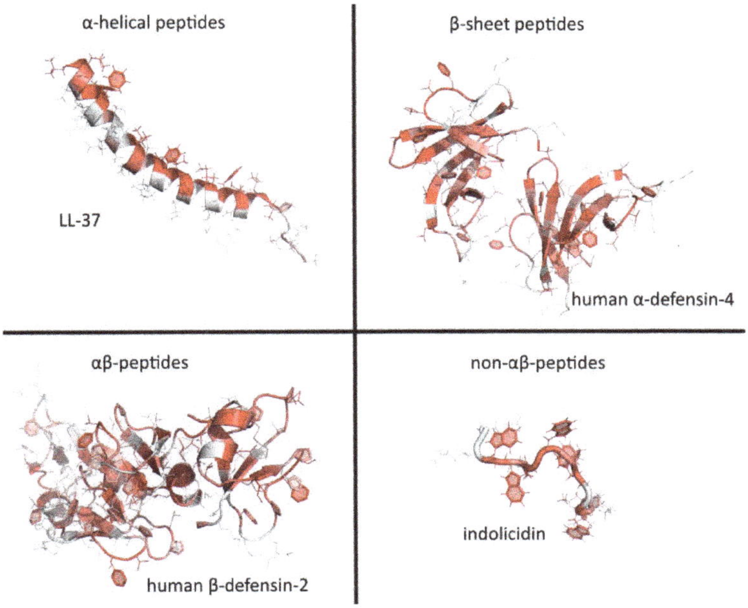

Figure 4. An overview of major structural classes of antimicrobial peptides. The structures taken as an example were solved either by nuclear magnetic resonance spectroscopy or X-ray diffraction and coordinates were downloaded from Protein Data Bank (PDB) (https://www.rcsb.org/) [77]. PDB IDs: LL-37 (2k6o), human α-defensin-4 (1zmm), human β-defensin-2 (1fd3), and indolicidin (1g89). Visualization was done using PyMOL 1.8 [78] and amino acids colored according to a normalized Eisenberg hydrophobicity scale (light grey—polar, red—hydrophobic) [79].

The "α−helical" peptides, which are among the most studied and, consequently, the better understood, generally exhibit little or no structuring in a bulk aqueous solution, but only adopt this defined secondary structure in the presence of a bacterial membrane or some other type of anisotropic environment (sodium dodecyl sulfate micelles or water/trifluoroethanol mixtures) [46]. The adoption of this active structure is aided by:

- the presence of helix-stabilizing residues distributed throughout the sequence (e.g., Leu, Ala, Lys),
- the clustering hydrophobic residues on one side of the helix when it forms, which allows insertion into the membrane bilayer (or, seen the other way round, the presence of a lipid layer that induces appropriately distributed hydrophobic residues to cluster into a well-defined sector of the helix by interacting with it),
- salt-bridging between oppositely charged residues placed next to each other when the helix forms (normally, but not necessarily, when these residues are spaced three or four positions apart [80]).

It should be noted that increased helix stability correlates with increased potency, but only to a certain extent [81]. In fact, an increased propensity to transit from a coiled to helical conformation on passing from bulk solution to the membrane surface will generally positively affect activity. On the other hand, an increased propensity for helix formation in bulk solution will result in 'sticky' molecules (due to formation of a hydrophobic sector exposed to an aqueous environment), that tend to oligomerize or interact with other hydrophobic molecular surfaces. In the first case, this will alter the mode of the membrane interaction, which inevitably affects activity. In the second case, it leads to sequestration, with a negative effect on potency [72,82].

The size of helical peptides (sequence length) is also an important feature, especially in the context of peptide activity, since a minimum of seven to eight amino acids are needed to form an amphipathic structure [71] with separate hydrophobic and hydrophilic faces. One of the shortest peptides (only 10 residues) that was reported to have antibacterial activity, PGLa-H, was isolated from the skin of the African clawed frog *Xenopus laevis*, and is capable of adopting an α−helical conformation in an anisotropic environment [83]. Peptide length was initially thought to be particularly important when helical peptides were thought to form barrel-stave pores in the membrane (see below). At least 22 residues are required for them to span a canonical lipid bilayer, whereas the more extended peptides could do so with as few as eight residues [71]. However, very few AMPs have been found to form barrel-stave pores.

It is difficult to directly relate the length of a membrane-active peptide to its cytotoxicity. The bee venom-derived melittin becomes significantly less toxic when shortened from the original 26 residues to only the 15 C-terminal residues [84], but this is more likely related to the removal of a section with features favoring membrane insertion, rather than just reducing the size. In fact, by conversely doubling the size of the previously mentioned 10-residue PGLa-H in a tandem repeat resulted in a peptide with significantly greater potency against bacteria without greatly affecting toxicity toward host cells, so that the effect can be selective [85]. Therefore, it is wrong to conclude that the simple shortening or extending of a certain sequence will result in more favorable properties. It may be more appropriate to alter the length, while at the same time, maintaining an appropriate balance between hydrophobic and polar residues, and not drastically affect hydrophobicity or charge (see below). In any case, natural AMPs generally have relatively short sequences (normally under 50 residues) with the majority of known peptides in the 10-30 residue range [45].

With respect to other structural types, one well understood group are the proline-rich AMPs that, due to the abundance of proline residues, adopt an extended, likely polyproline type-II, conformation [86]. This type of AMP acts principally by translocating into bacterial cells without damaging their membranes (see below) using specific transport systems to then hit intracellular targets. Shortening these peptides affects both of these functionally essential aspects, and it has been reported that a minimum length is required for efficient antimicrobial activity [86]. In this case, requirements for the presence of specific motifs in the sequence may constrain how and to what extent a peptide can be shortened [87].

Trp-rich peptides, which are generally also Arg-rich, appear to enter bacteria via a sort of self-promoted uptake, without membrane disruption (see below), to then inactivate internal targets [88–90]. Trp residues are aromatic but the indole ring has a dual hydrophobic/hydrophilic nature and tends to partition at the membrane water interface. While the Arg residues undergo electrostatic interactions with membrane surface components, the guanidinium group also contributes to cell penetration. These distinctive properties make very short Trp-rich peptides active.

Defensins have size constraints that are determined by the fact that their scaffold is braced by at least three disulfide bonds, and participating Cys residues can be located towards the termini, so that shortening knocks bridges out with an inevitable effect on structure and function. In any case, some attempts have been made to design 'mini-defensins' by maintaining only part of the scaffold and only one disulfide bridge [91,92]. Therefore, appreciable antimicrobial activity is maintained. Several studies suggest that disulfide bridges are dispensable, so that linear analogues or fragments may appear to maintain or even increase activity with respect to the native peptides [92–97]. One study reported that the reducing environment of the colon may break disulphide bridges in hBD1, which renders it a more a potent antimicrobial agent [98]. This very likely alters the target and mode of action with respect to the native defensin, which is suggested by the fact that the IDR activity of defensins on host cells, which also likely depends on membrane interaction but in a different manner, can be significantly reduced in linearized peptides while the AMP activity is not [93,94]. Furthermore, oligomerization plays a significant role in the mode of action of many defensins, by determining how they interact with the membrane, and is likely to be very sensitive to such drastic structural alteration as knocking out disulfide bridges.

4.2. Charge and Hydrophobicity

The net charge of known natural AMPs varies widely from cationic (most often) to anionic (rare), which ranges from +16 to −6 [80,99–101]. The vast majority of identified peptides have an intermediate net positive charge (centering around +6) that can be directly correlated with peptide potency and selectivity. There seems to be an optimum charge span for activity, so that higher or lower values outside this range can result in reduced activity toward bacterial cells and/or increased toxicity toward host cells. The relationship between these parameters and function has again been most extensively probed in helical peptides. Dathe et al. [102] have shown that increasing the charge of magainin analogs above +5 resulted in both increased hemolysis and loss of antimicrobial potency. Giangaspero et al. [81] came to a similar conclusion when varying the charge of helical peptides, which, otherwise, had relatively similar mean hydrophobicity, amphipathicity, and helicity. The decreased amicrobial potency was, in part, attributed to a reduced propensity for helix formation due to the increased charge density (clustering of positive charges in one sector of the helix leads to repulsion). More recent findings suggest it could also result from repulsion between highly charged peptides at the membrane surface, which leads to a lower concentration of membrane-adsorbed peptides [103]. In principle, the distribution of positively charged residues should not correlate with peptide potency, which should only depend on the overall peptide charge [81,104]. However, it can make a significant difference if it affects the formation of helix-stabilizing salt bridges, as observed in artificial variants of the human helical peptide LL-37 [73].

In a similar way, there seems to be an optimal hydrophobicity window for peptides to have an optimal balance between antimicrobial activity and host cell toxicity. On average, AMPs contain approximately 50% hydrophobic residues. The overall hydrophobicity affects the peptide's capacity to partition into the lipid bilayer, and can be directly correlated with both potency and host cell toxicity [105]. Increasing or decreasing this property outside the optimal range can result in a decrease of antimicrobial activity and an increase in blood cell lysis, not necessarily accompanied by improved antimicrobial activity [81,105,106]. In fact, an increased hydrophobicity can result in reduced antimicrobial activity if it promotes self-association, for the same reasons as an excessive stabilization of the helical structure (see above). This impedes access to the bacterial membrane and, therefore, lowers the concentration of the peptide actually impacting it [104,106].

4.3. Amphipathicity and Structural Stability in Helical Peptides

Overall, hydrophobicity is one of the key properties related to biological activity of any given AMP sequence. Unlike the charge, where its distribution is not necessarily correlated with potency (see above), the arrangement of polar and hydrophobic residues (~50% in AMPs) into an amphipathic structure sets some constraints on the primary structure, and it plays a key role in peptide activity more than the hydrophobicity itself [107]. However, what exactly is the amphipathicity? This property refers to the topographic distribution of hydrophobic and polar residues within the peptide sequence, which results in a more or less accentuated spatial separation in the active AMP structure. For a helical conformation, this occurs if polar/charged and hydrophobic residues cluster on opposite sides (of a hypothetical cylinder around which the helix is wound, see how this relates to the structure in the human helical AMP LL-37, in Figure 5). It can be numerically quantified in terms of the hydrophobic moment (μH) [79]. While an α-helix is one of the simplest and more efficient ways to generate an amphipathic structure, other types of active conformations such as in β−sheet peptides can also adopt an amphipathic arrangement, to a greater or lesser extent. However, it is more difficult to both visualize and quantify this [80,108].

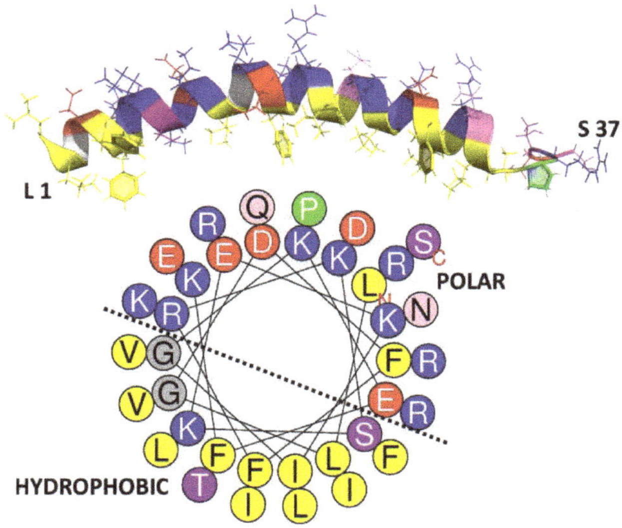

Figure 5. Secondary structure and helical wheel projection of human cathelicidin LL-37. The structure and projection were, respectively, obtained from PDB [77] (ID: 2k6o) and HeliQuest [109]. The residues were colored according to their hydrophobicity with ~40% hydrophobic and 60% polar amino acids in an appreciable amphipathic arrangement. Hydrophobic (yellow and green), polar charged [red (−) and blue (+)], polar uncharged (light to dark purple), and glycine (grey).

Amphipathicity aids activity of helical peptides since it allows them to sink their hydrophobic faces into the membrane bilayer, which is an essential step leading to membrane disruption. It must be correctly tuned for an optimal balance between anti-bacterial potency and host cell toxicity. In general, the hydrophobic moment in helical AMPs is around 60% of the maximum possible value. Increasing it above this value does not greatly increase potency but can significantly increase toxicity [81,82].

Lastly, helicity is the propensity of an AMP to adopt a helical structure. As discussed above, it plays a significant role in the antibacterial activity, and, in general, it seems to correlate more with the toxicity toward host cells than antimicrobial potency, in a manner that relates to its effect on oligomerization. It can be reduced by incorporating *D*-amino acids into the peptide sequence, without greatly affecting potency. However, this can narrow the activity spectrum. As reported by Papo et al. [110], replacing up to a third of *L*-amino acids with their *D*-enantiomers resulted in peptides devoid of haemolytic activity that maintained an appreciable antibacterial potency, especially against Gram-negative bacteria. Furthermore, they are protected from proteolytic degradation, which should increase the bioavailability of such synthetic peptides.

5. Mode of Action

The mechanism of action of numerous AMPs has been extensively studied. Experiments have often been carried out with artificial membranes, typically large or giant unilamellar vesicles, and less frequently on microbial cells, using fluorescent dyes and labeled peptides. In any case, a widely accepted subdivision of AMPs, according to their mode of action, is *i*) membrane active and *ii*) non-lytic [111]. Some AMPs can act upon bacteria using both of these two major mechanisms, and sometimes switching from one to the other, depending on the peptide concentration, the membrane characteristics of a particular bacterial species, or its growth phase [112].

5.1. Disrupting Bacterial Cytoplasmic Membrane Integrity—A Primary Inactivation Mechanism

The term "membrane permeabilizing" peptides (MPP) [75] is more general than the often, and sometimes inappropriately used, "pore-forming" peptides [113–115]. Considering the complexity of lytic mechanism(s) of membrane-active AMPs (which is not restricted only to "pore-forming"), it is more appropriate (see Figure 6). An MPP must initially partition into a membrane and, therefore, be amphipathic for at least part of its structure, i.e., it must have some form of "interfacial activity." This, nonetheless, allows for remarkable structural diversity, which results in functional diversity, so that some MPPs are active only against a narrow spectrum of microbial species while others have a very broad spectrum of activity. Wimley's group has recently pointed out that the process of membrane permeabilization should not be considered as being simply due to a series of stochastic events involving one or more peptide molecules, nor should it be ascribed to a well-defined sequence of events. In other words, it requires neither discrete events nor the formation of static molecular entities. It is better described as a "mechanistic landscape" that varies depending on the experimental conditions and variables such as peptide concentration, bilayer lipid composition, temperature, ionic strength, and pH [75].

Results from a recent molecular dynamics (MD) simulation case study on maculatin, isolated from the skin glands of a green-eyed tree frog *Litoria genimaculata* [116], are in line with the proposed scenario. This particular peptide has been reported to act by pore formation [117]. Simulation work done by Wang et al. [118] showed that these pores continuously form and dissociate in the membrane. Moreover, the architecture of the pores varies, and is dominated by hexamers, heptamers, and octamers, with peptide molecules having a strong but not absolute preference for an antiparallel peptide orientation in the oligomers. Remarkably, the assembly of maculatin into pores seems to be driven by the successive addition of peptide molecules to an already existing transmembrane inserted helix to form a growing oligomer. Therefore, the translocation of the polar side chains of the incoming peptide is 'catalyzed' by the polar face of the already inserted peptide(s).

It follows that, even for very well-studied AMPs, the molecular mechanisms of membrane permeabilization cannot have been completely elucidated, and many questions remain [75,119]. Yet, some common steps can be inferred to occur at the bacterial membrane, which eventually leads to its disruption. This includes: i) initial attraction of the AMP to the membrane surface and interaction, ii) adoption of an active conformation, iii) insertion into the bilayer and concentration dependent accumulation, and iv) (in some cases) self-association/oligomerization [111]. The order in which these three steps occur will significantly affect the type of membrane lesion.

Most α–helical AMPs do not adopt this conformation in a bulk solution (see above) so that the initial interaction at the membrane level occurs between a positively charged peptide coil and the negatively charged phospholipid head-groups in the bacterial membrane surface [80,121]. This allows redistribution of hydrophobic sidechains so that they can interact with the membrane acyl chains leading to adsorption into the membrane, which, in turn, induces conformational changes in the peptide (to a stable amphipathic helix) that maximize these interactions [122], and this allows a deeper insertion into the lipid bilayer, which alters its structure. In other words, it follows the steps i), ii), iii), iv) in that order. By contrast, β–sheet peptides already have a stable, disulphide-braced amphipathic conformation that is maintained on membrane insertion. This is likely the case for pre-formed helical peptides, like LL-37, which are stabilized by internal salt-bridges [123]. In both cases, the pre-formed structures favor oligomerization [72,124], so these peptides may follow a different order for the steps ii), iv), i), iii).

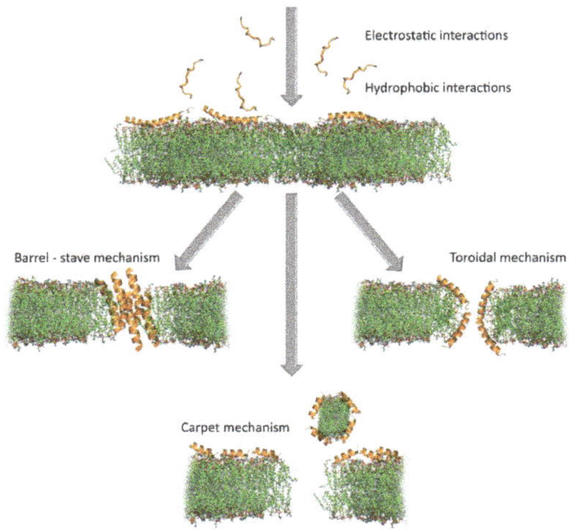

Figure 6. Proposed mechanisms of action of membrane permeabilizing peptides. Barrel-stave mechanism (alamethicin, PDB ID: 1amt), toroidal mechanism (magainin 2, PDB ID: 2 mag), carpet mechanism (aurein 1.2, PDB ID: 1vm5). Peptide structures were chosen from PDB [77] while taking their specific modes of action into account. The lipid bilayer was downloaded from the CHARMM-GUI.org website [120] (green: lipid tail, red, and blue: lipid head) and the manually created pores are only indicative. Visualization carried out using PyMOL 1.8 [78]. Note that not all interactions include pore formation and, for figure clarity, those are not included in this presentation (see above).

Once on the membrane, a critical peptide concentration is, in any case, required to induce membrane lysis, which can occur by different mechanisms (see Figure 6). Several different mechanisms have been proposed to lead to membrane lesions, which involve more or less well-defined molecular entities.

- Historically, the first proposed mechanism was for a certain number of peptide molecules to assemble and flip from a parallel to a perpendicular orientation with respect to the membrane surface, to form barrel-stave pores. The amphipathic structure would allow their hydrophobic surface to interact with the membrane lipids and hydrophilic regions to line the core of the channels, which promotes lateral peptide-peptide interactions. This mechanism, however, has turned out to be rare, and seems to apply to a limited number of AMPs, such as pardaxin [125] and non-proteinogenic alamethicin [126].
- In a second, less organized model, peptides remain aligned perpendicularly to the membrane surface, with the hydrophobic region inserted in among the acyl chains. On accumulation, this causes the bilayer itself to cavitate so that the hydrophilic region of the peptides line a wormhole or toroidal pore. Re-oriented phospholipid head groups also line the pore so that precise peptide-peptide interactions, or even a defined number of participating molecules, are not required, which makes it much more permissive for diverse primary structures than the barrel-stave pore. Such behavior is reported for the helical peptides magainin 2 [111] and aurein 2.2 [127]. These pores are reported to have relatively short lifetimes and can collapse, which allows the constituent AMPs to gather on the inside membrane bilayer surface, or it can extend and combine to lead to membrane micellization. For the bee toxin peptide melittin, for example, MD simulations suggest that toroidal pores are quite disordered and follow the latter pathway [128,129].
- In a third, even less organized model, peptides concentrate on and coat the surface to lead to micellar structures involving limited areas of the lipid bilayer, which, on removal from the membrane, leave large lesions behind. This non-specific, detergent-like mechanism does not

necessarily require discrete pore formation but just surface accumulation, so it has been called the carpet model. It has been proposed alternatively for magainin and aurein 1.2 [130,131].

The last two mechanisms are not necessarily mutually exclusive, but could fit into Wimley's "mechanistic landscape," which occurs at different peptide concentrations, in a membrane-dependent manner. Other less disruptive mechanisms have also been proposed for AMPs, and include membrane thinning, depolarization or fusion, electroporation, and targeting of specific phospholipids [74,80,119]. In any case, they are all attempts to simplify mechanisms that are extremely complex and dependent on a number of variable conditions, in an attempt to make them more comprehensible. This poses the risk of limiting the mode of action of membrane permeabilizing peptides to a few "main categories" considered separately. But how realistic are these proposed models? And how reliably do they explain AMP behavior? Years of research carried out mostly on very simplified membrane models (e.g., liposomes) [75], have shown that the mode of action can vary substantially with very subtle modifications in a lipid-to-peptide ratio or membrane surface charge, even for a given AMP [132].

In summary, although we are far from a complete picture of how even the best studied peptides act, it is safe to say that AMPs, likely act in vivo using several possible membrane-disrupting mechanisms, with time frames and to extents that depend on environmental conditions. Any given permeabilizing model may, therefore, solve part of the puzzle but is unlikely because it does not provide the entire solution. This lack of dependence on specific interactors or defined mechanistic pathways may have contributed to the relatively low incidence of bacterial resistance to AMPs, despite many millions of years of continuous exposure to them. Bacteria can counteract them by altering the surface properties (mainly charge) in different ways, but this is a metabolically expensive, and, therefore, a transient form of induced resistance [133].

5.2. Non-Lytic Intracellular or Extracellular Mechanisms of Action

Some AMPs do not rely on a directly membranolytic mechanism, but act on extracellular or intracellular targets [74,86,112,134,135]. These act on the outside, disrupting septation or cell-wall biogenesis to impede cell division and weaken the structural integrity of the cell, or it can pass through the bacterial cytoplasmic membrane, without necessarily disrupting it, and inactivate specific metabolically essential components inside the cell.

With respect to the latter type, Trp-rich AMPs have been reported to enter bacteria by direct translocation, which is a process that has some aspects in common with pore formation, but without resulting in cell lysis [88–90]. Instead, the proline-rich AMPs enter susceptible bacterial cells using specific membrane transport proteins [86,136]. Even helical AMPs could, in principle, internalize into bacteria, without apparent membrane lysis, simply through the rapid formation/collapse of pores. Regardless of the peptide uptake mechanism, and according to a slightly modified Le at al. classification [135], internally and externally acting AMPs can be classified into six groups depending of their specific targets, which is listed below.

5.3. Nucleic Acid Biosynthesis and Metabolism Inhibitors

This group of peptides is represented by the helical buforin II and Trp-rich indolicidin [137,138]. Buforin I, which is the parent peptide to buforin II (a 21 amino acid fragment), is homologues to the N-terminal fragment of the DNA-binding protein histone H2A [139]. Some variants of buforin II have shown affinity toward double stranded nucleic acids [140], while designed analogues were found to have a greater binding affinity for RNA [141]. Indolicidin, which is a peptide of bovine origin belonging to the cathelicidin family, has been found to act both by disrupting the bacterial membrane and by inhibiting DNA synthesis, or, more specifically, inactivating DNA topoisomerase [138,142,143].

5.4. Inhibitors of Protein Biosynthesis and Folding

Bovine cathelicidin Bac7, which is a 60 amino acid long peptide isolated from bovine neutrophils, interferes with complex machinery involved in protein synthesis. Its activity seems to reside at the N-terminus, so that a 35-long fragment, Bac7$_{1-35}$, is fully active and has been shown to inhibit protein translation by targeting ribosome subunits, without affecting DNA synthesis or transcription. This specifically inhibits the process of protein synthesis [135,144–147]. Other proline-rich peptides with a similar mode of action include PR-39, the porcine orthologue [148], and unrelated apidaecin-type peptides isolated in honeybees, hornets, and wasps [149]. Apart from Pro-rich peptides, CP10A, which is a synthetic indolicidin derivative in which proline has been substituted with alanine, is an example of a short, tryptophan-rich helical peptide that, in addition to membranolytic properties, has DNA-binding affinities, and also acts by disrupting protein metabolism [150].

Proline-rich peptides are also reported to exert antimicrobial activities by interfering with protein-folding. Pyrrhocoricin, apidaecin, drosocin, and Bac71-35 all inhibit the major bacterial heat shock protein DnaK, and, in some cases, disrupt its ATPase activity [147,151–153]. They prevent DnaK from refolding misfolded proteins, and apidaecin has been shown to also inhibit the associated chaperonin GroEL [134,135]. Furthermore, these peptides bind stereospecifically to their bacterial target, which are inactive toward the human counterpart chaperone Hsp70 [151]. Other proline-rich AMPs with the same mode of action are the insect abaecin, and redesigned oncocin [135].

5.5. Inhibitors of Bacterial Proteases

Some AMPs, like histatin-5, have been reported to inhibit both host-secreted and bacterium-secreted proteases [154]. Dysregulation of these enzymes is associated with oral diseases such as periodontitis. By competitively inhibiting the bacterial cysteine proteinase clostripain, produced by *Clostridium histolyticum*, whose infections cause gas gangrene [155], histatins-5 and other peptides of this kind have been proposed as a potential therapeutic to reduce extracellular matrix degradation caused by bacterial or dysregulated host proteases. These AMPs reduce virulence and are antimicrobial.

5.6. Cell Division Inhibitors

CRAMP, the mouse helical cathelicidin orthologous to human LL-37, is a potent membranolytic expected of a helical AMP [156]. A CRAMP fragment has also been reported to interfere with the septation process, most likely by inhibiting bacterial cytokinesis [134]. This fragment was identified due to a significant sequence similarity to a bacterial peptide that regulates septation by interacting with its machinery [157]. C18G, which is a C-terminal, α-helical fragment of platelet factor IV, was also found to inhibit cell division by strongly stimulating the PhoQ/PhoP signaling system. This, in turn, results in increased synthesis of QueE, which is an enzyme that inhibits septation by interacting with the divisome [158].

Human α−defensin-5 targets different cell mechanisms, and also interferes with cell division processes, as shown by extensive elongation of peptide treated bacteria [159]. For similar reasons, AMPs of different origin such as bacterial microcin J25, insect diptericin, and the cathelicidins indolicidin and PR-39 have also been proposed to interfere with cell division processes [135]. However, the precise mechanism(s) involved have not been elucidated.

5.7. Cell Wall Biosynthesis Inhibitors

The bacterial cell wall, being essential for the cell's structural integrity and survival, is the target for numerous antibiotics in current use [160]. It consists of alternating β−1,4-linked N-acetylglucosamine and N-acetylmuramic acid crosslinked with peptide chains [161]. Lipid II is a crucial component of the cell wall synthesis process, since it is the shuttle carrier that transports disaccharide-pentapeptide building blocks across the membrane to be incorporated into the existing cell wall structure [162]. A number of AMPs, including the bacterial lantibiotic peptides mersacidin and nisin [163,164], and the fungal defensins plectasin and copsin [165,166] target this molecule in different ways, to disrupt cell-wall biogenesis. These AMPs, therefore, act externally to disrupt peptidoglycan integrity analogously to β−lactam antibiotics like penicillin or glycopeptides like vancomycin.

5.8. Lipopolysaccharide (LPS)-Binding Peptides

These peptides specifically act on this major structural and functional component of the outer membrane that covers the surface of Gram-negative bacteria. LPS can be released during bacterial cell division or death and induce a variety of inflammatory effects in animals, which leads to sepsis, and this may occur as a result of using antibiotics to treat Gram-negative infections. At the moment, there is limited treatment for patients with septic shock, which most often results in death [167,168]. AMPs that can bind to LPS may disrupt the outer membrane, which affects the cell's structural integrity and reduces survival. They can make the cytoplasmic membrane more accessible to other AMPs/antibiotics that have difficulty in passing through the outer membrane, and can also help sequester and clear LPS, which reduces its pro-inflammatory effects.

On the other hand, the LPS layer can actively neutralize the activity of AMPs by inducing their self-association or aggregation and sequestering them [169]. This has been observed for the frog peptides temporins A and B from *Rana temporaria*. However, Rosenfeld et al. [170] showed a synergic effect between these peptides and temporin L, which prevents their LPS-mediated oligomerization and markedly improves their activity. Another way to restore activity is by introducing a boomerang motif (GWKRKRFG) at their C-terminus, which results in hybrid peptides no longer susceptible to LPS-induced aggregation [171]. Furthermore, melittin-cecropin hybrid peptide with two additional positive charges at the C-terminus proved to be effective in traversing the LPS layer [172]. In any case, amphipathicity and a high proportion of cationic residues in the AMP sequence seem to be important properties for the broad-spectrum LPS-binding peptides [173].

6. Strategies for Identifying or Designing New AMPs

6.1. Crude but Effective–Extraction and Assay-Guided Isolation

In the past, identification of novel AMPs involved handling of several specimens from the same species to obtain small amounts of active peptides. Initial tissue homogenization was followed by peptide extraction and the crude peptide was isolated in several steps, mainly by using chromatographic techniques. In some cases, the animals were pretreated with electric shocks or noradrenaline, or were exposed to bacterial infection, to stimulate AMP production [174,175]. Potential AMPs were then isolated by assay-guided fractionation and the sequence determined using different techniques, including Edman degradation and mass spectrometry. Magainin, which is one of the first frog peptides to be identified, was isolated in this manner, as were penaeidins, pleurocidin, and some mollusk cysteine-rich peptides, among others [176–179]. Several human peptides were also identified in this manner, from epithelial cells and plasma [180–183]. Some potential AMPs have been identified by analyzing lysates from proteins, and even common food sources, in particular whey. Theolier et al. [184] have recently reported six new peptide fragments from β-lactoglobulin and one fragment from α-lactalbumin derived by peptic cleavage of whey protein isolate, which all had antibacterial properties.

This approach is, therefore, evidently successful, but is also very time-consuming, and produces rather low yields. It can also raise ethical questions of animal protection, especially considering rare and endangered species. Lastly, it misses AMPs that are not constitutively produced or whose expression cannot be stimulated.

6.2. Make the Most of the Growing Abundance of Omics Data

The rapid development and plummeting cost of sequencing techniques (next generation sequencing or NGS), combined with efficient and relatively cheap solid phase synthesis techniques, has opened the possibility of mining for valuable sequence information hidden in the genome, and functional testing, without the necessity of isolating polypeptides. For example, frog peptides have been identified by isolating total RNA and reverse transcribing the mRNA based on the 3′ poly-A tail. A cDNA library was constructed by using appropriate vectors and the positive clones selected and analyzed by nucleotide sequencing. This allowed the identification of several novel temporins, which were then either synthesized or obtained using "classical methods" including isolation and purification, before activity testing and confirmation using amino acid analyses [175]. A similar procedure led to the discovery of several peptides in the pickerel frog, *Rana palustris* [185], of clavanins from tunicate hemocytes [186], of protegrins from porcine leukocytes [187] and of penaeidins from Indian white shrimp *Fenneropenaeus indicus* [188], among others. More recently, *in silico* analyses of cDNA data in EST databases [189] led to the discovery of trichoplaxin, which is a placozoan AMP from *Trichoplax adhaerens* [190].

Improvement in NGS techniques and analysis pipelines, as well as the abundance of publicly available genomic and transcriptomic data, has led to the development of high-throughput techniques for simultaneous identification of potential AMPs. Kim et al. [191] reported a de novo transcriptome analysis of the American cockroach *Periplaneta americana,* which leads to the discovery of 86 putative antimicrobial peptides out of which 21 were experimentally verified for this activity. A similar approach was used for the identification of novel AMP sequences in the grasshopper *Oxya chinensis sinuosa* [192]. A novel method has recently been successfully developed for simultaneous identification of AMPs in different frog species [193]. By utilizing highly conserved signal regions of the peptide precursors (see Section 3) to design forward degenerate primers and correlating with transcriptomic and proteomic data available in public databases, ~130 different potential AMPs were identified, of which 29 were novel sequences (see Figure 7). The same procedure could, in principle, be applied to other organisms that have AMP gene families with comparable properties, i.e., a conserved signal peptide region associated with hypervariable mature peptide regions.

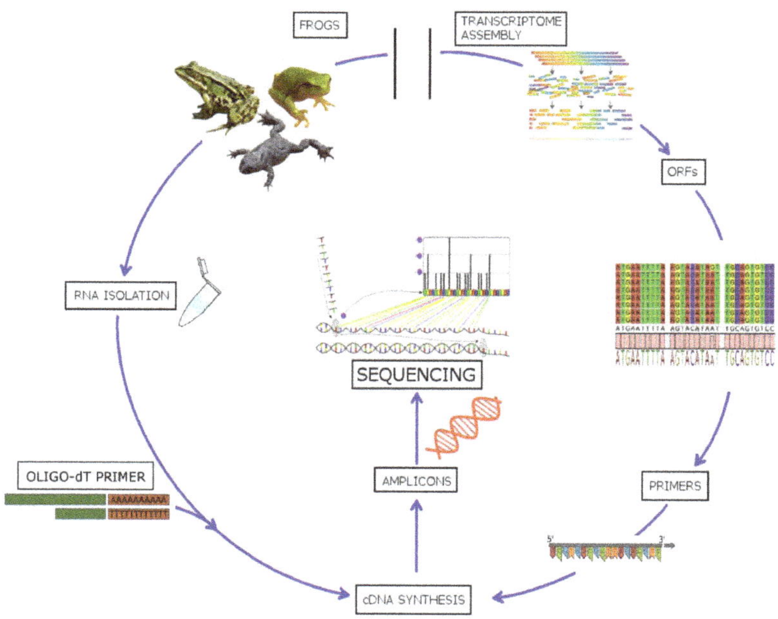

Figure 7. Schematic representation of the targeted DNA sequencing method. Figure modified from Rončević et al. [193] and reprinted under Creative Commons Attribution 4.0 International License (http://creativecommons.org/licenses/by/4.0/).

Brand et al. [194], on the other hand, developed a procedure for screening and identifying "intragenic antimicrobial peptides," which are bioactive fragments from larger proteins, based on specific physico-chemical properties, by finding eight novel peptides with different antibacterial potency. This experimental method was developed to identify peptide sets with membranolytic effects in model membranes [195]. This is a complementary step to the previously developed bioinformatic pipeline. Yi et al. [196] screened seven previously assembled genomic and transcriptomic datasets in the amphibious mudskippers and, based on sequence similarity, identified ~500 novel peptide sequences with the correct characteristics, by opening new pathways for AMP discovery. A similar procedure led to the analysis of gill transcriptomes from 87 ray-finned fish species, which leads to the successful identification of some novel AMPs [197].

Another interesting approach, which combines "bioreactor" AMP synthesis and high-throughput sequencing, was reported by Tucker et al. [198]. This leads to the identification of several thousand potential AMP sequences. To this purpose, the Surface Localized Antimicrobial displaY (SLAY) techniques were developed, inducing bacteria to express and self-test a random 20mer peptide library constructed using a codons subset, which remain tethered to a protein on the bacterial membrane surface. Therefore, bacteria expressing bactericidal or bacteriostatic peptides are depleted from the population, so that a comparison of high-throughput DNA sequencing of plasmid libraries before and after induction of expression leads to the identification of potential antimicrobial hits. A similar technique for improving AMP potency had been proposed several years ago by researchers at Novozymes, termed the "suicide expression system" (SES), but for soluble peptides. This is a cis-acting system based on induced mutation of bacterially expressed, but tightly controlled peptides, that are then secreted in increasing amounts until they result in the death of the producer strain, and was adapted from previously reported systems [199]. A trans-acting peptide screening system (TAPS) was also developed in which peptides expressed and secreted by one bacterium are screened against other bacterial species. This type of system was used to optimize the sequence of the fungal defensin plectasin for development as an antibiotic [165,200].

6.3. Quantitative Structure-Activity Relationship (QSAR)—From Virtual to Novel AMPs

The above-mentioned methods have proven to be effective in identifying putative AMPs, sometimes suggesting several sequences to select from, but provide no indication as to the eventual potency of their activity toward bacteria or their toxicity toward host cells. Rational design of artificial peptides and redesign of natural peptides, based on various physico-chemical properties associated with potency and/or selectivity (e.g., net charge, amphipathicity, structuring propensity, tendency for self-aggregation, etc.), has, however, provided a body of data that can be used to predict functional characteristics from the sequence, especially, but not only, for linear helical AMPs [201,202].

QSAR approaches may include virtual screening studies where the biophysical properties of known active peptides are used to construct molecular descriptors that are associated with different functional aspects. These descriptors are then used to link a novel sequence to its likely biological activity [203]. The main assumption is that a mathematical function can be developed that correctly links physico-chemical properties (e.g., net charge and amphipathicity) with an observable outcome [e.g., minimal inhibitory concentration (MIC) values]. Typically, a number of molecular descriptors are created by linking physico-chemical properties with the biological effect in a training set of peptides where the former are measurable and the latter are experimentally determined. This is followed by a statistical analysis to determine which descriptor (combination of parameters) provides a predicted functional value that correlates best with the experimentally-determined values. The QSAR model is then validated on an external (testing) set of peptides (see Figure 8) [204,205].

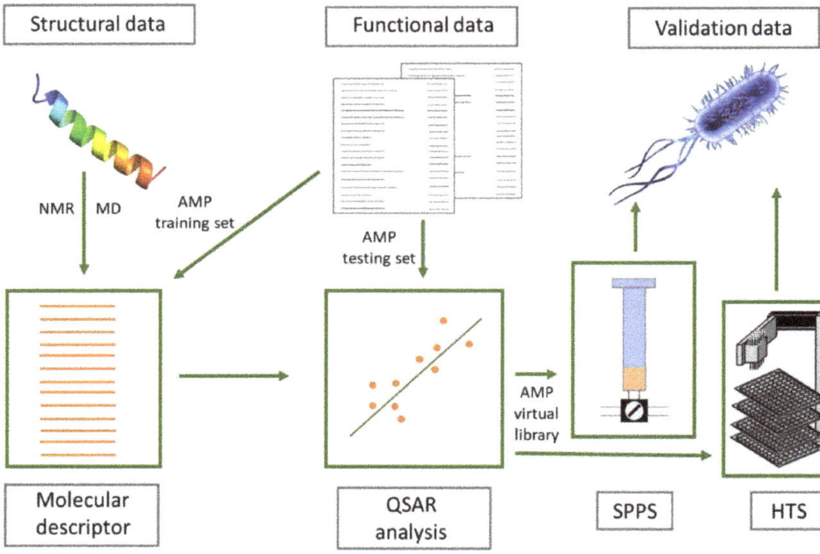

Figure 8. General overview of a QSAR method leading to design and validate novel AMPs. Structural data can be collected experimentally or predicted computationally (e.g., by MD). Functional data can be obtained from the literature or previous characterization campaigns to create specific databases. The best correlation between molecular descriptors and activity is determined based on statistical analysis, which allows us to propose new optimized sequences (putative AMP virtual library). These must then be either synthesized by solid phase peptide synthesis (SPPS) for in vitro validation and/or used for high-throughput screening (HTS) biological assays such as SLAY, SES, or TAPS (The 3D structure of magainin 2 was downloaded from PDB database (ID: 2 mag) and prepared using PyMOL 1.8).

Recurrent neural networks have successfully been used to develop such algorithms for de novo design of AMPs [206]. However, these models generate sequences without necessarily providing a quantitative prediction of antimicrobial activity. Witten and Witten [207] have improved on such models by creating a convolutional neural network, which was trained on a large set of peptides with known MIC values. This approach proved to be successful in designing AMPs, in which two have appreciable antimicrobial activity that has been experimentally validated.

An alternative approach is to combine QSAR with knowledge-based selection criteria for filtering putative AMP sequences. Adepantins were designed in this way, based on descriptors extracted from frog AMPs for which robust data on both MIC against *E. coli* and hemolytic activity were available. They proved to be remarkably selective toward some Gram-negative species [208]. The D-descriptor developed in this method has been recently adapted to the Mutator tool (http://split4.pmfst.hr/mutator/), which is a method that allows in silico re-design of peptide sequences to potentially improve selectivity. Dadapins were designed in this manner [209], by applying a strong filtering process on a devoted AMP database. In this case, we select for activity against Gram-positive bacteria, which is followed by optimization using the Mutator. They were shown to have high selectivity indices and comparable activities against Gram-positive and Gram-negative strains [210].

QSAR predictors are normally based on 2D-models, but, more recently, 3D-descriptors have also been developed [211,212]. This has, in part, been possible due to improved MD simulations on AMP/model membrane systems, which are used to optimize the starting 3D structure models, since experimentally determined 3D structures of AMPs are still rather limited [212].

6.4. MD Simulations—Seeing is Believing

This type of in silico approach has been used most often with helical, membrane–active AMPs [213,214], even though other types of membrane-active AMPs have also been studied [215,216]. Although they must always be considered critically and subsequently verified experimentally, MD simulations have the advantage of providing valuable information at the atomistic level, which is normally beyond experimental determination. This was the case for maculatin (see Section 5.1) where simulation data shed new light on the mode of action of this known lytic peptide [118]. Likewise, our group has recently used MD simulations to elucidate the mode of action of kiadins, which is another class of *in silico* designed peptides with membrane activity [217]. MD simulations can consider different trajectories to the membrane surface. For example, they can probe how the orientation of the peptide on its approach (e.g., how the angle of approach or peptide sector facing the membrane surface) affects the initial binding step, and subsequent insertion into the lipid bilayer (unpublished data). Simulation data can provide considerable insight on the mode of action of some peptides, which improves the existing models used for AMP prediction [218], but can also be used for a de novo design of peptides with desirable characteristics. Recently, an all-atom, simulation-guided design process, introduced by Chen et al. [219], has proven to be successful in producing a pore-forming AMP starting from a 14-residue polyleucine.

The main constraint with MD simulations is the computational time that is required to observe any given step in the permeabilization process [see steps i) to iv) in Section 5.1] [220,221]. Steps i) (binding) and iii) (insertion, but mostly in the specific conditions, such as at elevated temperature or when a pulling force is applied) are observed in a relatively short time-range. Therefore, these steps are amenable to an all-atom approach, but steps ii) and iv) (structuring and aggregation/pore-formation) can require considerably longer time frames and computational power to simulate. For this reason, simulations most often consider the AMP already in its active structure (which may not be realistic) and resorts to coarse-grain models when dealing with processes such as pore formation or translocation [222,223].

7. Therapeutic Potential

Since their initial discovery, AMPs have often been indicated as potential leads for the development of novel therapeutic agents for treating microbial infections. This has been one of the driving forces behind AMP research. However, the transition from in vitro to in vivo and translational applications of molecules derived from AMPs have proven to be very difficult [224,225]. Compared to "classical" antibiotic treatment, AMPs should have several advantages, especially when they are multimodal (can hit different microbial targets simultaneously), multifunctional (can directly inactivate microbes but also stimulate defense against them), fast acting, bactericidal, and can have accessory anti-inflammatory and/or healing activities [226]. However, they have evolved to act in a precisely orchestrated manner and are difficult to deliver as exogenous drugs. Furthermore, bacteria can fight back and interfere with AMP activity through proteolytic processing, active efflux, biofilm formation, and exopolymers entrapment, as well as by reducing their surface charge [227–229]. Especially worrisome is biofilm formation, since such bacteria can display up to 1000-fold higher resistance compared to planktonic bacteria due to the interaction of AMPs with specific components of the extracellular biofilm polymers [230]. Additionally, AMPs generally elicit relatively low levels of transient resistance compared to "classical" antibiotics. Several AMPs, such as the membrane-active gH625 and its analogue gH625-GCGKKKK, have been reported to be active against biofilm-forming bacteria [231]. In addition, Berditsch et al. [232] have recently reported a synergistic effect of two cyclic peptide antibiotics, polymyxin B and gramicidin S, against multidrug-resistant strains and biofilms of *Pseudomonas aeruginosa*. In this respect, AMPs can counteract biofilms in different ways such as by preventing their formation, and/or inactivating sessile bacteria, or modulating quorum sensing or twitching motility [233].

Despite years of trials, there are still some major obstacles to overcome before clinical application. Potent antimicrobial activity in AMPs is often accompanied by toxicity toward host cells. Although cationic AMPs preferentially target the negatively charged bacterial membrane, it has proven to be difficult to sufficiently reduce toxicity toward host cells, which can be significant, especially for helical AMPs. Melittin, for instance, is cytotoxic at comparable concentrations to those conferring antimicrobial activity [234]. These issues can, in principle, be resolved by designing peptides with more favorable physico-chemical properties (see above). Brevinin-1EMa analogues were less haemolytic when Ala residues replaced Leu, to reduce hydrophobicity [235]. A similar effect was observed when hydrophobicity was reduced at the N-terminus of mastoparan-X peptides [236]. In general, however, there is a trade-off between reduced toxicity and reduced potency. Other approaches used to reduce peptide toxicity include nanoencapsulation, which was the case with P34 [237]. It is also important to note that unwanted side effects may still be hidden, due to the large preponderance of in vitro over in vivo experiments involving AMPs. For example, it is well known that "classical" antibiotics impact gut microbiota, disturbing the well-adjusted and vital symbiosis between intestinal flora and the host [238]. It is likely that AMPs will display some toxicity toward the indigenous microflora, and partly disrupt its protective functions [239]. The host's own AMPs likely play a significant role in maintaining the tightly regulated homeostasis of the microbiota, which exogenous AMPs could alter [233].

A second major concern is peptide stability under physiological conditions (i.e., in the presence of serum, salt, pH variations, and proteolytic enzymes), or rapid clearance, which can result in unfavorable pharmacokinetics. This is especially the case with linear peptides, which are easily attacked by host proteases and peptidases [239], and particularly problematic if the peptides are administered systemically [240]. For this reason, AMPs are predominantly being considered for topical applications. The stability can, however, be enhanced by peptide cyclization (linking the C- and N-terminus), which prevents the proteases interaction peptide due to steric hindrance, or introduction of *D*-isomers or unnatural amino acids into the peptide sequence, which makes it unfit for enzyme degradation [241]. However, this is expensive and precludes biosynthesis. Nanoencapsulation can improve peptide stability while, at the same time, reducing toxicity [237]. Another strategy is to PEGylate peptides (link them to polyethylene glycol), which has also been shown to increase bioavailability by reducing renal clearance [242]. Very recently, an interesting continuous subcutaneous delivery method has been tested for another proline-rich AMP, Api137, which shows that it improved efficacy in an in vivo model of infection [243].

Lastly, high manufacturing costs represent another major obstacle for wider use of peptide antibiotics [54,239]. Production of one gram of such a drug by means of solid phase chemical synthesis can cost several hundred dollars. Therefore, there is a need for less expensive production platforms such as, for instance, biosynthesis in fungal, bacterial, or even plant expression systems. In recent years, several attempts have been made in this field. However, none has hitherto proven to be commercially feasible. An exception is the fungal expression system used to successfully obtain sufficient amounts of plectasin for development [200].

Despite these obstacles, approximately 20 AMPs are at various stages of clinical trials at the moment, with the majority intended for topical applications [244]. They include cyclic and linear AMPs, such as a twelve residue histatin derivative, P-113, the magainin derivative MSI-78 (pexiganan, which failed to gain FDA approval for its original use), the twelve residue indolicidin derivative omiganan, the arenicin-3 analog AA139, the cyclic protegrin I analog murepavadin, and others [239,244]. There are still years of trials before such peptides are accepted for clinical use by governing agencies. It is, however, safe to say that AMPs remain one of the more promising compounds to provide new classes of antibiotics, despite a somewhat less pronounced enthusiasm than in the past.

8. Conclusions

In the past 30 years, numerous studies have been carried out on AMPs, with most of them trying to answer to the question posed in the title of this review. In this case, it was our intention to focus not just on what we know, but also on some open issues surrounding the world of AMPs. It is hard to deny we are in the pre-post-antibiotic era, fully aware of the new, imminent threat to human health with no adequate solutions on the horizon. Can AMPs come, at least in part, to the rescue and justify years of motivated work by so many researchers? We are now well acquainted with how their structural and physico-chemical properties affect their activity but are still lacking a sufficiently deep knowledge of their modes of actions as well as the response of bacterial and host cells among them. This seems to be the major bottleneck in translating design and in vitro to in vivo efficacy as well as successful clinical applications. A limiting factor may be focusing too narrowly on pieces of the puzzle instead of adopting a wider perspective. Therefore, the question remains open—the potential is there but will we be able to tap it in a useful timeframe?

Author Contributions: Conceptualization, T.R.; funding acquisition, T.R. and J.P.; visualization, T.R.; writing—original draft, T.R., J.P. and A.T.; writing, review & editing, T.R., J.P. and A.T. All authors read and approved the final version of the manuscript.

Funding: This research received no external funding.

Acknowledgments: The authors would like to thank to Prof. Larisa Zoranić from the Faculty of Science, University of Split for her valuable comments regarding the MD section of the paper.

Conflicts of Interest: The authors declare no conflict of interest.

Abbreviations

AMPs	antimicrobial peptides
FDA	Food and Drug Administration
HDP	host defense peptides
HTS	high-throughput screening
IDR	innate defense regulatory
MD	molecular dynamics
MPP	membrane permeabilizing peptides
NGS	next-generation sequencing
NRPSs	non-ribosomal peptide synthetases
PDB	Protein Data Bank
QSAR	quantitative structure-activity relationship
SPPS	solid phase peptide synthesis
SLAY	Surface Localized Antimicrobial displaY
SES	suicide expression system
TAPS	trans-acting peptide screening system
WHO	World Health Organization

References

1. Gould, K. Antibiotics: From prehistory to the present day. *J. Antimicrob. Chemother.* **2016**, *71*, 572–575. [CrossRef] [PubMed]
2. Aminov, R.I. A Brief History of the Antibiotic Era: Lessons Learned and Challenges for the Future. *Front. Microbiol.* **2010**, *1*, 134. [CrossRef] [PubMed]
3. Bassett, E.J.; Keith, M.S.; Armelagos, G.J.; Martin, D.L.; Villanueva, A.R. Tetracycline-Labeled Human Bone from Ancient Sudanese Nubia (A.D. 350). *Sci. New Ser.* **1980**, *209*, 1532–1534. [CrossRef] [PubMed]
4. Cook, M.; Molto, E.; Anderson, C. Fluorochrome labelling in roman period skeletons from dakhleh oasis, Egypt. *Am. J. Phys. Anthropol.* **1989**, *80*, 137–143. [CrossRef]
5. Cui, L.; Su, X. Discovery, mechanisms of action and combination therapy of artemisinin. *Expert Rev. Anti-Infect. Ther.* **2009**, *7*, 999–1013. [CrossRef]
6. Zaffiri, L.; Gardner, J.; Toledo-Pereyra, L.H. History of Antibiotics. From Salvarsan to Cephalosporins. *J. Investig. Surg.* **2012**, *25*, 67–77. [CrossRef]
7. Nicolaou, K.C.; Rigol, S. A brief history of antibiotics and select advances in their synthesis. *J. Antibiot.* **2018**, *71*, 153–184. [CrossRef]
8. Landecker, H. Antibiotic Resistance and the Biology of History. *Body Soc.* **2016**, *22*, 19–52. [CrossRef]

9. Schatz, A.; Bugle, E.; Waksman, S.A. Streptomycin, a Substance Exhibiting Antibiotic Activity Against Gram-Positive and Gram-Negative Bacteria.*†. *Proc. Soc. Exp. Biol. Med.* **1944**, *55*, 66–69. [CrossRef]
10. Kresge, N.; Simoni, R.D.; Hill, R.L. Selman Waksman: The Father of Antibiotics. *J. Biol. Chem.* **2004**, *279*, e7.
11. Prestinaci, F.; Pezzotti, P.; Pantosti, A. Antimicrobial resistance: A global multifaceted phenomenon. *Pathog. Glob. Health* **2015**, *109*, 309–318. [CrossRef] [PubMed]
12. Draenert, R.; Seybold, U.; Grützner, E.; Bogner, J.R. Novel antibiotics: Are we still in the pre–post-antibiotic era? *Infection* **2015**, *43*, 145–151. [CrossRef] [PubMed]
13. Eliopoulos, G.M.; Meka, V.G.; Gold, H.S. Antimicrobial Resistance to Linezolid. *Clin. Infect. Dis.* **2004**, *39*, 1010–1015.
14. Gleason, P.P.; Shaughnessy, A.F. Telithromycin (Ketek) for Treatment of Community-Acquired Pneumonia. *AFP* **2007**, *76*, 1857.
15. Dixit, D.; Madduri, R.P.; Sharma, R. The role of tigecycline in the treatment of infections in light of the new black box warning. *Expert Rev. Anti-Infect. Ther.* **2014**, *12*, 397–400. [CrossRef] [PubMed]
16. Organization, W.H. Executive summary: The selection and use of essential medicines 2019. In Proceedings of the Report of the 22nd WHO Expert Committee on the Selection and Use of Essential Medicines: WHO Headquarters, Geneva, Switzerland, 1–5 April 2019.
17. Begg, E.J.; Barclay, M.L. Aminoglycosides—50 years on. *Br. J. Clin. Pharmacol.* **1995**, *39*, 597–603. [PubMed]
18. Jelić, D.; Antolović, R. From Erythromycin to Azithromycin and New Potential Ribosome-Binding Antimicrobials. *Antibiotics* **2016**, *5*, 29. [CrossRef]
19. Levine, D.P. Vancomycin: A History. *Clin. Infect. Dis.* **2006**, *42*, S5–S12. [CrossRef]
20. Nelson, M.L.; Levy, S.B. The history of the tetracyclines. *Ann. N. Y. Acad. Sci.* **2011**, *1241*, 17–32. [CrossRef]
21. Kumazawa, J.; Yagisawa, M. The history of antibiotics: The Japanese story. *J. Infect. Chemother.* **2002**, *8*, 125–133. [CrossRef]
22. Moellering, R.C. MRSA: The first half century. *J. Antimicrob. Chemother.* **2012**, *67*, 4–11. [CrossRef]
23. Ventola, C.L. The Antibiotic Resistance Crisis. *Pharm. Ther.* **2015**, *40*, 277–283.
24. Liu, Y.-Y.; Wang, Y.; Walsh, T.R.; Yi, L.-X.; Zhang, R.; Spencer, J.; Doi, Y.; Tian, G.; Dong, B.; Huang, X.; et al. Emergence of plasmid-mediated colistin resistance mechanism MCR-1 in animals and human beings in China: A microbiological and molecular biological study. *Lancet Infect. Dis.* **2016**, *16*, 161–168. [CrossRef]
25. Wang, X.; Wang, Y.; Zhou, Y.; Li, J.; Yin, W.; Wang, S.; Zhang, S.; Shen, J.; Shen, Z.; Wang, Y. Emergence of a novel mobile colistin resistance gene, mcr-8, in NDM-producing Klebsiella pneumoniae. *Emerg. Microbes Infect.* **2018**, *7*, 122. [CrossRef] [PubMed]
26. Carroll, L.M.; Gaballa, A.; Guldimann, C.; Sullivan, G.; Henderson, L.O.; Wiedmann, M. Identification of Novel Mobilized Colistin Resistance Gene mcr-9 in a Multidrug-Resistant, Colistin-Susceptible Salmonella enterica Serotype Typhimurium Isolate. *mBio* **2019**, *10*, e00853-e19. [CrossRef] [PubMed]
27. Lesch, J.E. *The First Miracle Drugs: How the Sulfa Drugs Transformed Medicine*; Oxford University Press: Oxford, NK, USA, 2007; ISBN 978-0-19-518775-5.
28. Fleming, A. On the antibacterial action of cultures of a Penicillium, with special reference to their use in the isolation of B. influenzae. *Br. J. Exp. Pathol.* **1929**, *10*, 226–236. [CrossRef]
29. Emmerson, A.M. The quinolones: Decades of development and use. *J. Antimicrob. Chem.* **2003**, *51*, 13–20. [CrossRef] [PubMed]
30. Orange Book: Approved Drug Products with Therapeutic Equivalence Evaluations. Available online: https://www.accessdata.fda.gov/scripts/cder/ob/index.cfm (accessed on 27 September 2019).
31. Drugs@FDA: FDA Approved Drug Products. Available online: https://www.accessdata.fda.gov/scripts/cder/daf/index.cfm (accessed on 1 October 2019).
32. Gall, Y.M.; Konashev, M.B. The discovery of Gramicidin S: The Intellectual Transformation of G.F. Gause from Biologist to Researcher of Antibiotics and on its Meaning for the Fate of Russian Genetics. *Hist. Philos. Life Sci.* **2001**, *23*, 137–150.
33. Eliopoulos, G.M.; Huovinen, P. Resistance to Trimethoprim-Sulfamethoxazole. *Clin. Infect. Dis.* **2001**, *32*, 1608–1614. [CrossRef]
34. De Groot, M.C.H.; van Puijenbroek, E.P. Clindamycin and taste disorders. *Br. J. Clin. Pharmacol.* **2007**, *64*, 542–545. [CrossRef]
35. Papp-Wallace, K.M.; Endimiani, A.; Taracila, M.A.; Bonomo, R.A. Carbapenems: Past, Present, and Future. *Antimicrob. Agents Chemother.* **2011**, *55*, 4943–4960. [CrossRef] [PubMed]

36. Ordooei Javan, A.; Shokouhi, S.; Sahraei, Z. A review on colistin nephrotoxicity. *Eur. J. Clin. Pharmacol.* **2015**, *71*, 801–810. [CrossRef] [PubMed]
37. Dhariwal, A.K.; Tullu, M.S. Colistin: Re-emergence of the "forgotten" antimicrobial agent. *J. Postgrad. Med.* **2013**, *59*, 208. [PubMed]
38. Paterson, D.L.; Harris, P.N.A. Colistin resistance: A major breach in our last line of defence. *Lancet Infect. Dis.* **2016**, *16*, 132–133. [CrossRef]
39. WHO | Global Priority List of Antibiotic-Resistant Bacteria to Guide Research, Discovery, and Development of New Antibiotics. Available online: http://www.who.int/medicines/publications/global-priority-list-antibiotic-resistant-bacteria/en/ (accessed on 18 December 2018).
40. Golkar, Z.; Bagasra, O.; Pace, D.G. Bacteriophage therapy: A potential solution for the antibiotic resistance crisis. *J. Infect. Dev. Cties.* **2014**, *8*, 129–136. [CrossRef]
41. Baptista, P.V.; McCusker, M.P.; Carvalho, A.; Ferreira, D.A.; Mohan, N.M.; Martins, M.; Fernandes, A.R. Nano-Strategies to Fight Multidrug Resistant Bacteria—"A Battle of the Titans". *Front. Microbiol.* **2018**, *9*, 1441. [CrossRef]
42. Mahlapuu, M.; Håkansson, J.; Ringstad, L.; Björn, C. Antimicrobial Peptides: An Emerging Category of Therapeutic Agents. *Front. Cell. Infect. Microbiol.* **2016**, *6*, 194. [CrossRef]
43. Nijnik, A.; Hancock, R. Host defence peptides: Antimicrobial and immunomodulatory activity and potential applications for tackling antibiotic-resistant infections. *Emerg. Health Threats J.* **2009**, *2*, 7078. [CrossRef]
44. Cederlund, A.; Gudmundsson, G.H.; Agerberth, B. Antimicrobial peptides important in innate immunity. *FEBS J.* **2011**, *278*, 3942–3951. [CrossRef]
45. Fan, L.; Sun, J.; Zhou, M.; Zhou, J.; Lao, X.; Zheng, H.; Xu, H. DRAMP: A comprehensive data repository of antimicrobial peptides. *Sci. Rep.* **2016**, *6*, 24482. [CrossRef]
46. Zhang, L.; Gallo, R.L. Antimicrobial peptides. *Curr. Biol.* **2016**, *26*, R14–R19. [CrossRef] [PubMed]
47. Hilchie, A.L.; Wuerth, K.; Hancock, R.E.W. Immune modulation by multifaceted cationic host defense (antimicrobial) peptides. *Nat. Chem. Biol.* **2013**, *9*, 761–768. [CrossRef] [PubMed]
48. Waghu, F.H.; Barai, R.S.; Gurung, P.; Idicula-Thomas, S. CAMPR3: A database on sequences, structures and signatures of antimicrobial peptides. *Nucleic Acids Res.* **2016**, *44*, D1094–D1097. [CrossRef] [PubMed]
49. Novkovic, M.; Simunic, J.; Bojovic, V.; Tossi, A.; Juretic, D. DADP: The database of anuran defense peptides. *Bioinformatics* **2012**, *28*, 1406–1407. [CrossRef] [PubMed]
50. Hassan, M.; Kjos, M.; Nes, I.F.; Diep, D.B.; Lotfipour, F. Natural antimicrobial peptides from bacteria: Characteristics and potential applications to fight against antibiotic resistance. *J. Appl. Microbiol.* **2012**, *113*, 723–736. [CrossRef] [PubMed]
51. Papagianni, M. Ribosomally synthesized peptides with antimicrobial properties: Biosynthesis, structure, function, and applications. *Biotechnol. Adv.* **2003**, *21*, 465–499. [CrossRef]
52. Finking, R.; Marahiel, M.A. Biosynthesis of Nonribosomal Peptides. *Ann. Rev. Microbiol.* **2004**, *58*, 453–488. [CrossRef]
53. Walsh, C.T.; O'Brien, R.V.; Khosla, C. Nonproteinogenic Amino Acid Building Blocks for Nonribosomal Peptide and Hybrid Polyketide Scaffolds. *Angew. Chem. Int. Ed.* **2013**, *52*, 7098–7124. [CrossRef]
54. Hancock, R.E.W.; Sahl, H.-G. Antimicrobial and host-defense peptides as new anti-infective therapeutic strategies. *Nat. Biotechnol.* **2006**, *24*, 1551–1557. [CrossRef]
55. Mogi, T.; Kita, K. Gramicidin S and polymyxins: The revival of cationic cyclic peptide antibiotics. *Cell. Mol. Life Sci.* **2009**, *66*, 3821–3826. [CrossRef]
56. Arnison, P.G.; Bibb, M.J.; Bierbaum, G.; Bowers, A.A.; Bugni, T.S.; Bulaj, G.; Camarero, J.A.; Campopiano, D.J.; Challis, G.L.; Clardy, J.; et al. Ribosomally synthesized and post-translationally modified peptide natural products: Overview and recommendations for a universal nomenclature. *Nat. Prod. Rep.* **2013**, *30*, 108–160. [CrossRef] [PubMed]
57. Lai, Y.; Gallo, R.L. AMPed up immunity: How antimicrobial peptides have multiple roles in immune defense. *Trends Immunol.* **2009**, *30*, 131–141. [CrossRef] [PubMed]
58. Patrzykat, A.; Gallant, J.W.; Seo, J.-K.; Pytyck, J.; Douglas, S.E. Novel Antimicrobial Peptides Derived from Flatfish Genes. *Antimicrob. Agents Chemother.* **2003**, *47*, 2464–2470. [CrossRef] [PubMed]
59. Broekaert, W.F.; Cammue, B.P.A.; De Bolle, M.F.C.; Thevissen, K.; De Samblanx, G.W.; Osborn, R.W.; Nielson, K. Antimicrobial Peptides from Plants. *Crit. Rev. Plant Sci.* **1997**, *16*, 297–323. [CrossRef]

60. Lu, Y.; Li, J.; Yu, H.; Xu, X.; Liang, J.; Tian, Y.; Ma, D.; Lin, G.; Huang, G.; Lai, R. Two families of antimicrobial peptides with multiple functions from skin of rufous-spotted torrent frog, *Amolops loloensis*. *Peptides* **2006**, *27*, 3085–3091. [CrossRef]
61. Conlon, J.M.; Sonnevend, Á.; Patel, M.; Al-Dhaheri, K.; Nielsen, P.F.; Kolodziejek, J.; Nowotny, N.; Iwamuro, S.; Pál, T. A family of brevinin-2 peptides with potent activity against Pseudomonas aeruginosa from the skin of the Hokkaido frog, Rana pirica. *Regul. Pept.* **2004**, *118*, 135–141. [CrossRef]
62. Ma, Y.; Liu, C.; Liu, X.; Wu, J.; Yang, H.; Wang, Y.; Li, J.; Yu, H.; Lai, R. Peptidomics and genomics analysis of novel antimicrobial peptides from the frog, Rana nigrovittata. *Genomics* **2010**, *95*, 66–71. [CrossRef]
63. Xu, X.; Li, J.; Han, Y.; Yang, H.; Liang, J.; Lu, Q.; Lai, R. Two antimicrobial peptides from skin secretions of Rana grahami. *Toxicon* **2006**, *47*, 459–464. [CrossRef]
64. Rholam, M.; Fahy, C. Processing of peptide and hormone precursors at the dibasic cleavage sites. *Cell. Mol. Life Sci.* **2009**, *66*, 2075–2091. [CrossRef]
65. Von Heijne, G. The signal peptide. *J. Membr. Biol.* **1990**, *115*, 195–201. [CrossRef]
66. Morrison, G.M.; Semple, C.A.M.; Kilanowski, F.M.; Hill, R.E.; Dorin, J.R. Signal Sequence Conservation and Mature Peptide Divergence Within Subgroups of the Murine β-Defensin Gene Family. *Mol. Biol. Evol.* **2003**, *20*, 460–470. [CrossRef] [PubMed]
67. Amiche, M.; Seon, A.A.; Pierre, T.N.; Nicolas, P. The dermaseptin precursors: A protein family with a common preproregion and a variable C-terminal antimicrobial domain. *FEBS Lett.* **1999**, *456*, 352–356. [CrossRef]
68. Zasloff, M. Antimicrobial peptides of multicellular organisms. *Nature* **2002**, *415*, 389–395. [CrossRef] [PubMed]
69. König, E.; Bininda-Emonds, O.R.P. Evidence for convergent evolution in the antimicrobial peptide system in anuran amphibians. *Peptides* **2011**, *32*, 20–25. [CrossRef]
70. Wang, G. Post-translational Modifications of Natural Antimicrobial Peptides and Strategies for Peptide Engineering. *Curr. Biotechnol.* **2014**, *1*, 72–79.
71. Bahar, A.; Ren, D. Antimicrobial Peptides. *Pharmaceuticals* **2013**, *6*, 1543–1575. [CrossRef]
72. Xhindoli, D.; Pacor, S.; Guida, F.; Antcheva, N.; Tossi, A. Native oligomerization determines the mode of action and biological activities of human cathelicidin LL-37. *Biochem. J.* **2014**, *457*, 263–275. [CrossRef]
73. Pacor, S.; Guida, F.; Xhindoli, D.; Benincasa, M.; Gennaro, R.; Tossi, A. Effect of targeted minimal sequence variations on the structure and biological activities of the human cathelicidin LL-37. *Pept. Sci.* **2018**, *110*, e24087. [CrossRef]
74. Nguyen, L.T.; Haney, E.F.; Vogel, H.J. The expanding scope of antimicrobial peptide structures and their modes of action. *Trends Biotechnol.* **2011**, *29*, 464–472. [CrossRef]
75. Guha, S.; Ghimire, J.; Wu, E.; Wimley, W.C. Mechanistic Landscape of Membrane-Permeabilizing Peptides. *Chem. Rev.* **2019**, *119*, 6040–6085. [CrossRef]
76. Wang, G. *Antimicrobial Peptides: Discovery, Design and Novel Therapeutic Strategies*, 2nd ed.; CABI: Omaha, NE, USA, 2017; ISBN 978-1-78639-039-4.
77. Berman, H.M.; Westbrook, J.; Feng, Z.; Gilliland, G.; Bhat, T.N.; Weissig, H.; Shindyalov, I.N.; Bourne, P.E. The Protein Data Bank. *Nucleic Acids Res.* **2000**, *28*, 235–242. [CrossRef] [PubMed]
78. Delano, W. The PyMOL Molecular Graphics System. Available online: http://www.pymol.org (accessed on 9 May 2017).
79. Eisenberg, D.; Schwarz, E.; Komaromy, M.; Wall, R. Analysis of membrane and surface protein sequences with the hydrophobic moment plot. *J. Mol. Biol.* **1984**, *179*, 125–142. [CrossRef]
80. Tossi, A.; Sandri, L.; Giangaspero, A. Amphipathic, α-helical antimicrobial peptides. *Pept. Sci.* **2000**, *55*, 4–30. [CrossRef]
81. Giangaspero, A.; Sandri, L.; Tossi, A. Amphipathic α helical antimicrobial peptides. *Eur. J. Biochem.* **2001**, *268*, 5589–5600. [CrossRef] [PubMed]
82. Zelezetsky, I.; Pacor, S.; Pag, U.; Papo, N.; Shai, Y.; Sahl, H.-G.; Tossi, A. Controlled alteration of the shape and conformational stability of α-helical cell-lytic peptides: Effect on mode of action and cell specificity. *Biochem. J.* **2005**, *390*, 177–188. [CrossRef]
83. Hou, F.; Li, J.; Pan, P.; Xu, J.; Liu, L.; Liu, W.; Song, B.; Li, N.; Wan, J.; Gao, H. Isolation and characterisation of a new antimicrobial peptide from the skin of Xenopus laevis. *Int. J. Antimicrob. Agents* **2011**, *38*, 510–515. [CrossRef]

84. Subbalakshmi, C.; Nagaraj, R.; Sitaram, N. Biological activities of C-terminal 15-residue synthetic fragment of melittin: Design of an analog with improved antibacterial activity. *FEBS Lett.* **1999**, *448*, 62–66. [CrossRef]
85. Rončević, T.; Gajski, G.; Ilić, N.; Goić-Barišić, I.; Tonkić, M.; Zoranić, L.; Simunić, J.; Benincasa, M.; Mijaković, M.; Tossi, A.; et al. PGLa-H tandem-repeat peptides active against multidrug resistant clinical bacterial isolates. *BBA Biomembr.* **2017**, *1859*, 228–237. [CrossRef]
86. Scocchi, M.; Tossi, A.; Gennaro, R. Proline-rich antimicrobial peptides: Converging to a non-lytic mechanism of action. *Cell. Mol. Life Sci.* **2011**, *68*, 2317–2330. [CrossRef]
87. Guida, F.; Benincasa, M.; Zahariev, S.; Scocchi, M.; Berti, F.; Gennaro, R.; Tossi, A. Effect of Size and N-Terminal Residue Characteristics on Bacterial Cell Penetration and Antibacterial Activity of the Proline-Rich Peptide Bac7. *J. Med. Chem.* **2015**, *58*, 1195–1204. [CrossRef]
88. Chan, D.I.; Prenner, E.J.; Vogel, H.J. Tryptophan- and arginine-rich antimicrobial peptides: Structures and mechanisms of action. *BBA Biomembr.* **2006**, *1758*, 1184–1202. [CrossRef]
89. Shagaghi, N.; Palombo, E.A.; Clayton, A.H.A.; Bhave, M. Archetypal tryptophan-rich antimicrobial peptides: Properties and applications. *World, J. Microbiol. Biotechnol.* **2016**, *32*, 31. [CrossRef]
90. Mishra, A.K.; Choi, J.; Moon, E.; Baek, K.-H. Tryptophan-Rich and Proline-Rich Antimicrobial Peptides. *Molecules* **2018**, *23*, 815. [CrossRef] [PubMed]
91. Scudiero, O.; Nigro, E.; Cantisani, M.; Colavita, I.; Leone, M.; Mercurio, F.A.; Galdiero, M.; Pessi, A.; Daniele, A.; Salvatore, F.; et al. Design and activity of a cyclic mini-β-defensin analog: A novel antimicrobial tool. *Int. J. Nanomed.* **2015**, *10*, 6523–6539.
92. Antcheva, N.; Morgera, F.; Creatti, L.; Vaccari, L.; Pag, U.; Pacor, S.; Shai, Y.; Sahl, H.-G.; Tossi, A. Artificial β-defensin based on a minimal defensin template. *Biochem. J.* **2009**, *421*, 435–447. [CrossRef] [PubMed]
93. Wu, Z.; Hoover, D.M.; Yang, D.; Boulègue, C.; Santamaria, F.; Oppenheim, J.J.; Lubkowski, J.; Lu, W. Engineering disulfide bridges to dissect antimicrobial and chemotactic activities of human β-defensin 3. *Proc. Natl. Acad. Sci. USA* **2003**, *100*, 8880–8885. [CrossRef]
94. Taylor, K.; Barran, P.E.; Dorin, J.R. Structure–activity relationships in β-defensin peptides. *Pept. Sci.* **2008**, *90*, 1–7. [CrossRef]
95. Krishnakumari, V.; Rangaraj, N.; Nagaraj, R. Antifungal Activities of Human Beta-Defensins HBD-1 to HBD-3 and Their C-Terminal Analogs Phd1 to Phd3. *Antimicrob. Agents Chemother.* **2009**, *53*, 256–260. [CrossRef]
96. Klüver, E.; Schulz-Maronde, S.; Scheid, S.; Meyer, B.; Forssmann, W.-G.; Adermann, K. Structure–Activity Relation of Human β-Defensin 3: Influence of Disulfide Bonds and Cysteine Substitution on Antimicrobial Activity and Cytotoxicity. *Biochemistry* **2005**, *44*, 9804–9816. [CrossRef]
97. Reynolds, N.L.; De Cecco, M.; Taylor, K.; Stanton, C.; Kilanowski, F.; Kalapothakis, J.; Seo, E.; Uhrin, D.; Campopiano, D.; Govan, J.; et al. Peptide Fragments of a β-Defensin Derivative with Potent Bactericidal Activity. *Antimicrob. Agents Chemother.* **2010**, *54*, 1922–1929. [CrossRef]
98. Schroeder, B.O.; Wu, Z.; Nuding, S.; Groscurth, S.; Marcinowski, M.; Beisner, J.; Buchner, J.; Schaller, M.; Stange, E.F.; Wehkamp, J. Reduction of disulphide bonds unmasks potent antimicrobial activity of human β-defensin 1. *Nature* **2011**, *469*, 419–423. [CrossRef] [PubMed]
99. Tossi, A.; Sandri, L. Molecular Diversity in Gene-Encoded, Cationic Antimicrobial Polypeptides. *Curr. Pharm. Des.* **2002**, *8*, 743–761. [CrossRef] [PubMed]
100. Harris, F.; Dennison, S.; Phoenix, D. Anionic Antimicrobial Peptides from Eukaryotic Organisms. *Curr. Protein Pept. Sci.* **2009**, *10*, 585–606. [CrossRef] [PubMed]
101. Wang, Z.; Wang, G. APD: The Antimicrobial Peptide Database. *Nucleic. Acids Res.* **2004**, *32*, D590–D592. [CrossRef]
102. Dathe, M.; Nikolenko, H.; Meyer, J.; Beyermann, M.; Bienert, M. Optimization of the antimicrobial activity of magainin peptides by modification of charge. *FEBS Lett.* **2001**, *501*, 146–150. [CrossRef]
103. López Cascales, J.J.; Zenak, S.; García de la Torre, J.; Lezama, O.G.; Garro, A.; Enriz, R.D. Small Cationic Peptides: Influence of Charge on Their Antimicrobial Activity. *ACS Omega* **2018**, *3*, 5390–5398. [CrossRef]
104. Yin, L.M.; Edwards, M.A.; Li, J.; Yip, C.M.; Deber, C.M. Roles of Hydrophobicity and Charge Distribution of Cationic Antimicrobial Peptides in Peptide-Membrane Interactions. *J. Biol. Chem.* **2012**, *287*, 7738–7745. [CrossRef]
105. Rosenfeld, Y.; Lev, N.; Shai, Y. Effect of the Hydrophobicity to Net Positive Charge Ratio on Antibacterial and Anti-Endotoxin Activities of Structurally Similar Antimicrobial Peptides. *Biochemistry* **2010**, *49*, 853–861. [CrossRef]

106. Chen, Y.; Guarnieri, M.T.; Vasil, A.I.; Vasil, M.L.; Mant, C.T.; Hodges, R.S. Role of Peptide Hydrophobicity in the Mechanism of Action of α-Helical Antimicrobial Peptides. *Antimicrob. Agents Chemother.* **2007**, *51*, 1398–1406. [CrossRef]
107. Fernández-Vidal, M.; Jayasinghe, S.; Ladokhin, A.S.; White, S.H. Folding amphipathic helices into membranes: Amphiphilicity trumps hydrophobicity. *J. Mol. Biol.* **2007**, *370*, 459–470. [CrossRef]
108. Edwards, I.A.; Elliott, A.G.; Kavanagh, A.M.; Zuegg, J.; Blaskovich, M.A.T.; Cooper, M.A. Contribution of Amphipathicity and Hydrophobicity to the Antimicrobial Activity and Cytotoxicity of β-Hairpin Peptides. *ACS Infect. Dis.* **2016**, *2*, 442–450. [CrossRef] [PubMed]
109. Gautier, R.; Douguet, D.; Antonny, B.; Drin, G. HELIQUEST: A web server to screen sequences with specific α-helical properties. *Bioinformatics* **2008**, *24*, 2101–2102. [CrossRef] [PubMed]
110. Papo, N.; Oren, Z.; Pag, U.; Sahl, H.-G.; Shai, Y. The Consequence of Sequence Alteration of an Amphipathic α-Helical Antimicrobial Peptide and Its Diastereomers. *J. Biol. Chem.* **2002**, *277*, 33913–33921. [CrossRef] [PubMed]
111. Lee, T.H.; N Hall, K.; Aguilar, M.I. Antimicrobial Peptide Structure and Mechanism of Action: A Focus on the Role of Membrane Structure. *Curr. Top. Med. Chem.* **2015**, *16*, 25–39. [CrossRef] [PubMed]
112. Hale, J.D.; Hancock, R.E. Alternative mechanisms of action of cationic antimicrobial peptides on bacteria. *Expert Rev. Anti-Infect. Ther.* **2007**, *5*, 951–959. [CrossRef] [PubMed]
113. Pillong, M.; Hiss, J.A.; Schneider, P.; Lin, Y.-C.; Posselt, G.; Pfeiffer, B.; Blatter, M.; Müller, A.T.; Bachler, S.; Neuhaus, C.S.; et al. Rational Design of Membrane-Pore-Forming Peptides. *Small* **2017**, *13*, 1701316. [CrossRef]
114. Wiedman, G.; Fuselier, T.; He, J.; Searson, P.; Hristova, K.; Wimley, W.C. A Novel Functional Class of Pore-Forming Peptides. *Biophys. J.* **2014**, *106*, 85a–86a. [CrossRef]
115. Brogden, K.A. Antimicrobial peptides: Pore formers or metabolic inhibitors in bacteria? *Nat. Rev. Microbiol.* **2005**, *3*, 238–250. [CrossRef]
116. Rozek, T.; Waugh, R.J.; Steinborner, S.T.; Bowie, J.H.; Tyler, M.J.; Wallace, J.C. The Maculatin peptides from the skin glands of the tree frogLitoria genimaculata: A comparison of the structures and antibacterial activities of Maculatin 1.1 and Caerin 1.1. *J. Pept. Sci.* **1998**, *4*, 111–115. [CrossRef]
117. Sani, M.-A.; Whitwell, T.C.; Gehman, J.D.; Robins-Browne, R.M.; Pantarat, N.; Attard, T.J.; Reynolds, E.C.; O'Brien-Simpson, N.M.; Separovic, F. Maculatin 1.1 Disrupts Staphylococcus aureus Lipid Membranes via a Pore Mechanism. *Antimicrob. Agents Chemother.* **2013**, *57*, 3593–3600. [CrossRef]
118. Wang, Y.; Chen, C.H.; Hu, D.; Ulmschneider, M.B.; Ulmschneider, J.P. Spontaneous formation of structurally diverse membrane channel architectures from a single antimicrobial peptide. *Nat. Commun.* **2016**, *7*, 13535. [CrossRef] [PubMed]
119. Sani, M.-A.; Separovic, F. How Membrane-Active Peptides Get into Lipid Membranes. *Acc. Chem. Res.* **2016**, *49*, 1130–1138. [CrossRef] [PubMed]
120. Jo, S.; Kim, T.; Im, W. Automated Builder and Database of Protein/Membrane Complexes for Molecular Dynamics Simulations. *PLoS ONE* **2007**, *2*, e880. [CrossRef] [PubMed]
121. Melo, M.N.; Ferre, R.; Castanho, M.A.R.B. Antimicrobial peptides: Linking partition, activity and high membrane-bound concentrations. *Nat. Rev. Microbiol.* **2009**, *7*, 245–250. [CrossRef] [PubMed]
122. Dathe, M.; Wieprecht, T. Structural features of helical antimicrobial peptides: Their potential to modulate activity on model membranes and biological cells. *BBA Biomembr.* **1999**, *1462*, 71–87. [CrossRef]
123. Xhindoli, D.; Pacor, S.; Benincasa, M.; Scocchi, M.; Gennaro, R.; Tossi, A. The human cathelicidin LL-37—A pore-forming antibacterial peptide and host-cell modulator. *BBA Biomembr.* **2016**, *1858*, 546–566. [CrossRef]
124. Chairatana, P.; Nolan, E.M. Molecular Basis for Self-Assembly of a Human Host-Defense Peptide That Entraps Bacterial Pathogens. *J. Am. Chem. Soc.* **2014**, *136*, 13267–13276. [CrossRef]
125. Rapaport, D.; Shai, Y. Interaction of fluorescently labeled pardaxin and its analogues with lipid bilayers. *J. Biol. Chem.* **1991**, *266*, 23769–23775.
126. Andrew Woolley, G.; Wallace, B.A. Model ion channels: Gramicidin and alamethicin. *J. Membr. Biol.* **1992**, *129*, 109–136. [CrossRef]
127. Cheng, J.T.J.; Hale, J.D.; Elliot, M.; Hancock, R.E.W.; Straus, S.K. Effect of Membrane Composition on Antimicrobial Peptides Aurein 2.2 and 2.3 From Australian Southern Bell Frogs. *Biophys. J.* **2009**, *96*, 552–565. [CrossRef]

128. Yang, L.; Harroun, T.A.; Weiss, T.M.; Ding, L.; Huang, H.W. Barrel-Stave Model or Toroidal Model? A Case Study on Melittin Pores. *Biophys. J.* **2001**, *81*, 1475–1485. [CrossRef]
129. Sengupta, D.; Leontiadou, H.; Mark, A.E.; Marrink, S.-J. Toroidal pores formed by antimicrobial peptides show significant disorder. *BBA Biomembr.* **2008**, *1778*, 2308–2317. [CrossRef] [PubMed]
130. Bechinger, B. Detergent-like properties of magainin antibiotic peptides: A 31P solid-state NMR spectroscopy study. *BBA Biomembr.* **2005**, *1712*, 101–108. [CrossRef] [PubMed]
131. Fernandez, D.I.; Brun, A.P.L.; Whitwell, T.C.; Sani, M.-A.; James, M.; Separovic, F. The antimicrobial peptide aurein 1.2 disrupts model membranes via the carpet mechanism. *Phys. Chem. Chem. Phys.* **2012**, *14*, 15739–15751. [CrossRef] [PubMed]
132. Manzini, M.C.; Perez, K.R.; Riske, K.A.; Bozelli, J.C.; Santos, T.L.; da Silva, M.A.; Saraiva, G.K.V.; Politi, M.J.; Valente, A.P.; Almeida, F.C.L.; et al. Peptide:lipid ratio and membrane surface charge determine the mechanism of action of the antimicrobial peptide BP100. Conformational and functional studies. *BBA Biomembr.* **2014**, *1838*, 1985–1999. [CrossRef] [PubMed]
133. Nuri, R.; Shprung, T.; Shai, Y. Defensive remodeling: How bacterial surface properties and biofilm formation promote resistance to antimicrobial peptides. *BBA Biomembr.* **2015**, *1848*, 3089–3100. [CrossRef]
134. Scocchi, M.; Mardirossian, M.; Runti, G.; Benincasa, M. Non-Membrane Permeabilizing Modes of Action of Antimicrobial Peptides on Bacteria. *Curr. Top. Med. Chem.* **2015**, *16*, 76–88. [CrossRef]
135. Le, C.-F.; Fang, C.-M.; Sekaran, S.D. Intracellular Targeting Mechanisms by Antimicrobial Peptides. *Antimicrob. Agents Chemother.* **2017**, *61*, e02340-16. [CrossRef]
136. Mattiuzzo, M.; Bandiera, A.; Gennaro, R.; Benincasa, M.; Pacor, S.; Antcheva, N.; Scocchi, M. Role of the Escherichia coli SbmA in the antimicrobial activity of proline-rich peptides. *Mol. Microbiol.* **2007**, *66*, 151–163. [CrossRef]
137. Park, C.B.; Kim, H.S.; Kim, S.C. Mechanism of Action of the Antimicrobial Peptide Buforin II: Buforin II Kills Microorganisms by Penetrating the Cell Membrane and Inhibiting Cellular Functions. *Biochem. Biophys. Res. Commun.* **1998**, *244*, 253–257. [CrossRef]
138. Subbalakshmi, C.; Sitaram, N. Mechanism of antimicrobial action of indolicidin. *FEMS Microbiol. Lett.* **1998**, *160*, 91–96. [CrossRef] [PubMed]
139. Park, C.B.; Yi, K.-S.; Matsuzaki, K.; Kim, M.S.; Kim, S.C. Structure–activity analysis of buforin II, a histone H2A-derived antimicrobial peptide: The proline hinge is responsible for the cell-penetrating ability of buforin II. *Proc. Natl. Acad. Sci. USA* **2000**, *97*, 8245–8250. [CrossRef] [PubMed]
140. Uyterhoeven, E.T.; Butler, C.H.; Ko, D.; Elmore, D.E. Investigating the nucleic acid interactions and antimicrobial mechanism of buforin II. *FEBS Lett.* **2008**, *582*, 1715–1718. [CrossRef] [PubMed]
141. Hao, G.; Shi, Y.-H.; Tang, Y.-L.; Le, G.-W. The intracellular mechanism of action on *Escherichia coli* of BF2-A/C, two analogues of the antimicrobial peptide Buforin 2. *J. Microbiol.* **2013**, *51*, 200–206. [CrossRef] [PubMed]
142. Hsu, C.-H.; Chen, C.; Jou, M.-L.; Lee, A.Y.-L.; Lin, Y.-C.; Yu, Y.-P.; Huang, W.-T.; Wu, S.-H. Structural and DNA-binding studies on the bovine antimicrobial peptide, indolicidin: Evidence for multiple conformations involved in binding to membranes and DNA. *Nucleic Acids Res.* **2005**, *33*, 4053–4064. [CrossRef]
143. Marchand, C.; Krajewski, K.; Lee, H.-F.; Antony, S.; Johnson, A.A.; Amin, R.; Roller, P.; Kvaratskhelia, M.; Pommier, Y. Covalent binding of the natural antimicrobial peptide indolicidin to DNA abasic sites. *Nucleic Acids Res.* **2006**, *34*, 5157–5165. [CrossRef]
144. Mardirossian, M.; Grzela, R.; Giglione, C.; Meinnel, T.; Gennaro, R.; Mergaert, P.; Scocchi, M. The Host Antimicrobial Peptide Bac71-35 Binds to Bacterial Ribosomal Proteins and Inhibits Protein Synthesis. *Chem. Biol.* **2014**, *21*, 1639–1647. [CrossRef]
145. Graf, M.; Mardirossian, M.; Nguyen, F.; Seefeldt, A.C.; Guichard, G.; Scocchi, M.; Innis, C.A.; Wilson, D.N. Proline-rich antimicrobial peptides targeting protein synthesis. *Nat. Prod. Rep.* **2017**, *34*, 702–711. [CrossRef]
146. Krizsan, A.; Volke, D.; Weinert, S.; Sträter, N.; Knappe, D.; Hoffmann, R. Insect-Derived Proline-Rich Antimicrobial Peptides Kill Bacteria by Inhibiting Bacterial Protein Translation at the 70 S Ribosome. *Angew. Chem. Int. Ed.* **2014**, *53*, 12236–12239. [CrossRef]
147. Kragol, G.; Lovas, S.; Varadi, G.; Condie, B.A.; Hoffmann, R.; Otvos, L. The Antibacterial Peptide Pyrrhocoricin Inhibits the ATPase Actions of DnaK and Prevents Chaperone-Assisted Protein Folding. *Biochemistry* **2001**, *40*, 3016–3026. [CrossRef]

148. Ho, Y.-H.; Shah, P.; Chen, Y.-W.; Chen, C.-S. Systematic Analysis of Intracellular-targeting Antimicrobial Peptides, Bactenecin 7, Hybrid of Pleurocidin and Dermaseptin, Proline–Arginine-rich Peptide, and Lactoferricin B, by Using *Escherichia coli* Proteome Microarrays. *Mol. Cell. Proteom.* **2016**, *15*, 1837–1847. [CrossRef] [PubMed]
149. Casteels, P.; Romagnolo, J.; Castle, M.; Casteels-Josson, K.; Erdjument-Bromage, H.; Tempst, P. Biodiversity of apidaecin-type peptide antibiotics. Prospects of manipulating the antibacterial spectrum and combating acquired resistance. *J. Biol. Chem.* **1994**, *269*, 26107–26115. [PubMed]
150. Friedrich, C.L.; Rozek, A.; Patrzykat, A.; Hancock, R.E.W. Structure and Mechanism of Action of an Indolicidin Peptide Derivative with Improved Activity against Gram-positive Bacteria. *J. Biol. Chem.* **2001**, *276*, 24015–24022. [CrossRef] [PubMed]
151. Otvos, L.; O, I.; Rogers, M.E.; Consolvo, P.J.; Condie, B.A.; Lovas, S.; Bulet, P.; Blaszczyk-Thurin, M. Interaction between Heat Shock Proteins and Antimicrobial Peptides. *Biochemistry* **2000**, *39*, 14150–14159. [CrossRef]
152. Chesnokova, L.S.; Slepenkov, S.V.; Witt, S.N. The insect antimicrobial peptide, L-pyrrhocoricin, binds to and stimulates the ATPase activity of both wild-type and lidless DnaK. *FEBS Lett.* **2004**, *565*, 65–69. [CrossRef]
153. Scocchi, M.; Lüthy, C.; Decarli, P.; Mignogna, G.; Christen, P.; Gennaro, R. The Proline-rich Antibacterial Peptide Bac7 Binds to and Inhibits in vitro the Molecular Chaperone DnaK. *Int. J. Pept. Res. Ther.* **2009**, *15*, 147–155. [CrossRef]
154. Gusman, H.; Travis, J.; Helmerhorst, E.J.; Potempa, J.; Troxler, R.F.; Oppenheim, F.G. Salivary Histatin 5 Is an Inhibitor of Both Host and Bacterial Enzymes Implicated in Periodontal Disease. *Infect. Immun.* **2001**, *69*, 1402–1408. [CrossRef]
155. Gusman, H.; Grogan, J.; Kagan, H.M.; Troxler, R.F.; Oppenheim, F.G. Salivary histatin 5 is a potent competitive inhibitor of the cysteine proteinase clostripain. *FEBS Lett.* **2001**, *489*, 97–100. [CrossRef]
156. Gallo, R.L.; Kim, K.J.; Bernfield, M.; Kozak, C.A.; Zanetti, M.; Merluzzi, L.; Gennaro, R. Identification of CRAMP, a Cathelin-related Antimicrobial Peptide Expressed in the Embryonic and Adult Mouse. *J. Biol. Chem.* **1997**, *272*, 13088–13093. [CrossRef]
157. Handler, A.A.; Lim, J.E.; Losick, R. Peptide inhibitor of cytokinesis during sporulation in *Bacillus subtilis*. *Mol. Microbiol.* **2008**, *68*, 588–599. [CrossRef]
158. Yadavalli, S.S.; Carey, J.N.; Leibman, R.S.; Chen, A.I.; Stern, A.M.; Roggiani, M.; Lippa, A.M.; Goulian, M. Antimicrobial peptides trigger a division block in Escherichia coli through stimulation of a signalling system. *Nat. Commun.* **2016**, *7*, 12340. [CrossRef] [PubMed]
159. Chileveru, H.R.; Lim, S.A.; Chairatana, P.; Wommack, A.J.; Chiang, I.-L.; Nolan, E.M. Visualizing Attack of Escherichia coli by the Antimicrobial Peptide Human Defensin 5. *Biochemistry* **2015**, *54*, 1767–1777. [CrossRef] [PubMed]
160. Koch, A.L. Bacterial Wall as Target for Attack: Past, Present, and Future Research. *Clin. Microbiol. Rev.* **2003**, *16*, 673–687. [CrossRef] [PubMed]
161. Scheffers, D.-J.; Pinho, M.G. Bacterial Cell Wall Synthesis: New Insights from Localization Studies. *Microbiol. Mol. Biol. Rev.* **2005**, *69*, 585–607. [CrossRef] [PubMed]
162. Chugunov, A.; Pyrkova, D.; Nolde, D.; Polyansky, A.; Pentkovsky, V.; Efremov, R. Lipid-II forms potential "landing terrain" for lantibiotics in simulated bacterial membrane. *Sci. Rep.* **2013**, *3*, 1678. [CrossRef]
163. Brötz, H.; Bierbaum, G.; Reynolds, P.E.; Sahl, H.-G. The Lantibiotic Mersacidin Inhibits Peptidoglycan Biosynthesis at the Level of Transglycosylation. *Eur. J. Biochem.* **1997**, *246*, 193–199. [CrossRef]
164. Hsu, S.-T.D.; Breukink, E.; Tischenko, E.; Lutters, M.A.G.; de Kruijff, B.; Kaptein, R.; Bonvin, A.M.J.J.; van Nuland, N.A.J. The nisin–lipid II complex reveals a pyrophosphate cage that provides a blueprint for novel antibiotics. *Nat. Struct. Mol. Biol.* **2004**, *11*, 963–967. [CrossRef]
165. Schneider, T.; Kruse, T.; Wimmer, R.; Wiedemann, I.; Sass, V.; Pag, U.; Jansen, A.; Nielsen, A.K.; Mygind, P.H.; Raventós, D.S.; et al. Plectasin, a fungal defensin, targets the bacterial cell wall precursor Lipid II. *Science* **2010**, *328*, 1168–1172. [CrossRef]
166. Essig, A.; Hofmann, D.; Münch, D.; Gayathri, S.; Künzler, M.; Kallio, P.T.; Sahl, H.-G.; Wider, G.; Schneider, T.; Aebi, M. Copsin, a Novel Peptide-based Fungal Antibiotic Interfering with the Peptidoglycan Synthesis. *J. Biol. Chem.* **2014**, *289*, 34953–34964. [CrossRef]
167. Cohen, J. The immunopathogenesis of sepsis. *Nature* **2002**, *420*, 885–891. [CrossRef]
168. Oberholzer, A.; Oberholzer, C.; Moldawer, L.L. Sepsis syndromes: Understanding the role of innate and acquired immunity. *Shock* **2001**, *16*, 83–96. [CrossRef] [PubMed]

169. Papo, N.; Shai, Y. A Molecular Mechanism for Lipopolysaccharide Protection of Gram-negative Bacteria from Antimicrobial Peptides. *J. Biol. Chem.* **2005**, *280*, 10378–10387. [CrossRef] [PubMed]
170. Rosenfeld, Y.; Barra, D.; Simmaco, M.; Shai, Y.; Mangoni, M.L. A Synergism between Temporins toward Gram-negative Bacteria Overcomes Resistance Imposed by the Lipopolysaccharide Protective Layer. *J. Biol. Chem.* **2006**, *281*, 28565–28574. [CrossRef] [PubMed]
171. Mohanram, H.; Bhattacharjya, S. Resurrecting Inactive Antimicrobial Peptides from the Lipopolysaccharide Trap. *Antimicrob. Agents Chemother.* **2014**, *58*, 1987–1996. [CrossRef]
172. Piers, K.L.; Brown, M.H.; Hancock, R.E. Improvement of outer membrane-permeabilizing and lipopolysaccharide-binding activities of an antimicrobial cationic peptide by C-terminal modification. *Antimicrob. Agents Chemother.* **1994**, *38*, 2311–2316. [CrossRef]
173. Sun, Y.; Shang, D. Inhibitory Effects of Antimicrobial Peptides on Lipopolysaccharide-Induced Inflammation. *Mediat. Inflamm.* **2015**, *2015*, 167572. [CrossRef]
174. Methods in Molecular Biology. In *Antimicrobial Peptides*; Giuliani, A.; Rinaldi, A.C. (Eds.) Humana Press: Totowa, NJ, USA, 2010; Volume 618, ISBN 978-1-60761-593-4.
175. Simmaco, M.; Mignogna, G.; Canofeni, S.; Miele, R.; Mangoni, M.L.; Barra, D. Temporins, Antimicrobial Peptides from the European Red Frog Rana temporaria. *Eur. J. Biochem.* **1996**, *242*, 788–792. [CrossRef]
176. Zasloff, M. Magainins, a class of antimicrobial peptides from Xenopus skin: Isolation, characterization of two active forms, and partial cDNA sequence of a precursor. *Proc. Natl. Acad. Sci. USA* **1987**, *84*, 5449–5453. [CrossRef]
177. Cole, A.M.; Weis, P.; Diamond, G. Isolation and Characterization of Pleurocidin, an Antimicrobial Peptide in the Skin Secretions of Winter Flounder. *J. Biol. Chem.* **1997**, *272*, 12008–12013. [CrossRef]
178. Charlet, M.; Chernysh, S.; Philippe, H.; Hetru, C.; Hoffmann, J.A.; Bulet, P. Innate Immunity: Isolation of several cysteine-rich antimicrobial peptides from the blood of a mollusc, *Mytilus edulis*. *J. Biol. Chem.* **1996**, *271*, 21808–21813. [CrossRef]
179. Destoumieux, D.; Bulet, P.; Loew, D.; Van Dorsselaer, A.; Rodriguez, J.; Bachère, E. Penaeidins, a New Family of Antimicrobial Peptides Isolated from the Shrimp *Penaeus vannamei* (Decapoda). *J. Biol. Chem.* **1997**, *272*, 28398–28406. [CrossRef] [PubMed]
180. Harder, J.; Schröder, J.-M. Psoriatic scales: A promising source for the isolation of human skin-derived antimicrobial proteins. *J. Leukoc. Biol.* **2005**, *77*, 476–486. [CrossRef] [PubMed]
181. Tang, Y.-Q.; Yeaman, M.R.; Selsted, M.E. Antimicrobial Peptides from Human Platelets. *Infect. Immun.* **2002**, *70*, 6524–6533. [CrossRef] [PubMed]
182. Harder, J.; Bartels, J.; Christophers, E.; Schröder, J.-M. Isolation and Characterization of Human β-Defensin-3, a Novel Human Inducible Peptide Antibiotic. *J. Biol. Chem.* **2001**, *276*, 5707–5713. [CrossRef]
183. Lande, R.; Chamilos, G.; Ganguly, D.; Demaria, O.; Frasca, L.; Durr, S.; Conrad, C.; Schröder, J.; Gilliet, M. Cationic antimicrobial peptides in psoriatic skin cooperate to break innate tolerance to self-DNA: Innate immunity. *Eur. J. Immunol.* **2015**, *45*, 203–213. [CrossRef]
184. Théolier, J.; Hammami, R.; Labelle, P.; Fliss, I.; Jean, J. Isolation and identification of antimicrobial peptides derived by peptic cleavage of whey protein isolate. *J. Funct. Foods* **2013**, *5*, 706–714. [CrossRef]
185. Zhou, M.; Wang, L.; Owens, D.E.; Chen, T.; Walker, B.; Shaw, C. Rapid identification of precursor cDNAs encoding five structural classes of antimicrobial peptides from pickerel frog (Rana palustris) skin secretion by single step "shotgun" cloning. *Peptides* **2007**, *28*, 1605–1610. [CrossRef]
186. Zhao, C.; Liaw, L.; Lee, I.H.; Lehrer, R.I. cDNA cloning of Clavanins: Antimicrobial peptides of tunicate hemocytes. *FEBS Lett.* **1997**, *410*, 490–492. [CrossRef]
187. Zhao, C.; Liu, L.; Lehrer, R.I. Identification of a new member of the protegrin family by cDNA cloning. *FEBS Lett.* **1994**, *346*, 285–288.
188. Shanthi, S.; Vaseeharan, B. cDNA cloning, characterization and expression analysis of a novel antimicrobial peptide gene penaeidin-3 (Fi-Pen3) from the haemocytes of Indian white shrimp Fenneropenaeus indicus. *Microbiol. Res.* **2012**, *167*, 127–134. [CrossRef]
189. DbEST—Database for "Expressed Sequence Tags" | Nature Genetics. Available online: https://www.nature.com/articles/ng0893-332 (accessed on 25 February 2019).
190. Simunić, J.; Petrov, D.; Bouceba, T.; Kamech, N.; Benincasa, M.; Juretić, D. Trichoplaxin—A new membrane-active antimicrobial peptide from placozoan cDNA. *BBA Biomembr.* **2014**, *1838*, 1430–1438. [CrossRef] [PubMed]

191. Kim, I.-W.; Lee, J.H.; Subramaniyam, S.; Yun, E.-Y.; Kim, I.; Park, J.; Hwang, J.S. De Novo Transcriptome Analysis and Detection of Antimicrobial Peptides of the American Cockroach *Periplaneta americana* (Linnaeus). *PLoS ONE* **2016**, *11*, e0155304. [CrossRef] [PubMed]
192. Kim, I.-W.; Markkandan, K.; Lee, J.H.; Subramaniyam, S.; Yoo, S.; Park, J.; Hwang, J.S. Transcriptome Profiling and In Silico Analysis of the Antimicrobial Peptides of the Grasshopper Oxya chinensis sinuosa. *J. Microbiol. Biotechnol.* **2016**, *26*, 1863–1870. [CrossRef] [PubMed]
193. Rončević, T.; Gerdol, M.; Spazzali, F.; Florian, F.; Mekinić, S.; Tossi, A.; Pallavicini, A. Parallel identification of novel antimicrobial peptide sequences from multiple anuran species by targeted DNA sequencing. *BMC Genom.* **2018**, *19*, 827. [CrossRef] [PubMed]
194. Brand, G.D.; Magalhães, M.T.Q.; Tinoco, M.L.P.; Aragão, F.J.L.; Nicoli, J.; Kelly, S.M.; Cooper, A.; Bloch, C., Jr. Probing Protein Sequences as Sources for Encrypted Antimicrobial Peptides. *PLoS ONE* **2012**, *7*, e45848. [CrossRef]
195. Brand, G.D.; Ramada, M.H.S.; Genaro-Mattos, T.C.; Bloch, C. Towards an experimental classification system for membrane active peptides. *Sci. Rep.* **2018**, *8*, 1194. [CrossRef]
196. Yi, Y.; You, X.; Bian, C.; Chen, S.; Lv, Z.; Qiu, L.; Shi, Q. High-Throughput Identification of Antimicrobial Peptides from Amphibious Mudskippers. *Mar. Drugs* **2017**, *15*, 364. [CrossRef]
197. Yi, Y.; Lv, Y.; You, X.; Chen, J.; Bian, C.; Huang, Y.; Xu, J.; Deng, L.; Shi, Q. High throughput screening of small immune peptides and antimicrobial peptides from the Fish-T1K database. *Genomics* **2019**, *111*, 215–221. [CrossRef]
198. Tucker, A.T.; Leonard, S.P.; DuBois, C.D.; Knauf, G.A.; Cunningham, A.L.; Wilke, C.O.; Trent, M.S.; Davies, B.W. Discovery of Next-Generation Antimicrobials through Bacterial Self-Screening of Surface-Displayed Peptide Libraries. *Cell* **2018**, *172*, 618–628. [CrossRef]
199. Raventos, D.; Taboureau, O.; Mygind, P.; Nielsen, J.; Sonksen, C.; Kristensen, H. Improving on Natures Defenses: Optimization & High Throughput Screening of Antimicrobial Peptides. *Comb. Chem. High Throughput Screen.* **2005**, *8*, 219–233.
200. Mygind, P.H.; Fischer, R.L.; Schnorr, K.M.; Hansen, M.T.; Sönksen, C.P.; Ludvigsen, S.; Raventós, D.; Buskov, S.; Christensen, B.; De Maria, L.; et al. Plectasin is a peptide antibiotic with therapeutic potential from a saprophytic fungus. *Nature* **2005**, *437*, 975–980. [CrossRef] [PubMed]
201. Zelezetsky, I.; Pag, U.; Sahl, H.-G.; Tossi, A. Tuning the biological properties of amphipathic α-helical antimicrobial peptides: Rational use of minimal amino acid substitutions. *Peptides* **2005**, *26*, 2368–2376. [CrossRef] [PubMed]
202. Jiang, Z.; Vasil, A.I.; Hale, J.D.; Hancock, R.E.W.; Vasil, M.L.; Hodges, R.S. Effects of net charge and the number of positively charged residues on the biological activity of amphipathic α-helical cationic antimicrobial peptides. *Pept. Sci.* **2008**, *90*, 369–383. [CrossRef]
203. Wang, Y.; Ding, Y.; Wen, H.; Lin, Y.; Hu, Y.; Zhang, Y.; Xia, Q.; Lin, Z. QSAR Modeling and Design of Cationic Antimicrobial Peptides Based on Structural Properties of Amino Acids. *Comb. Chem. High Throughput Screen.* **2012**, *15*, 347–353. [CrossRef] [PubMed]
204. Taboureau, O. Methods for Building Quantitative Structure–Activity Relationship (QSAR) Descriptors and Predictive Models for Computer-Aided Design of Antimicrobial Peptides. In *Antimicrobial Peptides*; Giuliani, A., Rinaldi, A.C., Eds.; Humana Press: Totowa, NJ, USA, 2010; Volume 618, pp. 77–86. ISBN 978-1-60761-593-4.
205. Veerasamy, R.; Rajak, H.; Jain, A.; Sivadasan, S.; Varghese, C.P.; Agrawal, R.K. Validation of QSAR Models -Strategies and Importance. 9. *Int. J. Drug Des. Dis.* **2011**, *3*, 511–519.
206. Müller, A.T.; Hiss, J.A.; Schneider, G. Recurrent Neural Network Model for Constructive Peptide Design. *J. Chem. Inf. Model.* **2018**, *58*, 472–479. [CrossRef]
207. Witten, J.; Witten, Z. Deep learning regression model for antimicrobial peptide design. *BioRxiv* **2019**. [CrossRef]
208. Ilić, N.; Novković, M.; Guida, F.; Xhindoli, D.; Benincasa, M.; Tossi, A.; Juretić, D. Selective antimicrobial activity and mode of action of adepantins, glycine-rich peptide antibiotics based on anuran antimicrobial peptide sequences. *BBA Biomembr.* **2013**, *1828*, 1004–1012. [CrossRef]
209. Kamech, N.; Vukičević, D.; Ladram, A.; Piesse, C.; Vasseur, J.; Bojović, V.; Simunić, J.; Juretić, D. Improving the Selectivity of Antimicrobial Peptides from Anuran Skin. *J. Chem. Inf. Model.* **2012**, *52*, 3341–3351. [CrossRef]

210. Rončević, T.; Vukičević, D.; Krce, L.; Benincasa, M.; Aviani, I.; Maravić, A.; Tossi, A. Selection and redesign for high selectivity of membrane-active antimicrobial peptides from a dedicated sequence/function database. *BBA Biomembr.* **2019**, *1861*, 827–834. [CrossRef]
211. Haney, E.F.; Brito-Sánchez, Y.; Trimble, M.J.; Mansour, S.C.; Cherkasov, A.; Hancock, R.E.W. Computer-aided Discovery of Peptides that Specifically Attack Bacterial Biofilms. *Sci. Rep.* **2018**, *8*, 1871. [CrossRef] [PubMed]
212. Liu, S.; Bao, J.; Lao, X.; Zheng, H. Novel 3D Structure Based Model for Activity Prediction and Design of Antimicrobial Peptides. *Sci. Rep.* **2018**, *8*, 11189. [CrossRef] [PubMed]
213. Leontiadou, H.; Mark, A.E.; Marrink, S.J. Antimicrobial Peptides in Action. *J. Am. Chem. Soc.* **2006**, *128*, 12156–12161. [CrossRef] [PubMed]
214. Reißer, S.; Strandberg, E.; Steinbrecher, T.; Elstner, M.; Ulrich, A.S. Best of Two Worlds? How MD Simulations of Amphiphilic Helical Peptides in Membranes Can Complement Data from Oriented Solid-State NMR. *J. Chem. Theory Comput.* **2018**, *14*, 6002–6014. [CrossRef] [PubMed]
215. Lee, J.; Jung, S.W.; Cho, A.E. Molecular Insights into the Adsorption Mechanism of Human β-Defensin-3 on Bacterial Membranes. *Langmuir* **2016**, *32*, 1782–1790. [CrossRef] [PubMed]
216. Lai, P.-K.; Kaznessis, Y.N. Insights into Membrane Translocation of Protegrin Antimicrobial Peptides by Multistep Molecular Dynamics Simulations. *ACS Omega* **2018**, *3*, 6056–6065. [CrossRef] [PubMed]
217. Rončević, T.; Vukičević, D.; Ilić, N.; Krce, L.; Gajski, G.; Tonkić, M.; Goić-Barišić, I.; Zoranić, L.; Sonavane, Y.; Benincasa, M.; et al. Antibacterial Activity Affected by the Conformational Flexibility in Glycine–Lysine Based α-Helical Antimicrobial Peptides. *J. Med. Chem.* **2018**, *61*, 2924–2936. [CrossRef]
218. Fjell, C.D.; Hiss, J.A.; Hancock, R.E.W.; Schneider, G. Designing antimicrobial peptides: Form follows function. *Nat. Rev. Drug Discov.* **2012**, *11*, 37–51. [CrossRef]
219. Chen, C.H.; Starr, C.G.; Troendle, E.; Wiedman, G.; Wimley, W.C.; Ulmschneider, J.P.; Ulmschneider, M.B. Simulation-Guided Rational *de Novo* Design of a Small Pore-Forming Antimicrobial Peptide. *J. Am. Chem. Soc.* **2019**, *141*, 4839–4848. [CrossRef]
220. Tieleman, D.P. Antimicrobial Peptides in the Cross Hairs of Computer Simulations. *Biophys. J.* **2017**, *113*, 1–3. [CrossRef]
221. Ulmschneider, J.P.; Ulmschneider, M.B. Molecular Dynamics Simulations Are Redefining Our View of Peptides Interacting with Biological Membranes. *Acc. Chem. Res.* **2018**, *51*, 1106–1116. [CrossRef] [PubMed]
222. Marrink, S.J.; Tieleman, D.P. Perspective on the Martini model. *Chem. Soc. Rev.* **2013**, *42*, 6801–6822. [CrossRef] [PubMed]
223. Parton, D.L.; Akhmatskaya, E.V.; Sansom, M.S.P. Multiscale Simulations of the Antimicrobial Peptide Maculatin 1.1: Water Permeation through Disordered Aggregates. *J. Phys. Chem. B* **2012**, *116*, 8485–8493. [CrossRef] [PubMed]
224. Wimley, W.C.; Hristova, K. Antimicrobial Peptides: Successes, Challenges and Unanswered Questions. *J. Membr. Biol.* **2011**, *239*, 27–34. [CrossRef] [PubMed]
225. Jenssen, H.; Hamill, P.; Hancock, R.E.W. Peptide Antimicrobial Agents. *Clin. Microbiol. Rev.* **2006**, *19*, 491–511. [CrossRef]
226. Gordon, Y.J.; Romanowski, E.G.; McDermott, A.M. A Review of Antimicrobial Peptides and Their Therapeutic Potential as Anti-Infective Drugs. *Curr. Eye Res.* **2005**, *30*, 505–515. [CrossRef]
227. Joo, H.-S.; Fu, C.-I.; Otto, M. Bacterial strategies of resistance to antimicrobial peptides. *Philos. Trans. R. Soc. B Biol. Sci.* **2016**, *371*, 20150292. [CrossRef]
228. Bechinger, B.; Gorr, S.-U. Antimicrobial Peptides: Mechanisms of Action and Resistance. *J. Dent. Res.* **2017**, *96*, 254–260. [CrossRef]
229. Andersson, D.I.; Hughes, D.; Kubicek-Sutherland, J.Z. Mechanisms and consequences of bacterial resistance to antimicrobial peptides. *Drug Resist. Updat.* **2016**, *26*, 43–57. [CrossRef]
230. Otto, M. Bacterial Evasion of Antimicrobial Peptides by Biofilm Formation. In *Antimicrobial Peptides and Human Disease*; Shafer, W.M., Ed.; Springer: Berlin/Heidelberg, Germany, 2006; Volume 306, pp. 251–258. ISBN 978-3-540-29915-8.
231. De Alteriis, E.; Lombardi, L.; Falanga, A.; Napolano, M.; Galdiero, S.; Siciliano, A.; Carotenuto, R.; Guida, M.; Galdiero, E. Polymicrobial antibiofilm activity of the membranotropic peptide gH625 and its analogue. *Microb. Pathog.* **2018**, *125*, 189–195. [CrossRef]

232. Berditsch, M.; Jäger, T.; Strempel, N.; Schwartz, T.; Overhage, J.; Ulrich, A.S. Synergistic Effect of Membrane-Active Peptides Polymyxin B and Gramicidin S on Multidrug-Resistant Strains and Biofilms of Pseudomonas aeruginosa. *Antimicrob. Agents Chemother.* **2015**, *59*, 5288–5296. [CrossRef] [PubMed]
233. Ostaff, M.J.; Stange, E.F.; Wehkamp, J. Antimicrobial peptides and gut microbiota in homeostasis and pathology. *EMBO Mol. Med.* **2013**, *5*, 1465–1483. [CrossRef] [PubMed]
234. Maher, S.; McClean, S. Investigation of the cytotoxicity of eukaryotic and prokaryotic antimicrobial peptides in intestinal epithelial cells in vitro. *Biochem. Pharmacol.* **2006**, *71*, 1289–1298. [CrossRef]
235. Won, H.-S.; Kang, S.-J.; Choi, W.-S.; Lee, B.-J. Activity Optimization of an Undecapeptide Analogue Derived from a Frog-Skin Antimicrobial Peptide. *Mol. Cells* **2011**, *31*, 49–54. [CrossRef] [PubMed]
236. Henriksen, J.R.; Etzerodt, T.; Gjetting, T.; Andresen, T.L. Side Chain Hydrophobicity Modulates Therapeutic Activity and Membrane Selectivity of Antimicrobial Peptide Mastoparan-X. *PLoS ONE* **2014**, *9*, e91007. [CrossRef] [PubMed]
237. Brandelli, A. Nanostructures as Promising Tools for Delivery of Antimicrobial Peptides. Available online: http://www.eurekaselect.com/99018/article (accessed on 26 February 2019).
238. Yoon, M.Y.; Yoon, S.S. Disruption of the Gut Ecosystem by Antibiotics. *Yonsei Med. J.* **2018**, *59*, 4–12. [CrossRef] [PubMed]
239. Kang, S.-J.; Park, S.J.; Mishig-Ochir, T.; Lee, B.-J. Antimicrobial peptides: Therapeutic potentials. *Expert Rev. Anti-Infect. Ther.* **2014**, *12*, 1477–1486. [CrossRef] [PubMed]
240. Deslouches, B.; Islam, K.; Craigo, J.K.; Paranjape, S.M.; Montelaro, R.C.; Mietzner, T.A. Activity of the De Novo Engineered Antimicrobial Peptide WLBU2 against Pseudomonas aeruginosa in Human Serum and Whole Blood: Implications for Systemic Applications. *Antimicrob. Agents Chemother.* **2005**, *49*, 3208–3216. [CrossRef]
241. Strömstedt, A.A.; Pasupuleti, M.; Schmidtchen, A.; Malmsten, M. Evaluation of Strategies for Improving Proteolytic Resistance of Antimicrobial Peptides by Using Variants of EFK17, an Internal Segment of LL-37. *Antimicrob. Agents Chemother.* **2009**, *53*, 593–602. [CrossRef]
242. Benincasa, M.; Zahariev, S.; Pelillo, C.; Milan, A.; Gennaro, R.; Scocchi, M. PEGylation of the peptide Bac7(1–35) reduces renal clearance while retaining antibacterial activity and bacterial cell penetration capacity. *Eur. J. Med. Chem.* **2015**, *95*, 210–219. [CrossRef]
243. Knappe, D.; Schmidt, R.; Adermann, K.; Hoffmann, R. Continuous Subcutaneous Delivery of Proline-Rich Antimicrobial Peptide Api137 Provides Superior Efficacy to Intravenous Administration in a Mouse Infection Model. *Front. Microbiol.* **2019**, *10*, 2283. [CrossRef] [PubMed]
244. Sierra, J.M.; Fusté, E.; Rabanal, F.; Vinuesa, T.; Viñas, M. An overview of antimicrobial peptides and the latest advances in their development. *Expert Opin. Biol. Ther.* **2017**, *17*, 663–676. [CrossRef] [PubMed]

International Journal of *Molecular Sciences*

MDPI

Article

Intragenic Antimicrobial Peptide Hs02 Hampers the Proliferation of Single- and Dual-Species Biofilms of *P. aeruginosa* and *S. aureus*: A Promising Agent for Mitigation of Biofilm-Associated Infections

Lucinda J. Bessa [1,*], Julia R. Manickchand [2], Peter Eaton [1], José Roberto S. A. Leite [3], Guilherme D. Brand [2] and Paula Gameiro [1]

1 LAQV/Requimte, Departamento de Química e Bioquímica, Faculdade de Ciências da, Universidade do Porto, 4050-313 Porto, Portugal

2 Laboratório de Síntese e Análise de Biomoléculas, Instituto de Química, Universidade de Brasília, UnB, Brasília DF 70910-900, Brasil

3 Núcleo de Pesquisa em Morfologia e Imunonologia Aplicada, NuPMIA, Área de Morfologia, Faculdade de Medicina, FM, Universidade de Brasília, UnB, Brasília DF 70910-900, Brasil

* Correspondence: lucinda.bessa@fc.up.pt; Tel.: +351-220-402-585

Received: 17 June 2019; Accepted: 15 July 2019; Published: 23 July 2019

Abstract: *Pseudomonas aeruginosa* and *Staphylococcus aureus* are two major pathogens involved in a large variety of infections. Their co-occurrence in the same site of infection has been frequently reported and is linked to enhanced virulence and difficulty of treatment. Herein, the antimicrobial and antibiofilm activities of an intragenic antimicrobial peptide (IAP), named Hs02, which was uncovered from the human unconventional myosin 1H protein, were investigated against several *P. aeruginosa* and *S. aureus* strains, including multidrug-resistant (MDR) isolates. The antibiofilm activity was evaluated on single- and dual-species biofilms of *P. aeruginosa* and *S. aureus*. Moreover, the effect of peptide Hs02 on the membrane fluidity of the strains was assessed through Laurdan generalized polarization (GP). Minimum inhibitory concentration (MIC) values of peptide Hs02 ranged from 2 to 16 μg/mL against all strains and MDR isolates. Though Hs02 was not able to hamper biofilm formation by some strains at sub-MIC values, it clearly affected 24 h preformed biofilms, especially by reducing the viability of the bacterial cells within the single- and dual-species biofilms, as shown by confocal laser scanning microscopy (CLSM) and atomic force microscopy (AFM) images. Laurdan GP values showed that Hs02 induces membrane rigidification in both *P. aeruginosa* and *S. aureus*. Peptide Hs02 can potentially be a lead for further improvement as an antibiofilm agent.

Keywords: polymicrobial biofilms; intragenic antimicrobial peptide; Hs02; *Pseudomonas aeruginosa*; *Staphylococcus aureus*

1. Introduction

Most types of infections, especially chronic infections, are of polymicrobial origin; that is, two or more microorganisms are present and play a role themselves in the infection [1–3]. Within the polymicrobial context, various bacterial species communicate, cooperate, and compete with each other [4]. Polymicrobial infections consisting of two or more bacterial pathogens are common in wounds and in cystic fibrosis lung infections and usually are biofilm-associated, which can greatly hamper the efficacy of treatment and thus of healing [5–7].

Staphylococcus aureus and *Pseudomonas aeruginosa* are two important bacterial pathogens that are known to have developed complex interactions in such chronic polymicrobial infections [4,8–10]. Though an antagonistic relationship was initially considered to occur between these two species, recently, *P. aeruginosa* and *S. aureus* have been isolated from the same site of infection, with evidence that both pathogens seem to worsen the infection's evolution [11–13]. According to Hotterbeekx et al. [4], there are several studies demonstrating that the copresence of *P. aeruginosa* and *S. aureus* in vivo is linked to worse disease outcomes and to a delay in the healing process.

Interspecies interactions differ between the planktonic and biofilm modes of growth [4,14]. Bacteria in biofilms exhibit different gene expression, growth rates, behaviors, and appearances to those that are in the planktonic state [15], and it is recognized that biofilms are more tolerant to antibiotics than planktonic cells [16]. Moreover, *S. aureus* interaction with *P. aeruginosa* within a biofilm can alter *S. aureus'* susceptibility to different antibiotics [9,17].

Antimicrobial peptides (AMPs) have been isolated from a variety of organisms, such as bacteria, reptiles, plants, and mammals and can also be chemically synthesized [18]. AMPs are regarded as promising alternatives to conventional antibiotics [18,19], presenting several mechanisms of action, and usually, more than one mechanism is present simultaneously [20,21]. There are also AMPs with the ability to specifically affect biofilms by inhibiting the biofilm formation or killing preformed biofilms [19, 22,23]. It has been demonstrated that proteins from varied organisms contain encrypted structural elements with significant physicochemical similarity to AMPs, termed intragenic antimicrobial peptides (IAPs) [24,25]. This opens an exciting opportunity to screen organism genomes for encrypted bioactive molecules with therapeutic potential. Once identified and synthesized as individual entities, IAPs may share biological activities with AMPs, such as broad and direct antimicrobial activity against susceptible and resistant bacterial strains, capacity to act as anti-inflammatory agents, to exert chemoattractant effect in immune cells, and also to disrupt the formation of biofilms [10,22,26]. The peptide Hs02 was identified as an internal fragment of the unconventional myosin 1H protein from a collection of human proteins using the software Kamal, which was originally developed to probe IAPs in plant proteins [24,25,27]. Hs02 was demonstrated to fold as an amphiphilic α-helix upon membrane interaction and to present potent antimicrobial and anti-inflammatory activity [27]. On that basis, in this study, it was aimed to evaluate the antibiofilm activity of the peptide against *P. aeruginosa* and *S. aureus* single- and dual-species biofilms formed by multidrug-resistant isolates.

2. Results and Discussion

2.1. Antibacterial Activity of Peptide Hs02

Peptide Hs02 exhibited antibacterial activity against both Gram-positive and Gram-negative strains, including multidrug-resistant clinical isolates (Table 1). Minimum inhibitory concentration (MIC) and minimum bactericidal concentration (MBC) values ranged from 2 to 16 μg/mL (1 to 8.2 μM), which are low values and thus indicative of effective antimicrobial activity. These results confirm the previously reported antimicrobial activity of the peptide [27]. Like many other AMPs, peptide Hs02 is bactericidal and has a broad-spectrum action [28].

Table 1. Minimum inhibitory concentration (MIC) and minimum bactericidal concentration (MBC) values of peptide Hs02 against several susceptible and multidrug-resistant strains.

	Strains	MIC μg/mL (μM)	MBC μg/mL (μM)
Reference strains	***E. coli* ATCC 25922**	4 (2.0)	4 (2.0)
	***P. aeruginosa* ATCC 27853**	8 (4.1)	8 (4.1)
	***S. aureus* ATCC 25923**	8 (4.1)	8 (4.1)
	***E. faecalis* ATCC 29212**	16 (8.2)	16 (8.2)
***E. coli* strains**	***E. coli* TBX1/1 (S)**	4 (2.0)	4 (2.0)
	***E. coli* TBX2/3 (S)**	2 (1.0)	2 (1.0)
	Ec1-SA1 (R)	4 (2.0)	4 (2.0)
	EC001 (R)	4 (2.0)	4 (2.0)

Table 1. *Cont.*

Strains		MIC μg/mL (μM)	MBC μg/mL (μM)
P. aeruginosa strains	**PAO1** [(S)]	8 (4.1)	8 (4.1)
	PA007 [(S)]	8 (4.1)	8 (4.1)
	PA008 [(S)]	8 (4.1)	8 (4.1)
	PA006 [(R)]	4 (2.0)	4 (2.0)
	Pa4 [(R)]	4 (2.0)	4 (2.0)
	PA002 [(R)]	16 (8.2)	16 (8.2)
	PA004 [(R)]	8 (4.1)	8 (4.1)
	Pa3 [(R)]	4 (2.0)	4 (2.0)
S. aureus strains	**Sa1** [(R)]	8 (4.1)	8 (4.1)
	SA007 [(R)]	4 (2.0)	4 (2.0)
	Sa3 [(R)]	8 (4.1)	8 (4.1)
E. faecalis strain	**Ef1** [(R)]	4 (2.0)	4 (2.0)

[(S)] Susceptible strain; [(R)] multidrug-resistant strain.

2.2. *Antibiofilm Activity of Peptide Hs02*

2.2.1. Peptide Hs02 Did Not Inhibit Biofilm Formation

No biofilm was formed by any of the isolates in the presence of the peptide at its MIC, as expected. Nonetheless, peptide Hs02 did not inhibit biofilm formation by the PA004, PA008, and Sa1 isolates in presence of subinhibitory concentrations, since the biofilm formed was similar or greater than the control biofilm (grown in the absence of peptide Hs02) (Figure 1). For the isolate SA007, however, at the sub-MIC levels of 0.5× and 0.25× MIC, the biofilm formed was significantly reduced in comparison to the control.

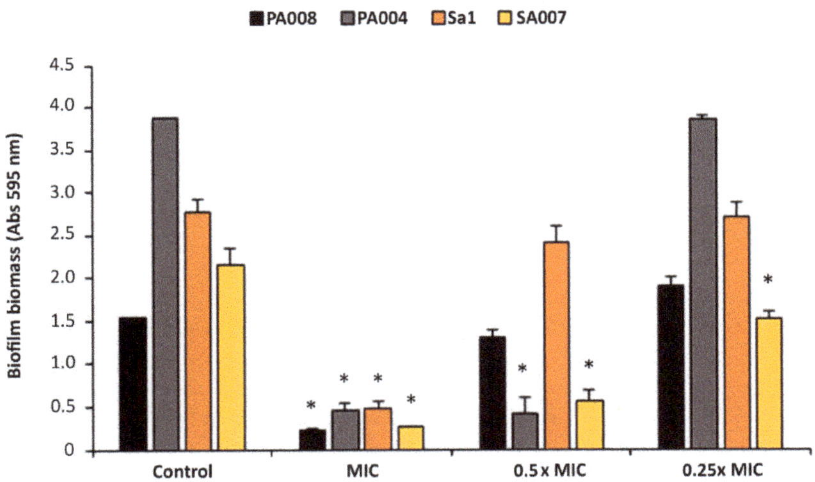

Figure 1. Biomass quantification through the crystal violet assay of biofilms formed by two *P. aeruginosa* and two methicillin-resistant *S. aureus* (MRSA) isolates in presence of peptide Hs02. Biofilms were formed in the presence of different concentrations (ranging from the MIC to 0.25× MIC) of peptide Hs02. Two independent experiments were performed in triplicate. Error bars represent SD. Statistically significant differences in comparison to the control ($p < 0.05$) are marked with an asterisk (*). Abs: Absorbance.

2.2.2. Peptide Hs02 Hampered the Proliferation of 24 h Biofilms

Peptide Hs02 had a more marked effect on 24 h performed biofilms (either single- and dual-species), by decreasing their proliferation at concentrations of 8× MIC. The biofilm proliferation was recorded by measuring the optical density at 600 nm (OD_{600}) of the planktonic phase of the treated and nontreated biofilms after 24 h (Figure 2). The single-species biofilms of *P. aeruginosa* isolates PA002 and Pa3 were less affected by peptide Hs02 than those of *S. aureus*, which were clearly affected for all three strains—SA007, Sa3, and Sa1. Two out of the three dual-species biofilms (PA002+Sa3 and Pa3+Sa1) suffered an intermediate reduction when treated with the peptide.

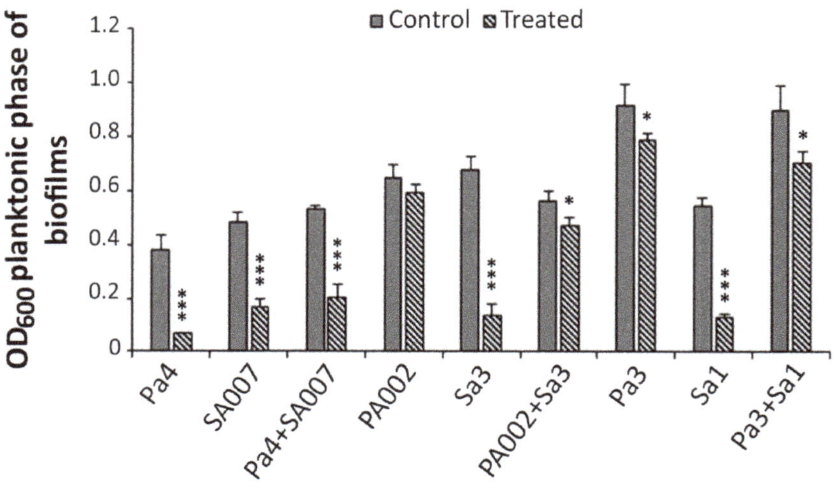

Figure 2. Effect of peptide Hs02 on the proliferation of 24-h preformed biofilms. The OD_{600} of planktonic phases of biofilms was measured and used to infer the biofilm proliferation. Pa4 and SA007 biofilms were treated with 8× MIC (32 µg/mL in both cases) of peptide Hs02; the dual-species biofilm of Pa4+SA007 was also treated with the same concentration. PA002 and Sa3 biofilms were treated with 8× MIC; that is, 128 µg/mL and 64 µg/mL, respectively, while the dual-species biofilm PA002+Sa3 was treated with 128 µg/mL. Pa3 and Sa1 biofilms were treated with 8× MIC; that is, 40 and 80 µg/mL, respectively, while the dual-species biofilm Pa3+Sa1 was treated with 80 µg/mL. Two independent experiments were performed in triplicate. Error bars represent SD. Statistically significant differences in comparison to the control are highlighted for $p < 0.001$ (***) or for $0.01 \leq p < 0.05$ (*). OD_{600}: optical density at 600 nm.

2.2.3. Peptide Hs02-Treated Biofilms Presented Reduced Viability

By using confocal laser scanning microscopy (CLSM), it could be observed that the peptide-treated biofilms maintained overall structure in comparison to the respective nontreated ones; however, the viable cells within the treated biofilms were reduced to some extent in all cases (Figure 3). As it is generally accepted that AMPs interact with the cytoplasmic membrane, causing membrane rupture that will eventually lead to cell lysis [29], one could expect to see more red (nonviable) cells in the peptide Hs02-treated biofilms, as was indeed observed. The peptide effect was quite striking in the case of both *P. aeruginosa* biofilms assayed, Pa4 and PA002, where more than 90% of the cells within the biofilm were red.

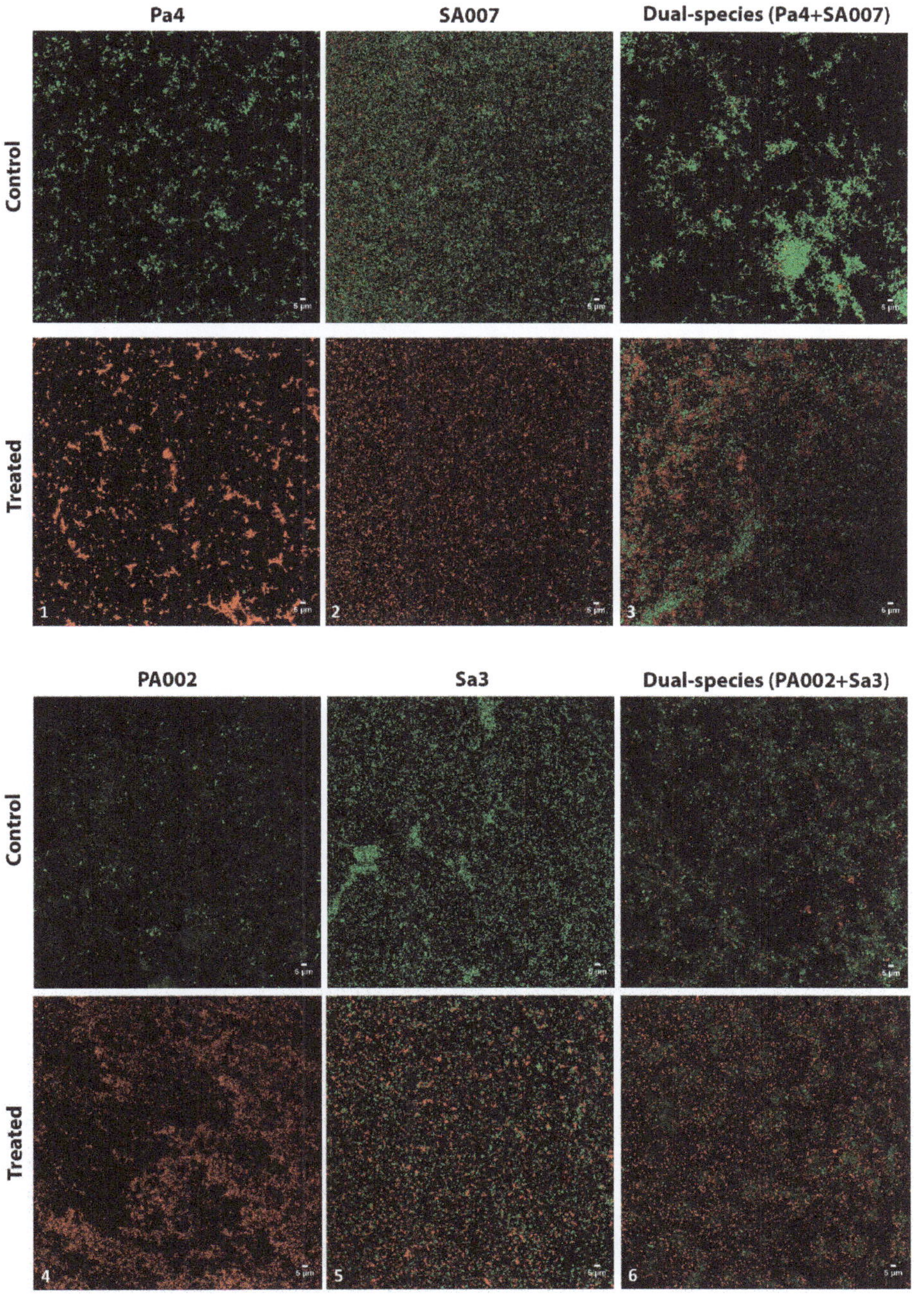

Figure 3. Confocal laser scanning microscopy (CLSM) images of single- and dual-species multidrug-resistant (MDR) *P. aeruginosa* and methicillin-resistant *S. aureus* (MRSA) biofilms. Biofilms were grown for 24 h and then treated for further 24 h with peptide Hs02 (**1–6**). Control images: no antimicrobial added. **1**, **2**, and **3**: 8× MIC peptide Hs02 (32 µg/mL); **4**: 8× MIC (128 µg/mL); **5**: 8× MIC (64 µg/mL). **6**: 128 µg/mL.

The dual-species biofilm of PA002+Sa3 was also analyzed by atomic force microscopy (AFM) (Figure 4). As we can see, the peptide-treated biofilm presents some undamaged cells, but clearly more damaged ones ("dead cells") in comparison to the nontreated biofilm. Again, we saw more damaged PA002 cells (in the form of empty membranes or "ghosts") than Sa3 cells, which in general looked quite intact in the AFM images. It is also worth noting that in the AFM images, the structure of the control biofilms seemed more compact compared to those treated with peptide Hs02. These results corroborate those obtained by the live/dead staining.

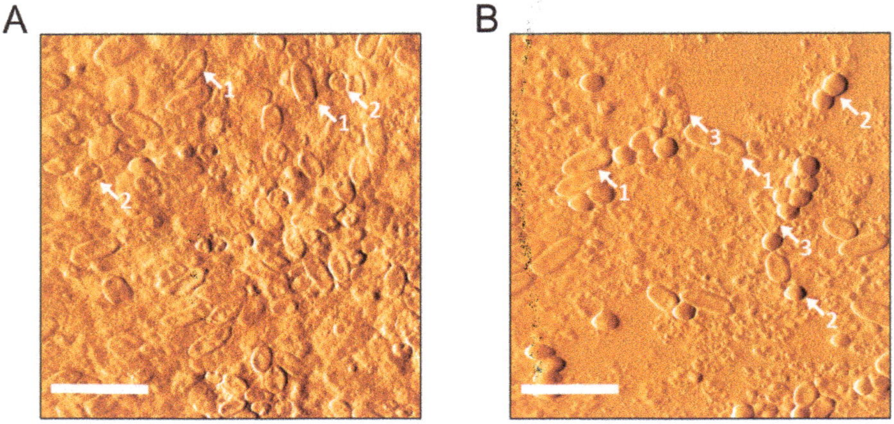

Figure 4. AFM images of the dual-species biofilm of PA002+Sa3. Biofilms were grown for 24 h and then treated for further 24 h with no peptide Hs02 (**A**) or with 128 μg/mL of the peptide (**B**). Scale bars correspond to 5 μm. Some features discussed in the text are indicated as follows: 1: undamaged PA002 cells; 2: undamaged Sa3 cells; 3: damaged PA002 cells.

2.3. Peptide Hs02 Decreased Bacterial Membrane Fluidity

The effect of peptide Hs02 on bacterial cytoplasmic membrane fluidity was assessed after calculating the Laurdan generalized polarization (GP). Laurdan is a polarity-sensitive fluorescent probe used to detect changes in the general membrane fluidity; an inverse relationship exists between Laurdan GP values and the degree of cytoplasmic membrane lipid order (i.e., lower GP values equate to greater membrane fluidity) [30–32].

As shown in Table 2, peptide Hs02 induced an increase in Laurdan GP values, reflecting a shift towards rigidification. This is in accordance with data that others [33] have obtained regarding the lipopeptide antibiotic daptomycin, which is active against Gram-positive pathogens. Daptomycin is a last-resort antibiotic for the treatment of infections caused by multidrug-resistant Gram-positive pathogens, such as MRSA, and is one of the few peptide antibiotics that can be administered intravenously [34]. Accordingly, others [35,36] have also reported that some AMPs can reduce the membrane fluidity of *S. aureus*. The antimicrobial hexapeptide MP196 and the cyclic β-sheet peptide gramicidin S also increased the Laurdan GP values in *S. aureus* and *Bacillus subtilis*, indicating that they also reduced the membrane fluidity [37].

Table 2. Laurdan generalized polarization (GP) values obtained for different bacterial strains in the presence and absence of peptide Hs02 (in concentrations equal to 0.5× MIC, MIC, and 2× MIC). GP values allow the measurement of membrane fluidity.

	Control	0.5× MIC	MIC	2× MIC
***P. aeruginosa* ATCC 27853**	0.068 ± 0.002	0.077 ± 0.003 *	0.123 ± 0.004 *	0.155 ± 0.003 *
***S. aureus* ATCC 25923**	0.007 ± 0.000	0.008 ± 0.000 *	0.014 ± 0.002 *	0.038 ± 0.002 *
PA002	0.015 ± 0.001	0.060 ± 0.003 *	0.118 ± 0.003 *	0.151 ± 0.002 *
PA004	0.065 ± 0.002	0.122 ± 0.002 *	0.165 ± 0.001 *	0.205 ± 0.002 *
Sa1	0.088 ± 0.003	0.057 ± 0.030	0.115 ± 0.002 *	0.148 ± 0.002 *
Sa3	0.117 ± 0.001	0.110 ± 0.004 *	0.093 ± 0.003 *	0.123 ± 0.001 *

Statistically significant differences in comparison to the control ($p < 0.05$) are marked with an asterisk (*).

The increase in Laurdan GP values was more marked in *P. aeruginosa* strains than in those of *S. aureus*. Moreover, as we can observe, membrane fluidity seems to be more variable among *S. aureus* strains than among *P. aeruginosa* strains. In the control condition, the Laurdan GP value for *S. aureus* ATCC 25923 was significantly lower than that obtained for the MRSA strains, Sa1 and Sa3, which was in accordance with previous reports [30].

3. Materials and Methods

3.1. Hs02 Peptide

The peptide Hs02, primary structure KWAVRIIRKFIKGFIS-NH_2, derived from the protein NP_001094891.3, was chemically synthesized using the Fmoc/t-butyl strategy [38] as described elsewhere [27]. For further details of the peptide purification and mass spectrometric analysis to confirm purity and primary structure, please refer to the literature [27].

3.2. Bacterial Strains and Growth Conditions

A range of susceptible and resistant strains were used in this study. Susceptible strains included *Pseudomonas aeruginosa* ATCC 27853, *P. aeruginosa* PAO1, *Escherichia coli* ATCC 25922, *Staphylococcus aureus* ATCC 25923, and *Enterococcus faecalis* ATCC 29212, as well as food isolates of *E. coli* (TBX1/1 and TBX2/3) and clinical isolates of *P. aeruginosa* (PA007 and PA008). Multidrug-resistant (MDR) clinical isolates of *P. aeruginosa* (Pa3, Pa4, PA002, PA004, and PA006) and of *S. aureus* (Sa1, Sa3, and SA007) were also used in this study. The antimicrobial resistance profile of all multidrug-resistant isolates is shown in Table A1 (see Appendix A). These bacteria were grown on Mueller–Hinton agar (MH; Liofilchem s.r.l., Roseto degli Abruzzi (Te), Italy) from stock cultures. MH plates were incubated at 37 °C for 20 h and then used to prepare fresh cultures for each experimental in vitro assay.

3.3. MIC and MBC Determination

The minimum inhibitory concentration (MIC) values of peptide Hs02 against the above-mentioned strains were determined by the broth microdilution method using cation-adjusted Mueller–Hinton broth (CAMHB; Sigma-Aldrich, Saint Louis, USA), and following the Clinical and Laboratory Standards Institute (CLSI) guidelines [39]. The minimum bactericidal concentration (MBC) was determined as reported by Bessa et al. [40].

3.4. Biofilm Formation Inhibition Assay

The effect of peptide Hs02 at concentrations equal to the MIC, 0.5× MIC, and 0.25× MIC on biofilm formation by isolates Sa1, SA007, PA004, and PA008 was assessed using the crystal violet assay as described previously [40]. Biofilms were formed in 96-well microtiter plates using tryptic soy broth (TSB; Liofilchem s.r.l., Roseto degli Abruzzi, Italy) and a starting inoculum of 1×10^6 colony-forming units (CFU)/mL.

3.5. Performed Biofilm Treatment Assay

The effect of peptide Hs02 on 24-h established single-species biofilms of *P. aeruginosa* (Pa4-SA2, PA002, and PA004) and *S. aureus* (Sa1, Sa3, and SA007) and on 24-h dual-species biofilms (*P. aeruginosa* and *S. aureus*) was investigated. The dual-species combinations studied were: Pa3+Sa1, Pa4+SA007, and PA002+Sa3. Prior to choosing these combinations, a few other combinations were previously studied; nevertheless, in some of those mixed-culture biofilms grown for 24 h without treatment, *P. aeruginosa* clearly outgrew *S. aureus*, and therefore, such combinations were not selected.

Briefly, biofilms were allowed to form for 24 h in 96-well microtiter plates, and then the planktonic phases were gently removed and the wells were rinsed and filled with concentrations of peptide Hs02 equal to 8× MIC. The OD_{600} was measured at time 0 h (just after the treatment with peptide Hs02) and time 24 h, after incubation for 24 h at 37 °C. Lower OD_{600} in the treated biofilms correlates to a reduction in the biofilm proliferation.

3.6. Visualization of Biofilms by CLSM

Single- and dual-species biofilms (Pa4 and SA007, PA002 and Sa3) grown for 24 h were formed on μ-Dishes 35 mm high, with ibidi polymer coverslips (ibidi GmbH, Germany), from a starting inoculum of 1×10^6 CFU/mL in TSB. After 24 h, biofilms were rinsed with phosphate-buffered saline (PBS) and treated with a concentration of peptide Hs02 (8× MIC) for another 24 h. Control biofilms were formed in the same way but not treated with peptide. All biofilms were then rinsed and stained using the live/dead staining BacLight bacterial viability kit (Molecular Probes, Thermo Fisher Scientific, USA). Biofilms were examined by a laser scanning confocal system Leica TCS SP5 II (Leica Microsystems, Germany), equipped with (i) an inverted microscope, Leica DMI6000-CS, using a HC PL APO CS 63× /1.30 glycerin 21 °C objective and the lasers diode 405 nm and DPSS561 561 nm, and (ii) the LAS AF software. Two to three independent experiments were performed for CLSM visualization.

3.7. Visualization of Biofilms by AFM

The dual-species biofilm of PA002+Sa3 was grown for 24 h on Thermanox circular (15 mm diameter) plastic coverslips (Thermo Scientific, NY, USA) placed in 35 mm diameter polystyrene plates. After 24 h, biofilms were rinsed with PBS and treated with 128 μg/mL of peptide Hs02 and incubated for a further 24 h at 37 °C. The respective control biofilm was formed in the same way but in the absence of the peptide. All biofilms were rinsed three times with 1 mM of phosphate buffer and air-dried before AFM imaging. Samples were scanned with a TT-AFM from AFMWorkshop in air in vibrating mode. A 50 μm scanner and 300 kHz silicon cantilevers (ACT, AppNano) were used. Images were processed using Gwyddion 2.47 software. Two independent experiments were performed for AFM visualization.

3.8. Membrane Fluidity Assessment by Laurdan Generalized Polarization (GP)

The membrane fluidity of *P. aeruginosa* (ATCC 27852, PA002, and PA004) and *S. aureus* (ATCC 25923, Sa1, and Sa3) strains in the presence and absence of peptide Hs02 was determined by assessing the Laurdan generalized polarization (GP) as previously described [30,31] with some modifications. Briefly, fresh colonies were used to inoculate nutrient broth (NB; Liofilchem s.r.l., Roseto degli Abruzzi, Italy) to obtain cell suspensions with an OD_{600} of 0.4. Several aliquots of 1.5 mL of these bacterial suspensions were taken and centrifuged (9000 rpm, 8 min). For each strain, the bacterial pellets obtained were then resuspended in 1.5 mL of NB (in duplicate, to serve as controls; one to be unlabeled and the other labeled with Laurdan) and NB containing 0.5× MIC, MIC, or 2× MIC of peptide Hs02. These new bacterial suspensions were incubated at 37 °C for 3 h. Afterwards, they were centrifuged (9000 rpm, 8 min) and cells were washed twice in 15 mM Tris–HCl buffer (pH 7.4) and finally resuspended in 10 μM of Laurdan (from a 2 mM stock solution in dimethylformamide). Each suspension was incubated in the dark at 37 °C with shaking (500 rpm) for 1.5 h. Aliquots of 1 mL were transferred to a 1 cm quartz cuvette, and Laurdan emission spectra were obtained in a Varian Cary Eclipse fluorescence spectrofluorometer (Agilent Technologies, Santa Clara, California, USA) at an excitation wavelength of 350 nm using emission wavelengths from 410 to 550 nm. The temperature was set at 37.0 ± 0.1 °C. The excitation GP was calculated using the following equation:

$$GP = (I_{440} - I_{490})/(I_{440} + I_{490}) \quad (1)$$

where I_{440} and I_{490} are fluorescence intensities at 440 and 490 nm, respectively.

3.9. Statistical Analysis

The assay to assess the membrane fluidity by Laurdan generalized polarization was performed in three independent experiments, with the results being expressed as mean values ± standard deviation.

The biofilm formation and the preformed biofilm treatment assays were carried out in two independent experiments, with each experiment being performed in triplicate.

The results regarding the biofilm formation were expressed as mean values ± standard deviation. The statistical significance of differences between controls and experimental groups was evaluated using the Student's *t*-test. *P*-values of < 0.05 were considered statistically significant.

4. Conclusions

By taking together the results from the MIC and MBC values and the live/dead staining, we can attribute a bactericidal action to the peptide Hs02, likely due to a direct effect on the bacterial cells by disrupting the cytoplasmic membrane. Moreover, the ability of the peptide to decrease the membrane fluidity of both *P. aeruginosa* and *S. aureus* strains also suggests a membrane-targeting antibacterial mechanism.

Peptide Hs02 hampered the proliferation and decreased the viability of single- and dual-species biofilms of two major pathogens, *P. aeruginosa* and *S. aureus*, confirming its potential as a lead for development towards an antibiofilm agent in complex cases involving polymicrobial biofilms.

Author Contributions: Conceptualization, L.J.B., J.R.M., J.R.S.A.L., and G.D.B.; Formal analysis, L.J.B.; Funding acquisition, J.R.M. and P.G.; Investigation, L.J.B, P.E., and G.D.B.; Project administration, L.J.B.; Resources, P.G.; Visualization, J.R.M. and J.R.S.A.L.; Writing—original draft, L.J.B. and G.D.B.; Writing—review & editing, P.E. and P.G.

Funding: This research was funded by UID/QUI/50006/2019 with funding from FCT/MCTES through national funds, and also the FAP-DF grant 0193.000866/2015.

Acknowledgments: The authors wish to express their gratitude to Laboratório de Espectrometria de Massa—Embrapa Recursos Genéticos e Biotecnologia for the access to the mass spectrometry facilities.

Conflicts of Interest: The authors declare no conflict of interest. The funders had no role in the design of the study; in the collection, analyses, or interpretation of data; in the writing of the manuscript; or in the decision to publish the results.

Abbreviations

AFM	Atomic force microscopy
AMP	Antimicrobial peptide
CFU	Colony-forming units
CLSM	Confocal laser scanning microscopy
GP	Generalized polarization
IAP	Intragenic antimicrobial peptide
MBC	Minimum bactericidal concentration
MDR	Multidrug-resistant
MIC	Minimum inhibitory concentration
MRSA	Methicillin-resistant *Staphylococcus aureus*
PBS	Phosphate-buffered saline
TSB	Tryptic soy broth

Appendix A

Table A1. Antimicrobial resistance pattern of the multidrug-resistant isolates used in this study.

Isolate	Antimicrobial resistance pattern
Pa3	ATM, CIP, FEP, GEN
Pa4	ATM, CAZ, CIP, FEP, IPM
PA002	AMK, CIP, COL, GEN, TOB
PA004	CIP, GEN, IPM, TOB, TZP
PA006	AMK, CAZ, CIP, COL, FEP, GEN, IPM, TOB
Sa1	AMC, AMP, CIP, FOX, TET
Sa3	AMC, AMP, CIP, FOX, IPM
SA007	CIP, CLI, ERI, GEN, LEV, MOX, OXA

AMC: amoxicillin/clavulanic acid; AMK: amikacin; AMP: ampicillin; ATM: aztreonam; CAZ: ceftazidime; CLI: clindamycin; CIP: ciprofloxacin; COL: colistin; ERI: erythromycin; FEP: cefepime; FOX: cefoxitin; GEN: gentamicin; IPM: imipenem; LEV: levofloxacin; MOX: moxifloxacin; OXA: oxacillin; TET: tetracycline; TOB: tobramycin; TZP: piperacillin/tazobactam.

References

1. Stacy, A.; McNally, L.; Darch, S.E.; Brown, S.P.; Whiteley, M. The biogeography of polymicrobial infection. *Nat. Rev. Microbiol.* **2016**, *14*, 93–105. [CrossRef] [PubMed]
2. Kong, E.F.; Tsui, C.; Kucharíková, S.; Andes, D.; Van Dijck, P.; Jabra-Rizk, M.A. Commensal Protection of Staphylococcus aureus against Antimicrobials by Candida albicans Biofilm Matrix. *mBio* **2016**, *7*. [CrossRef] [PubMed]
3. García-Pérez, A.N.; De Jong, A.; Junker, S.; Becher, D.; Chlebowicz, M.A.; Duipmans, J.C.; Jonkman, M.F.; Van Dijl, J.M. From the wound to the bench: Exoproteome interplay between wound-colonizing Staphylococcus aureus strains and co-existing bacteria. *Virulence* **2018**, *9*, 363–378. [CrossRef] [PubMed]
4. Hotterbeekx, A.; Kumar-Singh, S.; Goossens, H.; Malhotra-Kumar, S. In vivo and In vitro Interactions between Pseudomonas aeruginosa and Staphylococcus spp. *Front. Cell. Infect. Microbiol.* **2017**, *7*, 106. [CrossRef] [PubMed]
5. Bowler, P.G. Antibiotic resistance and biofilm tolerance: A combined threat in the treatment of chronic infections. *J. Wound Care* **2018**, *27*, 273–277. [CrossRef] [PubMed]
6. Burmølle, M.; Ren, D.; Bjarnsholt, T.; Sørensen, S.J. Interactions in multispecies biofilms: Do they actually matter? *Trends Microbiol.* **2014**, *22*, 84–91. [CrossRef] [PubMed]
7. Wolcott, R.; Costerton, J.; Raoult, D.; Cutler, S. The polymicrobial nature of biofilm infection. *Clin. Microbiol. Infect.* **2013**, *19*, 107–112. [CrossRef]
8. Bessa, L.J.; Fazii, P.; Di Giulio, M.; Cellini, L. Bacterial isolates from infected wounds and their antibiotic susceptibility pattern: Some remarks about wound infection. *Int. Wound J.* **2015**, *12*, 47–52. [CrossRef]
9. Radlinski, L.; Rowe, S.E.; Kartchner, L.B.; Maile, R.; Cairns, B.A.; Vitko, N.P.; Gode, C.J.; Lachiewicz, A.M.; Wolfgang, M.C.; Conlon, B.P. Pseudomonas aeruginosa exoproducts determine antibiotic efficacy against Staphylococcus aureus. *PLoS Boil.* **2017**, *15*, e2003981. [CrossRef]
10. Pfalzgraff, A.; Brandenburg, K.; Weindl, G. Antimicrobial Peptides and Their Therapeutic Potential for Bacterial Skin Infections and Wounds. *Front. Pharmacol.* **2018**, *9*, 281. [CrossRef]
11. Sagel, S.D.; Gibson, R.L.; Emerson, J.; McNamara, S.; Burns, J.L.; Wagener, J.S.; Ramsey, B.W.; Inhaled Tobramycin in Young Children Study Group; Cystic Fibrosis Foundation Therapeutics Development Network. Impact of Pseudomonas and Staphylococcus infection on inflammation and clinical status in young children with cystic fibrosis. *J. Pediatr.* **2009**, *154*, 183–188. [CrossRef] [PubMed]
12. Hubert, D.; Reglier-Poupet, H.; Sermet-Gaudelus, I.; Ferroni, A.; Le Bourgeois, M.; Burgel, P.-R.; Serreau, R.; Dusser, D.; Poyart, C.; Coste, J. Association between Staphylococcus aureus alone or combined with Pseudomonas aeruginosa and the clinical condition of patients with cystic fibrosis. *J. Cyst. Fibros.* **2013**, *12*, 497–503. [CrossRef] [PubMed]

13. Dalton, T.; Dowd, S.E.; Wolcott, R.D.; Sun, Y.; Watters, C.; Griswold, J.A.; Rumbaugh, K.P. An in vivo polymicrobial biofilm wound infection model to study interspecies interactions. *PLoS ONE* **2011**, *6*, e27317. [CrossRef] [PubMed]
14. O'Brien, S.; Fothergill, J.L. The role of multispecies social interactions in shaping Pseudomonas aeruginosa pathogenicity in the cystic fibrosis lung. *FEMS Microbiol. Lett.* **2017**, *364*, 128. [CrossRef] [PubMed]
15. Flemming, H.-C.; Wingender, J.; Szewzyk, U.; Steinberg, P.; Rice, S.A.; Kjelleberg, S. Biofilms: An emergent form of bacterial life. *Nat. Rev. Genet.* **2016**, *14*, 563–575. [CrossRef] [PubMed]
16. Koo, H.; Allan, R.N.; Howlin, R.P.; Stoodley, P.; Hall-Stoodley, L. Targeting microbial biofilms: Current and prospective therapeutic strategies. *Nat. Rev. Genet.* **2017**, *15*, 740–755. [CrossRef] [PubMed]
17. Beaudoin, T.; Yau, Y.C.W.; Stapleton, P.J.; Gong, Y.; Wang, P.W.; Guttman, D.S.; Waters, V. Staphylococcus aureus interaction with Pseudomonas aeruginosa biofilm enhances tobramycin resistance. *NPJ Biofilms Microbiomes* **2017**, *3*, 25. [CrossRef]
18. Kim, M.K.; Kang, H.K.; Ko, S.J.; Hong, M.J.; Bang, J.K.; Seo, C.H.; Park, Y. Mechanisms driving the antibacterial and antibiofilm properties of Hp1404 and its analogue peptides against multidrug-resistant Pseudomonas aeruginosa. *Sci. Rep.* **2018**, *8*, 1763. [CrossRef]
19. Chung, P.Y.; Khanum, R. Antimicrobial peptides as potential anti-biofilm agents against multidrug-resistant bacteria. *J. Microbiol. Immunol. Infect.* **2017**, *50*, 405–410. [CrossRef]
20. Bechinger, B.; Gorr, S.U. Antimicrobial peptides: Mechanisms of action and resistance. *J. Dent. Res.* **2017**, *96*, 254–260. [CrossRef]
21. Kumar, P.; Kizhakkedathu, J.N.; Straus, S.K. Antimicrobial Peptides: Diversity, Mechanism of Action and Strategies to Improve the Activity and Biocompatibility In Vivo. *Biomolecules* **2018**, *8*, 4. [CrossRef] [PubMed]
22. Pletzer, D.; Coleman, S.R.; Hancock, R.E. Anti-biofilm peptides as a new weapon in antimicrobial warfare. *Curr. Opin. Microbiol.* **2016**, *33*, 35–40. [CrossRef] [PubMed]
23. Di Luca, M.; Maccari, G.; Maisetta, G.; Batoni, G. BaAMPs: The database of biofilm-active antimicrobial peptides. *Biofouling* **2015**, *31*, 193–199. [CrossRef] [PubMed]
24. Brand, G.D.; Magalhães, M.T.Q.; Tinoco, M.L.P.; Aragão, F.J.L.; Nicoli, J.; Kelly, S.M.; Cooper, A.; Bloch, C. Probing Protein Sequences as Sources for Encrypted Antimicrobial Peptides. *PLoS ONE* **2012**, *7*, e45848. [CrossRef] [PubMed]
25. Ramada, M.H.S.; Brand, G.D.; Abrão, F.Y.; Oliveira, M.; Filho, J.L.C.; Galbieri, R.; Gramacho, K.P.; Prates, M.V.; Bloch, C. Encrypted Antimicrobial Peptides from Plant Proteins. *Sci. Rep.* **2017**, *7*, 13263. [CrossRef] [PubMed]
26. Giuliani, A.; Pirri, G.; Rinaldi, A.C. Antimicrobial peptides: The LPS connection. *Methods Mol. Biol.* **2010**, *618*, 137–154.
27. Brand, G.D.; Ramada, M.H.S.; Manickchand, J.R.; Correa, R.; Ribeiro, D.J.S.; Michele, A.; Santos, M.A.; Vasconcelos, A.G.; Abrão, F.Y.; Prates, M.V.; et al. Intragenic antimicrobial peptides (IAPs) from human proteins with potent antimicrobial and anti-inflammatory activity. *PLoS ONE*, in press.
28. Monserrat-Martinez, A.; Gambin, Y.; Sierecki, E. Thinking Outside the Bug: Molecular Targets and Strategies to Overcome Antibiotic Resistance. *Int. J. Mol. Sci.* **2019**, *20*, 1255. [CrossRef]
29. Travkova, O.G.; Moehwald, H.; Brezesinski, G. The interaction of antimicrobial peptides with membranes. *Adv. Colloid Interface Sci.* **2017**, *247*, 521–532. [CrossRef]
30. Bessa, L.J.; Ferreira, M.; Gameiro, P. Evaluation of membrane fluidity of multidrug-resistant isolates of Escherichia coli and Staphylococcus aureus in presence and absence of antibiotics. *J. Photochem. Photobiol. B Boil.* **2018**, *181*, 150–156. [CrossRef]
31. Bessa, L.J.; Ferreira, M.; Gameiro, P. Data on Laurdan spectroscopic analyses to compare membrane fluidity between susceptible and multidrug-resistant bacteria. *Data Brief* **2018**, *21*, 128–132. [CrossRef] [PubMed]
32. Scheinpflug, K.; Krylova, O.; Strahl, H. Measurement of cell membrane fluidity by Laurdan GP: Fluorescence spectroscopy and microscopy. *Methods Mol. Biol.* **2017**, *1520*, 159–174. [PubMed]
33. Müller, A.; Wenzel, M.; Strahl, H.; Grein, F.; Saaki, T.N.V.; Kohl, B.; Siersma, T.; Bandow, J.E.; Sahl, H.G.; Schneider, T.; et al. Daptomycin inhibits cell envelope synthesis by interfering with fluid membrane microdomains. *Proc. Natl. Acad. Sci. USA* **2016**, *113*, E7077–E7086. [CrossRef] [PubMed]
34. Baltz, R.H.; Miao, V.; Wrigley, S.K. Natural products to drugs: Daptomycin and related lipopeptide antibiotics. *Nat. Prod. Rep.* **2005**, *22*, 717–741. [CrossRef] [PubMed]

35. Shireen, T.; Singh, M.; Das, T.; Mukhopadhyay, K. Differential Adaptive Responses of Staphylococcus aureus to In Vitro Selection with Different Antimicrobial Peptides. *Antimicrob. Agents Chemother.* **2013**, *57*, 5134–5137. [CrossRef] [PubMed]
36. Yoon, Y.; Lee, H.; Lee, S.; Kim, S.; Choi, K.-H. Membrane fluidity-related adaptive response mechanisms of foodborne bacterial pathogens under environmental stresses. *Food Res. Int.* **2015**, *72*, 25–36. [CrossRef]
37. Wenzel, M.; Vischer, N.; Strahl, H.; Hamoen, L. Assessing Membrane Fluidity and Visualizing Fluid Membrane Domains in Bacteria Using Fluorescent Membrane Dyes. *Bio-Protoc.* **2018**, *8*, 1–26. [CrossRef]
38. Chan, W.C.; White, P. Basic procedures. In *Fmoc Solid Phase Peptide Synthesis: A Practical Approach*; Chan, W.C., White, P., Eds.; Oxford University Press: Oxford, UK, 2000.
39. Clinical and Laboratory Standards Institute (CLSI). *Methods for Dilution Antimicrobial Susceptibility Tests for Bacteria that Grow Aerobically*, 11th ed.; CLSI Standard M07: Wayne, PA, USA, 2012.
40. Bessa, L.J.; Eaton, P.; Dematei, A.; Plácido, A.; Vale, N.; Gomes, P.; Delerue-Matos, C.; Leite, J.R.S.; Gameiro, P. Synergistic and antibiofilm properties of ocellatin peptides against multidrug-resistant Pseudomonas aeruginosa. *Futur. Microbiol.* **2018**, *13*, 151–163. [CrossRef]

International Journal of
Molecular Sciences

MDPI

Review

Expression and Function of Host Defense Peptides at Inflammation Sites

Suhanya V. Prasad, Krzysztof Fiedoruk, Tamara Daniluk, Ewelina Piktel and Robert Bucki *

Department of Medical Microbiology and Nanobiomedical Engineering, Medical University of Bialystok, Mickiewicza 2c, Bialystok 15-222, Poland; suhanyavp@gmail.com (S.V.P.); krzysztof.fiedoruk@umb.edu.pl (K.F.); tamara.daniluk@umb.edu.pl (T.D.); ewelina.piktel@wp.pl (E.P.)

* Correspondence: buckirobert@gmail.com; Tel.: +48-85-7485483

Received: 12 November 2019; Accepted: 19 December 2019; Published: 22 December 2019

Abstract: There is a growing interest in the complex role of host defense peptides (HDPs) in the pathophysiology of several immune-mediated inflammatory diseases. The physicochemical properties and selective interaction of HDPs with various receptors define their immunomodulatory effects. However, it is quite challenging to understand their function because some HDPs play opposing pro-inflammatory and anti-inflammatory roles, depending on their expression level within the site of inflammation. While it is known that HDPs maintain constitutive host protection against invading microorganisms, the inducible nature of HDPs in various cells and tissues is an important aspect of the molecular events of inflammation. This review outlines the biological functions and emerging roles of HDPs in different inflammatory conditions. We further discuss the current data on the clinical relevance of impaired HDPs expression in inflammation and selected diseases.

Keywords: host defense peptides; human antimicrobial peptides; defensins; cathelicidins; inflammation; anti-inflammatory; pro-inflammatory

---

1. Introduction

The human body is in a constant state of conflict with the unseen microbial world that threatens to disrupt the host cell function and colonize the body surfaces. The immune system has an arsenal of destructive mechanisms to neutralize the toxic effect of the microbial pathogens. It functions through two layers of defense systems: The innate system and the more intricate adaptive immune system, which closely communicate with each other [1]. Each of those systems form a complex network of immune cells, signaling molecules, and regulatory pathways. Inflammation is a reaction of the host immune system that acts to eliminate the source of inflammatory stimulus, ranging from pathogens to burn injuries [2]. Although microbial infections largely initiate the events of inflammation, we must note that inflammation is also a hallmark feature of various autoimmune, cancer, and systemic diseases [3]. Inflammation is a highly coordinated biochemical sequence of events that commences with the rapid migration of leukocytes to the site of infection, followed by adequate blood supply that transports different inflammatory mediators that control the course of the immune response [4]. However, while the initial events of inflammation are constructive and beneficial to the host, incompetent inflammatory resolution mechanisms, along with inefficient elimination of foreign bodies or pathogens and cellular debris, prompt the onset of chronic inflammation.

Antimicrobial peptides (AMPs) such as defensins and cathelicidins represent a vital part of the human immune system due to their broad spectrum activity against pathogenic bacteria, fungi, protists, and enveloped viruses [5]. Furthermore, in recent years, a growing number of studies have recognized these peptides as potent immune modulators, implicated in multiple pro- and anti-inflammatory responses through (1) neutralization of bacterial toxins, (2) chemoattraction and activation of immune cells, (3) initiation of adaptive immunity, (4) neovascularization and wound healing, as well as (5) anti- or pro-tumor activity [6]. Actually, AMPs interact with innate and adaptive immune receptors, such as pattern recognition (PRRs) or chemokine receptors (CCRs), as well as inflammasomes and their complement systems, creating a link between innate and adaptive immunity [7–9]. In addition, AMPs may regulate fundamental cellular processes, such as differentiation, proliferation, and programmed cell death, e.g., by stimulating growth factor receptors or as complexes with the host nucleic acids [10,11], hence they resemble cytokines and growth factors. Overall, these activities, by controlling inflammation and/or accelerating repair process of the infected site, appear to support the direct microbicidal function of AMPs in resolving an infection. Therefore, the term host defense peptides (HDPs) was coined to encompass their pleiotropic nature, and association with both infectious as well as non-infectious inflammatory responses [12], although these two terms are used interchangeably. In the latter context, HDPs fit into the definition of "alarmins" or "danger signals", i.e., various endogenous molecules collectively known as DAMPs (damage associated molecular patterns), which are released from damaged or dying cells and initiate a diverse range of physiological and pathophysiological functions [13–15].

At present, HDPs are perceived as multifunctional agents that coordinate diverse immune surveillance functions necessary to maintain homeostasis (Figure 1) [16]. However, if their production is out of the physiological range, they may contribute to an undesirable inflammation in response to local (e.g., periodontal, respiratory, intestinal, and skin) and systemic (e.g., sepsis) infections. They might also function as pathophysiological events of inflammatory diseases, cancers, and even psychiatric disorders (Table 1) [17–23]. In addition, HDPs have been indicated as potential biomarkers in numerous infectious and non-infectious diseases [24].

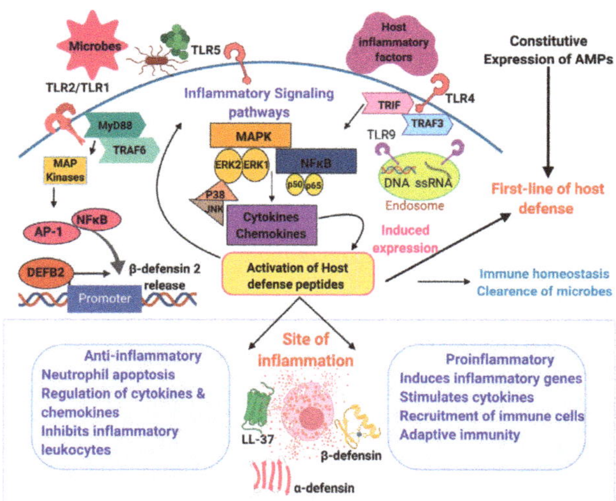

Figure 1. Illustration of the distinct role of host defense peptides at the sites of inflammation. Specific Toll-like receptors TLR-2, TLR-4, TLR-5, and TLR-6 are expressed on the plasma membrane of immune cells, non-immune cells, and intracellular compartments. TLR-7 and TLR-9 within endosomes participate in the host recognition of microbial cellular components and bind to host internal factors. The receptors, along with their adaptor proteins MyD88 and TRIF, can initiate the inflammatory signaling pathways. The host defense peptides promote innate immunity against various pathogens and maintain the immune system homeostasis. These peptides can also be induced in addition to their constitutive expression by the transcriptional modulatory factors NF-κB, AP-1, and intracellular release of cytokines and chemokines. They actively participate in coordinating the host immune signaling mechanisms during inflammation. Furthermore, HDPs can display both pro- and anti-inflammatory properties that may protect against the responses of inflammatory diseases. Abbreviations: AP-1, activator protein; NF-kB, nuclear factor kappa-light-chain enhancer of activated B cells; MAPK, mitogen-activated protein kinase; ERK1,2, extracellular signal-related kinases.

That being said, expanding our knowledge regarding the molecular mechanisms behind expression, processing, and mutual interactions of HDPs with other immune system components is crucial to better understand inflammation processes, and to develop new methods of anti-inflammatory treatment. Certainly, it is a challenging and long-term task, since a single antimicrobial peptide, e.g., human cathelicidin LL-37, may interact with dozens of proteins/receptors and subsequently engage hundreds of secondary effector proteins, as well as modify expression of >900 genes [6]. Considering this, the purpose of this review is to evaluate and summarize recent discoveries considering the functional expression and protective attributes of HDPs/AMPs in the acute inflammatory phase and the detrimental effects of their recruitment in chronic inflammation. Both in vitro and in vivo studies connecting the underlying mechanisms governing the immunoregulatory role of these peptides in the inflammatory microenvironment will be discussed.

2. Overview of Human Antimicrobial Peptides

Antimicrobial peptides are widely distributed in all living organisms, representing ancient and primary defense molecules, e.g., innate immune mechanisms conferred by the antimicrobial peptides in insects usually devoid of adaptive responses [25]. Discovery of defensins in rabbit leucocytes, lactoferrin in cow milk, and lysozyme in human saliva are among the first reports of animal-originated antimicrobial molecules, which paved the way for further identification and understanding of the physiological function of other antimicrobial peptides and proteins [26]. At present, 2272 peptides derived from animals, including ~130 of human origin (Figure 2), are recorded in the antimicrobial peptide database (http://aps.unmc.edu/AP), a comprehensive source of naturally existing families of antimicrobial peptides from all form of kingdoms of life [25].

Antimicrobial peptides and proteins contain a short chain of about 12–100 amino acids (Figure 2), and are classified according to their conformational structure (α, β, αβ, and non-αβ), amino acid motifs, and expression pattern [25,27]. For example, the major human AMPs, cathelicidin LL-37 and defensins, are characterized by α-helical and β-sheet structure, respectively. Furthermore, the latter are divided into α- and β-defensins based on the configuration of the disulfide bonds between six cysteine residues. AMPs are characterized by positive charge and substantial proportion (typically 50%) of hydrophobic residues, thus they are also known as cationic antimicrobial peptides (CAPs). However, at present, some negatively charged peptides are also classified as AMPs, e.g., human β-defensin DEFB118, psoriasin, or α-synuclein (Figure 2). Nevertheless, this amphiphilic–cationic organization allows them to selectively associate, and in turn disrupt, highly negatively charged microbial membranes. Hence, it explains their broad spectrum of activity, encompassing all cellular pathogens and enveloped viruses. Additionally, the cationic nature of AMPs may possibly facilitate, via electrostatic forces, their interactions with diverse host receptors, which are behind the immunomodulatory potential of these peptides [28].

Certain AMPs, e.g., cathelicidins, are produced as inactive pro-peptides and must be proteolytically processed for activity. It is noteworthy that this may generate multiple length variants characterized by diverse antimicrobial or immunomodulatory properties. Therefore, the presence of the appropriate proteases and their level is an important factor in regulating the function of the AMPs. Another important activity-related issue is that microbicidal action of AMPs is considerably suppressed by the physiological conditions present in some compartments of the body, including high salt, carbonate, lipoprotein, and polysaccharide concentrations [29–33].

Table 1. Host defense peptides—antimicrobial and immunomodulatory functions and disease association.

Class of Host Defense Peptides	Host Defense Peptides	Gene	Chromosome Location	Site of Expression	Biological Function	Dysregulated Expression of HDPs in Diseases	References
Cathelicidins	LL-37	*CAMP*	3p21.31	Innate immune cells Gut epithelial cells Respiratory system Salivary glands Skin	Wound healing and tissue repair LPS neutralization Recruitment of neutrophils Dendritic cells activation Intestinal barrier integrity Antiviral activity	Chronic intestinal infection↓ Systemic sclerosis↓ Chronic obstructive pulmonary disease↓ Psychological stress (murine CRAMP)↓ Atherosclerosis↑ Psoriasis↑	[18,19,23], [34–37]
α-defensins	HNP-1 HNP-2 HNP-3 HNP-4 HD-5 HD-6	*DEFA1* *DEFA3* *DEFA4* *DEFA5* *DEFA6*	8p23.1	Bone marrow Polymorphonuclear leukocytes Salivary glands Oronasal cavity and nasal mucosa Gastrointestinal and urinary tract Intestinal Paneth cells Bronchial cells Female reproductive system	Chemoattractant Phagocytosis induction Microbicidal activity Gut microbiota homeostasis Antifungal activity	Crohn's disease↓ Graft-versus-host disease↓ Sepsis↑ Coronary heart disease↑ Systemic lupus erythematosus↑ Periodontal infections↓ Colorectal cancer↑	[16,17,21] [38–40]
β-defensins	hBD-1 hBD-2 hBD-3 hBD-4	*DEFB1* *DEFB2* *DEFB3* *DEFB4*	8p 23.1-p23.2 8p23.1-p22 8p23 8p23	Epithelial and blood cells Skin Gut epithelium Respiratory tract Bone marrow Epidermal keratinocytes Gingival epithelium Small intestine	Innate immune defense Wound healing Cytokine enhancement Dendritic cell modulation Neutrophil recruitment Pro-inflammatory mediator Antimicrobial activity	Oral squamous cell carcinoma↓ Liver cancer and colorectal cancer↓ Periodontitis↓ Asthma↑ Esophageal and cervical cancer↑ Interleukin-17A-mediated psoriasis↑ Ulcerative colitis↑ Chronic obstructive pulmonary disorder↑	[20,22], [41–47]
Histatins	His1His3His5	*HTN1* *HTN3* *HTN3* *	4q13.3	Salivary glands	Oral health Wound healing	Aqueous deficient dry eye disease↓ Oral candidiasis↓	[48,49]
RNases	RNase 7	*RNASE7*	14q11.2	Skin Genito-urinary tract	Immunomodulatory	Allergic rhinitis↓ Urinary tract infections↑	[50,51]

Abbreviations: HNP-1, human neutrophil peptide 1; HNP-2, human neutrophil peptide 2; HNP-3, human neutrophil peptide 3; HNP-4, human neutrophil peptide 4; HD-5, human defensin 5; HD-6, human defensin 6; hBD-1, beta-defensin 1; hBD-2, beta-defensin 2; hBD-3, beta-defensin 3; hBD-4, beta-defensin 4; His1, histatin-1; His3, histatin-3; His5, histatin-5; RNase 7, ribonuclease 7; HTN3*, proteolytic variant of HTN3; ↑, upregulated; ↓, downregulated; HDPs, host defense peptides; LPS, lipopolysaccharides.

The sensitivity to environmental factors of these peptides was well illustrated by the inability to reproduce the protective role of insect-derived AMPs, such as drosocin, in a mouse model [52]. Briefly, the authors explained this difference by an unusually high degradation rate of such peptides in mammalian sera (human and mouse) in comparison to insect hemolymph. In contrast, physiological conditions have no impact on the immunomodulatory properties of AMPs, such as chemoattraction or activation of immune cells. In addition, the antimicrobial activity of AMPs estimated in vitro, i.e., MIC (minimal inhibitory concentration) values, is usually observed at micromolar concentrations which are significantly higher than the physiological concentrations of these peptides. For instance, the concentration of LL-37 or β-defensins is less than 2 μg/mL at mucosal sites, and the MIC of LL-37 in vitro against *Escherichia coli* is more than 32 μg/mL [10], whereas modulation of immune responses by AMPs occurs at nanomolar levels [53]. Therefore, it is possible that the other biological functions of AMPs, e.g., as alarmins, may play more prominent roles than their direct microbicidal effects in combating invading pathogens in vivo [6,10,53]. Indeed, several synthetic AMP derivatives, known as innate defense regulator (IDR) peptides, are characterized by potent immunomodulatory activities [54].

In fact, certain human AMPs such as the histone protein H2A (known as buforin I) or ribosomal protein S30 (known as ubiquicidin) were initially known from non-antimicrobial functions, before their antimicrobial potential was recognized. In addition, around 20% of human AMPs (Figure 2) are chemokines, which as cationic and amphipathic molecules are characterized by antimicrobial activity [55]. In addition, the specific chemokine receptor CCR6, expressed by dendritic cells and T cells, is utilized also by human β-defensin-2 peptide [56], supporting the hypothesis that AMPs create a bridge between innate and adaptive immune system.

AMPs protect all human body sites that are continually exposed to microbes, like the skin and mucous membranes, since they are produced by multiple immune and epithelial cells (Table 1). Their expression may be constitutive and some cells (e.g., neutrophils) store a high number of AMPs, or the expression is induced by various microbial or the host stimuli. As a consequence, each tissue has its own profile of different AMPs that may vary significantly depending on the actual host condition. It is tempting to name it as a "peptidiome" using an analogy to microbiome bacteria within a given body habitat. Therefore, (1) a synergism of AMPs activity, supported by their (2) accumulation, e.g., in neutrophil extracellular traps (NETs) (see below), as well as (3) enhanced expression, may explain the insufficient microbicidal concentration issue observed at the basal physiological background. On the other hand, this effect may just be a derivative of inadequate in vitro MIC testing methods. For example, Dorschner et al. [57] showed that cultivation of bacteria, like *Staphylococcus aureus* and *Escherichia coli*, in a medium mimicking the mammalian ionic environment, i.e., carbonate-containing solutions, causes changes in their cell wall thickness and an altered gene expression pattern, that in turn increased susceptibility to AMPs. Furthermore, it is possible that in the skin or inside phagocytic cells, i.e., body niches where the level of the AMPs-inhibiting factors is minor, this antagonism is not significant.

2.1. Human Defensin and Cathelicidin (LL-37) Peptides

Defensins are cysteine-rich peptides classified based on configuration of the disulfide bonds between six cysteine residues into α-, β-, and θ-defensins; however, in humans, the latter exist only as pseudogenes [58]. From the evolutionary perspective, β-defensins are the common ancestor of all vertebrate defensins, and α-defensins are mammalian-specific genes co-located with β-defensin ones on adjacent loci on human chromosome 8p22–p23 [59,60]. Human α-defensins are produced mainly by neutrophils; hence, they are known as human neutrophil peptides 1–4 (HNP-1, HNP-2, HNP-3, and HNP-4), as well as by Paneth cells of the small intestine (HD5 and HD6) (Table 1) [61,62]. Interestingly, the four HNPs are encoded by three genes, since HNP-2 is a truncated variant of HNP-1 or HNP-3 peptides, lacking the first alanine or aspartic acid residue, respectively [63]. HNP1–4 are constitutively expressed and stored in azurophil granules, where they constitute more than 30% of the protein content; however, HNP-4 is the least abundant [61,64].

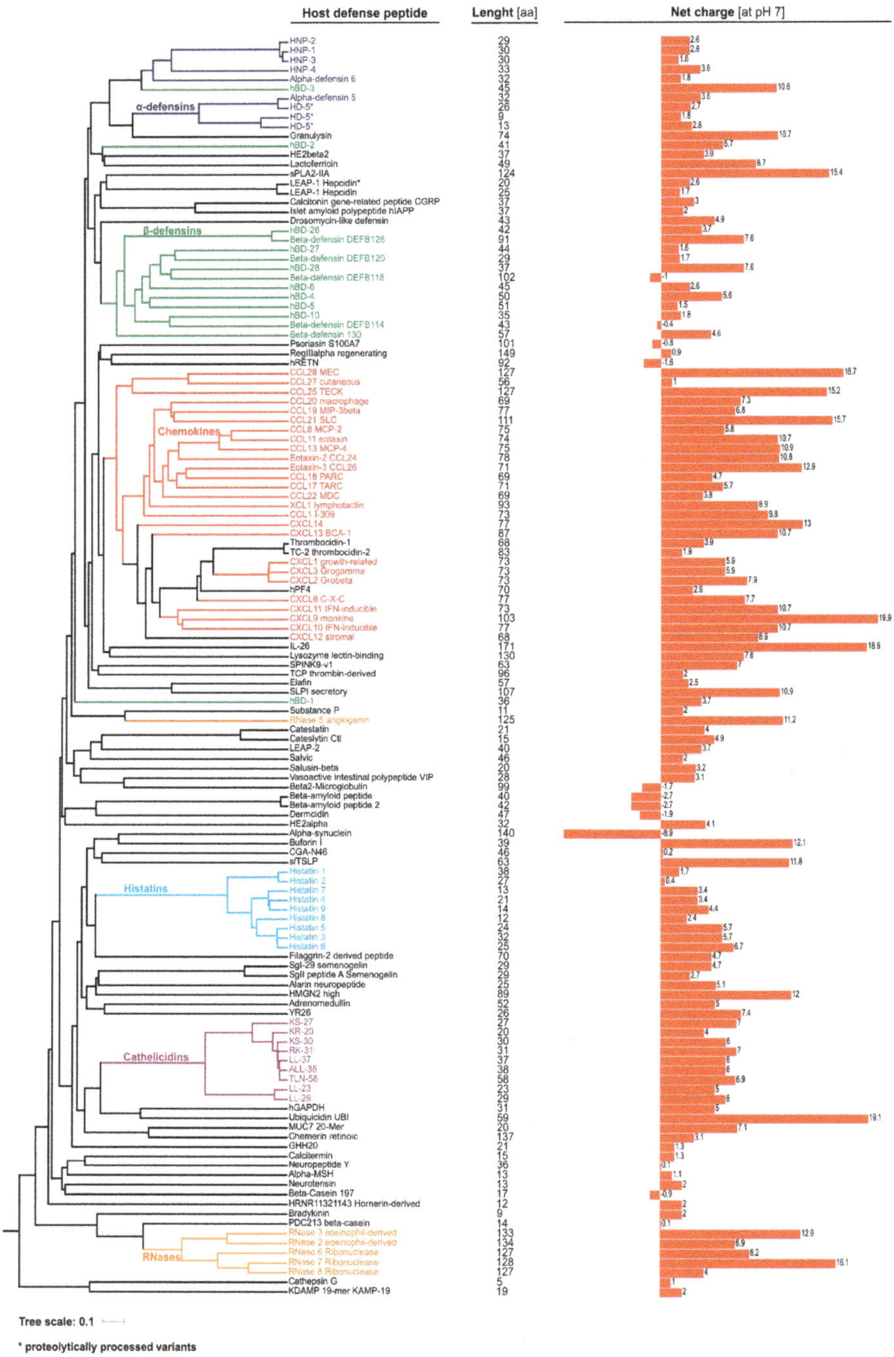

Figure 2. Comparison of human host defense peptides (n = 133) curated in antimicrobial peptide database (http://aps.unmc.edu/AP; accessed in September 2019). The dendrogram was built based on amino acid sequence alignment using MAFFT aligner (https://mafft.cbrc.jp) [65], visualized and annotated with Archaeopteryx [66] and iTOL [67], respectively. Net charge values of the peptides (at pH = 7.0) were estimated using Protein Calculator https://pepcalc.com/protein-calculator.php [68].

In contrast, at least 17 β-defensins (hBDs) have been described, yet hBD1–hBD4 are the best studied [69]. They are produced by various epithelial and mucosal cells, thus protecting body sites directly exposed to microbes, such as respiratory, intestinal, and genitourinary tracts, as well as skin (Table 1), where their expression may be constitutive or inducible. For example, expression of the *hBD-1* gene is essentially constitutive, whereas expression of the *hBD2-4* genes is infection-related or triggered by host-derived stimuli [70]. Remarkably, the microbicidal effect of hBD-3 peptide is not weakened in the presence of the physiological salt concentration found in mucus, which enables it to have a substantially strong anti-HIV effect [71]. In addition, ß-defensin genes *(DEFB4, DEFB103,* and *DEFB104*) have a high degree of copy-number variation (CNV), ranging from 2 to 12 copies per diploid genome [72], which affects their expression level.

Cathelicidins were named based on a conserved cathelin-like domain connected with a C-terminal antimicrobial domain, and are produced mainly by leucocytes and epithelial cells [73,74]. In the human genome, only one cathelicidin gene (*CAMP*) is present. Nevertheless, as the result of proteolytic cleavage by various proteases of its product, i.e., hCAP-18 (human cationic antimicrobial protein 18 kDa), several cathelicidin peptide variants are generated (Figure 2). In detail, in the first step, hCAP-18 is processed by protease 3 to the full-length active peptide LL-37 (leucine–leucine 37 aa), which in turn is cleaved into shorter variants by tissue-specific proteases. In the skin, serine proteases from the kallikrein family, SCTE (stratum corneum tryptic enzyme; kalikrein 5) and SCCE (stratum corneum chymotryptic enzyme; kalikrein 7), generate peptides KS30, KS22, LL29, and RK31 and KR20, respectively [75]. In fact, in the skin, LL-37 accounts for less than 20% of all cathelicidin variants. Interestingly, KS30, KS22, and LL29 are characterized by stronger antimicrobial activity, but lack of chemotactic properties. On the other hand, RK31 and KR20 peptides possess weak antibacterial but strong antifungal activity. Recently, also the TLN-58 variant, possibly generated by neutrophil elastase (ELA2), has been found in the skin palmoplantar pustulosis (PPP) vesicles [76]. Furthermore, since hCAP-18 is present in semen, a longer, by an additional alanine residue, peptide ALL-38 is produced as the result of action of prostate-derived protease, gastricsin, under acid vaginal pH conditions.

2.2. Other Host Antimicrobial Peptides

Besides the classical antimicrobial peptides, there is an array of small proteins regulating immunomodulatory and antimicrobial functions against a broad range of pathogens. For instance, histatin, lysozyme, hepcidin, thrombocidin-1, neuropeptide α-MSH, RNase 7, RNase 5, and dermcidin are inherently expressed in specific tissues and cells (Table 1). Briefly, histatins 1, 3, and 5 belong to a family of salivary peptides that help to maintain the human oral mucosa, along with the β-defensins. An elevated expression of histatin 5 is detected in the saliva of children with a high level of dental cavities harboring specific bacterial species, such as *Streptococcus mutans, S. sanguinis, S. mitis,* as well as *Lactobacillus rhamnosus* in the oral environment [77,78]. In contrast, RNAse 7 is abundantly found within specialized uroepithelial cells in bladder lining, ureters, and kidneys, protecting the urinary system from invading microbes. This peptide exhibits a significant role in maintaining a bacteria-free bladder, as it inhibits the microbial activity of various drug-resistant microbes, including *Klebsiella pneumoniae, Pseudomonas aeruginosa,* and vancomycin-resistant *Enterococcus faecium* [79]. Another important antimicrobial peptide synthesized in the liver is hepcidin. While its primary function involves maintenance of iron absorption and transport, hepcidin also exhibits strong antimicrobial activity. During inflammatory conditions, hepcidin mRNA expression is highly stimulated by the cytokines IL-6, IL-1α and IL-1β, which modulates host response [80].

3. Role of Host Defense Peptides in Inflammation

Over the years, our view on antimicrobial peptides (AMPs) has evolved from just endogenous antibiotics into multifunctional agents (HDPs), which execute their antimicrobial tasks at the same time as participating in a pro-inflammatory response and, if required, mediating its suppression. Currently, HDPs are perceived as factors contributing either to efficient clearance of infections or resolution of the infected sites. To illustrate, these peptides not only attract immune cells, e.g., neutrophils, but also by blocking apoptosis prolong their lifespan, and in turn phagocytic functions [81]. On the other hand, HDPs may function as a "molecular brake" on macrophage-driven inflammation to maximize eradication of pathogens with minimal adverse effects on surrounding tissues [82].

Furthermore, HDPs are essential for proper host–microbiota interactions. In this context, HDPs serve as a buffer, maintaining immune homeostasis via neutralization of pro-inflammatory MAMPs, e.g., lipopolysaccharides (LPS) and lipoteichoic acid (LTA), constantly released by microbiota, as well as a factor shaping its composition, hence protecting from dysbiosis [83]. For example, LL-37 inhibits the expression of specific pro-inflammatory genes up-regulated by NF-κB in the presence of LPS, unlike to LPS-induced genes which antagonize inflammation and certain chemokine genes classically considered pro-inflammatory [84]. On the other hand, the microbiota are a key factor in stimulating production of HDPs, as supported in a classical experiment by Mangoni et al. [85], showing that the presence of HDPs in frog skin (*Rana esculenta*) is microbiota-dependent, and frogs living in a sterile, i.e., the microbiota-free, environment do not synthesize antimicrobial peptides.

Importantly, HDPs inhibit not only pro-inflammatory action of exogenous PAMPs, but also endogenous ones, like DAMPs (also known as "alarmins" or "danger signals"), which are expressed in stressed or dying cells and convey alarm signals to the immune system, including those responsible for autoimmune disorders [13]. For instance, in the skin cathelicidin peptides block release of cytokines induced by the alarmin hyaluronan, thus their low expression may be a risk factor of the development of atopic dermatitis [86]. Initially, the term DAMPs involves factors from various cell/tissue compartments, such as extracellular matrix (hyaluronan, heparan sulfate, eDNA), cytoplasm (heat shock proteins: HSP60, HSP70, HSP90, HSP27; calcium-binding proteins: S100A8, S100A9, S100A12; β-Galactoside binding lectins: Galectin-1, Galectin-3; and uric acid), mitochondria (mitochondrial DNA, ATP, N-formylated peptides), and other subcellular organelles (HMGB1, IL-33, IL-1α, Calreticulin). However, currently, several HDPs, e.g., cathelicidins and α- and β-defensins, are also classified as alarmins [14], which in fact are frequently co-expressed with DAMPs (Figure 3).

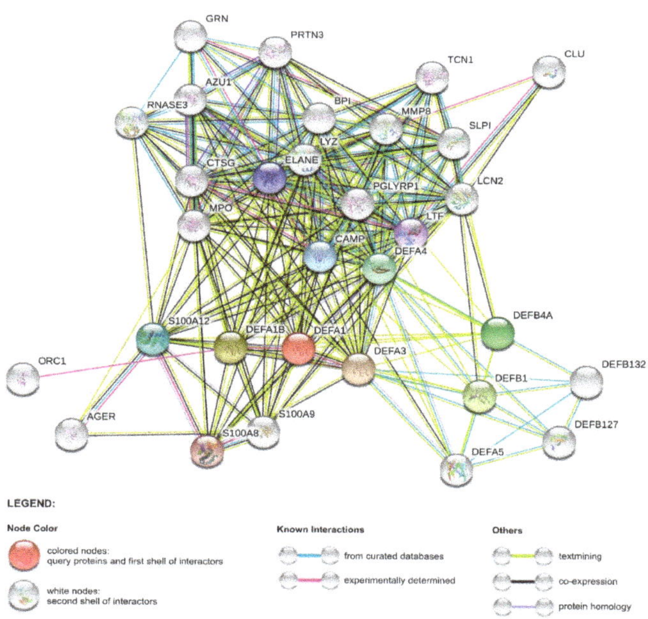

Figure 3. Network of interactions between α-defensin 1 gene (*DEFA-1*) and other human proteins (the network was obtained from STRING v11 database) [87]. Represented proteins are central to antimicrobial

and immunomodulatory activities. Abbreviations: AGER, advanced glycosylation end-product-specific receptor; BPI, bactericidal permeability increasing protein; CAMP, cathelicidin antimicrobial peptide; CLU, clusterin; CTSG, cathepsin; DEFA3, defensin alpha 3; DEFA4, defensin alpha 4; DEFA5, defensin alpha 5; DEFA1B, defensin alpha 1B; DEFB4A, defensin beta 4A; DEFB1, defensin beta 1; DEFB132, defensin beta 132; DEFB127, defensin beta 127; ELANE, neutrophil elastase; GRN, granulin precursor; LCN2, lipocalin 2; LTF, lactotransferrin; LYZ, lysozyme; MPO, myeloperoxidase; MMP8, matrix metallopeptidase 8; ORC1, origin recognition complex subunit 1; PGLYRP1, peptidoglycan recognition protein 1; PRTN3, proteinase 3; RNASE3, ribonuclease A family member 3; SLP I, secretory leukocyte peptidase inhibitor; S100A8, S100 calcium binding protein A8; S100A9, S100 calcium binding protein A9; S100A12, S100 calcium binding protein A12; TCN1, transcobalamin 1.

Indeed, a link between HDPs and multiple autoinflammatory diseases such as skin disease (atopic dermatitis, psoriasis, rosacea) or microbiota-related ones, e.g., IBD (Crohn's disease, colitis ulcerosa), acne vulgaris, and periodontitis, has been established by several studies (see below). An enhancement of Th17 response by HDPs may serve as an example. Briefly, HDPs efficiently attract Th17 (T helper 17 cells), which in turn secrete pro-inflammatory cytokines, IL-17A, IL-17F, IL-21, and IL-22, responsible for mounting mucosal defense against pathogenic microbes in the respiratory or intestinal tract. For instance, IL-17A and IL-22 work synergistically to induce certain β-defensins hBD-1, hBD-3, and hBD-4 in both human and primary mouse gastric epithelial cells (GEC) and gastroids co-cultured with *Helicobacter pylori* [88]. On the other hand, an elevated level of Th17 cells has been connected with various autoimmune diseases, such as systemic lupus erythematosus, rheumatoid arthritis, or psoriasis [89].

Also, genetically-mediated deficiency/excess of HDPs, gene sequence polymorphisms, as well as disturbed expression may be a risk factor in inflammatory diseases. For instance, Hollox et al. [90] showed a significant association between higher genomic copy numbers for β-defensin genes, ranging from 2 to 7 copies, and the relative risk of developing psoriasis. Likewise, a lower the *hBD-2* gene copy number in the β-defensin locus predisposes to colonic Crohn's disease [91]. Recently, experimental evidence has highlighted the genetic association between the clinical phenotype of sepsis and *DEFA-1/DEFA-3* copy number. Transgenic mice models were engineered to produce a high gene copy number of *DEFA-1/DEFA-3*, which manipulated the outcome of sepsis progression [92]. The consequential effect was compared to the low gene copy number wild-type mice models, in that the former showed chronic inflammation, endothelial cell damage, vascular leakage, severe organ injury, and mortality. Thus, treatment of patients with sepsis can be challenging due to the underlying individual genetic associations. However, further research is needed to obtain conclusive data. In addition, single nucleotide polymorphisms (SNPs) of the *hBD-1* gene was connected with the pathogenesis of inflammatory bowel diseases and chronic gastritis [93], as well as oral diseases [94].

In general, expression of HDPs is enhanced during infection or inflammation through transcription factors initialized by pro-inflammatory cytokines or signaling pathways associated with activation of PRRs, e.g., Toll-like receptors (TLRs). For instance, promoter regions of α- and β-defensin genes contain binding sites for major cellular transcription factors, notably nuclear factor κB (NF-κB) and activator protein 1 (AP-1) (Figure 1). It should be noted that NF-κB also plays a crucial role in the pathogenesis of Crohn's disease, along with many other pro-inflammatory molecules that modulate the hBD-2 expression [95], as well as in triggering its production (and IL-6) in severe sepsis [96]. Moreover, the expression of genes encoding LL-37 (and hBD-2) is modulated by vitamin D3 via binding with specific DNA sequences in their promoters, the so-called vitamin D response elements (VDRE) [97]. Additionally, a recent in silico analysis identified a wide range of transcription factors which possibly bind and modulate the gene transcription of many antimicrobial peptides and proteins, such as LL-37, RNAse1, CCL18, CXCL14, and HTN1 [98].

In line with this, it has been established that DNA methylation of the CpG sites in the 5′ flanking region of the *hBD-1* gene contributes to its deficiency in patients with atopic dermatitis [99]. Furthermore, point mutations in the promoter region of *hBD-1* explain a cancer-specific loss of this peptide in 90% and 82% of renal cell carcinomas and prostate [100]. Thus, *hBD-1* was suggested as a potential tumor suppressor gene for urological cancers. Also, in oral squamous cell carcinoma (OSCC), hBD-1 appears to have anti-tumor properties, while *hBD-2* and *hBD-3* might be proto-oncogenes [101].

Nonetheless, the relation between HDPs and inflammation is not always straightforward, and either their deficiency or overproduction, as well as a balance between pro- or anti-inflammatory effects, may contribute to the pathological inflammatory response. For instance, in atopic dermatitis (AD), despite severe skin inflammation, the expression level of major skin HDPs, dermcidin, LL-37, hBD-2, and hBD-3, is not increased, hence patients with AD are more prone to skin infections and have altered skin colonization patterns. By contrast, in psoriasis, expression of LL-37, hBD-2, and hBD-3 is elevated, hence skin infection is rare. Nevertheless, LL-37 and hBDs are considered as a major driving force of inflammation in psoriasis by mechanisms involving increased production of IFN-α and activation of pDCs, respectively [102]. In addition, these peptides stimulating degranulation of mast cells and increasing production of the pruritogenic cytokine IL-31 may escalate itching (pruritus) manifestation [103]. However, it has been recently observed that LL-37 may also act as an anti-inflammatory agent by blocking the release of inflammatory cytokine IL-1β, depending on its concentration. Interestingly, this observation possibly explains the mechanism underlying the paradoxical effectiveness of vitamin D3, i.e., inducer of LL-37 expression, in treatment of psoriasis [11]. Similarly, hBD-3 may inhibit inflammation by inducing expression of anti-inflammatory cytokine IL-37 in keratinocytes [104]. An elevated level of cathelicidin is also observed in other inflammatory skin conditions, namely rosacea and palmoplantar pustulosis, but instead of the native form of LL-37, its proteolytically cleaved variants drive the inflammation [76,105].

Considering the above, the final contribution of HDPs to inflammatory processes is a derivative of multiple variables related to their expression, processing, concentration, and combination, as well as reciprocal interactions with the remaining components of the host immune system. For instance, pro- and anti-inflammatory effects of cathelicidin LL-37 are concentration-dependent, i.e., the former is visible at >20 μg/mL, whereas the latter at 1–5 μg/mL. Similarly, at the concentration range 1–100 ng/mL, β-defensins can act as chemokines only, since other immunomodulatory functions are not visible [10]. Furthermore, there are more than 20 antimicrobial peptides in the skin characterized by different expression levels (Figure 4) [106]. Therefore, the relationship between HDPs and inflammation appears to be strongly context-dependent (e.g., inflammation site, type of cells, or stimuli), and as such, it should be analyzed on a multidimensional level, rather than as a single action of individual peptides. Otherwise, valid conclusions regarding the ultimate role of HDPs in inflammation may be difficult to draw. In the next paragraph, we discuss mechanisms behind anti- and pro-inflammatory actions of HDPs.

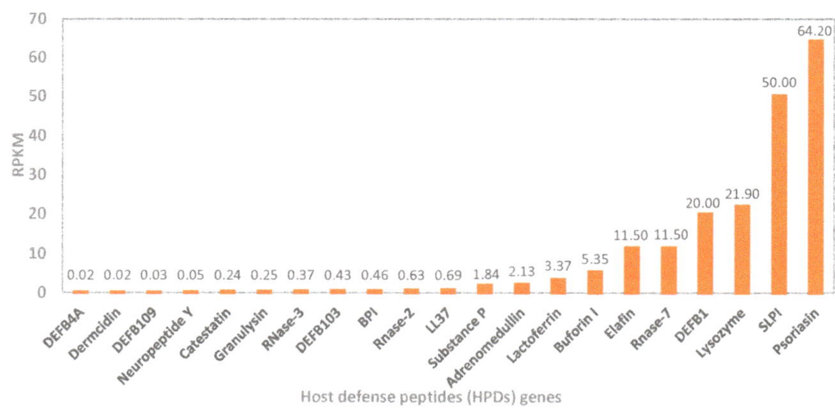

Figure 4. Relative expression of selected host defense peptide (HDP) genes in the human skin, based on RNA-seq analysis of tissue samples from 95 human individuals performed by Fagerberg et al. [107]. The data were obtained from the GenBank Bio Project: PRJEB4337 (RPKM, reads per kilobase per million reads).

4. Molecular Mechanisms of Anti- and Pro-Inflammatory Action of HDPs

The immunomodulatory potential of HDPs is strictly connected with their ability to recruit and activate immune and non-immune cells, as well as a direct or indirect impact on their fate, including maturation, differentiation, degranulation, or apoptosis [10]. This is mediated through interaction with a wide range of membrane-bound and intracellular receptors, followed by stimulation of their downstream signaling pathways. So far, HDPs have been recognized to interact with the following receptors: (1) Pattern recognition receptors (PRRs), (2) purinergic G-protein coupled receptors (formyl peptide receptor like-1), (3) P2X7 receptor, (4) MRGPRX2, (5) chemokine receptors (commonly known CCR2, CCR6), (6) epidermal growth factor receptors (tyrosine kinases), (7) integrin family receptors (macrophage-1 antigen), nucleotide oligomerization domain (NOD) proteins, and NODlike receptors (NLRs), and their number is still growing [108,109]. Hence, HDPs modulate immune responses using the same receptors as MAMPs/PAMPs and DAMPs [110,111]. To illustrate, β-defensins attract cells by interaction mainly with CCR2 and CCR6 receptors, and exert their "alarmin" activity, e.g., induction of cytokine production, via TLRs, EGFR, GPCR, and MrgX2 ones; however, both activities may overlap in one receptor.

It should be noted that TLR receptors may be the root cause of certain HDP-associated diseases. For instance, in individuals with rosacea, a higher expression of TLR-2 sensitizes the facial skin to microbes or environmental stimuli. Under these conditions, enhanced expression of kallikrein-5 proteinase is observed in keratinocytes, and ultimately affects production of cathelicidin peptides, which drives inflammation and abnormal growth of blood vessels [105]. Moreover, the tumor-suppressing effect of hBD-1 is associated with its ability to modulate epidermal growth factor and human epidermal growth factor receptor 2 (EGFR/HER2)-associated signaling pathways [112]. Finally, hBD-3, through deactivation of TLR-4 and TLR-2, may reduce the adverse immune reaction initialized by NF-κB in response to LPS [113].

Furthermore, β-defensins and cathelicidins, in the same manner as PAMPs (e.g., LPS) and DAMPs (e.g., heat shock antigens Hsp60 and Hsp70), are ligands of Toll-like receptor 4 (TLR-4). However, the resulting outcome of the receptor's stimulation may be different for these molecules. For instance, unlike LPS, hBD-3 does not induce production of IL-10, which is an important anti-inflammatory cytokine, e.g., via suppressing function of antigen-presenting cells (APCs), suggesting that hBD-3 can shift the immune response toward pro-inflammatory direction [114]. Similarly, hBD-2 via TLR-4 leads to maturation of dendritic cells (DCs), which consequently exhibit Th1-polarized responses, such as the production of pro-inflammatory cytokines IL-12, IL-1α, IL-1β, and IL-6, which may possibly counter suppressive action of microbial factors by generating more robust host inflammatory and Th1 responses [115]. In contrast, cathelicidin is considered as an inhibitor of TLR-4, and thus can antagonize with other TLR-4 ligands released during skin injury, e.g., hyaluronan [10].

Therefore, HPDs joining properties of MAMPs/PAMPs and DAMPs may operate as central nodes in a network that coordinates immune response to infections as well as non-infectious insults. For instance, it has been shown that synergistic action of MAMPs/PAMPs and DAMPs is necessary for synthesis and subsequent secretion of pro-inflammatory cytokine IL-1β [13]. It is important to note that a lack of IL-1β results in high susceptibility to infections, but its overproduction causes uncontrolled inflammation and tissue damage via T cell-mediated autoinflammatory response [116]. Indeed, overproduction of IL-1β was observed in patients with inflammatory bowel disease, and has been connected with deficiency of α-defensins that serve as regulators of IL-1β maturation [117]. As aforementioned, also in psoriasis, LL-37 may act as an inhibitor of the IL-1β release in keratinocytes by blocking activation of the cytosolic DNA-sensing signaling AIM2, i.e., cytosolic receptor for dsDNA. Hence, cytoplasmic DNA appears to contribute to the pathogenesis of psoriasis via activation of IL-1β in keratinocytes by AIM2-mediated inflammasomes [11].

In this context, it is relevant to mention the relationship of HDPs and self-nucleic acids, and its impact on inflammation. Under normal homeostatic conditions, the host-derived nucleic acids released from damaged and dying cells do not mediate inflammatory responses because of the systematic regulation and physiological location of nucleic acid sensing TLR7/9. However, several studies have shown that HDPs may disturb immune tolerance to self-nucleic acids, and in turn significantly enhance cell responses—in particular, plasmacytoid dendritic cells (pDCs) [118]. In fact, in the skin, this mechanism has been identified as an important initiator of psoriasis development, where LL-37 and defensins are able to condense self-DNA into particles, which are internalized by pDCs, inducing robust IFN-α response via activation of the TLR-9 signaling pathway [119,120]. This enhances production of large amounts of type I IFN, leading to the functional activation of myeloid dendritic cells (mDCs), monocytes, NK cells, keratinocytes, as well as Th1/Th17 differentiation, which further increase the pro-inflammatory, e.g., IFN-γ, IL-22, and IL-17, cytokine expression [120,121]. In line with this, a novel mechanism of nucleic acid recognition by LL-37 utilizing cell surface RNA scavenger receptors (SRs) has been described [122], which results is enhanced clathrin-dependent endocytosis, facilitating the overproduction of inflammatory cytokines and chemokines. Recently also RNase7 was found to utilize plasmacytoid dendritic cell (pDC) TLR-9 signaling mode of IFN-α activation even more strongly than LL-37, emphasizing its crucial role in autoimmune inflammatory skin diseases [123]. Interestingly, other antimicrobial peptides expressed in the skin, such as psoriasin, elafin, or hBD-1, lack the ability of interacting with the host nucleic acids, which may be related to their lower net charge (Figure 2) [120].

Another interesting consequence of interactions between HDPs and the host nucleic acids is a novel wound healing mechanism, where LL-37 may alter wound repair by modifying the responses to dsRNAs released as a result of skin injury. In detail, LL-37 enhances endosomal uptake of non-coding double stranded RNA in TLR-3-mediated mechanism, that results in activation of several important wound repair growth factors, including fibroblast growth factor (FGF2), and heparin binding EGF-like growth factor (HBEGF) from the dermal keratinocytes and fibroblasts [124]. Inhibition of LL-37/dsRNA relation may contribute to the development of hyperproliferation-based diseases, like psoriasis, whereas its augmentation can lead to increased wound regeneration in pathological conditions of abnormal wound repair (e.g., diabetic ulcers).

Finally, LL-37 actively participates in neutrophil extracellular trap (NET) formation via disruption of the nuclear membrane and promotes their stability [125]. NETs are structures composed of decondensed chromatin and multiple enzymes (elastase, myeloperoxidase, gelatinase, etc.) and proteins, including antimicrobial ones. Thus, NETs act as a mechanical barrier that entraps and subsequently reduces spreading of pathogens and/or their toxic products into the host tissues, where the antimicrobial activity of HPDs is boosted by their accumulation and combination [126]. Moreover, Stephan et al. have shown that complexes of LL-37/DNA formed inside human macrophages may participate in defense against intracellular bacteria, e.g., mycobacteria [127]. Accordingly, a recent study investigated the therapeutic potential of LL-37 in modulating macrophage-mediated excessive inflammatory responses. It was found that LL-37 reduced the severity of tuberculosis by rapidly enhancing the anti-inflammatory cytokine TGF-β, IL-10, and prostaglandin E from the infected macrophages [128]. However, further studies regarding the exogenous effect of LL-37 in severe pulmonary tuberculosis are warranted. Interestingly, administration of vitamin D3 or another potent inducer of LL-37, i.e., 4-phenyl butyrate (PBA), may be an alternative treatment method of tuberculosis [129].

5. Deregulations of HDPs Expression in Selected Diseases

5.1. Periodontal Diseases

An imbalanced unhealthy oral microbiota ushers the entry of various cariogenic, periodontal microbes which engenders oral biofilm formation and periodontal diseases such as gingivitis and periodontitis. The oral epithelial tissues, mainly the gingival epithelium, play a significant role in resisting the colonization of unfavorable oral pathogens. These tissues readily secrete beta-defensin peptides, as well as histatins, which are the major host defense proteins of the saliva that maintain homeostasis of oral microbiota [130]. A significant correlation is observed between elevated levels of hBD-2, hBD-4, and HNP4 in the oral mucosal epithelial cells of both adults and children with the development of dental caries. They are considered as important clinical biomarkers of periodontal diseases and dental caries. It has been shown that there is a declined expression of beta-defensin 1 mRNA gene in the inflamed gingival tissues and periodontal structures. Conversely, chronic cases of periodontitis manifest an elevated expression of the *hBD-1* gene [131,132]. The severity of periodontal diseases and dental plaques grows, along with a heightened expression of hBD-2 and histatin-5. Higher activity of pro-inflammatory cytokines modulates the progression of the infection, which further stimulates the production of the defensin peptides through various transcription factors [133]. Recently, 89 patients were monitored according to their periodontal status in relation to other clinical parameters [134]. This study identified an increase in the salivary production of hBD-2, triggered by inflammatory processes and pathogen derived metabolites that can be considered as a possible diagnostic biomarker s of periodontal diseases.

Healing of periodontal lesions is initiated by various growth factors, pro-inflammatory mediators, and antimicrobial peptides accumulating at the infected site. In detail, a complex network of highly specialized growth factors, namely, insulin-like growth factor (IGF1, IGF2), transforming growth factor (TGF-α, TGF-β), epidermal growth factor, and platelet-derived growth factor, coordinates the reparative process by rapid differentiation of keratinocytes and fibroblasts [135]. These growth factors also assist the wound healing mechanism by influencing the gene expression pattern of antimicrobial peptides that typically participate in the epithelial cell proliferation, migration, and inhibition of colonizing microbial pathogens at the site of injury. While it is established that wounding influences the expression of HDPs, not all of them function the same way. Recent reports have highlighted the distinct immune responses triggered within the wounded gingival epithelial cells (GECs) and gingival fibroblasts (HGFs) upon treatment with IGF1 and TGF-α. These growth factors enabled efficient wound closure and differently modulated the expression of hBD-2, CCL20, IL-1, and IL-8. The findings indicate that hBD-2 was exclusively enhanced in the gingival epithelial cells measured at set time points of 6 h and 24 h post-wounding, particularly in those cells associated with the keratinocyte differentiation marker involucrin. Additionally, hBD-2 along with CCL20, IL-1, and IL-8 control the invasion of bacterial microbes and impact the neutrophil defense mechanisms [136]. Contrarily, the wounded gingival fibroblasts (HGFs) witnessed a substantially low expression pattern of hBD-2 and CCL20, with or without growth factor treatment, that was suggested as a mechanism protecting fibroblast overgrowth into the epithelial wound [136]. fibroblast overgrowth into the epithelial wound.

5.2. Inflammatory Lung Diseases

Cystic fibrosis (CF) is a life-limiting disease characterized by recurrent respiratory infections and inflammation, connected to altered composition and volume of the airway surface liquid (ASL). For instance, a reduced bicarbonate HCO_3^- secretion resulting in a decrease of airway surface pH (average 6.8–7.5) was observed. Interestingly, it was also found that the acidic pH weakened the action of LL-37 and hBD-3 against invasive *Staphylococcus aureus* and *Pseudomonas aeruginosa* infections by affecting their structural net charge. Therefore, it could be suggested that a similar mechanism of acidic pH-reduced antimicrobial activity may occur in other inflammatory conditions taking place in cerebral spinal fluid, peritoneal fluid, and pleural fluid. Additionally, it has been noted that high ionic strength (Na^+, K^+, Cl^+) may impair the antimicrobial activity of hBD-2, lysozyme, and lactoferrin [137]. Moreover, CF patients suffer from viscous sputum that accumulates and obstructs their airways. The thick mucus is characterized by heterogenous complex aggregates of DNA and F-actin filaments derived from leukocytes that have encountered necrotic death. Thus, the antimicrobial function of LL-37, lysozyme, lactoferrin, and hBD-3 released in the respiratory airways is substantially hindered as they stabilize DNA/F-actin bundles. Additionally, neutralization of the immune function of neutrophil protease and IL-8 take place during DNA/F-actin bundles formation [138,139]. It is also worthwhile to underline that abundant secretion of cysteine cathepsins from the macrophages hinders the functional expression of hBD-2.

Chronic obstructive pulmonary disease (COPD), bronchitis, and asthma are all characterized by inflammation that develops as a consequence of pro-inflammatory mediator secretion. Immune cells distributed throughout the lungs are responsible for sudden exacerbations associated with the production of cytokines, oxidative stress, and protease secretion, including caspases, neutrophil elastase, and matrix metalloproteinases. One study reports the enhanced expression of hBD-2 in the distal airway epithelial cells of COPD patients, but a rather diminished expression of hBD-2 in the central airways, despite the exaggerated expression of TLR-4 receptors [140]. This distinct variation was found to be in correlation with exposure to cigarette smoking. While it is evident that every cell in the body requires ATP for its biological function of energy production and retention, little is known about its possible involvement in the immune system response to bacterial infection and inflammation. In a *P. aeruginosa*-infected rat model, ATP administration led to rapid stimulation of hBD-2 production. The mechanism of ATP action involved NADPH family of oxidases (DUOX 1) via ion channel receptors P2X, P2Y activation, and regulation of multiple signaling pathways ERK1/2 and NF-κB [141]. The released defensin peptide was found to control the inflammatory processes underlying the acute infection of pneumonia by suppression of TNF-α and IL-6. Furthermore, another study detected the potent ability of the IL-17 family of cytokines in the induction mechanism of the *hBD-2* gene. Typically, most of the immune cells, including T helper cells, macrophages, dendritic cells, and natural killer cells, secrete IL-17 family of cytokines. These cytokines act in concordance with the tumor necrosis factor and IL-1 to promote the induction of other inflammatory mediators production, which individually or collectively can stimulate the secretion of beta-defensins via the activation of various signaling pathways [142,143]. For example, the stimulatory functions of IL-17 in the airway epithelial cells promoting transcription of the *hBD-2* gene through the action of JAK and NF-κB signaling have been reported [144]. Moreover, IL-17 has an impact on other cytokines such as IL-1α, IL-β, IL-6, IL-7, and TNF-α, which contribute to the production of hBD-2. On the other hand, the alveolar macrophages and dendritic cells consistently maintain the release of IL-22. According to a recent study, in which alveolar epithelial cells (A549) were screened for the abundant display of IL-22 receptors and subjected to treatment with different doses of IL-22, an increase of *hBD-2* mRNA transcript synthesis via the STAT3 pathway was observed [145]. Thus, this study revealed a new immunomodulatory role of IL-22 in stimulation of the lung defensins in response to exposure to pathogenic bacteria and viruses. Interestingly, hBD-1 has also emerged as a clinical biomarker of COPD and other inflammatory lung diseases, such as asthma [146]. However, in this case, an altered expression of hBD-1 may be aggravated by gene copy number variations.

5.3. *Inflammatory Bowel Diseases*

The human defensins 5 and 6 (HD5, HD6) are particularly important in preserving the homeostatic equilibrium of the enteric mucosa layer, exhibiting different effects against the essential inducers of their secretion, i.e., various products of the Gram-positive and Gram-negative bacteria [147]. To illustrate, there is a remarkable reduction in the expression levels of HD5 and HD6 by the Paneth cells in inflammatory bowel conditions such as Crohn's disease. This shift in expression could be attributed to the cause by genetic changes in the NOD2 receptor [148]. Furthermore, a recent study suggests the possibility of using other HDPs, such as the level of fecal HNP, as a non-invasive biomarker of intestinal inflammation in patients suffering from colitis ulcerosa [149].

The human beta-defensins are also naturally expressed in the epithelial cells of the gastric mucosa and extensively participate in host defense against *Helicobacter pylori* colonization, a bacterium present in a high proportion (~80%) of people throughout the world [150]. Multilevel signaling pathways promote the molecular mechanism of induction of beta-defensins in response to the initial stages of *H. pylori* infection. In addition, it has been shown that the phosphorylation of a serine residue of EGFR may modulate the release of hBD-3 [151], revealingan underlying interdependent relation between the stimulated transforming growth factor β-activated kinase-1 (TAK1), p38α pathway, and phosphorylation of EGFR receptor in the amplified release of hBD-3 in the gastric mucosa involved in *H. pylori* infection.

6. Conclusion

The multifunctional host defense peptides provide a link between innate and adaptive immunity against different microorganisms and contribute to inflammation of infected sites. Depending on the cell type and extracellular environment, some of these peptides exert contrasting functions, wherein they promote or suppress inflammatory processes. A strongly compromised action of host defense peptides against intruders and delayed resolution of inflammatory mediators underlies the development of inflammation in different diseases. Some of these peptides may serve as potential clinical biomarkers for a wide range of inflammatory diseases. Evidently, antimicrobial regulation is crucial to limit the exacerbation of inflammatory signaling molecules. While various factors govern the release of HDPs, any dysregulation can favor an imbalanced feedback mechanism between the host-induced anti-inflammatory and pro-inflammatory processes. In summary, a deeper understanding of the diverse functional roles of HDPs in the body's physiological response to inflammation and disease is crucial and represents the first approach to develop new therapeutic strategies based on HDPs aimed at resolving the progression of inflammatory diseases and strengthening the host barrier defenses.

Author Contributions: Conceptualization, writing, and original draft preparation, S.V.P.; conceptualization, writing, figure preparation, and review and editing, K.F.; writing and figure preparation, T.D.; conceptualization and writing, E.P.; conceptualization and review and editing, R.B. All authors have read and agreed to the published version of the manuscript.

Funding: This work was financially supported by the National Science Center, Poland under Grant: UMO-2018/30/M/NZ6/00502 (to RB). This work was conducted within a project which received funding from the European Union's Horizon 2020 research and innovation programme under the Marie Skłodowska-Curie grant agreement No. 754432 and the Polish Ministry of Science and Higher Education, from financial resources for science in 2018–2023 granted for the implementation of an international co-financed project.

Conflicts of Interest: The authors declare no conflict of interest. The funders had no role in the design of the study, in the collection, analyses, or interpretation of data; in the writing of the manuscript; or in the decision to publish.

Abbreviations

AMPs	Antimicrobial peptides
ATP	Adenosine triphosphate
BM	Bone morphogenetic protein
CAMP	Cathelicidin antimicrobial peptide
CCR2	C–C chemokine receptor type 2
CCR6	C–C chemokine receptor type 6
CNV	Copy number variation
COPD	Chronic obstructive pulmonary disease
DAMPs	Damage-associated molecular patterns
DC	Dendritic cell
DUOX1	Dual oxidase 1
EGFR	Epidermal growth factor receptor
ELA2	Neutrophil elastase 2
FGF2	Fibroblast growth factor
GEC	Gingival epithelial cell
HBEGF	Heparin binding EGF like growth factor
HER2	Human epidermal growth factor receptor 2
HGFs	Human gingival fibroblasts
HMGB1	High mobility group box 1
IDR	Innate defense regulator

IGF	Insulin like growth factor
IL-1α	Interleukin-1α
IL-33	Interleukin-33
JAK	Janus Kinase
LPS	Lipopolysaccharides
LTA	Lipoteichoic acid
MAPK	Mitogen-activated protein kinase
MIC	Minimum inhibitory concentration
MIP–3	Macrophage inflammatory protein-3 alpha
MRGPRX2	Mas-related G-protein coupled receptor member X2
NETs	Neutrophil extracellular traps
NF-κB	Nuclear factor kappa-light-chain-enhancer of activated B cells
NOD2	Nucleotide-binding oligomerization domain
OSCC	Oral squamous cell carcinoma
P2X7	Purinoceptor 7
SNP	Single nucleotide polymorphism
STAT	Signal transducer and activator of transcription

References

1. Mills, C.D.; Ley, K.; Buchmann, K.; Canton, J. Sequential Immune Responses: The Weapons of Immunity. *J. Innate Immun.* **2015**, *7*, 443–449. [CrossRef] [PubMed]
2. Chen, L.; Deng, H.; Cui, H.; Fang, J.; Zuo, Z.; Deng, J.; Li, Y.; Wang, X.; Zhao, L. Inflammatory responses and inflammation-associated diseases in organs. *Oncotarget* **2018**, *9*, 7204–7218. [CrossRef] [PubMed]
3. Libby, P. Inflammatory mechanisms: The molecular basis of inflammation and disease. *Nutr. Rev.* **2007**, *65*, S140–S146. [CrossRef]
4. Muller, W.A. Getting leukocytes to the site of inflammation. *Vet. Pathol.* **2013**, *50*, 7–22. [CrossRef] [PubMed]
5. Gupta, S.; Bhatia, G.; Sharma, A.; Saxena, S. Host defense peptides: An insight into the antimicrobial world. *J. Oral. Maxillofac. Pathol.* **2018**, *22*, 239–244. [CrossRef]
6. Hancock, R.E.; Haney, E.F.; Gill, E.E. The immunology of host defence peptides: Beyond antimicrobial activity. *Nat. Rev. Immunol.* **2016**, *16*, 321–334. [CrossRef]
7. Kopp, Z.A.; Jain, U.; Van Limbergen, J.; Stadnyk, A.W. Do antimicrobial peptides and complement collaborate in the intestinal mucosa? *Front. Immunol.* **2015**, *6*, 17. [CrossRef]
8. Bhat, S.; Song, Y.H.; Lawyer, C.; Milner, S.M. Modulation of the complement system by human beta-defensin 2. *J. Burns Wounds* **2007**, *5*, e10.
9. Lee, E.Y.; Lee, M.W.; Wong, G.C.L. Modulation of toll-like receptor signaling by antimicrobial peptides. *Semin. Cell Dev. Biol.* **2019**, *88*, 173–184. [CrossRef]
10. Lai, Y.; Gallo, R.L. AMPed up immunity: How antimicrobial peptides have multiple roles in immune defense. *Trends Immunol.* **2009**, *30*, 131–141. [CrossRef]
11. Dombrowski, Y.; Schauber, J. Cathelicidin LL-37: A defense molecule with a potential role in psoriasis pathogenesis. *Exp. Dermatol.* **2012**, *21*, 327–330. [CrossRef] [PubMed]
12. Haney, E.F.; Straus, S.K.; Hancock, R.E.W. Reassessing the Host Defense Peptide Landscape. *Front. Chem.* **2019**, *7*, 43. [CrossRef] [PubMed]
13. Saïd-Sadier, N.; Ojcius, D.M. Alarmins, inflammasomes and immunity. *Biomed. J.* **2012**, *35*, 437–449. [CrossRef] [PubMed]
14. Pouwels, S.D.; Heijink, I.H.; ten Hacken, N.H.; Vandenabeele, P.; Krysko, D.V.; Nawijn, M.C.; van Oosterhout, A.J. DAMPs activating innate and adaptive immune responses in COPD. *Mucosal. Immunol.* **2014**, *7*, 215–226. [CrossRef] [PubMed]
15. Roh, J.S.; Sohn, D.H. Damage-Associated Molecular Patterns in Inflammatory Diseases. *Immune. Netw.* **2018**, *18*, e27. [CrossRef]
16. Fruitwala, S.; El-Naccache, D.W.; Chang, T.L. Multifaceted immune functions of human defensins and underlying mechanisms. *Semin. Cell Dev. Biol.* **2019**, *88*, 163–172. [CrossRef]

17. Jourdain, M.L.; Pierrard, L.; Kanagaratnam, L.; Velard, F.; Sergheraert, J.; Lefèvre, B.; Gangloff, S.C.; Braux, J. Antimicrobial peptide gene expression in periodontitis patients: A pilot study. *J. Clin. Periodontol.* **2018**, *45*, 524–537. [CrossRef]
18. Uysal, P.; Simsek, G.; Durmus, S.; Sozer, V.; Aksan, H.; Yurt, S.; Cuhadaroglu, C.; Kosar, F.; Gelisgen, R.; Uzun, H. Evaluation of plasma antimicrobial peptide LL-37 and nuclear factor-kappaB levels in stable chronic obstructive pulmonary disease. *Int. J. Chron Obstruct. Pulmon. Dis* **2019**, *14*, 321–330. [CrossRef]
19. Muniz, L.R.; Knosp, C.; Yeretssian, G. Intestinal antimicrobial peptides during homeostasis, infection, and disease. *Front. Immunol.* **2012**, *3*, 310. [CrossRef]
20. Kolbinger, F.; Loesche, C.; Valentin, M.A.; Jiang, X.; Cheng, Y.; Jarvis, P.; Peters, T.; Calonder, C.; Bruin, G.; Polus, F.; et al. β-Defensin 2 is a responsive biomarker of IL-17A-driven skin pathology in patients with psoriasis. *J. Allergy Clin. Immunol.* **2017**, *139*, 923–932. [CrossRef]
21. Xie, G.H.; Chen, Q.X.; Cheng, B.L.; Fang, X.M. Defensins and sepsis. *Biomed. Res. Int.* **2014**, *2014*, 180109. [CrossRef] [PubMed]
22. Joly, S.; Compton, L.M.; Pujol, C.; Kurago, Z.B.; Guthmiller, J.M. Loss of human beta-defensin 1, 2, and 3 expression in oral squamous cell carcinoma. *Oral Microbiol. Immunol.* **2009**, *24*, 353–360. [CrossRef] [PubMed]
23. Aberg, K.M.; Radek, K.A.; Choi, E.H.; Kim, D.K.; Demerjian, M.; Hupe, M.; Kerbleski, J.; Gallo, R.L.; Ganz, T.; Mauro, T.; et al. Psychological stress downregulates epidermal antimicrobial peptide expression and increases severity of cutaneous infections in mice. *J. Clin. Invest.* **2007**, *117*, 3339–3349. [CrossRef] [PubMed]
24. Silva, O.N.; Porto, W.F.; Ribeiro, S.M.; Batista, I.; Franco, O.L. Host-defense peptides and their potential use as biomarkers in human diseases. *Drug Discov. Today* **2018**, *23*, 1666–1671. [CrossRef]
25. Wang, G.; Li, X.; Wang, Z. APD3: The antimicrobial peptide database as a tool for research and education. *Nucleic Acids Res.* **2016**, *44*, D1087–D1093. [CrossRef]
26. Pasupuleti, M.; Schmidtchen, A.; Malmsten, M. Antimicrobial peptides: Key components of the innate immune system. *Crit. Rev. Biotechnol.* **2012**, *32*, 143–171. [CrossRef]
27. Nakatsuji, T.; Gallo, R.L. Antimicrobial peptides: Old molecules with new ideas. *J. Invest. Dermatol.* **2012**, *132*, 887–895. [CrossRef]
28. Semple, F.; Dorin, J.R. beta-Defensins: Multifunctional modulators of infection, inflammation and more? *J. Innate Immun.* **2012**, *4*, 337–348. [CrossRef]
29. Sieprawska-Lupa, M.; Mydel, P.; Krawczyk, K.; Wójcik, K.; Puklo, M.; Lupa, B.; Suder, P.; Silberring, J.; Reed, M.; Pohl, J.; et al. Degradation of human antimicrobial peptide LL-37 by Staphylococcus aureus-derived proteinases. *Antimicrob. Agents Chemother.* **2004**, *48*, 4673–4679. [CrossRef]
30. Barańska-Rybak, W.; Sonesson, A.; Nowicki, R.; Schmidtchen, A. Glycosaminoglycans inhibit the antibacterial activity of LL-37 in biological fluids. *J. Antimicrob. Chemother.* **2006**, *57*, 260–265. [CrossRef]
31. Wang, G. Human antimicrobial peptides and proteins. *Pharmaceuticals* **2014**, *7*, 545–594. [CrossRef] [PubMed]
32. Tomita, T.; Hitomi, S.; Nagase, T.; Matsui, H.; Matsuse, T.; Kimura, S.; Ouchi, Y. Effect of ions on antibacterial activity of human beta defensin 2. *Microbiol. Immunol.* **2000**, *44*, 749–754. [CrossRef] [PubMed]
33. Goldman, M.J.; Anderson, G.M.; Stolzenberg, E.D.; Kari, U.P.; Zasloff, M.; Wilson, J.M. Human beta-defensin-1 is a salt-sensitive antibiotic in lung that is inactivated in cystic fibrosis. *Cell* **1997**, *88*, 553–560. [CrossRef]
34. Sun, L.; Wang, W.; Xiao, W.; Yang, H. The Roles of Cathelicidin LL-37 in Inflammatory Bowel Disease. *Inflamm. Bowel. Dis.* **2016**, *22*, 1986–1991. [CrossRef] [PubMed]
35. Dürr, U.H.; Sudheendra, U.S.; Ramamoorthy, A. LL-37, the only human member of the cathelicidin family of antimicrobial peptides. *Biochim. Biophys. Acta* **2006**, *1758*, 1408–1425. [CrossRef] [PubMed]
36. Scott, M.G.; Davidson, D.J.; Gold, M.R.; Bowdish, D.; Hancock, R.E. The human antimicrobial peptide LL-37 is a multifunctional modulator of innate immune responses. *J. Immunol.* **2002**, *169*, 3883–3891. [CrossRef]
37. Edfeldt, K.; Agerberth, B.; Rottenberg, M.E.; Gudmundsson, G.H.; Wang, X.B.; Mandal, K.; Xu, Q.; Yan, Z.Q. Involvement of the antimicrobial peptide LL-37 in human atherosclerosis. *Arterioscler. Thromb. Vasc. Biol.* **2006**, *26*, 1551–1557. [CrossRef]
38. Cerrillo, E.; Moret, I.; Iborra, M.; Ramos, D.; Busó, E.; Tortosa, L.; Sáez-González, E.; Nos, P.; Beltrán, B. Alpha-defensins (α-Defs) in Crohn's disease: Decrease of ileal α-Def 5 via permanent methylation and increase in plasma α-Def 1-3 concentrations offering biomarker utility. *Clin. Exp. Immunol.* **2018**, *192*, 120–128. [CrossRef]

39. Sthoeger, Z.M.; Bezalel, S.; Chapnik, N.; Asher, I.; Froy, O. High alpha-defensin levels in patients with systemic lupus erythematosus. *Immunology* **2009**, *127*, 116–122. [CrossRef]
40. Lisitsyn, N.A.; Bukurova, Y.A.; Nikitina, I.G.; Krasnov, G.S.; Sykulev, Y.; Beresten, S.F. Enteric alpha defensins in norm and pathology. *Ann. Clin. Microbiol. Antimicrob.* **2012**, *11*, 1. [CrossRef]
41. Bonamy, C.; Sechet, E.; Amiot, A.; Alam, A.; Mourez, M.; Fraisse, L.; Sansonetti, P.J.; Sperandio, B. Expression of the human antimicrobial peptide β-defensin-1 is repressed by the EGFR-ERK-MYC axis in colonic epithelial cells. *Sci. Rep.* **2018**, *8*, 18043. [CrossRef] [PubMed]
42. Ling, Y.M.; Chen, J.Y.; Guo, L.; Wang, C.Y.; Tan, W.T.; Wen, Q.; Zhang, S.D.; Deng, G.H.; Lin, Y.; Kwok, H.F. β-defensin 1 expression in HCV infected liver/liver cancer: An important role in protecting HCV progression and liver cancer development. *Sci. Rep.* **2017**, *7*, 13404. [CrossRef] [PubMed]
43. Dilek, F.; Emin, Ö.; Gültepe, B.; Yazıcı, M.; Çakır, E.; Gedik, A.H. Evaluation of nasal fluid β-defensin 2 levels in children with allergic rhinitis. *Turk. Pediatri. Ars.* **2017**, *52*, 79–84. [CrossRef] [PubMed]
44. Andresen, E.; Günther, G.; Bullwinkel, J.; Lange, C.; Heine, H. Increased expression of beta-defensin 1 (DEFB1) in chronic obstructive pulmonary disease. *PLoS ONE* **2011**, *6*, e21898. [CrossRef] [PubMed]
45. Shi, N.; Jin, F.; Zhang, X.; Clinton, S.K.; Pan, Z.; Chen, T. Overexpression of human β-defensin 2 promotes growth and invasion during esophageal carcinogenesis. *Oncotarget* **2014**, *5*, 11333–11344. [CrossRef] [PubMed]
46. Xu, D.; Zhang, B.; Liao, C.; Zhang, W.; Wang, W.; Chang, Y.; Shao, Y. Human beta-defensin 3 contributes to the carcinogenesis of cervical cancer via activation of NF-κB signaling. *Oncotarget* **2016**, *7*, 75902–75913. [CrossRef]
47. Abe, S.; Miura, K.; Kinoshita, A.; Mishima, H.; Miura, S.; Yamasaki, K.; Hasegawa, Y.; Higashijima, A.; Jo, O.; Sasaki, K.; et al. Copy number variation of the antimicrobial-gene, defensin beta 4, is associated with susceptibility to cervical cancer. *J. Hum. Genet.* **2013**, *58*, 250–253. [CrossRef]
48. Kalmodia, S.; Son, K.N.; Cao, D.; Lee, B.S.; Surenkhuu, B.; Shah, D.; Ali, M.; Balasubramaniam, A.; Jain, S.; Aakalu, V.K. Presence of Histatin-1 in Human Tears and Association with Aqueous Deficient Dry Eye Diagnosis: A Preliminary Study. *Sci. Rep.* **2019**, *9*, 10304. [CrossRef]
49. Khan, S.A.; Fidel, P.L.; Thunayyan, A.A.; Varlotta, S.; Meiller, T.F.; Jabra-Rizk, M.A. Impaired Histatin-5 Levels and Salivary Antimicrobial Activity against. *J. AIDS Clin. Res.* **2013**, *4*. [CrossRef]
50. Bogefors, J.; Kvarnhammar, A.M.; Millrud, C.R.; Georén, S.K.; Cardell, L.O. LEAP-2, LL-37 and RNase7 in tonsillar tissue: Downregulated expression in seasonal allergic rhinitis. *Pathog. Dis.* **2014**, *72*, 55–60. [CrossRef]
51. Spencer, J.D.; Schwaderer, A.L.; Wang, H.; Bartz, J.; Kline, J.; Eichler, T.; DeSouza, K.R.; Sims-Lucas, S.; Baker, P.; Hains, D.S. Ribonuclease 7, an antimicrobial peptide upregulated during infection, contributes to microbial defense of the human urinary tract. *Kidney Int.* **2013**, *83*, 615–625. [CrossRef] [PubMed]
52. Hoffmann, R.; Bulet, P.; Urge, L.; Otvos, L., Jr. Range of activity and metabolic stability of synthetic antibacterial glycopeptides from insects. *Biochim. Biophys. Acta* **1999**, *1426*, 459–467. [CrossRef]
53. Weinberg, A.; Jin, G.; Sieg, S.; McCormick, T.S. The yin and yang of human Beta-defensins in health and disease. *Front. Immunol.* **2012**, *3*, 294. [CrossRef] [PubMed]
54. Haney, E.F.; Hancock, R.E. Peptide design for antimicrobial and immunomodulatory applications. *Biopolymers* **2013**, *100*, 572–583. [CrossRef]
55. Yang, D.; Chen, Q.; Hoover, D.M.; Staley, P.; Tucker, K.D.; Lubkowski, J.; Oppenheim, J.J. Many chemokines including CCL20/MIP-3alpha display antimicrobial activity. *J. Leukoc. Biol.* **2003**, *74*, 448–455. [CrossRef]
56. Vongsa, R.A.; Zimmerman, N.P.; Dwinell, M.B. CCR6 regulation of the actin cytoskeleton orchestrates human beta defensin-2- and CCL20-mediated restitution of colonic epithelial cells. *J. Biol. Chem.* **2009**, *284*, 10034–10045. [CrossRef]
57. Dorschner, R.A.; Lopez-Garcia, B.; Peschel, A.; Kraus, D.; Morikawa, K.; Nizet, V.; Gallo, R.L. The mammalian ionic environment dictates microbial susceptibility to antimicrobial defense peptides. *FASEB J.* **2006**, *20*, 35–42. [CrossRef]
58. Venkataraman, N.; Cole, A.L.; Ruchala, P.; Waring, A.J.; Lehrer, R.I.; Stuchlik, O.; Pohl, J.; Cole, A.M. Reawakening retrocyclins: Ancestral human defensins active against HIV-1. *PLoS Biol.* **2009**, *7*, e95. [CrossRef]

59. Liu, L.; Zhao, C.; Heng, H.H.; Ganz, T. The human beta-defensin-1 and alpha-defensins are encoded by adjacent genes: Two peptide families with differing disulfide topology share a common ancestry. *Genomics* **1997**, *43*, 316–320. [CrossRef]
60. Schneider, J.J.; Unholzer, A.; Schaller, M.; Schäfer-Korting, M.; Korting, H.C. Human defensins. *J. Mol. Med.* **2005**, *83*, 587–595. [CrossRef]
61. Ganz, T.; Selsted, M.E.; Szklarek, D.; Harwig, S.S.; Daher, K.; Bainton, D.F.; Lehrer, R.I. Defensins. Natural peptide antibiotics of human neutrophils. *J. Clin. Invest.* **1985**, *76*, 1427–1435. [CrossRef] [PubMed]
62. Bevins, C.L. Paneth cell defensins: Key effector molecules of innate immunity. *Biochem. Soc. Trans.* **2006**, *34*, 263–266. [CrossRef] [PubMed]
63. Lehrer, R.I.; Lichtenstein, A.K.; Ganz, T. Defensins: Antimicrobial and cytotoxic peptides of mammalian cells. *Annu. Rev. Immunol.* **1993**, *11*, 105–128. [CrossRef] [PubMed]
64. Faurschou, M.; Kamp, S.; Cowland, J.B.; Udby, L.; Johnsen, A.H.; Calafat, J.; Winther, H.; Borregaard, N. Prodefensins are matrix proteins of specific granules in human neutrophils. *J. Leukoc. Biol.* **2005**, *78*, 785–793. [CrossRef] [PubMed]
65. Katoh, K.; Rozewicki, J.; Yamada, K.D. MAFFT online service: Multiple sequence alignment, interactive sequence choice and visualization. *Brief. Bioinform.* **2019**, *20*, 1160–1166. [CrossRef]
66. Czech, L.; Huerta-Cepas, J.; Stamatakis, A. A Critical Review on the Use of Support Values in Tree Viewers and Bioinformatics Toolkits. *Mol. Biol. Evol.* **2017**, *34*, 1535–1542. [CrossRef]
67. Letunic, I.; Bork, P. Interactive tree of life (iTOL) v3: An online tool for the display and annotation of phylogenetic and other trees. *Nucleic Acids Res.* **2016**, *44*, W242–W245. [CrossRef]
68. Lear, S.; Cobb, S.L. Pep-Calc.com: A set of web utilities for the calculation of peptide and peptoid properties and automatic mass spectral peak assignment. *J. Comput. Aided Mol. Des.* **2016**, *30*, 271–277. [CrossRef]
69. Suarez-Carmona, M.; Hubert, P.; Delvenne, P.; Herfs, M. Defensins: "Simple" antimicrobial peptides or broad-spectrum molecules? *Cytokine Growth Factor Rev.* **2015**, *26*, 361–370. [CrossRef]
70. Ganz, T. Defensins: Antimicrobial peptides of innate immunity. *Nat. Rev. Immunol* **2003**, *3*, 710–720. [CrossRef]
71. Dhople, V.; Krukemeyer, A.; Ramamoorthy, A. The human beta-defensin-3, an antibacterial peptide with multiple biological functions. *Biochim. Biophys. Acta* **2006**, *1758*, 1499–1512. [CrossRef] [PubMed]
72. Hollox, E.J.; Armour, J.A.; Barber, J.C. Extensive normal copy number variation of a beta-defensin antimicrobial-gene cluster. *Am. J. Hum. Genet.* **2003**, *73*, 591–600. [CrossRef] [PubMed]
73. Bucki, R.; Leszczynska, K.; Namiot, A.; Sokolowski, W. Cathelicidin LL-37: A multitask antimicrobial peptide. *Arch. Immunol Ther. Exp.* **2010**, *58*, 15–25. [CrossRef] [PubMed]
74. Kościuczuk, E.M.; Lisowski, P.; Jarczak, J.; Strzałkowska, N.; Jóźwik, A.; Horbańczuk, J.; Krzyżewski, J.; Zwierzchowski, L.; Bagnicka, E. Cathelicidins: Family of antimicrobial peptides. A review. *Mol. Biol. Rep.* **2012**, *39*, 10957–10970. [CrossRef] [PubMed]
75. Murakami, M.; Lopez-Garcia, B.; Braff, M.; Dorschner, R.A.; Gallo, R.L. Postsecretory processing generates multiple cathelicidins for enhanced topical antimicrobial defense. *J. Immunol.* **2004**, *172*, 3070–3077. [CrossRef]
76. Murakami, M.; Kameda, K.; Tsumoto, H.; Tsuda, T.; Masuda, K.; Utsunomiya, R.; Mori, H.; Miura, Y.; Sayama, K. TLN-58, an Additional hCAP18 Processing Form, Found in the Lesion Vesicle of Palmoplantar Pustulosis in the Skin. *J. Invest. Dermatol.* **2017**, *137*, 322–331. [CrossRef]
77. Gornowicz, A.; Tokajuk, G.; Bielawska, A.; Maciorkowska, E.; Jabłoński, R.; Wójcicka, A.; Bielawski, K. The assessment of sIgA, histatin-5, and lactoperoxidase levels in saliva of adolescents with dental caries. *Med. Sci. Monit.* **2014**, *20*, 1095–1100. [CrossRef]
78. Güncü, G.N.; Yilmaz, D.; Könönen, E.; Gürsoy, U.K. Salivary Antimicrobial Peptides in Early Detection of Periodontitis. *Front. Cell Infect. Microbiol.* **2015**, *5*, 99. [CrossRef]
79. Spencer, J.D.; Schwaderer, A.L.; Dirosario, J.D.; McHugh, K.M.; McGillivary, G.; Justice, S.S.; Carpenter, A.R.; Baker, P.B.; Harder, J.; Hains, D.S. Ribonuclease 7 is a potent antimicrobial peptide within the human urinary tract. *Kidney Int.* **2011**, *80*, 174–180. [CrossRef]
80. Lee, P.; Peng, H.; Gelbart, T.; Wang, L.; Beutler, E. Regulation of hepcidin transcription by interleukin-1 and interleukin-6. *Proc. Natl Acad Sci. USA* **2005**, *102*, 1906–1910. [CrossRef]
81. Nagaoka, I.; Suzuki, K.; Niyonsaba, F.; Tamura, H.; Hirata, M. Modulation of neutrophil apoptosis by antimicrobial peptides. *ISRN Microbiol.* **2012**, *2012*, 345791. [CrossRef] [PubMed]

82. Brook, M.; Tomlinson, G.H.; Miles, K.; Smith, R.W.; Rossi, A.G.; Hiemstra, P.S.; van 't Wout, E.F.; Dean, J.L.; Gray, N.K.; Lu, W.; et al. Neutrophil-derived alpha defensins control inflammation by inhibiting macrophage mRNA translation. *Proc. Natl. Acad. Sci. USA* **2016**, *113*, 4350–4355. [CrossRef] [PubMed]
83. Meade, K.G.; O'Farrelly, C. beta-Defensins: Farming the Microbiome for Homeostasis and Health. *Front. Immunol.* **2018**, *9*, 3072. [CrossRef] [PubMed]
84. Agier, J.; Efenberger, M.; Brzezinska-Blaszczyk, E. Cathelicidin impact on inflammatory cells. *Cent. Eur. J. Immunol.* **2015**, *40*, 225–235. [CrossRef] [PubMed]
85. Mangoni, M.L.; Miele, R.; Renda, T.G.; Barra, D.; Simmaco, M. The synthesis of antimicrobial peptides in the skin of Rana esculenta is stimulated by microorganisms. *FASEB J.* **2001**, *15*, 1431–1432. [CrossRef] [PubMed]
86. Morioka, Y.; Yamasaki, K.; Leung, D.; Gallo, R.L. Cathelicidin antimicrobial peptides inhibit hyaluronan-induced cytokine release and modulate chronic allergic dermatitis. *J. Immunol.* **2008**, *181*, 3915–3922. [CrossRef]
87. Szklarczyk, D.; Gable, A.L.; Lyon, D.; Junge, A.; Wyder, S.; Huerta-Cepas, J.; Simonovic, M.; Doncheva, N.T.; Morris, J.H.; Bork, P.; et al. STRING v11: Protein-protein association networks with increased coverage, supporting functional discovery in genome-wide experimental datasets. *Nucleic Acids Res.* **2019**, *47*, D607–D613. [CrossRef]
88. Dixon, B.R.; Radin, J.N.; Piazuelo, M.B.; Contreras, D.C.; Algood, H.M. IL-17a and IL-22 Induce Expression of Antimicrobials in Gastrointestinal Epithelial Cells and May Contribute to Epithelial Cell Defense against *Helicobacter pylori*. *PLoS ONE* **2016**, *11*, e0148514. [CrossRef]
89. Zambrano-Zaragoza, J.F.; Romo-Martinez, E.J.; Duran-Avelar Mde, J.; Garcia-Magallanes, N.; Vibanco-Perez, N. Th17 cells in autoimmune and infectious diseases. *Int. J. Inflam.* **2014**, *2014*, 651503. [CrossRef]
90. Hollox, E.J.; Huffmeier, U.; Zeeuwen, P.L.; Palla, R.; Lascorz, J.; Rodijk-Olthuis, D.; van de Kerkhof, P.C.; Traupe, H.; de Jongh, G.; den Heijer, M.; et al. Psoriasis is associated with increased beta-defensin genomic copy number. *Nat. Genet.* **2008**, *40*, 23–25. [CrossRef]
91. Fellermann, K.; Stange, D.E.; Schaeffeler, E.; Schmalzl, H.; Wehkamp, J.; Bevins, C.L.; Reinisch, W.; Teml, A.; Schwab, M.; Lichter, P.; et al. A chromosome 8 gene-cluster polymorphism with low human beta-defensin 2 gene copy number predisposes to Crohn disease of the colon. *Am. J. Hum. Genet.* **2006**, *79*, 439–448. [CrossRef] [PubMed]
92. Chen, Q.; Yang, Y.; Hou, J.; Shu, Q.; Yin, Y.; Fu, W.; Han, F.; Hou, T.; Zeng, C.; Nemeth, E.; et al. Increased gene copy number of *DEFA1/DEFA3* worsens sepsis by inducing endothelial pyroptosis. *Proc. Natl. Acad Sci. USA* **2019**, *116*, 3161–3170. [CrossRef] [PubMed]
93. Huang, Y.P.; Wang, T.Y.; Wang, W.; Sun, H.Z. Association between Genetic Polymorphisms in DEFB1 and Susceptibility to Digestive Diseases. *Med. Sci. Monit.* **2015**, *21*, 2240–2250. [CrossRef] [PubMed]
94. Polesello, V.; Zupin, L.; Di Lenarda, R.; Biasotto, M.; Ottaviani, G.; Gobbo, M.; Cecco, L.; Alberi, G.; Pozzato, G.; Crovella, S.; et al. Impact of DEFB1 gene regulatory polymorphisms on hBD-1 salivary concentration. *Arch. Oral Biol.* **2015**, *60*, 1054–1058. [CrossRef] [PubMed]
95. Atreya, I.; Atreya, R.; Neurath, M.F. NF-kappaB in inflammatory bowel disease. *J. Intern. Med.* **2008**, *263*, 591–596. [CrossRef]
96. Mookherjee, N.; Rehaume, L.M.; Hancock, R.E. Cathelicidins and functional analogues as antisepsis molecules. *Expert Opin. Ther. Targets* **2007**, *11*, 993–1004. [CrossRef]
97. Wang, T.T.; Nestel, F.P.; Bourdeau, V.; Nagai, Y.; Wang, Q.; Liao, J.; Tavera-Mendoza, L.; Lin, R.; Hanrahan, J.W.; Mader, S.; et al. Cutting edge: 1,25-dihydroxyvitamin D3 is a direct inducer of antimicrobial peptide gene expression. *J. Immunol.* **2004**, *173*, 2909–2912. [CrossRef]
98. Mandal, S.M.; Manna, S.; Mondal, S.; Ghosh, A.K.; Chakraborty, R. Transcriptional regulation of human defense peptides: A new direction in infection control. *Biol. Chem.* **2018**, *399*, 1277–1284. [CrossRef]
99. Noh, Y.H.; Lee, J.; Seo, S.J.; Myung, S.C. Promoter DNA methylation contributes to human β -defensin-1 deficiency in atopic dermatitis. *Anim. Cells Syst.* **2018**, *22*, 172–177. [CrossRef]
100. Sun, C.Q.; Arnold, R.; Fernandez-Golarz, C.; Parrish, A.B.; Almekinder, T.; He, J.; Ho, S.M.; Svoboda, P.; Pohl, J.; Marshall, F.F.; et al. Human beta-defensin-1, a potential chromosome 8p tumor suppressor: Control of transcription and induction of apoptosis in renal cell carcinoma. *Cancer Res.* **2006**, *66*, 8542–8549. [CrossRef]

101. Winter, J.; Pantelis, A.; Reich, R.; Martini, M.; Kraus, D.; Jepsen, S.; Allam, J.P.; Novak, N.; Wenghoefer, M. Human beta-defensin-1, -2, and -3 exhibit opposite effects on oral squamous cell carcinoma cell proliferation. *Cancer Invest.* **2011**, *29*, 196–201. [CrossRef] [PubMed]
102. Takahashi, T.; Gallo, R.L. The Critical and Multifunctional Roles of Antimicrobial Peptides in Dermatology. *Dermatol. Clin.* **2017**, *35*, 39–50. [CrossRef] [PubMed]
103. Niyonsaba, F.; Ushio, H.; Hara, M.; Yokoi, H.; Tominaga, M.; Takamori, K.; Kajiwara, N.; Saito, H.; Nagaoka, I.; Ogawa, H.; et al. Antimicrobial peptides human beta-defensins and cathelicidin LL-37 induce the secretion of a pruritogenic cytokine IL-31 by human mast cells. *J. Immunol.* **2010**, *184*, 3526–3534. [CrossRef] [PubMed]
104. Smithrithee, R.; Niyonsaba, F.; Kiatsurayanon, C.; Ushio, H.; Ikeda, S.; Okumura, K.; Ogawa, H. Human β-defensin-3 increases the expression of interleukin-37 through CCR6 in human keratinocytes. *J. Dermatol. Sci.* **2015**, *77*, 46–53. [CrossRef]
105. Yamasaki, K.; Gallo, R.L. Rosacea as a disease of cathelicidins and skin innate immunity. *J. Investig. Dermatol. Symp. Proc.* **2011**, *15*, 12–15. [CrossRef]
106. Schauber, J.; Gallo, R.L. Antimicrobial peptides and the skin immune defense system. *J. Allergy Clin. Immunol.* **2008**, *122*, 261–266. [CrossRef]
107. Fagerberg, L.; Hallström, B.M.; Oksvold, P.; Kampf, C.; Djureinovic, D.; Odeberg, J.; Habuka, M.; Tahmasebpoor, S.; Danielsson, A.; Edlund, K.; et al. Analysis of the human tissue-specific expression by genome-wide integration of transcriptomics and antibody-based proteomics. *Mol. Cell Proteomics* **2014**, *13*, 397–406. [CrossRef]
108. Suresh, R.; Mosser, D.M. Pattern recognition receptors in innate immunity, host defense, and immunopathology. *Adv. Physiol. Educ.* **2013**, *37*, 284–291. [CrossRef]
109. Skovbakke, S.L.; Holdfeldt, A.; Forsman, H.; Bylund, J.; Franzyk, H. The Role of Formyl Peptide Receptors for Immunomodulatory Activities of Antimicrobial Peptides and Peptidomimetics. *Curr. Pharm. Des.* **2018**, *24*, 1100–1120. [CrossRef]
110. Mogensen, T.H. Pathogen recognition and inflammatory signaling in innate immune defenses. *Clin. Microbiol. Rev.* **2009**, *22*, 240–273. [CrossRef]
111. Schaefer, L. Complexity of danger: The diverse nature of damage-associated molecular patterns. *J. Biol. Chem.* **2014**, *289*, 35237–35245. [CrossRef] [PubMed]
112. Sun, C.Q.; Arnold, R.S.; Hsieh, C.L.; Dorin, J.R.; Lian, F.; Li, Z.; Petros, J.A. Discovery and mechanisms of host defense to oncogenesis: Targeting the beta-defensin-1 peptide as a natural tumor inhibitor. *Cancer Biol. Ther.* **2019**, *20*, 774–786. [CrossRef] [PubMed]
113. Cui, D.; Lyu, J.; Li, H.; Lei, L.; Bian, T.; Li, L.; Yan, F. Human β-defensin 3 inhibits periodontitis development by suppressing inflammatory responses in macrophages. *Mol. Immunol.* **2017**, *91*, 65–74. [CrossRef] [PubMed]
114. Funderburg, N.; Lederman, M.M.; Feng, Z.; Drage, M.G.; Jadlowsky, J.; Harding, C.V.; Weinberg, A.; Sieg, S.F. Human -defensin-3 activates professional antigen-presenting cells via Toll-like receptors 1 and 2. *Proc. Natl. Acad. Sci. USA* **2007**, *104*, 18631–18635. [CrossRef] [PubMed]
115. Biragyn, A.; Ruffini, P.A.; Leifer, C.A.; Klyushnenkova, E.; Shakhov, A.; Chertov, O.; Shirakawa, A.K.; Farber, J.M.; Segal, D.M.; Oppenheim, J.J.; et al. Toll-like receptor 4-dependent activation of dendritic cells by beta-defensin 2. *Science* **2002**, *298*, 1025–1029. [CrossRef] [PubMed]
116. Carta, S.; Semino, C.; Sitia, R.; Rubartelli, A. Dysregulated IL-1β Secretion in Autoinflammatory Diseases: A Matter of Stress? *Front. Immunol.* **2017**, *8*, 345. [CrossRef] [PubMed]
117. Shi, J.; Aono, S.; Lu, W.; Ouellette, A.J.; Hu, X.; Ji, Y.; Wang, L.; Lenz, S.; van Ginkel, F.W.; Liles, M.; et al. A novel role for defensins in intestinal homeostasis: Regulation of IL-1beta secretion. *J. Immunol.* **2007**, *179*, 1245–1253. [CrossRef]
118. Tewary, P.; de la Rosa, G.; Sharma, N.; Rodriguez, L.G.; Tarasov, S.G.; Howard, O.M.; Shirota, H.; Steinhagen, F.; Klinman, D.M.; Yang, D.; et al. β-Defensin 2 and 3 promote the uptake of self or CpG DNA, enhance IFN-α production by human plasmacytoid dendritic cells, and promote inflammation. *J. Immunol.* **2013**, *191*, 865–874. [CrossRef]
119. Lande, R.; Gregorio, J.; Facchinetti, V.; Chatterjee, B.; Wang, Y.H.; Homey, B.; Cao, W.; Wang, Y.H.; Su, B.; Nestle, F.O.; et al. Plasmacytoid dendritic cells sense self-DNA coupled with antimicrobial peptide. *Nature* **2007**, *449*, 564–569. [CrossRef]

120. Lande, R.; Chamilos, G.; Ganguly, D.; Demaria, O.; Frasca, L.; Durr, S.; Conrad, C.; Schroder, J.; Gilliet, M. Cationic antimicrobial peptides in psoriatic skin cooperate to break innate tolerance to self-DNA. *Eur. J. Immunol.* **2015**, *45*, 203–213. [CrossRef]
121. McAleer, J.P.; Kolls, J.K. Mechanisms controlling Th17 cytokine expression and host defense. *J. Leukoc. Biol.* **2011**, *90*, 263–270. [CrossRef] [PubMed]
122. Takahashi, T.; Kulkarni, N.N.; Lee, E.Y.; Zhang, L.J.; Wong, G.C.L.; Gallo, R.L. Cathelicidin promotes inflammation by enabling binding of self-RNA to cell surface scavenger receptors. *Sci. Rep.* **2018**, *8*, 4032. [CrossRef] [PubMed]
123. Kopfnagel, V.; Wagenknecht, S.; Harder, J.; Hofmann, K.; Kleine, M.; Buch, A.; Sodeik, B.; Werfel, T. RNase 7 Strongly Promotes TLR9-Mediated DNA Sensing by Human Plasmacytoid Dendritic Cells. *J. Invest. Dermatol.* **2018**, *138*, 872–881. [CrossRef] [PubMed]
124. Adase, C.A.; Borkowski, A.W.; Zhang, L.J.; Williams, M.R.; Sato, E.; Sanford, J.A.; Gallo, R.L. Non-coding Double-stranded RNA and Antimicrobial Peptide LL-37 Induce Growth Factor Expression from Keratinocytes and Endothelial Cells. *J. Biol. Chem.* **2016**, *291*, 11635–11646. [CrossRef] [PubMed]
125. Neumann, A.; Vollger, L.; Berends, E.T.; Molhoek, E.M.; Stapels, D.A.; Midon, M.; Friaes, A.; Pingoud, A.; Rooijakkers, S.H.; Gallo, R.L.; et al. Novel role of the antimicrobial peptide LL-37 in the protection of neutrophil extracellular traps against degradation by bacterial nucleases. *J. Innate Immun.* **2014**, *6*, 860–868. [CrossRef]
126. Delgado-Rizo, V.; Martinez-Guzman, M.A.; Iniguez-Gutierrez, L.; Garcia-Orozco, A.; Alvarado-Navarro, A.; Fafutis-Morris, M. Neutrophil Extracellular Traps and Its Implications in Inflammation: An Overview. *Front. Immunol.* **2017**, *8*, 81. [CrossRef]
127. Stephan, A.; Batinica, M.; Steiger, J.; Hartmann, P.; Zaucke, F.; Bloch, W.; Fabri, M. LL37:DNA complexes provide antimicrobial activity against intracellular bacteria in human macrophages. *Immunology* **2016**, *148*, 420–432. [CrossRef]
128. Torres-Juarez, F.; Cardenas-Vargas, A.; Montoya-Rosales, A.; González-Curiel, I.; Garcia-Hernandez, M.H.; Enciso-Moreno, J.A.; Hancock, R.E.; Rivas-Santiago, B. LL-37 immunomodulatory activity during *Mycobacterium tuberculosis* infection in macrophages. *Infect. Immun.* **2015**, *83*, 4495–4503. [CrossRef]
129. Mily, A.; Rekha, R.S.; Kamal, S.M.; Arifuzzaman, A.S.; Rahim, Z.; Khan, L.; Haq, M.A.; Zaman, K.; Bergman, P.; Brighenti, S.; et al. Significant Effects of Oral Phenylbutyrate and Vitamin D3 Adjunctive Therapy in Pulmonary Tuberculosis: A Randomized Controlled Trial. *PLoS ONE* **2015**, *10*, e0138340. [CrossRef]
130. Khurshid, Z.; Naseem, M.; Sheikh, Z.; Najeeb, S.; Shahab, S.; Zafar, M.S. Oral antimicrobial peptides: Types and role in the oral cavity. *Saudi. Pharm. J.* **2016**, *24*, 515–524. [CrossRef]
131. Gursoy, U.K.; Könönen, E. Understanding the roles of gingival beta-defensins. *J. Oral Microbiol.* **2012**, 4. [CrossRef] [PubMed]
132. Costa, L.C.M.; Soldati, K.R.; Fonseca, D.C.; Costa, J.E.; Abreu, M.H.N.G.; Costa, F.O.; Zandim-Barcelos, D.L.; Cota, L.O.M. Gingival crevicular fluid levels of human beta-defensin 1 in individuals with and without chronic periodontitis. *J. Periodontal Res.* **2018**, *53*, 736–742. [CrossRef] [PubMed]
133. Jurczak, A.; Kościelniak, D.; Papież, M.; Vyhouskaya, P.; Krzyściak, W. A study on β-defensin-2 and histatin-5 as a diagnostic marker of early childhood caries progression. *Biol. Res.* **2015**, *48*, 61. [CrossRef] [PubMed]
134. Pereira, A.L.; Franco, G.C.; Cortelli, S.C.; Aquino, D.R.; Costa, F.O.; Raslan, S.A.; Cortelli, J.R. Influence of periodontal status and periodontopathogens on levels of oral human β-defensin-2 in saliva. *J. Periodontol.* **2013**, *84*, 1445–1453. [CrossRef] [PubMed]
135. Werner, S.; Grose, R. Regulation of wound healing by growth factors and cytokines. *Physiol. Rev.* **2003**, *83*, 835–870. [CrossRef]
136. Dommisch, H.; Winter, J.; Götz, W.; Miesen, J.; Klein, A.; Hierse, L.; Deschner, J.; Jäger, A.; Eberhard, J.; Jepsen, S. Effect of growth factors on antimicrobial peptides and pro-inflammatory mediators during wound healing. *Clin. Oral. Investig.* **2015**, *19*, 209–220. [CrossRef]
137. Abou Alaiwa, M.H.; Reznikov, L.R.; Gansemer, N.D.; Sheets, K.A.; Horswill, A.R.; Stoltz, D.A.; Zabner, J.; Welsh, M.J. pH modulates the activity and synergism of the airway surface liquid antimicrobials β-defensin-3 and LL-37. *Proc. Natl. Acad Sci. USA* **2014**, *111*, 18703–18708. [CrossRef]
138. Weiner, D.J.; Bucki, R.; Janmey, P.A. The antimicrobial activity of the cathelicidin LL37 is inhibited by F-actin bundles and restored by gelsolin. *Am. J. Respir. Cell Mol. Biol.* **2003**, *28*, 738–745. [CrossRef]

139. Bucki, R.; Byfield, F.J.; Janmey, P.A. Release of the antimicrobial peptide LL-37 from DNA/F-actin bundles in cystic fibrosis sputum. *Eur. Respir. J.* **2007**, *29*, 624–632. [CrossRef]
140. Liao, Z.; Dong, J.; Hu, X.; Wang, T.; Wan, C.; Li, X.; Li, L.; Guo, L.; Xu, D.; Wen, F. Enhanced expression of human β-defensin 2 in peripheral lungs of patients with chronic obstructive pulmonary disease. *Peptides* **2012**, *38*, 350–356. [CrossRef]
141. Dalcin, D.; Ulanova, M. The Role of Human Beta-Defensin-2 in *Pseudomonas aeruginosa* Pulmonary Infection in Cystic Fibrosis Patients. *Infect. Dis. Ther.* **2013**, *2*, 159–166. [CrossRef] [PubMed]
142. Archer, N.K.; Adappa, N.D.; Palmer, J.N.; Cohen, N.A.; Harro, J.M.; Lee, S.K.; Miller, L.S.; Shirtliff, M.E. Interleukin-17A (IL-17A) and IL-17F Are Critical for Antimicrobial Peptide Production and Clearance of *Staphylococcus aureus* Nasal Colonization. *Infect. Immun.* **2016**, *84*, 3575–3583. [CrossRef] [PubMed]
143. Altieri, A.; Piyadasa, H.; Recksiedler, B.; Spicer, V.; Mookherjee, N. Cytokines IL-17, TNF and IFN-γ Alter the Expression of Antimicrobial Peptides and Proteins Disparately: A Targeted Proteomics Analysis using SOMAscan Technology. *Vaccines* **2018**, *6*. [CrossRef] [PubMed]
144. Kusagaya, H.; Fujisawa, T.; Yamanaka, K.; Mori, K.; Hashimoto, D.; Enomoto, N.; Inui, N.; Nakamura, Y.; Wu, R.; Maekawa, M.; et al. Toll-like receptor-mediated airway IL-17C enhances epithelial host defense in an autocrine/paracrine manner. *Am. J. Respir Cell Mol. Biol.* **2014**, *50*, 30–39. [CrossRef]
145. Li, A.; Gan, Y.; Wang, R.; Liu, Y.; Ma, T.; Huang, M.; Cui, X. IL-22 Up-Regulates β-Defensin-2 Expression in Human Alveolar Epithelium via STAT3 but Not NF-κB Signaling Pathway. *Inflammation* **2015**, *38*, 1191–1200. [CrossRef]
146. Baines, K.J.; Wright, T.K.; Simpson, J.L.; McDonald, V.M.; Wood, L.G.; Parsons, K.S.; Wark, P.A.; Gibson, P.G. Airway β-Defensin-1 Protein Is Elevated in COPD and Severe Asthma. *Mediators Inflamm.* **2015**, *2015*, 407271. [CrossRef]
147. Nakamura, K.; Sakuragi, N.; Takakuwa, A.; Ayabe, T. Paneth cell α-defensins and enteric microbiota in health and disease. *Biosci. Microbiota Food Health* **2016**, *35*, 57–67. [CrossRef]
148. Wehkamp, J.; Salzman, N.H.; Porter, E.; Nuding, S.; Weichenthal, M.; Petras, R.E.; Shen, B.; Schaeffeler, E.; Schwab, M.; Linzmeier, R.; et al. Reduced Paneth cell alpha-defensins in ileal Crohn's disease. *Proc. Natl. Acad. Sci. USA.* **2005**, *102*, 18129–18134. [CrossRef]
149. Kanmura, S.; Hamamoto, H.; Morinaga, Y.; Oda, K.; Fujita, T.; Arima, S.; Nasu, Y.; Sasaki, F.; Hashimoto, S.; Taguchi, H.; et al. Fecal Human Neutrophil Peptide Levels Correlate with Intestinal Inflammation in Ulcerative Colitis. *Digestion* **2016**, *93*, 300–308. [CrossRef]
150. Pero, R.; Coretti, L.; Nigro, E.; Lembo, F.; Laneri, S.; Lombardo, B.; Daniele, A.; Scudiero, O. β-Defensins in the Fight against *Helicobacter pylori*. *Molecules* **2017**, *22*, 424. [CrossRef]
151. Muhammad, J.S.; Zaidi, S.F.; Zhou, Y.; Sakurai, H.; Sugiyama, T. Novel epidermal growth factor receptor pathway mediates release of human β-defensin 3 from *Helicobacter pylori*-infected gastric epithelial cells. *Pathog. Dis.* **2016**, *74*. [CrossRef]

International Journal of
Molecular Sciences

MDPI

Article

Osteostatin Inhibits Collagen-Induced Arthritis by Regulation of Immune Activation, Pro-Inflammatory Cytokines, and Osteoclastogenesis

Josep Nácher-Juan, María Carmen Terencio, María José Alcaraz * and María Luisa Ferrándiz *

Instituto Interuniversitario de Investigación de Reconocimiento Molecular y Desarrollo Tecnológico (IDM), Universitat Politècnica de València, Universitat de València, Av. Vicent A. Estellés s/n, 46100 Burjasot, Valencia, Spain

* Correspondence: maria.j.alcaraz@uv.es (M.J.A.); luisa.ferrandiz@uv.es (M.L.F.)

Received: 26 July 2019; Accepted: 5 August 2019; Published: 7 August 2019

Abstract: In chronic inflammatory joint diseases, such as rheumatoid arthritis, there is an important bone loss. Parathyroid hormone-related protein (PTHrP) and related peptides have shown osteoinductive properties in bone regeneration models, but there are no data on inflammatory joint destruction. We have investigated whether the PTHrP (107-111) C-terminal peptide (osteostatin) could control the development of collagen-induced arthritis in mice. Administration of osteostatin (80 or 120 μg/kg s.c.) after the onset of disease decreased the severity of arthritis as well as cartilage and bone degradation. This peptide reduced serum IgG2a levels as well as T cell activation, with the downregulation of RORγt+CD4+ T cells and upregulation of FoxP3+CD8+ T cells in lymph nodes. The levels of key cytokines, such as interleukin(IL)-1β, IL-2, IL-6, IL-17, and tumor necrosis factor-α in mice paws were decreased by osteostatin treatment, whereas IL-10 was enhanced. Bone protection was related to reductions in receptor activator of nuclear factor-κB ligand, Dickkopf-related protein 1, and joint osteoclast area. Osteostatin improves arthritis and controls bone loss by inhibiting immune activation, pro-inflammatory cytokines, and osteoclastogenesis. Our results support the interest of osteostatin for the treatment of inflammatory joint conditions.

Keywords: osteostatin; arthritis; inflammation; immune response; cartilage destruction; bone erosion

1. Introduction

Parathyroid hormone (PTH) and parathyroid hormone-related protein (PTHrP) show structural homology at the N-terminal region, which determines the interaction with the common PTH type 1 receptor. The bone anabolic properties of different PTHrP-derived peptides have been demonstrated in vitro and in vivo. In particular, the Food and Drug Administration have approved abaloparatide, which is an analogue that is based on the N-terminal 1-34 sequence of human PTHrP, for the treatment of postmenopausal women with osteoporosis at high risk for fracture. Several studies have demonstrated that both the N-terminal fragment and the PTH-unrelated C-terminal domain of PTHrP can enhance osteoblast proliferation and differentiation [1,2] and it can confer osteoinductive effects to implants in animal models of bone repair [3]. In addition, some C-terminal peptides have been found to reduce osteoclast activity in vitro, suggesting that the C-terminal PTHrP may act as a paracrine regulator of bone metabolism [4].

Rheumatoid arthritis (RA) is the most common chronic inflammatory joint disease. Autoimmunity responses precede joint inflammation and cellular activation to release a wide range of pro-inflammatory and catabolic mediators inducing the chronic inflammatory state and articular damage [5]. RA is characterized by an important alteration in bone homeostasis with an imbalance between bone resorption and formation [6]. Therefore, localized bone resorption and generalized bone loss are both associated to the progression of disease [7]. Bone damage arises from complex interactions that determine osteoclast precursor maturation that is mainly mediated by receptor activator of nuclear factor-κB ligand (RANKL), macrophage colony-stimulating factor and tumor necrosis factor-α (TNFα) pathways [5].

High levels of PTHrP in synovial fluid from RA patients are thought to participate in the regulation of different articular cells, such as subchondral bone osteoblasts, lymphocytes, or chondrocytes [8]. The production of PTHrP has been related to chondrogenesis and the inhibition of hypertrophic development of mesenchymal stem cells [9]. Indeed, PTHrP delays chondrocyte differentiation and it is involved in cartilage maturation [10]. Nevertheless, some report has suggested a pathogenic role for PTHrP that is related to synovial cell proliferation or osteoclast induction in arthritic articular tissues [11].

We have recently shown that PTHrP peptides are able to control senescence and inflammation in osteoarthritic osteoblasts. These properties were mainly associated to the C-terminal moiety of these agents [12]. No studies have yet addressed if these peptides could control joint inflammatory conditions. Therefore, we have explored whether PTHrP (107-111) (osteostatin), which includes the -Thr-Arg-Ser-Ala-Trp- sequence of the PTHrP C terminal fragment, exerts inhibitory effects on inflammation and joint degradation in the collagen-induced arthritis (CIA). This model exhibits morphological features that are similar to RA, including synovitis and erosion of cartilage and bone [13].

2. Results

2.1. *Effects of Osteostatin on the Progression of Arthritis*

As shown in Figure 1A, the severity of arthritis increased over time, until the end of the experiment. Mice that were treated with osteostatin (80 µg/kg or 120 µg/kg) showed a sustained reduction in this score, which was significant from days 34 to 40.

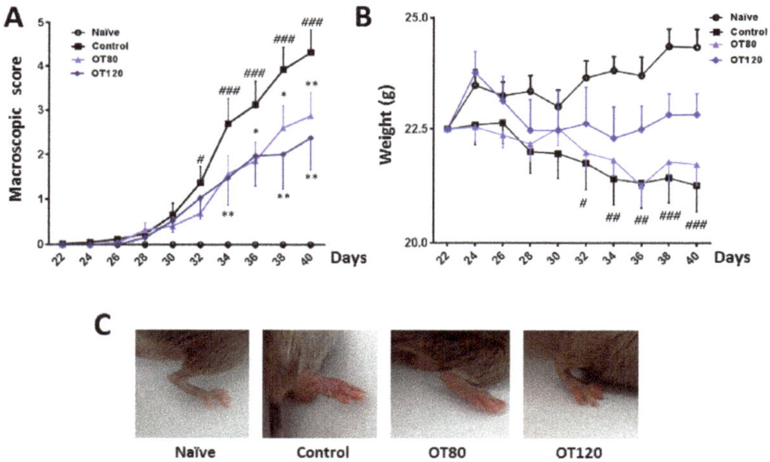

Figure 1. (**A**) Time course of arthritis macroscopic score. For each animal, a total score value was calculated after the second immunization on day 21. (**B**) Time course of body weight. (**C**) Representative images of mice hindpaw on day 40. Data are presented as mean ± SEM ($n = 8$ mice per group). # $p < 0.05$, ## $p < 0.01$, ### $p < 0.001$ versus naïve group; * $p < 0.05$, ** $p < 0.01$ versus control group. Two-way ANOVA (Bonferroni post-test).

Figure 1C shows representative images of these experimental groups. The progression of the arthritic process resulted in weight loss when compared with naïve animals, while the treatment with the highest dose of osteostatin tended to normalize this parameter (Figure 1B).

2.2. *Joint Histological Analysis*

Cartilage pathology and joint inflammation were assessed by joint histological analysis of ankle sections that were stained with haematoxylin/eosin or safranin O. Representative sections are shown in Figure 2A and the histological score in Figure 2B. Histopathological examination of ankle joints of arthritic control mice showed cartilage damage with chondrocyte death and proteoglycan depletion, besides synovial infiltrate and exudate. Treatment with both doses of osteostatin led to reductions in chondrocyte death, cartilage erosion, proteoglycan depletion, and synovial exudate, while a reduction in synovial infiltrate was observed at the highest dose.

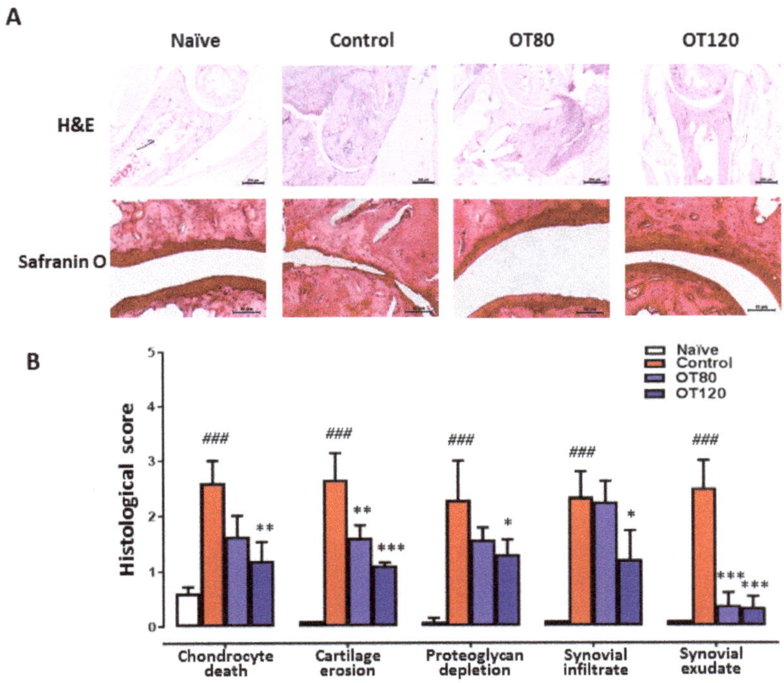

Figure 2. Histological analysis of ankle joints at day 40. (**A**) Haematoxylin and eosin (H&E) and safranin O stained sections (40× and 200×, respectively). Bar = 200 µm (H&E) and 50 µm (safranin O). (**B**) The histological scores are presented as mean ± S.D. ($n = 5$) and were analyzed using one-way ANOVA with Tukey's post-test. ### $p < 0.001$ versus naïve group; * $p < 0.05$, ** $p < 0.01$, *** $p < 0.001$ versus control group.

2.3. *Effects on Local Cytokine Levels and Myeloperoxidase Activity*

Next, we investigated whether osteostatin could regulate local cytokines. For this purpose, we measured by ELISA the levels of key cytokines present in paw homogenates at the end of the experiment. We found a significant reduction of interleukin(IL)-1β, TNFα, IL-6, and IL-2 by both doses of osteostatin, whereas no significant effect was observed on the chemotactic factor CXCL-1 (Figure 3). In addition, the levels of IL-17 and RANKL (osteoclast-differentiation factor) were decreased by both doses of osteostatin, although the results reached statistical significance for the 80 µg/kg dose. Notably, the levels of the anti-inflammatory cytokine IL-10 were significantly enhanced by both doses of osteostatin. Myeloperoxidase (MPO) activity was measured in paw homogenates. As shown in Figure 3, MPO was increased by arthritis induction, and this effect was significantly reduced by the highest dose of osteostatin.

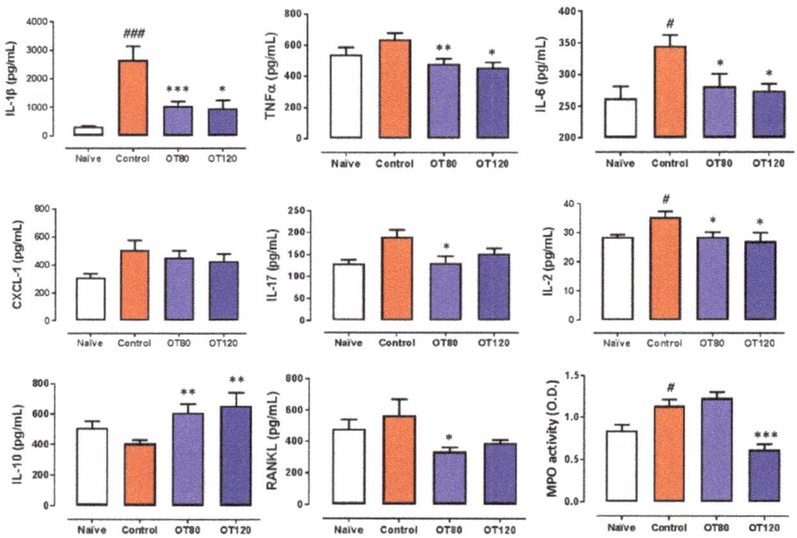

Figure 3. Levels of mediators in paw homogenates. Cytokines were measured by ELISA. Myeloperoxidase (MPO) activity was determined by spectrophotometry O.D. (optical density, 450 nm). Data presented as mean ± SEM. # $p < 0.05$, ### $p < 0.001$ versus naïve group; * $p < 0.05$, ** $p < 0.01$, *** $p < 0.001$ versus control group. One-way ANOVA (Tukey's post-test with $n = 8$).

2.4. Effects on Serum IgG2a and Bone Metabolism Biomarkers

As the severity of arthritis in this experimental model is positively correlated with the IgG autoantibody response to collagen II [14] and the presence of the IgG2a isotype [15], we wanted to investigate whether osteostatin can modify the serum levels of IgG2a. ELISA analyzed serum samples that were taken at the end of the study. Figure 4 shows that the IgG2a levels were significantly increased in arthritic control mice, while osteostatin treatment reduced them, although statistical significance was only reached for the 120 µg/kg dose. In addition, several biomarkers of bone metabolism were measured in serum. The wingless-related integration site (Wnt) signaling pathways play an essential role in regulating bone development and homeostasis of joints and the skeleton mass [16]. Arthritis induction did not modify osteocalcin (a biomarker of mature osteoblasts) levels, but enhanced the Wnt regulatory molecules sclerostin and Dickkopf-related protein 1 (DKK-1) (Figure 4). Our results indicate that DKK-1 levels were dose-dependently inhibited by osteostatin, whereas osteocalcin and sclerostin were not significantly modified.

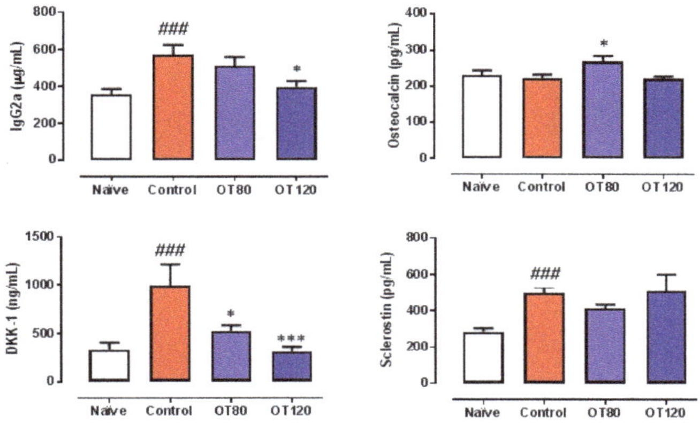

Figure 4. Serum levels of IgG2a and bone metabolism biomarkers Dickkopf-related protein 1 (DKK-1), sclerostin, and osteocalcin. IgG2a levels were measured by ELISA and bone metabolism biomarkers by multiplexing. Data are presented as mean ± SEM. One-way ANOVA with Tukey's post-test. ### $p < 0.001$, versus naïve group; * $p < 0.05$, *** $p < 0.001$ versus control group ($n = 8$).

2.5. Lymph Node T Cell Proliferation and Release of Cytokines

We next determined the effects of osteostatin on T cells that were present in lymph nodes. Figure 5 shows that T cell proliferation was stimulated in arthritic control mice versus naïve group. A lower proliferation was observed in osteostatin-treated animals, with a significant effect for the highest dose, which reduced proliferation to levels similar to those found in naïve animals. In addition, T cells from arthritic control mice released higher levels of the cytokines IL-2, IL-4, and interferon γ (IFNγ) as compared with non-arthritic animals. The treatment with osteostatin reduced the release of these cytokines with significant effects for IL-2 at the highest dose and for IL-4 and IFNγ at both doses.

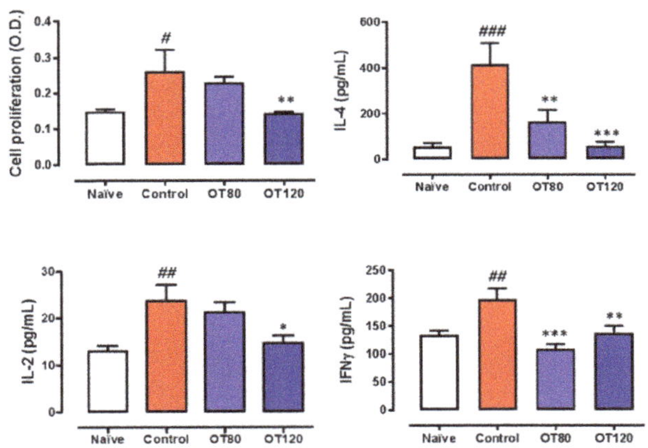

Figure 5. T cell proliferation assay and release of cytokines. Lymph node T cell proliferation was measured by bromdeoxiuridine incorporation. O.D. (optical density, 405 nm). ELISA measured levels of T cell differentiation cytokines in the conditioned media of the cell proliferation assay. Data are presented as mean ± SEM ($n = 5$), one-way ANOVA with Tukey's post-test. # $p < 0.05$, ## $p < 0.01$, ### $p < 0.001$ versus naïve group; * $p < 0.05$, ** $p < 0.01$, *** $p < 0.001$ versus control group.

The analysis of T cell populations revealed that arthritis induction increased the number of RORγt+CD4+ T cells, whereas Foxp3+CD4+ T cells and to a lower extent Foxp3+CD8+ T cells were reduced in arthritic controls when compared with naïve mice (Figure 6).

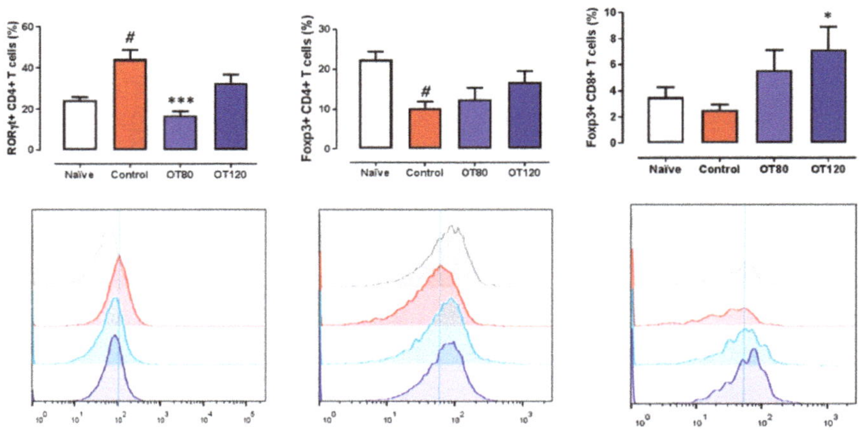

Figure 6. Flow cytometry analysis of lymph node T cell populations. FoxP3+ and RORγt+ T cell populations are represented. Data are mean ± SEM ($n = 5$), one-way ANOVA with Tukey's post-test. # $p < 0.05$ versus naïve group; * $p < 0.05$, *** $p < 0.001$ versus control group.

Osteostatin reverted the effect of CIA on RORγt+CD4+ T cells and tended to enhance Foxp3+CD4+ T cells. In addition, osteostatin increased the number of Foxp3+CD8+T cells, with statistically significant results for the 120 µg/kg dose.

2.6. Bone Degradation

After identifying histological improvements in ankle joint by osteostatin treatment, we checked whether this compound could modify arthritic bone alterations. Analysis of X-rays taken from the mice paws revealed that osteostatin treatment reduced the severity of radiographic changes. Figure 7A shows the quantification of bone density (front and hind paws) in Hounsfield units and representative images of hind paws. Arthritis induction significantly reduced bone density while osteostatin normalized these values. The differentiation of osteoclasts is a key process driving major erosive lesions in RA [17]. Osteoclast area was assessed by tartrate-resistant acid phosphatase (TRAP) staining of mice ankle sections. Arthritic control mice showed a significant enhancement of TRAP staining when compared with non-arthritic animals (Figure 7B). Of note, osteostatin dose-dependently reduced TRAP staining, which showed values that were similar to those of naïve mice in animals that were treated with the 120 µg/kg dose.

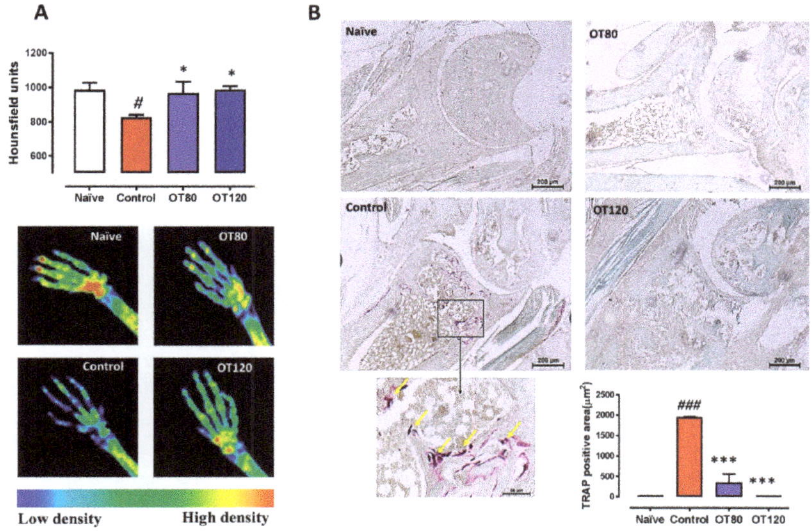

Figure 7. Bone density and osteoclast area analysis. (**A**) X-rays analysis of both front and hind limbs. Bone density was measured by analyzing a ROI (region of interest) around the wrist and ankle joints. Values are presented in Hounsfield units. Representative images of hind paws with color scale: red, high bone density; green/yellow, middle density; blue, low density (n= 5). (**B**) Tartrate-resistant acid phosphatase (TRAP) staining of mice ankle sections. The purple staining represent the TRAP positive area associated to osteoclasts activity. Bar = 200 µm and 50 µm. Data are presented as mean ± SEM (n = 5). One-way ANOVA with Tukey's post-test. # $p < 0.05$, ### $p < 0.001$ versus naïve group; * $p < 0.05$, *** $p < 0.001$ versus control group.

3. Discussion

In the present study, we show that osteostatin treatment ameliorates the severity of experimental arthritis with efficacy on clinical signs and structural alterations of CIA. Mice injected with collagen II develop arthritis that is similar in many respects to human RA characterized by a progressive breakdown of articular cartilage and bone erosion. Histological analysis revealed a significant improvement in inflammation, proteoglycan loss, and cartilage damage by osteostatin administration as compared with arthritic control mice. Pro-inflammatory cytokines are abundantly expressed in the arthritic joints of CIA mice as in human RA. In particular, TNFα plays a relevant role in the integration of the arthritic response, as demonstrated by the therapeutic efficacy of different strategies aimed at blocking this cytokine [18]. Our results indicate that osteostatin treatment reduces cell influx into the synovium, as well as the number of cells that were present in the joint cavity and the levels of key pro-inflammatory cytokines, such as TNFα, IL-1β, and IL-6. High levels of IL-10 are present in synovial tissues of RA patients, where it may have an immunoregulatory function [19,20]. Interestingly, osteostatin treatment significantly enhanced the local production of IL-10, which may be contribute to the control of disease severity as the deficiency of this cytokine exacerbates the CIA response [21].

Genetically linked autoimmunity to collagen II leads to the induction of arthritis and it may be determinant for the chronicity of the process and joint cartilage destruction [22,23]. We found that osteostatin is able to downregulate the production of anti-collagen II IgG2a, an autoantibody subclass that appears to be of particular importance in the CIA model [15]. The adaptive immune response and additional mechanisms play a role in the progression of CIA [24]. It has been suggested that collagen II-specific T cells act during the effector phase of arthritis, leading to perpetuation and exacerbation of disease [25]. CD4+ T cells differentiate into several subsets that are involved in the arthritic process [26]. In particular, IL-17-producing Th17 cells appear to play a predominant role in autoimmune arthritis [27]. Our data indicate that osteostatin affects T cell immunity in the CIA model. Therefore, this peptide may downregulate the activity of Th17 cells, as it reduced the number of RORγt+CD4+ T cells in the lymph node and the levels of IL-17 in the joints [28]. In addition, osteostatin counteracted the enhanced production of IL-2, IFNγ (Th1 activity), and IL-4 (Th2 activity) induced by the arthritic process. Accumulating evidence has revealed a key role of IL-17 in the production of pro-inflammatory cytokines [29] and the potentiation of their effects for cartilage degradation [30]. Furthermore, IL-17 is an important inducer of RANKL expression stimulating osteoclastogenesis and bone erosion in arthritis [31]. These results suggest that osteostatin might inhibit T cell activation and the subsequent generation of cytokines that play key roles in the pathogenesis of arthritis and joint degradation.

On the other hand, it has become evident that CD4+ regulatory T cells (Tregs) are involved in the control of clinical symptoms of CIA, cytokine production, effector T cell activity [32], and osteoclastogenesis [33,34]. As a result, the enhancement of Treg activity may be beneficial for the treatment of inflammation-induced bone loss [35]. In addition to CD4+ Tregs, CD8+ Tregs possess important immunosuppressive functions [36] and regulate Th17-mediated autoimmune diseases [37]. Interestingly, CD8+ Tregs are present in synovium from RA patients and they have the ability to decrease the activity of lesional T cells as well as the production of proinflammatory cytokines, suggesting a role in synovitis downregulation and potential therapeutic applications (reviewed in [38]). Thus, induced human CD8+Tregs have been shown to alleviate the severity of CIA and inhibit the mRNA expression of IL-17A and RANKL in mouse paw [39]. We found that osteostatin enhances CD8+Treg cells, which may contribute to the downregulation of the immune response to collagen II and the development of arthritis.

The control of bone destruction is a relevant objective in RA treatment. The erosion of subchondral and cortical bone is common in areas of synovial hyperplasia. Besides, bone destruction is associated to the presence of osteoclasts and RANKL expression [40]. X-rays analysis of mouse limbs showed that osteostatin significantly counteracted the bone density loss that is induced by the arthritic process. High levels of pro-inflammatory cytokines and mainly TNFα produced in inflammatory arthritis alter the resorption/formation cycle necessary for bone homeostasis. TNFα may contribute to the inhibition of osteoblast maturation and function acting directly on these cells [41] or through the modulation of the Wnt pathway. In fact, TNFα is a key inducer of the Wnt inhibitor DKK-1 in mouse inflammatory arthritis and in human RA [42]. In addition to the upregulation of RANKL expression by pro-inflammatory cytokines [43], it is known that TNFα cooperates with RANKL for osteoclast differentiation [41]. Furthermore, TNFα can stimulate osteoclast formation by RANKL-independent mechanisms [44]. The bone protective effect of osteostatin that is reported here was supported by reductions in osteoclast numbers and the inhibition of pro-inflammatory cytokines, such as TNFα and RANKL, suggesting that osteostatin is able to inhibit osteoclastogenesis in vivo, a critical process in the pathogenesis of joint damage. These results extend previous findings on the inhibitory effects of PTHrP C-terminal peptides on osteoclast function in vitro [4].

The Wnt pathway plays an important role in joint remodeling. It is generally accepted that members of the Wnt family are involved in the production of pro-inflammatory cytokines or catabolic enzymes in RA [45]. The inhibition of Wnt signaling may be responsible for the suppression of normal osteoblast function at sites of bone erosion in RA [42,46]. It is also known that DKK1 enhances RANKL expression and facilitates osteoclastogenesis [47]. Interestingly, the DKK1 levels are elevated in sera from early RA patients and correlate with disease activity [48]. Thus, recent studies have reported the reduction in DKK1 serum levels in RA patients in remission. In contrast, sclerostin levels do not correlate with disease activity and they do not change during remission [49]. In experimental arthritis, neutralization of DKK-1 with an antibody inhibited bone erosion without affecting inflammation [42], whereas the neutralization of sclerostin improved systemic bone loss, but did not affect disease severity or focal bone erosions in the CIA model [50]. However, this last approach worsened clinical outcome in models of TNFα-dependent inflammation, as sclerostin has an inhibitory effect on TNFα-induced p38 activation [51]. In line with these findings, we have observed a strong inhibitory effect of osteostatin on the serum levels of DKK-1 without a significant modification of sclerostin levels in CIA mice which suggests that DKK-1 downregulation may play a role in the inhibition of arthritic bone loss by osteostatin treatment.

Our data have shown the anti-arthritic properties of the oligopeptide osteostatin. It is interesting to note that oligopeptides may have some advantages for drug development when compared with higher molecular weight products, such as convenient synthesis and low immunogenicity, as well as favorable pharmacokinetics. However, some studies have reported that oligopeptides show an acceptable in vivo stability and integrity, good tissue penetration to exert their effects and a low toxicity. Additionally, they have gained wide attention for drug delivery systems and different modifications can be easily applied to improve their pharmacological profile [52]. All of these properties are relevant regarding potential therapeutic applications.

In conclusion, our results demonstrate the protective effects of osteostatin in a model of RA. This agent downregulated the immune and inflammatory responses resulting in reduced cartilage and bone destruction. The bone protective effects of osteostatin can be dependent on the inhibition of osteoclastogenesis, which may be mediated by the downregulation of key cytokines and DKK-1. Therefore, C-terminal PTHrP peptides provide an interesting approach for developing novel therapeutic opportunities for inflammatory joint conditions.

4. Materials and Methods

4.1. Animals

Male DBA/1 mice (Janvier, Le Genest-Saint-Isle, France) between 10 and 12 weeks of age (18–20 g) were used for all experiments. All the mice were housed in plastic cages (four per cage) with wood chips for bedding in a quiet room under controlled lighting (12 h day/night cycle) and temperature (22 ± 1 °C). Standard diet and water were provided ad libitum. All of the experiments were performed in accordance with European regulations for the handling and use of laboratory animals (Directive 2010/63/EU and Spanish R. D. 53/2013). The Institutional Animal Care and Use Committee (Comité de Etica de Experimentación Animal de la Universidad de Valencia, Spain) approved the protocols (number 2016/VSC/PEA/00053, 11 June 2016). All of the studies are reported in accordance with the ARRIVE guidelines for reporting experiments involving animals [53].

4.2. Induction of Arthritis

Arthritis was induced, as previously described [54]. Bovine type II collagen (2 mg/mL) (Chondrex Inc. Redmond, WA, USA) was emulsified in equal volumes of Freund's complete adjuvant (Thermofisher Scientific Inc. Waltham, MA, USA). On day 0, the mice were immunized by subcutaneous (s.c.) injection of the emulsion (100 µL) at the base of the tail. On day 21, animals received an intraperitoneal booster injection of collagen II (2 mg/mL, 100 µL) that was dissolved in phosphate buffered saline (PBS).

4.3. Experimental Groups and Treatment

Mice were randomly assigned to experimental groups: naïve group, which was not immunized ($n = 8$); control group, which was immunized but not treated ($n = 8$); OT80, which was immunized and treated with 80 µg/kg per day of osteostatin (Bachem AG, Bubendorf, Switzerland) in physiological saline ($n = 8$); and, OT120, which was immunized and treated with 120 µg/kg per day of osteostatin in physiological saline ($n = 8$). Doses, route, and frequency of administration were selected in preliminary experiments. The treated mice received 100 µL per day (s.c.) of each dose while naïve and control groups received 100 µL per day (s.c.) of physiological saline, after the onset of disease (day 28) for 13 days. On day 40, the mice were euthanized by cervical dislocation, lymph nodes were dissected, and limbs were surgically removed. Hind paws were processed for histologic analysis or homogenized for measurement of inflammatory mediators.

4.4. Arthritis Score

Joint inflammation was scored visually in each paw, while using a scale of 0–2, where 0-uninflamed, 1-mild, 1.5-marked, and 2–severe, as previously reported [54]. This macroscopic grading system assessed the extent of changes in redness, swelling, and ulceration of the paws. Scoring was performed every other day by two trained, independent observers (who were blinded with regard to experimental group), and weight measurements were also taken.

4.5. Histological Analysis

Ankles were kept in 4% paraformaldehyde in PBS (pH 7.4) for three weeks and decalcified with Osteosoft® (Merck KGaA, Darmstadt, Germany) for four weeks. Ankles were dehydrated and then embedded in paraffin. Lateral sections of tissue (7 μm) were obtained while using a Leica microtome and mounted on SuperFrost slides (Menzel-Gläser/Thermofisher Scientific Inc., Waltham, MA, USA). Histopathological changes in joints were measured using a scoring system [54]. Haematoxylin and eosin staining was performed to study joint inflammation. The severity of inflammation in the joints was scored on a scale of 0–3 (0 = no cells, 1 = mild cellularity, 2 = moderate cellularity, and 3 = maximal cellularity). To study proteoglycan depletion from the cartilage matrix, sections were stained with safranin O, followed by counterstaining with fast green. The depletion of proteoglycan was determined using an arbitrary scale of 0–3, ranging from normal, fully stained cartilage to destained cartilage that was fully depleted of proteoglycan. Cartilage erosion was scored by assigning a value of 0–3, depending on the integrity of the joint cartilage. Chondrocyte death was determined by counting the number of gaps in the cartilage without any chondrocyte. TRAP staining was performed in order to determine osteoclasts by measuring the TRAP-positive area. Scoring was performed in a blinded manner by two independent observers. The scores are the result of the mean of three sections from each mouse. The sections were examined under a light microscope DM IL LED (Leica®, Wetzlar, Germany) and pictures were taken with a camera Leica® (DFC 450 C).

4.6. Determination of Mediators in Paw Homogenates

Hind limbs were homogenized in liquid N_2 with 2 mL of A buffer pH 7.4 (10 mM HEPES, pH 8, 1 mM EDTA, 1 mM EGTA, 10 mM KCl, 1 mM dithiothreitol, 5 mM NaF, 1 mM Na_3VO_4, 1 mg/mL leupeptin, 0.1 mg/mL aprotinine, and 0.5 mM phenylmethylsulfonyl fluoride). The tissue homogenates were sonicated (3 × 10 s) and centrifuged at 12,000× *g*, 10 min. at 4 °C. Supernatants were removed and used for determinations. TNFα, IL-1β, and IL-17 were measured by ELISA (R&D Systems, Minneapolis, MN, USA) (range of detection of 32–2700 pg/mL, 25–2000 pg/mL and 10.9-700 pg/mL, respectively). CXCL-1 was determined by ELISA (Promokine, Heidelberg, Germany) (8–1000 pg/mL). IL-2 and IL-10 were measured with Th1/Th2 Mouse Uncoated ELISA kit (Invitrogen, Thermofisher Scientific Inc.) (2–200 pg/mL and 30-4000 pg/mL, respectively). The IL-6 levels were determined by IL-6 mouse ELISA kit (Invitrogen, Thermofisher Scientific Inc.) (7.8–500 pg/mL). RANKL concentrations were assessed by TRANCE mouse ELISA kit (Invitrogen, Thermofisher Scientific Inc.) (2.74–2000 pg/mL). MPO activity, a neutrophil marker, was measured by a spectrophotometric method, as previously reported [55].

4.7. Serum Determinations

Blood was collected in heparinized tubes on day 40. After centrifugation at 12,000× *g*, serum was separated. DKK-1, sclerostin, and osteocalcin were determined by Multiplex assay while using the Merck-Millipore kit (assay range 15–60,000 pg/mL, 3–12,000 pg/mL and 146–600,000 pg/mL, respectively) (Merck KGaA). IgG2a was measured by IgG2a mouse ELISA kit (Thermofisher Scientific Inc.) (assay range 0.614–150 ng/mL).

4.8. Lymph Node Cells Isolation, Proliferation Assay and Cytokine Determination

Inguinal, popliteal, brachial, and axillary lymph nodes were dissected, mechanically disaggregated, and incubated in DMEM (Dulbecco's Modified Eagle's Medium) that was supplemented with fetal bovine serum (FBS) (10%), streptomycin/penicillin (1%), collagenase (1.6 mg/mL), and DNAse (200 µg/mL) (Merck KGaA) for 40 min. at 37 °C. After filtration with a 40 µm strainer, 1 mL of DMEM was added and the suspension was centrifuged at 500× *g*, 4 °C, 10 min. Medium was removed and the cells were resuspended in DMEM supplemented with FBS, streptomycin/penicillin and DNAse, as previously indicated. Cells were incubated for 15 min. at 37 °C. Subsequently, 500 µL of DMEM was added and the suspension was centrifuged. Medium was removed and cells were resuspended in PBS for flow cytometry or in DMEM that was supplemented with FBS (10%), streptomycin/penicillin (1%) for bromdeoxiuridine proliferation assay. Lymph node cells resuspended in DMEM were seeded into 96-well plates at 2.5×10^5 cells/mL. At 24 h BrdU was added and the ELISA assay was performed at 48 h while using the cell proliferation ELISA BrdU from Merck KGaA. IL-2, IL-4, and IFNγ were measured by ELISA in the conditioned media of the cell proliferation assay with the Th1/Th2 Mouse Uncoated ELISA kit (Invitrogen, Thermofisher Scientific Inc.), with detection limits of 2–200 pg/mL, 4–500 pg/mL, and 15–2000 pg/mL, respectively.

4.9. Flow Cytometry

Cells that were obtained from lymph nodes were labeled with antibodies: antiCD3-Indo1(ref. 563565, lot 734861, clone 145-2C11), antiCD8-FITC (ref. 564422, lot 8037507, clone 53-6.7), antiRORγt-Pacific blue (ref. 562894, lot 8151861, clone Q31-378), and antiFoxP3-APC and antiCD4-PerCP-Cy5.5 (mouse Th17/Treg kit, ref. 51-9006647, lot 8067595) (BD Biosciences Europe, Madrid, Spain). Flow cytometry assays were performed with BD LSRFORTESSA (BD Biosciences). The software used was FACS DIVA 7.0 (BD Biosciences) and FlowJo v9 (BD Biosciences).

4.10. X-ray Analysis

Bone destruction in hind and front limbs was determined by X-ray analysis, which was carried out by microPET-CT (Albira) (Bruker, Billerica, MA, USA). A color scale represented bone density, where blue represents low density, yellow/green means middle density, and red high density. For numeric analysis, three-dimensional (3D) areas called region of interest (ROI) were used. These areas were exactly the same for all paws in 3 axis (X = 1.38 mm, Y = 2.94 mm, Z = 2.8 mm for hind limbs, and X = 2 mm, Y = 2 mm, and Z = 1 mm for front limbs) adjusted from the beginning of calcaneus, astragalus, and tarsus, until metatarsus for hind limbs and carpal bones for front limbs.

4.11. Statistical Analysis

Data are presented as mean ± S.D or S.E.M., with the number of individual values (*n*). Unless otherwise stated, duplicated determinations were done. Differences between experimental groups were tested by two-way ANOVA with Bonferroni post-test for the time course of arthritis macroscopic score and one-way ANOVA with Tukey's post-test for all other data. Statistical analyses were performed with GraphPad PRISM 5.0 (GraphPad software, San Diego, CA, USA). A value of $p < 0.05$ was considered to be significant.

Author Contributions: M.J.A., M.C.T. and M.L.F. were responsible for conception, design and interpretation of data. J.N.J. performed the experiments and analysed data. M.J.A. wrote the paper. All authors have full access to the data and take responsability for the integrity and accuracy of analysis.

Funding: This work has been funded by grant SAF2017-85806-R (Ministerio de Ciencia, Innovación y Universidades, Spain, FEDER). J. Nácher-Juan thanks Universitat de València, Spain, for a PhD fellowship (INV18-01-13-01).

Conflicts of Interest: The authors declare no conflict of interest.

References

1. de Castro, L.F.; Lozano, D.; Portal-Nuñez, S.; Maycas, M.; De la Fuente, M.; Caeiro, J.R.; Esbrit, P. Comparison of the skeletal effects induced by daily administration of PTHrP (1–36) and PTHrP (107–139) to ovariectomized mice. *J. Cell. Physiol.* **2012**, *227*, 1752–1760. [CrossRef] [PubMed]
2. de Gortazar, A.R.; Alonso, V.; Alvarez-Arroyo, M.V.; Esbrit, P. Transient exposure to PTHrP (107–139) exerts anabolic effects through vascular endothelial growth factor receptor 2 in human osteoblastic cells in vitro. *Calcif. Tissue Int.* **2006**, *79*, 360–369. [CrossRef] [PubMed]
3. Trejo, C.G.; Lozano, D.; Manzano, M.; Doadrio, J.C.; Salinas, A.J.; Dapia, S.; Gomez-Barrena, E.; Vallet-Regi, M.; Garcia-Honduvilla, N.; Bujan, J.; et al. The osteoinductive properties of mesoporous silicate coated with osteostatin in a rabbit femur cavity defect model. *Biomaterials* **2010**, *31*, 8564–8573. [CrossRef] [PubMed]
4. Fenton, A.J.; Martin, T.J.; Nicholson, G.C. Carboxyl-terminal parathyroid hormone-related protein inhibits bone resorption by isolated chicken osteoclasts. *J. Bone Miner. Res.* **1994**, *9*, 515–519. [CrossRef] [PubMed]
5. Firestein, G.S.; McInnes, I.B. Immunopathogenesis of Rheumatoid Arthritis. *Immunity* **2017**, *46*, 183–196. [CrossRef]
6. Scholtysek, C.; Kronke, G.; Schett, G. Inflammation-associated changes in bone homeostasis. *Inflamm. Allergy Drug Targets* **2012**, *11*, 188–195. [CrossRef] [PubMed]
7. Szentpetery, A.; Horvath, A.; Gulyas, K.; Petho, Z.; Bhattoa, H.P.; Szanto, S.; Szucs, G.; FitzGerald, O.; Schett, G.; Szekanecz, Z. Effects of targeted therapies on the bone in arthritides. *Autoimmun. Rev.* **2017**, *36*, 313–320. [CrossRef]
8. Kohno, H.; Shigeno, C.; Kasai, R.; Akiyama, H.; Iida, H.; Tsuboyama, T.; Sato, K.; Konishi, J.; Nakamura, T. Synovial fluids from patients with osteoarthritis and rheumatoid arthritis contain high levels of parathyroid hormone-related peptide. *J. Bone Miner. Res.* **1997**, *12*, 847–854. [CrossRef]
9. Fischer, J.; Dickhut, A.; Rickert, M.; Richter, W. Human articular chondrocytes secrete parathyroid hormone-related protein and inhibit hypertrophy of mesenchymal stem cells in coculture during chondrogenesis. *Arthritis Rheum.* **2010**, *62*, 2696–2706. [CrossRef]
10. Chen, X.; Macica, C.M.; Nasiri, A.; Broadus, A.E. Regulation of articular chondrocyte proliferation and differentiation by indian hedgehog and parathyroid hormone-related protein in mice. *Arthritis Rheum.* **2008**, *58*, 3788–3797. [CrossRef]
11. Horiuchi, T.; Yoshida, T.; Koshihara, Y.; Sakamoto, H.; Kanai, H.; Yamamoto, S.; Ito, H. The increase of parathyroid hormone-related peptide and cytokine levels in synovial fluid of elderly rheumatoid arthritis and osteoarthritis. *Endocr. J.* **1999**, *46*, 643–649. [CrossRef] [PubMed]
12. Platas, J.; Guillen, M.I.; Gomar, F.; Castejon, M.A.; Esbrit, P.; Alcaraz, M.J. Anti-senescence and Anti-inflammatory Effects of the C-terminal Moiety of PTHrP Peptides in OA Osteoblasts. *J. Gerontol. A Biol. Sci. Med. Sci.* **2016**, *172*, 624–631. [CrossRef] [PubMed]
13. Myers, L.K.; Rosloniec, E.F.; Cremer, M.A.; Kang, A.H. Collagen-induced arthritis, an animal model of autoimmunity. *Life Sci.* **1997**, *61*, 1861–1878. [CrossRef]
14. Williams, P.J.; Jones, R.H.; Rademacher, T.W. Correlation between IgG anti-type II collagen levels and arthritic severity in murine arthritis. *Autoimmunity* **1998**, *27*, 201–207. [CrossRef] [PubMed]
15. Watson, W.C.; Townes, A.S. Genetic susceptibility to murine collagen II autoimmune arthritis. Proposed relationship to the IgG2 autoantibody subclass response, complement C5, major histocompatibility complex (MHC) and non-MHC loci. *J. Exp. Med.* **1985**, *162*, 1878–1891. [CrossRef] [PubMed]
16. Lories, R.J.; Corr, M.; Lane, N.E. To Wnt or not to Wnt: The bone and joint health dilemma. *Nat. Rev. Rheumatol.* **2013**, *9*, 328–339. [CrossRef] [PubMed]
17. Schett, G.; Teitelbaum, S.L. Osteoclasts and Arthritis. *J. Bone Miner. Res.* **2009**, *24*, 1142–1146. [CrossRef] [PubMed]
18. Williams, R.O. Collagen-induced arthritis in mice: A major role for tumor necrosis factor-alpha. *Methods Mol. Biol.* **2007**, *361*, 265–284. [PubMed]
19. Cush, J.J.; Splawski, J.B.; Thomas, R.; McFarlin, J.E.; Schulze-Koops, H.; Davis, L.S.; Fujita, K.; Lipsky, P.E. Elevated interleukin-10 levels in patients with rheumatoid arthritis. *Arthritis Rheum.* **1995**, *38*, 96–104. [CrossRef]
20. Isomaki, P.; Luukkainen, R.; Saario, R.; Toivanen, P.; Punnonen, J. Interleukin-10 functions as an antiinflammatory cytokine in rheumatoid synovium. *Arthritis Rheum.* **1996**, *39*, 386–395. [CrossRef]

21. Finnegan, A.; Kaplan, C.D.; Cao, Y.; Eibel, H.; Glant, T.T.; Zhang, J. Collagen-induced arthritis is exacerbated in IL-10-deficient mice. *Arthritis Res. Ther.* **2003**, *5*, R18–R24. [CrossRef] [PubMed]
22. Ronnelid, J.; Lysholm, J.; Engstrom-Laurent, A.; Klareskog, L.; Heyman, B. Local anti-type II collagen antibody production in rheumatoid arthritis synovial fluid. Evidence for an HLA-DR4-restricted IgG response. *Arthritis Rheum.* **1994**, *37*, 1023–1029. [CrossRef] [PubMed]
23. Croxford, A.M.; Whittingham, S.; McNaughton, D.; Nandakumar, K.S.; Holmdahl, R.; Rowley, M.J. Type II collagen-specific antibodies induce cartilage damage in mice independent of inflammation. *Arthritis Rheum.* **2013**, *65*, 650–659. [CrossRef] [PubMed]
24. Plows, D.; Kontogeorgos, G.; Kollias, G. Mice lacking mature T and B lymphocytes develop arthritic lesions after immunization with type II collagen. *J. Immunol.* **1999**, *162*, 1018–1023. [PubMed]
25. Nandakumar, K.S.; Backlund, J.; Vestberg, M.; Holmdahl, R. Collagen type II (CII)-specific antibodies induce arthritis in the absence of T or B cells but the arthritis progression is enhanced by CII-reactive T cells. *Arthritis Res. Ther.* **2004**, *6*, R544–R550. [CrossRef] [PubMed]
26. Zhu, J.; Paul, W.E. Peripheral CD4+ T-cell differentiation regulated by networks of cytokines and transcription factors. *Immunol. Rev.* **2010**, *238*, 247–262. [CrossRef] [PubMed]
27. Lubberts, E. Th17 cytokines and arthritis. *Semin. Immunopathol.* **2010**, *32*, 43–53. [CrossRef] [PubMed]
28. Ito, Y.; Usui, T.; Kobayashi, S.; Iguchi-Hashimoto, M.; Ito, H.; Yoshitomi, H.; Nakamura, T.; Shimizu, M.; Kawabata, D.; Yukawa, N.; et al. Gamma/delta T cells are the predominant source of interleukin-17 in affected joints in collagen-induced arthritis, but not in rheumatoid arthritis. *Arthritis Rheum.* **2009**, *60*, 2294–2303. [CrossRef]
29. Lubberts, E.; Koenders, M.I.; Oppers-Walgreen, B.; Van Den, B.L.; Coenen-De Roo, C.J.; Joosten, L.A.; van den Berg, W.B. Treatment with a neutralizing anti-murine interleukin-17 antibody after the onset of collagen-induced arthritis reduces joint inflammation, cartilage destruction, and bone erosion. *Arthritis Rheum.* **2004**, *50*, 650–659. [CrossRef]
30. Koenders, M.I.; Marijnissen, R.J.; Devesa, I.; Lubberts, E.; Joosten, L.A.; Roth, J.; van Lent, P.L.; van de Loo, F.A.; van den Berg, W.B. Tumor necrosis factor-interleukin-17 interplay induces S100A8, interleukin-1beta, and matrix metalloproteinases, and drives irreversible cartilage destruction in murine arthritis: Rationale for combination treatment during arthritis. *Arthritis Rheum.* **2011**, *63*, 2329–2339. [CrossRef]
31. Lubberts, E.; Van Den, B.L.; Oppers-Walgreen, B.; Schwarzenberger, P.; Coenen-De Roo, C.J.; Kolls, J.K.; Joosten, L.A.; van den Berg, W.B. IL-17 promotes bone erosion in murine collagen-induced arthritis through loss of the receptor activator of NF-kappa B ligand/osteoprotegerin balance. *J. Immunol.* **2003**, *170*, 2655–2662. [CrossRef] [PubMed]
32. Cools, N.; Ponsaerts, P.; Van Tendeloo, V.F.; Berneman, Z.N. Regulatory T cells and human disease. *Clin. Dev. Immunol.* **2007**, *2007*, 89195. [CrossRef] [PubMed]
33. Kelchtermans, H.; Geboes, L.; Mitera, T.; Huskens, D.; Leclercq, G.; Matthys, P. Activated CD4+CD25+ regulatory T cells inhibit osteoclastogenesis and collagen-induced arthritis. *Ann. Rheum. Dis.* **2009**, *68*, 744–750. [CrossRef] [PubMed]
34. Notley, C.A.; McCann, F.E.; Inglis, J.J.; Williams, R.O. anti-CD3 therapy expands the numbers of CD4+ and CD8+ Treg cells and induces sustained amelioration of collagen-induced arthritis. *Arthritis Rheum.* **2010**, *62*, 171–178. [CrossRef] [PubMed]
35. Zaiss, M.M.; Frey, B.; Hess, A.; Zwerina, J.; Luther, J.; Nimmerjahn, F.; Engelke, K.; Kollias, G.; Hunig, T.; Schett, G.; et al. Regulatory T cells protect from local and systemic bone destruction in arthritis. *J. Immunol.* **2010**, *184*, 7238–7246. [CrossRef] [PubMed]
36. Yu, Y.; Ma, X.; Gong, R.; Zhu, J.; Wei, L.; Yao, J. Recent advances in CD8(+) regulatory T cell research. *Oncol. Lett.* **2018**, *15*, 8187–8194. [CrossRef] [PubMed]
37. Nakagawa, T.; Tsuruoka, M.; Ogura, H.; Okuyama, Y.; Arima, Y.; Hirano, T.; Murakami, M. IL-6 positively regulates Foxp3+CD8+ T cells in vivo. *Int. Immunol.* **2010**, *22*, 129–139. [CrossRef]
38. Filaci, G.; Fenoglio, D.; Indiveri, F. CD81 T regulatory/suppressor cells and their relationships with autoreactivity and autoimmunity. *Autoimmunity* **2011**, *44*, 51–57. [CrossRef]
39. Sun, J.; Yang, Y.; Huo, X.; Zhu, B.; Li, Z.; Jiang, X.; Xie, R.; Gao, L.; Sun, Y.; Fan, H.; et al. Efficient therapeutic function and mechanisms of human polyclonal CD8+CD103+Foxp3+ regulatory T cells on collagen-induced arthritis in mice. *J. Immunol. Res.* **2019**, *2019*, 8575407. [CrossRef]

40. Stolina, M.; Adamu, S.; Ominsky, M.; Dwyer, D.; Asuncion, F.; Geng, Z.; Middleton, S.; Brown, H.; Pretorius, J.; Schett, G.; et al. RANKL is a marker and mediator of local and systemic bone loss in two rat models of inflammatory arthritis. *J. Bone Miner. Res.* **2005**, *20*, 1756–1765. [CrossRef]
41. Nanes, M.S. Tumor necrosis factor-alpha: Molecular and cellular mechanisms in skeletal pathology. *Gene* **2003**, *321*, 1–15. [CrossRef]
42. Diarra, D.; Stolina, M.; Polzer, K.; Zwerina, J.; Ominsky, M.S.; Dwyer, D.; Korb, A.; Smolen, J.; Hoffmann, M.; Scheinecker, C.; et al. Dickkopf-1 is a master regulator of joint remodeling. *Nat. Med.* **2007**, *13*, 156–163. [CrossRef]
43. Nakashima, T.; Kobayashi, Y.; Yamasaki, S.; Kawakami, A.; Eguchi, K.; Sasaki, H.; Sakai, H. Protein expression and functional difference of membrane-bound and soluble receptor activator of NF-kappaB ligand: Modulation of the expression by osteotropic factors and cytokines. *Biochem. Biophys. Res. Commun.* **2000**, *275*, 768–775. [CrossRef]
44. Kobayashi, K.; Takahashi, N.; Jimi, E.; Udagawa, N.; Takami, M.; Kotake, S.; Nakagawa, N.; Kinosaki, M.; Yamaguchi, K.; Shima, N.; et al. Tumor necrosis factor alpha stimulates osteoclast differentiation by a mechanism independent of the ODF/RANKL-RANK interaction. *J. Exp. Med.* **2000**, *191*, 275–286. [CrossRef]
45. Sen, M. Wnt signalling in rheumatoid arthritis. *Rheumatology* **2005**, *44*, 708–713. [CrossRef]
46. Walsh, N.C. and Gravallese, E.M. Bone remodeling in rheumatic disease: A question of balance. *Immunol. Rev.* **2010**, *233*, 301–312. [CrossRef]
47. Braun, T.; Zwerina, J. Positive regulators of osteoclastogenesis and bone resorption in rheumatoid arthritis. *Arthritis Res. Ther.* **2011**, *13*, 235. [CrossRef]
48. Seror, R.; Boudaoud, S.; Pavy, S.; Nocturne, G.; Schaeverbeke, T.; Saraux, A.; Chanson, P.; Gottenberg, J.E.; Devauchelle-Pensec, V.; Tobon, G.J.; et al. Increased Dickkopf-1 in Recent-onset Rheumatoid Arthritis is a New Biomarker of Structural Severity. Data from the ESPOIR Cohort. *Sci. Rep.* **2016**, *6*, 18421. [CrossRef]
49. Kerschan-Schindl, K.; Ebenbichler, G.; Foeger-Samwald, U.; Leiss, H.; Gesslbauer, C.; Herceg, M.; Stummvoll, G.; Marculescu, R.; Crevenna, R.; Pietschmann, P. Rheumatoid arthritis in remission: Decreased myostatin and increased serum levels of periostin. *Wien. Klin. Wochenschr.* **2019**, *131*, 1–7. [CrossRef]
50. Marenzana, M.; Vugler, A.; Moore, A.; Robinson, M. Effect of sclerostin-neutralising antibody on periarticular and systemic bone in a murine model of rheumatoid arthritis: A microCT study. *Arthritis Res. Ther.* **2013**, *15*, R125. [CrossRef]
51. Wehmeyer, C.; Frank, S.; Beckmann, D.; Bottcher, M.; Cromme, C.; Konig, U.; Fennen, M.; Held, A.; Paruzel, P.; Hartmann, C.; et al. Sclerostin inhibition promotes TNF-dependent inflammatory joint destruction. *Sci. Transl. Med.* **2016**, *8*, 330ra35. [CrossRef]
52. Ji, Y.; Qiao, H.; He, J.; Li, W.; Chen, R.; Wang, J.; Wu, L.; Hu, R.; Duan, J.; Chen, Z. Functional oligopeptide as a novel strategy for drug delivery. *J. Drug. Target.* **2017**, *25*, 597–607. [CrossRef]
53. Kilkenny, C.; Browne, W.J.; Cuthill, I.C.; Emerson, M.; Altman, D.G. Improving bioscience research reporting: The ARRIVE guidelines for reporting animal research. *PLoS. Biol.* **2010**, *8*, e1000412. [CrossRef]
54. Maicas, N.; Ibañez, L.; Alcaraz, M.J.; Ubeda, A.; Ferrandiz, M.L. Prostaglandin D2 regulates joint inflammation and destruction in murine collagen-induced arthritis. *Arthritis Rheum.* **2012**, *64*, 130–140. [CrossRef]
55. Paya, M.; Terencio, M.C.; Ferrandiz, M.L.; Alcaraz, M.J. Involvement of secretory phospholipase A2 activity in the zymosan rat air pouch model of inflammation. *Br. J. Pharmacol.* **1996**, *117*, 1773–1779. [CrossRef]

International Journal of
Molecular Sciences

MDPI

Article

Alterations in Galanin-Like Immunoreactivity in the Enteric Nervous System of the Porcine Stomach Following Acrylamide Supplementation

Katarzyna Palus *, Krystyna Makowska and Jarosław Całka

Department of Clinical Physiology, Faculty of Veterinary Medicine, University of Warmia and Mazury in Olsztyn, Olsztyn, Poland, Oczapowskiego Str. 13, 10-718 Olsztyn, Poland

* Correspondence: katarzyna.palus@uwm.edu.pl; Tel.: +48-89-523-44-60

Received: 31 May 2019; Accepted: 4 July 2019; Published: 8 July 2019

Abstract: In recent years, a significant increase in the consumption of products containing large amounts of acrylamide (e.g., chips, fries, coffee), especially among young people has been noted. The present study was created to establish the impact of acrylamide supplementation, in tolerable daily intake (TDI) dose and a dose ten times higher than TDI, on the population of galanin-like immunoreactive (GAL-LI) stomach neurons in pigs. Additionally, in the present study, the possible functional co-operation of GAL with other neuroactive substances and their role in acrylamide intoxication was investigated. Using double-labelling immunohistochemistry, alterations in the expression of GAL were examined in the porcine stomach enteric neurons after low and high doses of acrylamide supplementation. Generally, upregulation in GAL-LI immunoreactivity in both myenteric and submucous plexuses was noted in all stomach fragments studied. Additionally, the proportion of GAL-expressing cell bodies simultaneously immunoreactive to vasoactive intestinal peptide (VIP), neuronal nitric oxide synthase (nNOS) and cocaine- and amphetamine- regulated transcript peptide (CART) also increased. The results suggest neurotrophic or/and neuroprotective properties of GAL and possible co-operation of GAL with VIP, nNOS, CART in the recovery processes in the stomach enteric nervous system (ENS) neurons following acrylamide intoxication.

Keywords: acrylamide; enteric nervous system; galanin; pig; stomach

1. Introduction

Galanin (GAL) is a 29 (or 30 in humans) amino acid peptide which has widespread distribution in the central and peripheral nervous systems, as well as in peripheral tissues of numerous species, including humans [1–4]. The occurrence of GAL was observed in both nerve cell bodies and nerve fibres located in different fragments of the gastrointestinal (GI) tracts of many species [2,5–7]. To date, three G-protein-coupled receptors known as galanin receptors 1, 2 and 3 (GAL-R1, GAL-R2 and GAL-R3) have been described in studies which examined tissues and organs [3]. GAL participates in many physiological functions in the GI tract, such as regulation of motility, hydrochloric acid secretion, exocrine function of the pancreas and intestinal absorption [1]. GAL may act in both an inhibitory or an excitatory role depending on the fragment of GI tract, the species and the experimental conditions [8]. Moreover, previous studies have indicated that GAL as a neuroprotective factor mediates survival or regeneration after neural injury and exerts anti-inflammatory activities [9,10]. Indeed, upregulation of GAL expression in neural structures of the GI tract was demonstrated during experimentally-induced and naturally occurring intestinal inflammation [2,11]. An enhanced percentage of GAL-like (GAL-LI) immunoreactive enteric nervous system (ENS) neurons was also observed in injuries of the digestive tract as well as in toxaemia [2,12].

One of the toxins occurring in food products is acrylamide (ACM). Acrylamide is formed at high temperature in a process known as the Maillard reaction. A high content of acrylamide has been found in many food products, such as biscuits, chips, coffee and cornflakes [13]. The literature in the field contains many papers dealing with the influence of acrylamide on living organisms. Namely, acrylamide-induced axonal damage and impaired retrograde transport resulting in symptoms of neuropathy were noted in both the central and peripheral nervous systems. [14]. Furthermore, acrylamide contributes to the creation of oxidative stress condition [15]. It is assumed that oxidative stress and an insufficient amount of antioxidants are responsible for the gastric mucosal damage [16]. The GI tract, the principal place of absorption of acrylamide, is thus directly exposed to the irritant effects of acrylamide, as well as systemic toxicity of its metabolite glycidamide [17]. According to the World Health Organization (WHO) data, the daily intake of acrylamide contained in food products ranges between 0.3 to 0.8 μg/kg of body weight [18]. It should be emphasized that the choice of the pig as a model animal, as well as young animals in the present experiment, was not accidental. The pig is an omnivorous animal, and the structure of the gastrointestinal tract and physiological processes occurring in it are similar to humans, which makes them a good animal model in biomedical research, especially concerning GI pathology [19,20]. In turn, products with a high content of acrylamide (like chips) are very popular, especially among young consumers, which makes them particularly vulnerable to their toxic effects.

On the other hand, it is known that both extrinsic (sympathetic, parasympathetic and sensory) as well as intrinsic (localized in the GI wall and forming the enteric nervous system (ENS)) neurons participate in the neural regulation of the digestive tract function. Based on autonomy, the huge number of neurons that create enteric plexuses, as well as numerous neurotransmitters synthesized and released by neurons, the ENS has been called the intestinal brain [21]. Moreover, the ENS is one of the first barriers of the organism against toxins in food. The ENS anatomy mainly depends on the animal species, as well as the part of the GI tract. In pigs and other large animals, in contrast to rodents, in the oesophagus and stomach, the ENS is arranged into two plexuses: The myenteric plexus (MP) and the submucous plexus (SP). While in the intestines, the SP is divided into two submucosal plexuses: The outer submucous plexus (OSP) and inner submucous plexus (ISP) [12,20]. Additionally, the ENS neurons show many adaptive changes, both functional and morphological, in response to inflammatory factors, injuries or toxins referred to as neural plasticity. One of them is alteration in the immunohistochemical phenotype of ENS neurons expressed as up- or down-regulation in the expression of neuroactive substances [2]. Although changes in the neurochemical profile of ENS neurons have been reported in many diseases and gastrointestinal toxicities, there is no data on the response of ENS neurons to acrylamide supplementation, particularly GAL-expression in stomach ENS neurons subjected to acrylamide toxicity. The present study was created to establish the impact of acrylamide supplementation, in tolerable daily intake (TDI) doses and a dose ten times higher than TDI, on the population of GAL-LI stomach neurons in pigs. Additionally, the possible functional co-operation of GAL with other neuroactive substances known from their neuroprotective features (VIP, nNOS, CART) and their role in acrylamide intoxication, was also investigated in the present study.

2. Results

During the present experiment, neurons displaying immunoreactivity to GAL were observed in each part of the stomach in animals from the control and experimental groups. The occurrence of GAL-LI cell bodies was detected in both the myenteric plexus (MP) and submucous plexus (SP) in all studied stomach fragments (Table 1, Figures 1 and 2). The most numerous populations of GAL-LI neurons with reference to all cells immunoreactive to protein gene product 9.5 (PGP 9.5) was found in the MP of the pylorus and cardia (25.14 ± 1.15% and 24.67 ± 0.67%, respectively) (Figure 1A,G). A slightly smaller number of GAL-positive cell bodies was noted in the corpus (19.71 ± 0.70%) (Figure 1D). In turn, in the SP, the largest group of GAL-LI nerve cells was found in the corpus (40.92 ± 0.84%), slightly smaller in the cardia (37.36 ± 0.71%), and the smallest was in the pylorus (36.87 ± 1.11%) (Figure 2A,D,G).

Table 1. GAL-LI immunoreactive neurons in various parts of the porcine stomach under physiological conditions (C group) and after low dose (LD group) or high dose (HD group) of acrylamide administration.

Part of the Stomach	Cardia			Corpus			Pylorus		
	C Group	LD Group	HD Group	C Group	LD Group	HD Group	C Group	LD Group	HD Group
MP	24.67 ± 0.67	29.89 ± 1.08 (**)	35.35 ± 0.64 (***)	19.71 ± 0.70	27.81 ± 1.23 (***)	33.55 ± 0.92 (***)	25.14 ± 1.15	33.59 ± 1.17 (***)	37.14 ± 0.43 (***)
SP	37.36 ± 0.71	40.43 ± 0.70 (**)	46.92 ± 0.42 (***)	40.92 ± 0.84	51.38 ± 1.28 (***)	58.53 ± 1.63 (***)	36.87 ± 1.11	40.37 ± 0.66 (*)	45.42 ± 0.58 (***)

MP- myenteric plexus, SP- submucous plexus. Significant differences were assessed with one-way analysis of variance (ANOVA) with Dunnett's test (* $p < 0.05$, ** $p < 0.01$, *** $p < 0.001$).

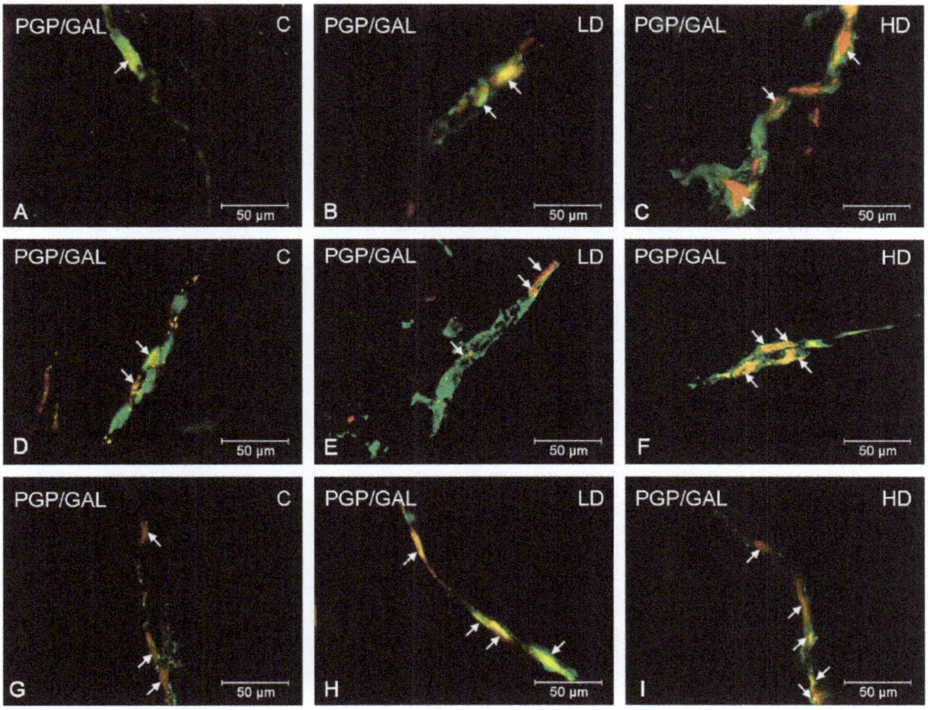

Figure 1. GAL-LI neurons in myenteric plexuses in the porcine stomach. The ENS neurons immunoreactive to protein gene-product 9.5 (PGP9.5)—used as a pan-neuronal marker and galanin (GAL) in myenteric plexuses in the porcine stomach under physiological condition (**A**,**D**,**G**), after low (**B**,**E**,**H**) and high doses (**C**,**F**,**I**) of acrylamide supplementation. Photographs A–C showing myenteric plexuses in the cardia, D–F—myenteric plexuses in the corpus and G–I—myenteric plexuses in the pylorus. All photographs have been created by digital superimposition of two colour channels (green for PGP 9.5 and red for GAL). Neurons immunoreactive to GAL are indicated with arrows.

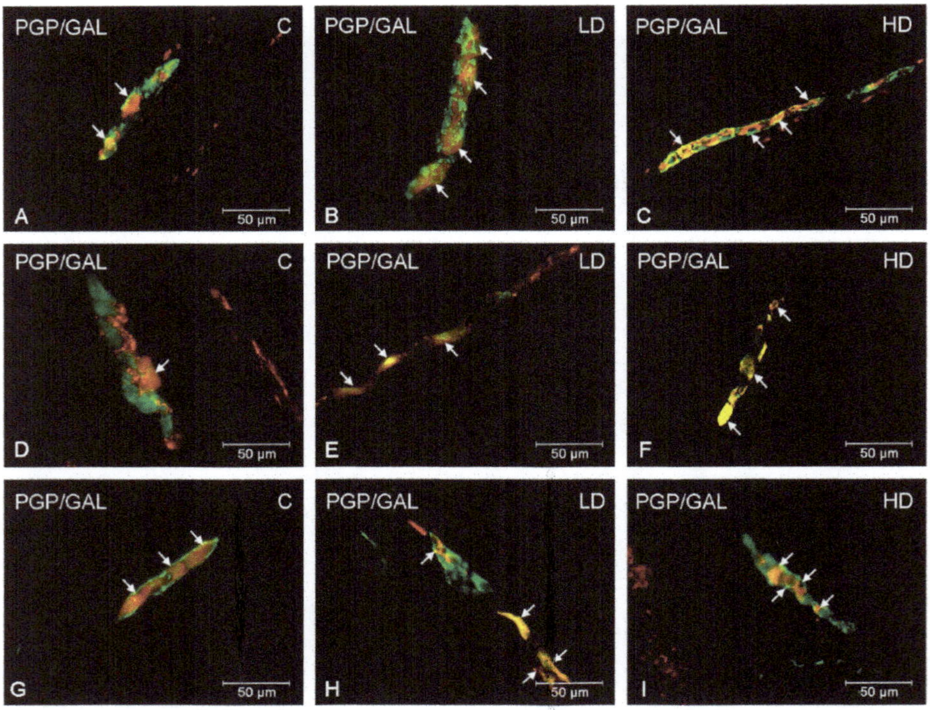

Figure 2. GAL-LI neurons in submucous plexuses in the porcine stomach. The ENS neurons immunoreactive to protein gene-product 9.5 (PGP9.5)—used as a pan-neuronal marker and galanin (GAL) in submucous plexuses in the porcine stomach under physiological condition (**A**,**D**,**G**), after low (**B**,**E**,**H**) and high doses (**C**,**F**,**I**) of acrylamide supplementation. Photographs A–C showing submucous plexuses in the cardia, D–F— submucous plexuses in the corpus and G–I— submucous plexuses in the pylorus. All photographs have been created by digital superimposition of two colour channels (green for PGP 9.5 and red for GAL). Neurons immunoreactive to GAL are indicated with arrows.

The supplementation of low and high doses of acrylamide affected the number of GAL-LI neurons in both the MP and SP in all studied fragments of the stomach (Table 1). The character of the observed changes depended on the type of enteric plexus and the segment of the stomach investigated. In particular, the most remarkable changes in the number of GAL-LI neurons, relative to control animals, was observed in the MP of the corpus in both experimental groups (an increase approximately 8 percentage points (pp) in the low dose (LD) and 14 pp in the high dose (HD) group, respectively) (Figure 1E,F). Similarly, in the SP of corpus, an increase in the expression of GAL was the most significant (approximately 11pp in the LD group and 18 pp in the HD group) (Figure 2E,F). A slightly smaller alteration was noted in the MP of the cardia (an increase of approximately 5 pp in the LD group and approximately 11 pp in the HD group) (Figure 1B,C) and pylorus (elevated approximately 8 pp in the LD group and 12 pp in the HD group) (Figure 1H,I). Comparable changes were observed in the gastric SP, where the increase in GAL-LI cell bodies was statistically significant (in the cardia, approximately 3 pp in LD group and 9 pp in the HD group; in the pylorus, approximately 4 in the LD group and 9 in the HD group, respectively) (Figure 2B,C,H,I).

In the present study, co-localization of GAL with other neuroactive substances studied (VIP, nNOS and CART) was noted within both stomach plexuses in animals under physiological conditions, as well as after low and high doses of acrylamide supplementation. The double-labelling immunofluorescence revealed extensive co-expression of GAL with VIP in the stomach ENS neurons (Table 2). Approximately half of the GAL-LI cell bodies in the MP were simultaneously immunoreactive to VIP (46.79 ± 2.18% in the cardia (Figure 3A), 43.20 ± 1.33% in the corpus and 47.53 ± 0.67% in the pylorus). Additionally, numerous MP neurons immunoreactive to GAL also showed the presence of nNOS (33.37 ± 0.79% in the cardia, 47.77 ± 1.22% in the corpus and 32.98 ± 0.51% in the pylorus (Figure 3D)) (Table 3). The degree of co-localization of GAL with CART in MP neurons was also high and amounted to 51.68 ± 0.85% in the cardia, 32.25 ± 1.12% in the corpus and 46.29 ± 1.30 in the pylorus (Figure 1G), respectively (Table 4). In turn, in SP neurons, the highest degree of co-localization with GAL was exhibited by CART (51.34 ± 1.02% in the cardia (Figure 4G), 32.41 ± 1.36% in the corpus and 51.30 ± 1.16% in the pylorus) (Table 4). A slightly smaller number of GAL-LI perikarya was simultaneously VIP-positive (48.20 ± 0.89% in the cardia, 39.32 ± 0.69% in the corpus (Figure 4A) and 48.36 ± 1.37% in the pylorus) (Table 2). Moreover, GAL-LI neurons in the SP were also immunoreactive to nNOS in all studied stomach fragments (in the cardia 35.17 ± 0.62% (Figure 4D); in the corpus 46.16 ± 1.15; in the pylorus 37.62 ± 1.51%) (Table 3).

Table 2. Co-localization of GAL with vasoactive intestinal polypeptide (VIP) in the enteric neurons in various parts of the porcine stomach under physiological conditions (C group) and after low dose (LD group) or high dose (HD group) of acrylamide administration.

GAL/VIP									
Part of the Stomach	**Cardia**			**Corpus**			**Pylorus**		
	C Group	**LD Group**	**HD Group**	**C Group**	**LD Group**	**HD Group**	**C Group**	**LD Group**	**HD Group**
MP	46.79 ± 2.18	50.28 ± 1.55	60.67 ± 2.34 (***)	43.20 ± 1.33	45.17 ± 0.73	51.49 ± 0.89 (***)	47.53 ± 0.67	50.57 ± 0.79	58.63± 1.72 (***)
SP	48.20 ± 0.89	52.51 ± 1.81	60.19 ± 2.06 (***)	39.32 ± 0.69	42.34 ± 0.78 (*)	49.27 ± 0.86 (***)	48.36 ± 1.37	51.48 ± 1.24	59.95 ± 1.16 (***)

MP- myenteric plexus, SP- submucous plexus. Significant differences were assessed with one-way analysis of variance (ANOVA) with Dunnett's test (* $p < 0.05$, ** $p < 0.01$, *** $p < 0.001$).

Table 3. Co-localization of GAL with nitric oxide synthase (nNOS) in the enteric neurons in various parts of the porcine stomach under physiological conditions (C group) and after low dose (LD group) or high dose (HD group) of acrylamide administration.

GAL/nNOS									
Part of the Stomach	**Cardia**			**Corpus**			**Pylorus**		
	C Group	**LD Group**	**HD Group**	**C Group**	**LD Group**	**HD Group**	**C Group**	**LD Group**	**HD Group**
MP	33.37 ± 0.79	39.28 ± 1.30 (**)	46.77 ± 1.20 (***)	47.77 ± 1.22	51.76 ± 1.46	56.81 ± 1.06 (***)	32.98 ± 0.51	39.50 ± 1.15 (**)	46.77 ± 1.38 (***)
SP	35.17 ± 0.62	39.92 ± 1.40 (*)	46.76 ± 0.86 (***)	46.16 ± 1.152	50.55 ± 1.23 (*)	57.23 ± 0.84 (***)	37.62 ± 1.51	42.32 ± 1.01 (*)	48.03 ± 1.14 (***)

MP- myenteric plexus, SP- submucous plexus. Significant differences were assessed with one-way analysis of variance (ANOVA) with Dunnett's test (* $p < 0.05$, ** $p < 0.01$, *** $p < 0.001$).

Table 4. Co-localization of GAL with cocaine- and amphetamine-regulated transcript (CART) in the enteric neurons in various parts of the porcine stomach under physiological conditions (C group) and after low dose (LD group) or high dose (HD group) of acrylamide administration.

GAL/CART									
Part of the Stomach	Cardia			Corpus			Pylorus		
	C Group	LD Group	HD Group	C Group	LD Group	HD Group	C Group	LD Group	HD Group
MP	51.68 ± 0.85	54.29 ± 1.52	66.74 ± 0.48 (***)	32.25 ± 1.12	36.39 ± 1.55	40.95 ± 0.76 (***)	46.29 ± 1.30	54.60 ± 0.51 (***)	56.81± 1.65 (***)
SP	51.34 ± 1.02	59.06 ± 1.69 (**)	63.62 ± 0.97 (***)	32.41 ± 1.36	34.98 ± 0.67	40.59 ± 1.46 (***)	51.30 ± 1.16	51.79 ± 1.24	54.97 ± 1.23

MP- myenteric plexus, SP- submucous plexus. Significant differences were assessed with one-way analysis of variance (ANOVA) with Dunnett's test (* $p < 0.05$, ** $p < 0.01$, *** $p < 0.001$).

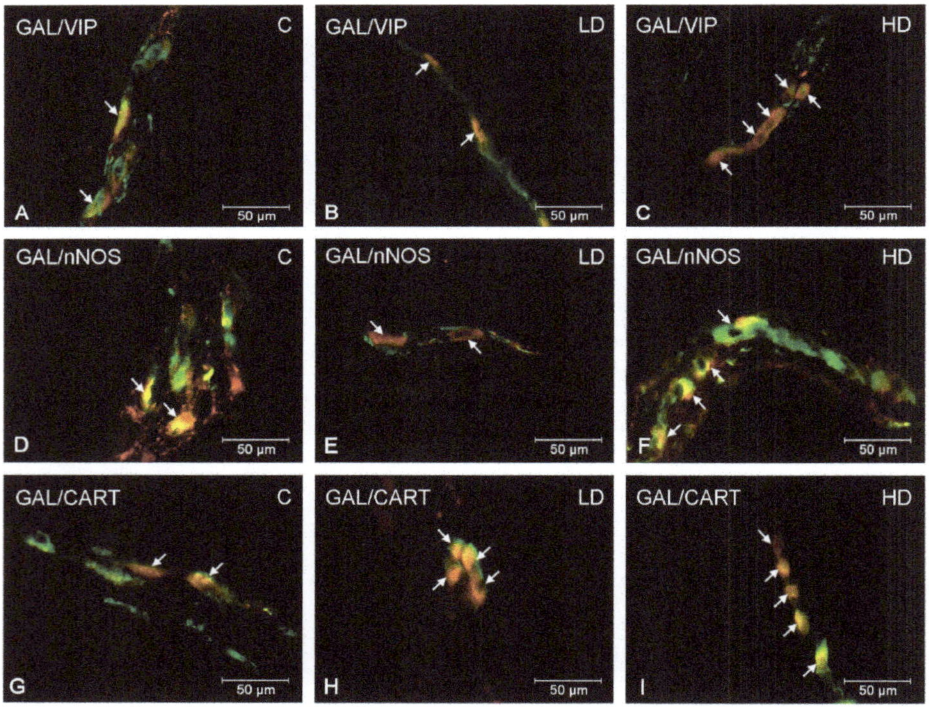

Figure 3. Co-localization of GAL with VIP, nNOS and CART in myenteric plexuses in the porcine stomach. The most representative images showing changes in the degree of co-localization of GAL with VIP in the myenteric plexuses of the cardia (**A**,**B**,**C**); GAL with nNOS in the myenteric plexuses of the pylorus (**D**,**E**,**F**); GAL with CART in the myenteric plexuses of the pylorus (**G**,**H**,**I**). Photographs A, D, G showing myenteric plexuses in the physiological conditions; B, E, H—after low and C, F, I after high doses of acrylamide administration. All photographs have been created by digital superimposition of two color channels (green for GAL and red for VIP, nNOS and CART respectively). Neurons immunoreactive to GAL and VIP or nNOS or CART are indicated with arrows.

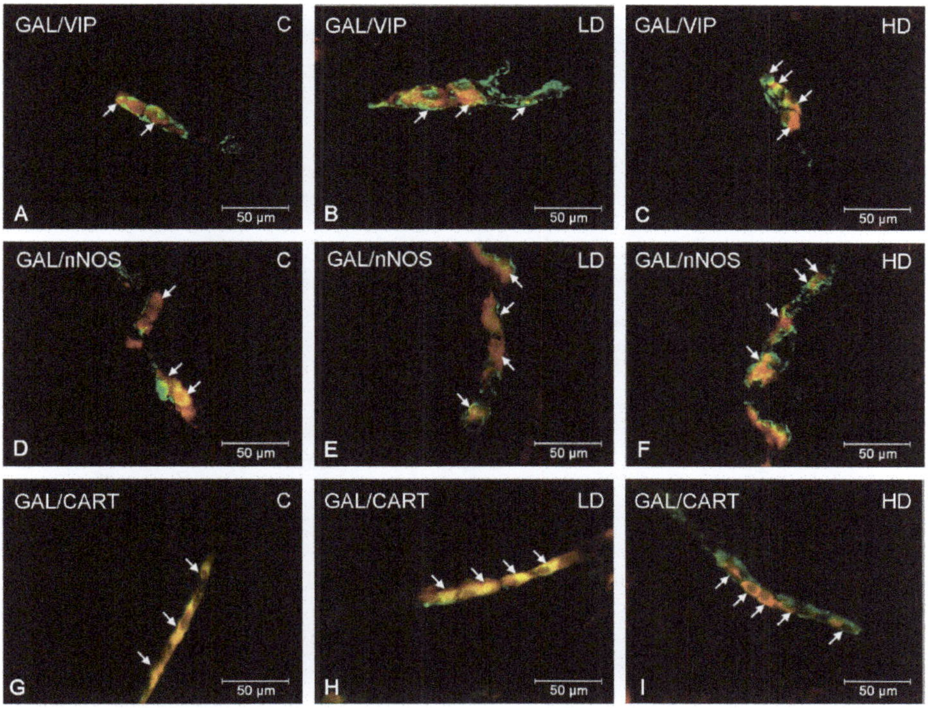

Figure 4. Co-localization of GAL with VIP, nNOS and CART in submucous plexuses in the porcine stomach. The most representative images showing changes in the degree of co-localization of GAL with VIP in the submucous plexuses of the corpus (**A**,**B**,**C**); GAL with nNOS in the submucous plexuses of the cardia (**D**,**E**,**F**); GAL with CART in the submucous plexuses of the cardia (**G**,**H**,**I**). Photographs A, D, G showing submucous plexuses in the physiological conditions; B, E, H—after low and C, F, I after high doses of acrylamide administration. All photographs have been created by digital superimposition of two color channels (green for GAL and red for VIP, nNOS and CART respectively). Neurons immunoreactive to GAL and VIP or nNOS or CART are indicated with arrows.

Acrylamide administration changed the number of GAL-LI enteric neurons simultaneously, immunoreactive to VIP, nNOS and CART in all stomach fragments studied (Tables 2–4). Both low and high doses of acrylamide substantially increased the number of SP neurons being simultaneously VIP- and GAL-positive in the stomach corpus (about 3 pp in the LD group and 10 pp in the HD group) (Figure 4B,C). However, in SP of the cardia and pylorus, significantly important alterations in VIP immunoreactivity in GAL-LI neurons were noted only in the HD group (about 12 pp in both stomach parts). In turn, in the MP, elevated numbers of GAL+/VIP+ perikarya were observed only in the HD group in all studied stomach fragments (about 14 pp in the cardia (Figure 3C), 8 pp in the corpus and 11 pp in the pylorus, respectively). In turn, nNOS-immunoreactivity in GAL-LI neurons increased the most in the pylorus and cardia where statistically relevant changes were noted in both experimental groups and stomach plexuses. In MP of the pylorus, an increase of approximately 7 pp in the LD group and 14 pp in the HD group were observed (Figure 3E,F), whereas in the cardia approximately 6 pp in the LD group and 13 pp in the HD group, respectively. In SP, in the cardia, the number of GAL+/nNOS+ neurons were elevated approximately 5 pp in the LD group and 12 pp in the HD group (Figure 4E,F), while in the pylorus the elevation was approximately 6 pp in the LD group and 10 pp in the HD group. On the other hand, in the corpus, significant changes in nNOS expression were observed in MP only in the HD group (an increase of approximately 9 pp). In turn, in the SP, similar to other parts of stomach in both LD and HD groups, the population of GAL+/nNOS+ neurons was elevated (about 4 and 11 pp, respectively). Furthermore, the expression level of CART in GAL-LI neurons was also significantly elevated. However, the character of observed changes depended on the part of the stomach as well as the kind of ENS plexus under investigation. In the MP of the cardia and corpus, only in the HD group was the increase statistically significant (approximately 15 pp in the cardia and 9 pp in the corpus), whereas changes were noticeable in the pylorus in both LD and HD group (approximately 8 pp and 10 pp, respectively) (Figure 3H,I). For SP, the most remarkable changes were noted in the cardia (an increase of approximately 8 pp in the LD group and 12 pp in the HD group) (Figure 4H,I). In the corpus, alterations in GAL+/CART+ neurons were observed only in the HD group (an increase of approximately 2 pp), although the changes observed in the pylorus were not statistically significant.

3. Discussion

During the present investigation, GAL- like immunoreactive neurons have been noted in both types of enteric plexuses within all stomach fragments studied. Previous studies have shown that GAL is distributed throughout the ENS of several species, including humans [2,8,11,12,22]. The occurrence of this peptide in enteric nervous structures (both neuronal cell and nerve fibers) as well as extrinsic neurons supplying the stomach was also confirmed [22–25]. The number of ENS neurons showing GAL-immunoreactivity clearly depends on the animal species and the part of the GI tract studied. Recent studies on the porcine stomach, confirmed by the results of the present experiment, have showed that GAL-positive neurons were present in a significant number of neurons in both the myenteric (MP) and submucous plexus (SP), with a slight predominance of SP [22,23]. It is well known that myenteric plexus mainly regulates the motility of the GI tract, whereas submucous plexus is responsible for the regulation of the secretion and conduction of pain reactions in the GI tract [26,27]. The above-mentioned distribution of GAL is well in line with its physiological function. In particular, GAL inhibits gastric acid and the pancreatic peptide secretions, modulates the motility of the GI tract and influences the release of other neurotransmitters [28,29]. However, the effect of its action clearly depends on both the animal species and the part of the GI tract. The inhibition of the motility in guinea pig ileum, canine pylorus and human intestines was reported [30–32]. In contrast, the opposite action was observed in mouse distal colon, rat stomach, as well as in the porcine and rabbit ileum [1,33,34].

Furthermore, in physiological conditions GAL-IR neurons in the ENS simultaneously showed immunoreactivity to other neurotransmitters and/or neuromodulators [35,36]. It is well known that neuroactive substances secreted in the same neurons may play similar or complementary roles [12,35]. During the present investigation, the co-localization of GAL with VIP, nNOS and CART was observed in both the SP and MP of the porcine stomach. VIP is known as a neuronal inhibitory factor within the digestive tract. It is involved in relaxation of smooth muscles, inhibits motor activity of the stomach and secretion of gastric juices but also stimulates mucus secretion, calcium ions and intestinal juice secretion [37,38]. The next substance was nNOS, an indicator of nitric oxide (NO). In the GI tract, NO inhibits smooth muscle contractility, regulates mesenteric and intestinal blood flow and participates in gut mucosal protection [39]. In turn, the physiological role of CART in the GI tract is not fully recognized. Ekblad et al. [40] suggested that CART is engaged in reducing gastric emptying, induction of stomach secretion and colonic motility. The co-localization of GAL with VIP, nNOS and CART observed in the present investigation, suggests the possible interaction of these substances in the regulation of physiological processes in the porcine stomach.

The present experiment, for the first time, has shown alterations in GAL-like immunoreactivity in the ENS neurons within the porcine stomach in response to acrylamide administration. The supplementation of TDI and ten times higher doses of acrylamide induced an increase in the number of GAL-positive neurons in both intramural plexuses (MP and SP) in all parts of the stomach under investigation. This is consistent with previous reports in which upregulated GAL-expression in the ENS structures was described in many pathological conditions within the GI tract, including inflammatory processes, tumors, neuronal damage and intoxication [2,12,22,41]. It is also in accordance with the neuronal plasticity phenomenon found in ENS neurons, defined as the ability to modify the chemical code of neurons by altering neuropeptide expression in response to changing environmental conditions, injures or other disorders [2,42].

A possible explanation of the changes observed in the present study may be the neurotoxic effects of acrylamide on the stomach ENS neurons. The results of experimental studies confirmed that acrylamide causes cell damage in the nervous and reproductive systems and leads to the occurrence of some types of cancer [43]. Particularly important, however, is its neurotoxic effects, which was examined, not only in experimental animals, but also in humans [43,44]. Until now, it has been shown that acrylamide causes swelling of distal axons leading to degeneration and, consequently, to the occurrence of peripheral neuropathy [45]. Further, acrylamide induces a neurotoxic effect by binding to cysteine-rich receptor proteins, causing nerve conduction disorders and disturbances in nerve function [43]. Although the influence of ACM on the peripheral nervous system is well known, knowledge of their impact on the ENS neurons is rather fragmentary. In particular, only our previous data have shown that even a low dose of acrylamide may induce changes in the number of enteric neurons positive to CART as well as VIP-, calcitonin gene-related peptide- and substance P–immunoreactive neuronal structures in various parts of the porcine GI tract [46,47]. It is suspected that the mechanism of neurotoxic action described in the peripheral nervous system is also similar in the case of ENS. However, this hypothesis needs to be confirmed by more detailed research. Nevertheless, the fluctuations in the number of GAL-LI neurons observed in this experiment are strongly supported by the fact that GAL is known to be an important neuroprotective factor. The association of elevated galanin expression with trophic and regenerative processes after peripheral nerve injury was described in both central and peripheral nervous systems [48]. Moreover, previous investigations have shown an increase in the number of GAL-LI structures in all intramural plexuses of the porcine descending colon under axotomy [2]. In view of the above-mentioned data, it may be speculated with a high probability that an increase in the number of GAL-LI neurons is associated with adaptive and neuroprotective changes of the enteric neurons in response to the neurotoxic effect of acrylamide. However, it is also possible that acrylamide led to damage of ENS neurons suggesting that cells lacking GAL are more vulnerable to acrylamide. In the literature, there is a lack of data describing the influence of acrylamide consumed in food products on the survival of the ENS neurons. However, an in vitro study on the cultured rats myenteric neurons did not cause death of neuronal cells but led to a decrease in number of axons [49]. It is therefore more likely that the increase in the number of GAL-LI neurons observed in the present study is the result of up-regulated synthesis of GAL in ENS neurons.

On the other hand, it is also very likely that the observed changes result from the pro-inflammatory properties of acrylamide. Chronic exposure to acrylamide in humans leads to fluctuations in the level of proinflammatory cytokines and enhanced levels of plasma C-reactive protein (CRP) [15]. Similarly, the inducible-form of nitric oxide synthase (iNOS) in the blood serum of albino mice was noted during acrylamide intoxication [50]. This corresponds well with the fact that GAL is commonly known to take part in the regulation of inflammatory processes. Accordingly, an increase in galanin expression has been noted in numerous animal models of inflammation, such as chemically-induced colitis [2] and enteric Salmonella infection [51]. Moreover, GAL modulates inflammatory responses by its effect on the secretion of cytokines, such as TNF-α, IL-1α and IL-8 [52]. Although the participation of GAL in inflammatory processes in the GI tract is unquestionable, further studies are necessary to confirm that changes observed in the course of acrylamide intoxication are associated with inflammation.

Furthermore, the present investigation, for the first time, demonstrated changes in the co-localization of GAL with other neuroactive substances in the stomach ENS neurons following acrylamide intoxication. The results showed that the proportion of GAL-expressing nerve cell bodies simultaneously immunoreactive to VIP, nNOS and CART increased. The changes were the most remarkable in the HD group. The present findings closely correlated with recent investigations in which neuroprotective properties of VIP, nNOS and CART in the ENS were demonstrated [35,40,46]. VIP is engaged in the regulation of inflammatory conditions, which closely interacts with the immune system, inhibits macrophage activity, supports Th2 cells and decreases T cell migration through intestinal Peyer's patches [53]. As an effect of its action, an elevated level of anti-inflammatory cytokines (mainly IL-10) and reduced synthesis of pro-inflammatory factors (such as TNF α, IL-6 and IL-12) were noted [54]. Moreover, VIP promotes the survival of cultured porcine myenteric neurons [55]. Similarly, nNOS leads to an increase in the survival rate of cultured neurons from adult rat colonic submucous ganglia [56]. An increased nNOS expression has also been reported in the ENS structures during extrinsic denervation [57] as well as in inflammatory bowel disease [58]. This confirms the important role of nNOS in neuronal plasticity. Additionally, the role of CART in neuroprotection was also discussed by many authors. In some studies, upregulated CART expression in the ENS structures were reported during hypertension, inflammatory conditions or intoxication [12,47,59,60]. The neurotrophic effect of CART was also demonstrated in the CNS [40]. The results of the present experiment, together with the above-mentioned data, point towards a neuroprotective role of CART. In view of the foregoing, it is very likely that changes in GAL-positive neurons reflect neuroprotective and recovery processes in the stomach ENS neurons following acrylamide intoxication. However, the mechanisms of these changes are unknown and further study is needed to clarify this issue.

4. Material and Methods

4.1. Animals and Experimental Procedures

The experimental procedure, including animal euthanasia, was approved by the Local Committee for Animal Experiments in Olsztyn (Approval No.: 11/2017). All possible efforts were made to minimize animal suffering. The experiment was conducted on 15 sexually immature female pigs of the Danish Landrace, approximately 8 weeks old and approximately 20 kg of body weight (b.w.). The gilts were kept under regular lighting conditions in a temperature-controlled environment. They were fed by commercial grain mixture and tap water ad libitum. After a period of acclimatization, the animals were randomly assigned into three groups: Control (C group), which were given empty gelatine capsules; a low dose group (LD group), animals given capsules with a tolerable daily intake (TDI) dose of acrylamide (0.5 μg/kg b.w./day); and a high dose group (HD group), animals given capsules with acrylamide in a dose ten times higher than TDI (5 μg/kg b.w./day). The capsules in all groups of animals were administered per os, once daily before the morning feed for 28 days. In order to determine the exact dose of acrylamide per animal, on the first day of the experiment, all animals were weighed, and then weighed once a week. After a period of supplementation, all gilts were treated with azaperone (Stresnil, Jansen Pharmaceutica N.V., Belgium, 4 mg/kg of body weight, i.m.) and after 15 min euthanized using a lethal dose of sodium pentobarbital (Morbital, Biowet Puławy, Puławy, Poland; 0.6 mL/kg of body weight, i.v.). Then, fragments of the stomach, the cardia, corpus and pylorus, were collected for further analysis. The tissues were fixed in 4% buffered paraformaldehyde (pH 7.4) for 1 h, rinsed in phosphate buffer (0.1 M, pH 7.4, at 4 °C) for 3 days with the exchange of the buffer every day, and then inserted into 18% phosphate buffered sucrose (at 4 °C) for 2 weeks. The frozen samples were cut using a Microm HM 560 cryostat (Carl Zeiss, Oberkochen, Germany) into 14 μm-thick sections and mounted on gelatine-coated slides.

4.2. Immunofluorescence Procedure

The frozen sections were then subjected to routine double-labelling immunofluorescence (as described previously by Palus et al. [61]). The sections were briefly air-dried at room temperature for 45 min and rinsed in 0.1 M phosphate-buffered saline (PBS, pH 7.4; 3 × 10 min). Subsequently, they were incubated with buffered blocking mixture (containing 10% horse serum and 0.1% bovine serum albumin in 0.1 M PBS, 1% Triton X-100, 0.05% Thimerosal and 0.01% sodium azide) for an hour in a humid chamber at room temperature. After rinsing in PBS (3 × 10 min), the samples were incubated overnight with a combination of antisera directed towards protein gene-product 9.5 (PGP 9.5;

mouse, cat. No. 7863-2004, Bio-Rad, Hercules, CA, USA, working dilution 1:1000, used here as a pan neuronal marker) and galanin (GAL, rabbit, cat. No. RIN7153, Peninsula, San Carlos, CA, USA, working dilution 1:3000), as well as GAL (guinea pig, cat. No. T-5036, Peninsula, San Carlos, CA, USA, working dilution 1:2000) and vasoactive intestinal peptide (VIP, rabbit, cat. No. 11428, Cappel, Aurora, OH, USA, working dilution 1:3000), cocaine- and amphetamine-regulated transcript peptide (CART, rabbit, cat. No. H-003-61, Phoenix Pharmaceuticals, Burlingame, CA, USA, working dilution 1:8000), neuronal nitric oxide synthase (nNOS, rabbit, cat. No. AB5380, Sigma-Aldrich, Saint Louis, MO, USA, working dilution 1:2000). On the next day, following subsequent rinsing in PBS (3×10 min), the sections were incubated with species-specific secondary antibodies (appropriate products: Alexa Fluor 488 (donkey anti-mouse IgG, cat. No. A21202, Invitrogen, Carlsbad, California, USA, working dilution 1:1000), Alexa Fluor 488 (donkey anti-guinea pig IgG, cat. No. A11073, Invitrogen, Carlsbad, California, USA, working dilution 1:1000) and Alexa Fluor 546 (goat anti-rabbit IgG, cat. No. A11010, Invitrogen, Carlsbad, California, USA, working dilution 1:1000) for 1 h at room temperature. Next, the washed sections (PBS, 3×10 min) were cover-slipped in carbonate-buffered glycerol (pH 8.6).

4.3. Negative Control

The standard controls, i.e., pre-absorption for the neuropeptide antisera with appropriate antigen: PGP 9.5 (AbD Serotec, Kidlington, UK), GAL (026-06, Phoenix Pharmaceutical), nNOS (N3033, Sigma, St Louis, MO, USA), VIP (064-24, Phoenix Pharmaceutical), and CART (Phoenix Pharmaceuticals, Burlingame, CA, USA) at a concentration of 20 µg/mL for 18 h at 37 °C and the omission, as well as the replacement of all primary antisera by non-immune sera, were performed to test immunohistochemical labelling. The above-mentioned controls completely eliminated labelling in the tissue.

4.4. Counting and Statistics

The stained preparations were then analyzed using an Olympus BX51 epi-fluorescence microscope and photographed with a digital camera connected to a PC and then processed with Olympus Cell F image-analysis software (Olympus, Tokyo, Japan). The number of the GAL-like immunoreactive (LI) enteric neurons was expressed as a percentage of the total number of PGP 9.5 positive perikarya. At least 500 PGP 9.5 labelled cell bodies in sections, at least 200 µm away from each other, were counted per animal in both the myenteric (MP) and submucosal plexus (SP) within each part of the stomach (cardia, corpus and pylorus). In order to determine the co-localization of GAL with other neuroactive substances (VIP, nNOS and CART), at least 100 GAL-positive cell bodies were investigated for immunoreactivity to particular neuronal factors. In these studies, GAL-positive neurons were considered as representing 100%. Only cells with well-visible nucleus were considered, pooled and expressed as a mean ± standard error of mean (SEM). The results from each group were analyzed statistically with Statistica 12 (StatSoft Inc., Tulsa, OK, USA). The significant differences were assessed with one-way analysis of variance (ANOVA) with Dunnett's test (* $p < 0.05$, ** $p < 0.01$, *** $p < 0.001$).

5. Conclusions

In summary, the present experiment revealed that supplementation of acrylamide triggers a reaction of the porcine stomach ENS neurons, expressed as an increased number of GAL-like immunoreactive neurons. Simultaneously, an increase in the expression of VIP, nNOS and CART in GAL-LI neurons was also observed. The observed alterations may be associated with adaptive and neuroprotective changes of the enteric neurons in response to the neurotoxic effect of acrylamide or may result from the inflammatory conditions that may accompany acrylamide supplementation. This suggests neurotrophic or/and neuroprotective properties of GAL and possible co-operation of GAL with VIP, nNOS, CART in the recovery processes within the ENS. Furthermore, the marked changes observed with exposure even to a low dose (TDI) of acrylamide raise doubts about the safe dose of acrylamide for consumers. In view of the growing demand for products containing large amounts of acrylamide, especially among children, the present study may be the starting point for further toxicological, pharmacological and clinical studies to ensure consumer safety.

Author Contributions: Conceptualization, K.P.; investigation, K.P. and K.M.; methodology, K.P.; writing—original draft, K.P.; writing—review and editing, J.C.

Funding: Funded by KNOW (Leading National Research Centre) Scientific Consortium "Healthy Animal—Safe Food", decision of Ministry of Science and Higher Education No. 05-1/KNOW2/2015 and the University of Warmia and Mazury in Olsztyn (statutory research): grant No 15.610.003-300.

Conflicts of Interest: The authors declare no conflicts of interest with respect to the publication of this manuscript.

References

1. Lang, R.; Gundlach, A.L.; Kofler, B. The galanin peptide family: receptor pharmacology, pleiotropic biological actions, and implications in health and disease. *Pharm. Ther.* **2007**, *115*, 177–207. [CrossRef] [PubMed]
2. Gonkowski, S.; Burliński, P.; Skobowiat, C.; Majewski, M.; Całka, J. Inflammation- and axotomy-induced changes in galanin-like immunoreactive (GAL-LI) nerve structures in the porcine descending colon. *Acta Vet. Hung.* **2010**, *58*, 91–103. [CrossRef] [PubMed]
3. Arciszewski, M.B.; Barabasz, S.; Całka, J. Immunohistochemical localization of galanin receptors (GAL-R1, GAL-R2, and GAL-R3) on myenteric neurons from the sheep and dog stomach. *Ann. Anat.* **2008**, *190*, 360–367. [CrossRef] [PubMed]
4. Kisfalvi Jr, I.; Burghardt, B.; Bálint, A.; Zelles, T.; Vizi, E.S.; Varga, G. Antisecretory effects of galanin and its putative antagonists M15, M35 and C7 in the rat stomach. *J. Physiol. Paris* **2000**, *94*, 37–42. [CrossRef]
5. Tatemoto, K.; Rökaeus, A.; Jörnvall, H.; McDonald, T.J.; Mutt, V. Galanin—A novel biologically active peptide from porcine intestine. *Febs Lett.* **1983**, *164*, 124–128. [CrossRef]
6. Hoyle, C.H.; Burnstock, G. Galanin-like immunoreactivity in enteric neurons of the human colon. *J. Anat.* **1989**, *166*, 23–33. [PubMed]
7. Ekblad, E.; Rökaeus, A.; Håkanson, R.; Sundler, F. Galanin nerve fibers in the rat gut: distribution, origin and projections. *Neuroscience* **1985**, *16*, 355–363. [CrossRef]
8. Anselmi, L.; Stella SLJr Lakhter, A.; Hirano, A.; Tonini, M.; Sternini, C. Galanin receptors in the rat gastrointestinal tract. *Neuropeptides* **2005**, *39*, 349–352. [CrossRef] [PubMed]
9. Liu, H.X.; Hökfelt, T. The participation of galanin in pain processing at the spinal level. *Trends Pharm. Sci.* **2002**, *23*, 468–474. [CrossRef]
10. Locker, F.; Lang, A.A.; Koller, A.; Lang, R.; Bianchini, R.; Kofler, B. Galanin modulates human and murine neutrophil activation in vitro. *Acta Physiol.* **2015**, *213*, 595–602. [CrossRef] [PubMed]
11. Pidsudko, Z.; Kaleczyc, J.; Wasowicz, K.; Sienkiewicz, W.; Majewski, M.; Zajac, W.; Lakomy, M. Distribution and chemical coding of intramural neurons in the porcine ileum during proliferative enteropathy. *J. Comp. Pathol.* **2008**, *138*, 23–31. [CrossRef]
12. Szymanska, K.; Makowska, K.; Gonkowski, S. The Influence of High and Low Doses of Bisphenol A (BPA) on the Enteric Nervous System of the Porcine Ileum. *Int. J. Mol. Sci.* **2018**, *19*, 917. [CrossRef] [PubMed]
13. Van Lancker, F.; Adams, A.; De Kimpe, N. Chemical modifications of peptides and their impact on food properties. *Chem. Rev.* **2011**, *111*, 7876–7903. [CrossRef] [PubMed]
14. Attoff, K.; Kertika, D.; Lundqvist, J.; Oredsson, S.; Forsby, A. Acrylamide affects proliferation and differentiation of the neural progenitor cell line C17.2 and the neuroblastoma cell line SH-SY5Y. *Toxicol. In Vitro* **2016**, *35*, 100–111. [CrossRef] [PubMed]
15. Naruszewicz, M.; Zapolska-Downar, D.; Kośmider, A.; Nowicka, G.; Kozłowska-Wojciechowska, M.; Vikström, A.S.; Törnqvist, M. Chronic intake of potato chips in humans increases the production of reactive oxygen radicals by leukocytes and increases plasma C-reactive protein: a pilot study. *Am. J. Clin. Nutr.* **2009**, *89*, 773–777. [CrossRef] [PubMed]
16. Bhattacharyya, A.; Chattopadhyay, R.; Mitra, S.; Crowe, S.E. Oxidative stress: an essential factor in the pathogenesis of gastrointestinal mucosal diseases. *Physiol. Rev.* **2014**, *94*, 329–354. [CrossRef] [PubMed]
17. Zödl, B.; Schmid, D.; Wassler, G.; Gundacker, C.; Leibetseder, V.; Thalhammer, T.; Ekmekcioglu, C. Intestinal transport and metabolism of acrylamide. *Toxicology* **2007**, *232*, 99–108. [CrossRef]

18. WHO. *Health Implications of Acrylamide in Food*; FAO/WHO: Geneva, Switzerland, 2002. Available online: http://apps.who.int/iris/handle/10665/42563 (accessed on 15 April 2019).
19. Verma, N.; Rettenmeier, A.W.; Schmitz-Spanke, S. Recent advances in the use of Sus scrofa (pig) as a model system for proteomic studies. *Proteomics* **2011**, *11*, 776–793. [CrossRef]
20. Brown, D.R.; Timmermans, J.P. Lessons from the porcine enteric nervous system. *Neurogastroenterol. Motil.* **2004**, *16*, 50–54. [CrossRef]
21. Furness, J.B.; Callaghan, B.P.; Rivera, L.R.; Cho, H.J. The enteric nervous system and gastrointestinal innervation: integrated local and central control. *Adv. Exp. Med. Biol.* **2014**, *817*, 39–71.
22. Bulc, M.; Palus, K.; Zielonka, Ł.; Gajęcka, M.; Całka, J. Changes in expression of inhibitory substances in the intramural neurons of the stomach following streptozotocin- induced diabetes in the pig. *World J. Gastroenterol.* **2017**, *23*, 6088–6099. [CrossRef] [PubMed]
23. Zalecki, M.; Sienkiewicz, W.; Franke-Radowiecka, A.; Klimczuk, M.; Kaleczyc, J. The Influence of Gastric Antral Ulcerations on the Expression of Galanin and GalR1, GalR2, GalR3 Receptors in the Pylorus with Regard to Gastric Intrinsic Innervation of the Pyloric Sphincter. *PLoS ONE.* **2016**, *11*, e0155658. [CrossRef] [PubMed]
24. Palus, K.; Całka, J. Neurochemical Plasticity of the Coeliac-Superior Mesenteric Ganglion Complex Neurons Projecting to the Prepyloric Area of the Porcine Stomach following Hyperacidity. *Neural Plast.* **2016**, *2016*, 8596214. [CrossRef]
25. Gańko, M.; Całka, J. Localization and chemical coding of the dorsal motor vagal nucleus (DMX) neurons projecting to the porcine stomach prepyloric area in the physiological state and after stomach partial resection. *J. Mol. Neurosci.* **2014**, *52*, 90–100. [CrossRef] [PubMed]
26. Huizinga, J.D.; Martz, S.; Gil, V.; Wang, X.Y.; Jimenez, M.; Parsons, S. Two independent networks of interstitial cells of cajal work cooperatively with the enteric nervous system to create colonic motor patterns. *Front. Neurosci.* **2011**, *5*, 93. [CrossRef]
27. Timmermans, J.P.; Hens, J.; Adriaensen, D. Outer submucous plexus: an intrinsic nerve network involved in both secretory and motility processes in the intestine of large mammals and humans. *Anat. Rec.* **2001**, *262*, 71–78. [CrossRef]
28. Schepp, W.; Prinz, C.; Tatge, C.; Håkanson, R.; Schusdziarra, V.; Classen, M. Galanin inhibits gastrin release from isolated rat gastric G-cells. *Am. J. Physiol.* **1990**, *258*, G596–G602. [CrossRef] [PubMed]
29. Delvaux, M.; Botella, A.; Fioramonti, J.; Frexinos, J.; Bueno, L. Galanin induces contraction of isolated cells from circular muscle layer of pig ileum. *Regul. Pept.* **1991**, *32*, 369–374. [CrossRef]
30. Sternini, C.; Anselmi, L.; Guerrini, S.; Cervio, E.; Pham, T.; Balestra, B.; Vicini, R.; Baiardi, P.; D'agostino, G.L.; Tonini, M. Role of galanin receptor 1 in peristaltic activity in the guinea pig ileum. *Neuroscience* **2004**, *125*, 103–112. [CrossRef] [PubMed]
31. Guerrini, S.; Raybould, H.E.; Anselmi, L.; Agazzi, A.; Cervio, E.; Reeve, J.R., Jr.; Tonini, M.; Sternini, C. Role of galanin receptor 1 in gastric motility in rat. *Neurogastroenterol. Motil.* **2004**, *16*, 429–438. [CrossRef]
32. Allescher, H.D.; Daniel, E.E.; Dent, J.; Fox, J.E. Inhibitory function of VIP-PHI and galanin in canine pylorus. *Am. J. Physiol.* **1989**, *256*, G789–G797. [CrossRef] [PubMed]
33. Bauer, F.E.; Zintel, A.; Kenny, M.J.; Calder, D.; Ghatei MABloom, S.R. Inhibitory effect of galanin on postprandial gastrointestinal motility and gut hormone release in humans. *Gastroenterology* **1989**, *97*, 260–264. [CrossRef]
34. Botella, A.; Delvaux, M.; Bueno, L.; Frexinos, J. Intracellular pathways triggered by galanin to induce contraction of pig ileum smooth muscle cells. *J. Physiol.* **1992**, *458*, 475–486. [CrossRef] [PubMed]
35. Makowska, K.; Rytel, L.; Lech, P.; Osowski, A.; Kruminis-Kaszkiel, E.; Gonkowski, S. Cocaine- and amphetamine-regulated transcript (CART) peptide in the enteric nervous system of the porcine esophagus. *C. R. Biol.* **2018**, *341*, 325–333. [CrossRef] [PubMed]
36. Mirabella, N.; Lamanna, C.; Assisi, L.; Botte, V.; Cecio, A. The relationships of nicotinamide adenine dinucleotide phosphate-d to nitric oxide synthase, vasoactive intestinal polypeptide, galanin and pituitary adenylate activating polypeptide in pigeon gut neurons. *Neurosci. Lett.* **2000**, *293*, 147–151. [CrossRef]
37. Biancani, P.; Walsh, J.H.; Behar, J. Vasoactive intestinal polypeptide. A neurotransmitter for lower esophageal sphincter relaxation. *J. Clin. Investig.* **1984**, *73*, 963–967. [CrossRef] [PubMed]
38. Wiley, J.W.; O'Dorisio, T.M.; Owyang, C. Vasoactive intestinal polypeptide mediates cholecystokinin-induced relaxation of the sphincter of Oddi. *J. Clin. Investig.* **1988**, *81*, 1920–1924. [CrossRef]

39. Nishio, H.; Hayashi, Y.; Terashima, S.; Takeuchi, K. Role of endogenous nitric oxide in mucosal defense of inflamed rat stomach following iodoacetamide treatment. *Life Sci.* **2006**, *79*, 1523–1530. [CrossRef]
40. Ekblad, E. CART in the enteric nervous system. *Peptides* **2006**, *27*, 2024–2030. [CrossRef]
41. Kozłowska, A.; Godlewski, J.; Majewski, M. Distribution Patterns of Cocaine- and Amphetamine-Regulated Transcript- and/or Galanin-Containing Neurons and Nerve Fibers Located in the Human Stomach Wall Affected by Tumor. *Int. J. Mol. Sci.* **2018**, *19*, 3357. [CrossRef]
42. Vasina, V.; Barbara, G.; Talamonti, L.; Stanghellini, V.; Corinaldesi, R.; Tonini, M.; De Ponti, F.; De Giorgio, R. Enteric neuroplasticity evoked by inflammation. *Auton. Neurosci.* **2006**, *126–127*, 264–272. [CrossRef] [PubMed]
43. Shipp, A.; Lawrence, G.; Gentry, R.; McDonald, T.; Bartow, H.; Bounds, J.; Macdonald, N.; Clewell, H.; Allen, B.; Van Landingham, C. Acrylamide: review of toxicity data and dose-response analyses for cancer and noncancer effects. *Crit. Rev. Toxicol.* **2006**, *36*, 481–608. [CrossRef] [PubMed]
44. Lo Pachin, R.M. The changing view of acrylamide neurotoxicity. *Toxicol. Vitro* **2010**, *25*, 573–579.
45. Ewaleifoh, O.; Trinh, M.; Griffin, J.W.; Nguyen, T. A novel system to accelerate the progression of nerve degeneration in transgenic mouse models of neuropathies. *Exp. Neurol.* **2012**, *237*, 153–159. [CrossRef] [PubMed]
46. Palus, K.; Bulc, M.; Całka, J. Changes in VIP-, SP- and CGRP- like immunoreactivity in intramural neurons within the pig stomach following supplementation with low and high doses of acrylamide. *Neurotoxicology* **2018**, *69*, 47–59. [CrossRef] [PubMed]
47. Palus, K.; Makowska, K.; Całka, J. Acrylamide-induced alterations in the cocaine- and amphetamine-regulated peptide transcript (CART)-like immunoreactivity within the enteric nervous system of the porcine small intestines. *Ann. Anat.* **2018**, *219*, 94–101. [CrossRef]
48. Mahoney, S.A.; Hosking, R.; Farrant, S.; Holmes, F.E.; Jacoby, A.S.; Shine, J.; Iismaa, T.P.; Scott, M.K.; Schmidt, R.; Wynick, D. The second galanin receptor GalR2 plays a key role in neurite outgrowth from adult sensory neurons. *J. Neurosci.* **2003**, *23*, 416–421. [CrossRef] [PubMed]
49. Lourenssen, S.; Miller, K.G.; Blennerhassett, M.G. Discrete responses of myenteric neurons to structural and functional damage by neurotoxins in vitro. *Am. J. Physiol. Gastrointest. Liver Physiol.* **2009**, *297*, G228–G239. [CrossRef]
50. Santhanasabapathy, R.; Vasudevan, S.; Anupriya, K.; Pabitha, R.; Sudhandiran, G. Farnesol quells oxidative stress, reactive gliosis and inflammation during acrylamide-induced neurotoxicity: Behavioral and biochemical evidence. *Neuroscience* **2015**, *308*, 212–227. [CrossRef] [PubMed]
51. Matkowskyj, K.; Royan, S.V.; Blunier, A.; Hecht, G.; Rao, M.; Benya, R.V. Age-dependent differences in galanin-dependent colonic fluid secretion after infection with Salmonella typhimurium. *Gut* **2009**, *58*, 1201–1206. [CrossRef] [PubMed]
52. Dallos, A.; Kiss, M.; Polyánka, H.; Dobozy, A.; Kemény, L.; Husz, S. Galanin receptor expression in cultured human keratinocytes and in normal human skin. *J. Peripher. Nerv. Syst.* **2006**, *11*, 156–164. [CrossRef] [PubMed]
53. Delgado, M.; Pozo, D.; Ganea, D. The significance of vasoactive intestinal peptide in immunomodulation. *Pharm. Rev.* **2004**, *56*, 249–290. [CrossRef] [PubMed]
54. Ganea, D.; Hooper, K.M.; Kong, W. The neuropeptide vasoactive intestinal peptide: direct effects on immune cells and involvement in inflammatory and autoimmune diseases. *Acta Physiol.* **2015**, *213*, 442–452. [CrossRef] [PubMed]
55. Arciszewski, M.B.; Ekblad, E. Effects of vasoactive intestinal peptide and galanin on survival of cultured porcine myenteric neurons. *Regul. Pept.* **2005**, *125*, 185–192. [CrossRef] [PubMed]
56. Lin, Z.; Sandgren, K.; Ekblad, E. Increased expression of nitric oxide synthase in cultured neurons from adult rat colonic submucous ganglia. *Auton. Neurosci.* **2004**, *114*, 29–38. [CrossRef] [PubMed]
57. Yunker, A.M.; Galligan, J.J. Extrinsic denervation increases myenteric nitric oxide synthase-containing neurons and inhibitory neuromuscular transmission in guinea pig. *J. Auton. Nerv. Syst.* **1998**, *71*, 148–158. [CrossRef]
58. Rychlik, A.; Gonkowski, S.; Nowicki, M.; Calka, J. Inflammatory bowel disease affects density of nitrergic nerve fibers in the mucosal layer of the canine gastrointestinal tract. *Can. J. Vet. Res.* **2017**, *81*, 129–136.

59. Gonkowski, S.; Burlinski, P.; Szwajca, P.; Całka, J. Changes in cocaine- and amphetamine-regulated transcript-like immunoreactive (CART-LI) nerve structures of the porcine descending colon during proliferative enteropathy. *Bull. Vet. Inst. Pulawy* **2012**, *56*, 199–203. [CrossRef]
60. Kasacka, I.; Piotrowska, Z. Evaluation of density and distribution of CART-immunoreactive structures in gastrointestinal tract of hypertensive rats. *Biofactors.* **2012**, *38*, 407–415. [CrossRef]
61. Palus, K.; Bulc, M.; Całka, J. Changes in Somatostatin-Like Immunoreactivity in the Sympathetic Neurons Projecting to the Prepyloric Area of the Porcine Stomach Induced by Selected Pathological Conditions. *BioMed Res. Int.* **2017**, *2017*, 9037476. [CrossRef]

International Journal of
Molecular Sciences

MDPI

Article

EDB-FN Targeted Peptide–Drug Conjugates for Use against Prostate Cancer

Shang Eun Park [1], Kiumars Shamloo [1], Timothy A. Kristedja [2], Shaban Darwish [1,3], Marco Bisoffi [1,2], Keykavous Parang [1] and Rakesh Kumar Tiwari [1,*]

1. Center for Targeted Drug Delivery, Department of Biomedical and Pharmaceutical Sciences, Chapman University School of Pharmacy, Harry and Diane Rinker Health Science Campus, 9401 Jeronimo Road, Irvine, CA 92618, USA
2. Biochemistry and Molecular Biology, Schmid College of Science and Technology, Chapman University, Orange, CA 92866, USA
3. Organometallic and Organometalloid Chemistry Department, Chemical Industries Research Division, National Research Centre, 33 EL Bohouth St. (former EL Tahrir st.) Dokki, Giza 12622, Egypt

* Correspondence: tiwari@chapman.edu; Tel.: +1-714-516-5483

Received: 30 May 2019; Accepted: 28 June 2019; Published: 4 July 2019

Abstract: Prostate cancer (PCa) is the most common malignancy in men and is the leading cause of cancer-related male mortality. A disulfide cyclic peptide ligand [CTVRTSADC] **1** has been previously found to target extra domain B of fibronectin (EDB-FN) in the extracellular matrix that can differentiate aggressive PCa from benign prostatic hyperplasia. We synthesized and optimized the stability of ligand **1** by amide cyclization to obtain [KTVRTSADE] **8** using Fmoc/tBu solid-phase chemistry. Optimized targeting ligand **8** was found to be stable in phosphate buffered saline (PBS, pH 6.5, 7.0, and 7.5) and under redox conditions, with a half-life longer than 8 h. Confocal microscopy studies demonstrated increased binding of ligand **8** to EDB-FN compared to ligand **1**. Therefore, we hypothesized that the EDB-FN targeted peptides (**1** and **8**) conjugated with an anticancer drug via a hydrolyzable linker would provide selective cytotoxicity to the cancer cells. To test our hypothesis, we selected both the normal prostate cell line, RWPE-1, and the cancerous prostate cell lines, PC3, DU-145, LNCaP, and C4-2, to evaluate the anticancer activity of synthesized peptide–drug conjugates. Docetaxel (Doce) and doxorubicin (Dox) were used as anticancer drugs. Dox conjugate **13** containing disulfide linkage showed comparable cytotoxicity versus Dox after 72 h incubation in all the cancer cell lines, whereas it was found to be less cytotoxic on RWPE-1, suggesting that it can act as a Dox prodrug. Doce conjugate **14** was found to be less cytotoxic in all the cell lines as compared to drug alone.

Keywords: antiproliferative assay; conjugation; docetaxel; doxorubicin; extra domain B; fibronectin; Fmoc/tBu; peptide–drug conjugate; prostate cancer; solid-phase synthesis; targeting

1. Introduction

Prostate cancer (PCa) is the most common form of cancer in males and the second leading cause of death after lung cancer in the United States. According to the American Cancer Society, it is also the first leading cause of cancer mortality in males aged 65 and older in the United States [1]. There are several treatment options for prostate cancer patients, which depend on the clinical risk assessment at the time of diagnosis, including tumor stage and grade, but also patient age and health. High-risk local disease is routinely addressed by local treatment with curative intent, such as radical prostatectomy followed by adjuvant radiation or androgen ablation therapy. Unfortunately, a fraction of patients (~20%) will progress to more aggressive disease, manifested by a metastatic spread of primary tumor cells, which in the United States results in a mortality rate of ~30,000 individuals per year [2]. Recurrent systemic disease, including metastatic castration resistance prostate cancer (mCRPC) after androgen ablation therapy, is typically treated with chemotherapy, most notably with docetaxel (Doce) and cabazitaxel [3,4]. The anthracycline doxorubicin (Dox) as a main chemotherapeutic drug for prostate cancer remains in the exploratory setting, while its synthetic chemical modifications, such as the anthracenedione (mitoxantrone) is in clinical use [5]. Regardless of the chemotherapeutic drug in use, whether in single or in combinatorial regimens, the major problem of off-target effects leads to substantial morbidity due to side effects. This has led to multiple lines of research into targeting strategies, including the use of liposomes and nanoparticles [6,7]. We have previously reported the synthesis of cell-penetrating peptide–drug conjugates for the delivery of anticancer drugs [8]. Targeted therapy could deliver the cytotoxic or therapeutic molecules to cancer cells, thereby minimizing exposure to normal cells. This promises to reduce the side effects associated with drug-based systemic therapies [9]. The use of tumor-specific or homing peptides as an example of targeted delivery in cancer was shown by Ruoslahti et al. [10]. A heightened efficiency of the delivery of a cytotoxic agent may also counteract the efflux of drugs due to multidrug resistance pumps. Hence, tumor targeting with enhanced selectivity should be more beneficial for an effective therapeutic outcome. Several ligands have been reported for targeting cytotoxic or chemotherapeutics molecules, such as monoclonal antibodies, peptide hormones, receptor-binding ligands, aptamers, oligosaccharides, vitamins, folic acids, nanoparticles, and DUPA [11]. However, most of these targeting ligands lack specificity for tumor cells, which in the clinical setting, is expected to lead to side effects due to their effect on non-malignant normal cells. Therefore, there is a need to develop more tumor-specific targeting modalities.

The extracellular matrix (ECM) is a microenvironment within all types of tissues and organs containing non-cellular components such as collagen, fibroblasts (which ARE cells), fibronectins (FNs), and blood vessels as a part of the basal lamina for the support of epithelial cells [12]. The ECM is responsible for adhesion, migration, cell-to-cell communication, homeostasis, growth, and proliferation of the cells, and thus contributes significantly to metastasis of tumors [13]. In malignant tumors, the microenvironment exhibits unusual overexpression of cancer-related proteins, such as the extra-domain B fibronectin (EDB-FN). EDB-FN is a high-molecular glycoprotein that mediates cell adhesion and migration [14]. EDB-FN has been reported to be found around new blood vessels of tumors and provides a promising specific biomarker for cancer [15–17]. Therefore, EDB-FN could be used for targeted therapy in PCa. For example, Zheng et al. showed the use of a specific peptide ligand [CTVRTSADC] **1**, which targets EDB-FN in PCa (Figure 1). This ligand was used in the diagnosis of later stages of PCa, differentiating them from benign hyperplasia. Later, ligand **1** was used as a targeted contrast agent in MRI application for PCa [18]. Therefore, peptide ligand **1** could be selected as a potential targeting ligand to develop targeted chemotherapeutics for PCa.

The rationale for the selection of peptides as targeting ligands is due to the fact that peptides are highly selective, safer, and tolerable agents in vivo. The specificity of the peptide ligand for its target could be used to deliver an anticancer drug to their target in cancer. Thus, we hypothesized that the conjugates of EDB-FN targeted peptide ligand with anticancer drugs containing a suitable linker will release the anticancer drug in the PCa and minimize side effects in normal cells. Here in this study, we report the synthesis and modification in peptide ligand **1**, developing peptide-anticancer drug conjugates, their characterization, stability, and in vitro assays against a panel of PCa and normal prostate cell lines.

2. Results and Discussion

2.1. Chemistry

We synthesized two different cyclic peptide EDB-FN targeting [CTVRTSADC] **1** and [KTVRTSADE] **8** (Figure 1) using solid phase peptide chemistry and used for the conjugation with anticancer drugs. Two anticancer drugs, Dox and Doce, were used for the conjugation. We adopted six carbon long aminohexanoic acid to provide spacer followed by cathepsin-B enzyme sensitive linker to release the drug moiety efficiently. Two types of dipeptide cathepsin-B enzyme sensitive linkers including valine and citrulline (VCit) and phenylalanine and lysine (FK) were adopted in the early phase of the study, but later a tetrapeptide linker with better release profile involving glycine, phenylalanine, leucine, and glycine (GFLG) was chosen. To facilitate drug attachment, hydrazone, disulfide, and thiomaleimide linkages were used. To provide a spacer for better recognition of the cathepsin-B sensitive linker, dipeptide of glycine was attached on hydrazone linkage, or additional cysteine was attached on the *N*-terminal to facilitate disulfide linkage.

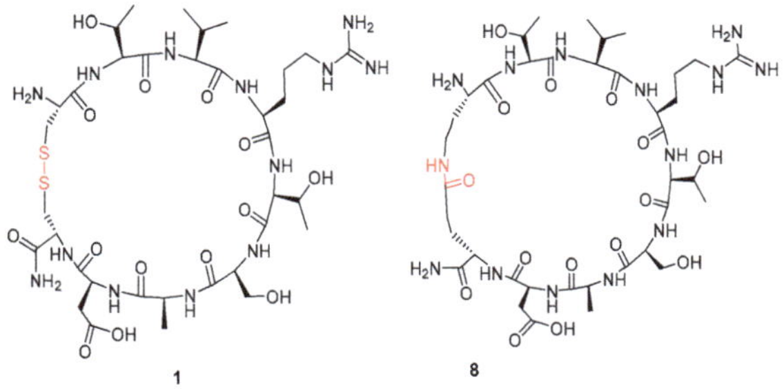

Figure 1. Chemical structures of peptide ligands **1** and **8**.

2.1.1. Synthesis of Targeting Ligand

Selected peptide ligand [CTVRTSADC] **1** contained a disulfide bridge between *C*- and *N*-terminal cysteine residues and was generated after the first-generation screening of phage display library bound to EDB-FN as previously reported [15]. Using this peptide, we designed the peptide-linker hydrazine-glutarate-GG-FK-C_6-[CTVRTSADC] **5** and hydrazine-glutarate-GG-VCit-C_6-[CTVRTSADC] **6** containing a cathepsin-sensitive linker (Scheme 1).

This disulfide linkage in targeting peptide **1** may be hydrolyzed in vivo due to the presence of glutathione or under physiological condition, which may reduce targeting efficiency of the ligand. Alternatively, we designed a more stable analog of ligand **1**, which contains lactam cyclization between lysine and glutamic side chain in place of cysteine amino acids. Therefore, two cyclic peptides [CTVRTSADC] **1** and [KTVRTSADE] **8** were synthesized using standard Fmoc/tBu solid-phase peptide synthesis (SPPS) on Rink amide resin as solid support. All the peptides were purified by preparative Reverse-Phase High-Performance Liquid Chromatography (RP-HPLC) and analyzed by Matrix-Assisted Laser Desorption/Ionization-Time of Flight (MALDI-TOF) mass spectroscopy (see supporting material). Most of the peptides showed an $[M + H]^+$ peak in the spectra, which is due to the presence of the protonated amino group. We also observed mass fragments presumably because of the breakdown of disulfide linkage to thiol (SH) groups when higher laser energy was used in the MALDI-TOF mass spectroscopy.

2.1.2. Synthesis of Peptide-Linker Conjugates

A spacer of six carbon (C_6) was included using Fmoc-aminohexanoic acid (Fmoc-Ahx-OH) in the design of peptide–drug conjugates to allow targeting ligand to recognize their binding motif. The release of drug from the conjugates was expected using a suitable linker. A cathepsin enzyme based linker was selected to allow the drug to be released at the site of the tumor [19]. Cathepsin-based linkers are dipeptides (FK, VCit) or tetrapeptides (GFLG), which release drug moiety efficiently from the conjugate as previously reported [20–22].

Scheme 1 depicts the synthesis of various cathepsin linkers in the peptide **1** cyclized through a disulfide linkage. A spacer of two glycine amino acids was added in the peptide sequence to avoid any steric hindrance for enzyme recognition of the linker. Furthermore, a hydrazine group was also included at the *N*-terminal to allow the selected model anticancer drug (Dox) to be conjugated using hydrazone linkage.

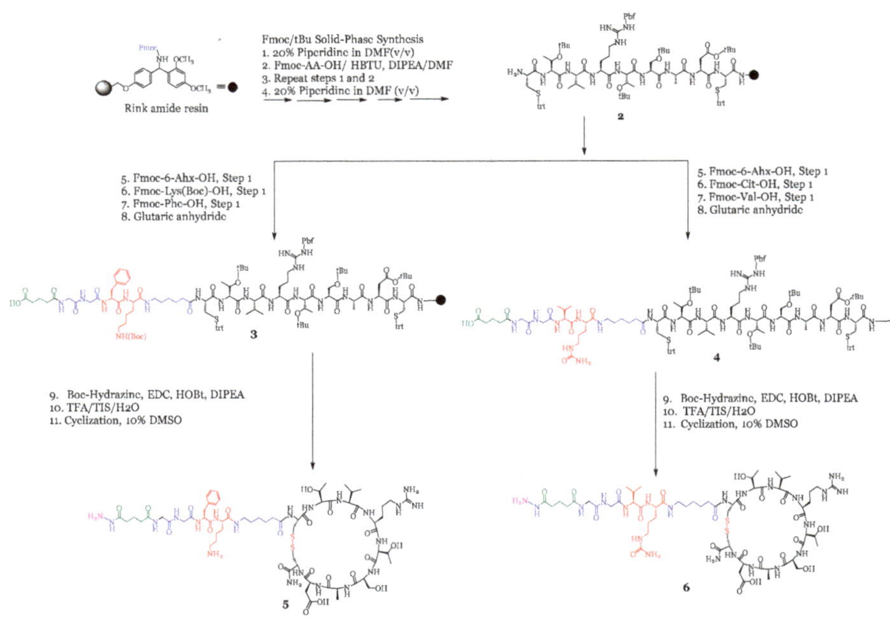

Scheme 1. Synthesis of hydrazine-glutarate-GG-FK-C_6-[CTVRTSADC] **5** and hydrazine-glutarate-GG-VCit-C_6-[CTVRTSADC] **6**.

Modified peptide ligand (KTVRTSADE) **8** was assembled on the resin with the modification of reported methodology of ligand **1** [18]. Scheme 2 depicts the synthesis of **8** that has lysine and glutamic acid residues instead of cysteine residues. Peptidyl resin went through lactam cyclization and conjugated with a linker. The orthogonal protected groups used during the synthesis for lysine and glutamic acid were 1-(4,4-dimethyl-2,6-dioxocyclohex-1-ylidene)ethyl (Dde) and 2-phenylisopropyl (Phipr), respectively. The protected linear version of peptide **8** was assembled on Rink amide resin followed by selective deprotection of Fmoc and Phipr group in the side chains. Lactam cyclization afforded side-chain cyclized intermediate peptide. After cleavage of a part of the resin, the calculated mass of **8** was 986.5145 $[M]^+$ that was consistent with experimental value 987.2699 $[M + H]^+$. Scheme 2 shows the synthesis of the linker, spacer, and targeting moiety in peptide **9**. Cysteine residue was incorporated at the *N*-terminal for functionalization in drug conjugation.

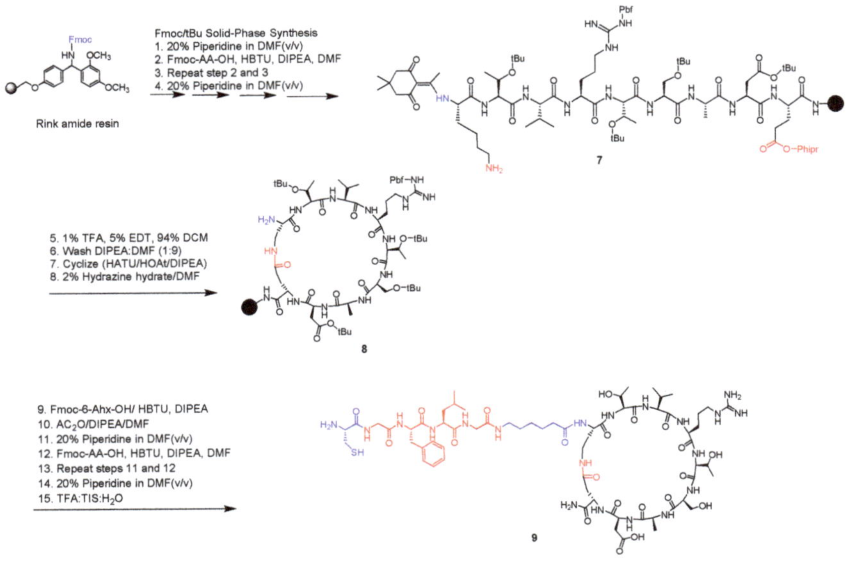

Scheme 2. Selected synthesis of C-GFLG-C_6-[KTVRTSADE] **9** containing lactam cyclization.

2.1.3. Conjugation of Anticancer Drugs with Peptide-Linker Conjugates

Doxorubicin (Dox) and docetaxel (Doce) were selected as model anticancer drugs for the conjugation with the synthesized linker-EDB peptides to form compound Dox-s-s-GFLG-C_6-[KTVRTSADE] **13** and Doce-βA-thioether-CGFLG-C_6-[KTVRTSADE] **14**. Dox and Doce have been widely used in chemotherapy against several cancers, including PCa [23]. The intermediate compound Dox-disulfide-pyridine was prepared by reaction of Dox with 2-iminothiolane followed by activation of the sulfhydryl group by reacting with dithiodipyridine at the room temperature (Scheme 3). Similarly, Doce was functionalized with Fmoc-β-alanine-OH to reduce the steric hindrance in the reaction with the linker-EDB peptide. The C2′ hydroxyl group in the Doce has been previously reported to be more reactive as compared to other hydroxyl groups in the compound. The other hydroxyl groups have shown to have low chemical reactivity due to the presence of more steric hindrance [24–26]. After Fmoc removal, the crude esterified Doce-β-alanine was purified using RP-HPLC. The purified compound showed one major peak with expected mass and was concentrated to use for the next reaction. A bifunctional linker sulfosuccinimidyl 4-(*N*-maleimidomethyl)cyclohexane-1-carboxylate) (sulfo-SMCC) was used to facilitate conjugation of Doce-β-alanine-NH_2 to form Doce-β-alanine-SMCC **11**, which was utilized in the conjugation with peptide **9** (Scheme 4). In another strategy, ketone group at the C13 position in the Dox was reacted under mildly acidic condition with the hydrazine group present at the EDB peptide-linker **6** to form hydrazone linkage in conjugate **12** as shown in Scheme 5.

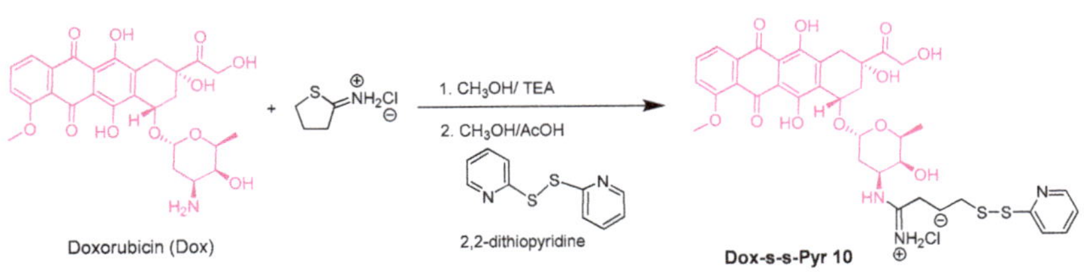

Scheme 3. Synthesis of Dox-disulfide-pyridyl **10**.

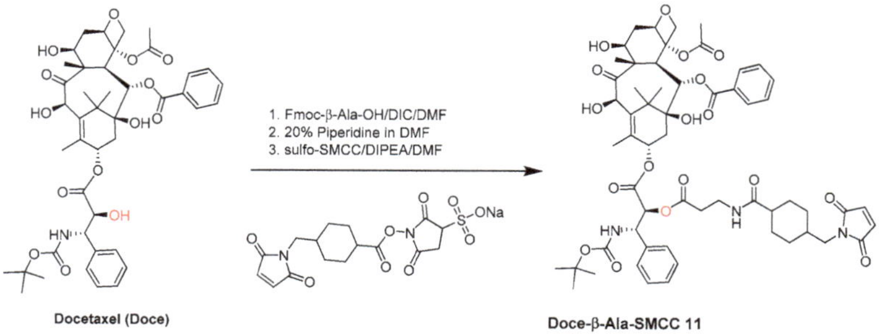

Scheme 4. Synthesis of Doce-βA-SMCC **11**.

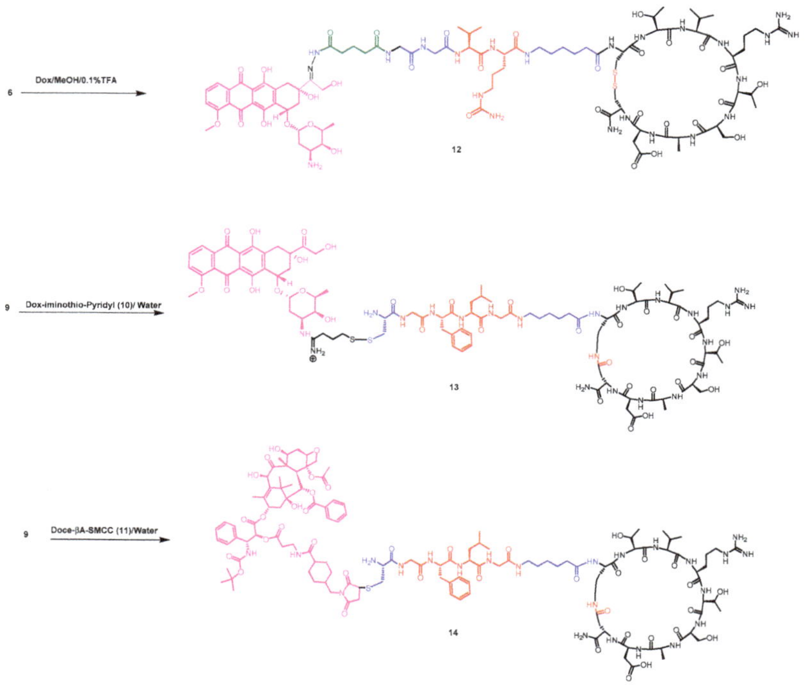

Scheme 5. Conjugation of Dox, Dox-SH, and Doce-βA-SMCC with the peptide targeting moiety through a hydrolyzable linker

2.1.4. Conjugation of Anticancer Drugs with Peptides

Three peptide–drug conjugated compounds (**12**, **13**, and **14**) were synthesized (Scheme 5). In the first reaction, Dox was reacted under a mild acidic condition with hydrazine-glutarate-GG-VCit-C_6-[CTVRTSADC] **6** for 20 h at room temperature to afford the Dox–hydrazone peptide conjugate **12** (Scheme 5). In the second attempt, Dox-activated disulfide compound **10** was reacted with C-GFLG-C_6-[KTVRTSADE] **9** to afford Dox-s-s-C-GFLG-C_6-[KTVRTSADE] **13**. In the third attempt, the Doce-β-alanine-SMCC **11** was reacted with sulfhydryl group in **9** using phosphate buffered saline (PBS, pH 8.0) containing 20 mM EDTA for 17 h to form Doce-thioether-CGFLG-C_6-[KTVRTSADE] **13**. All the conjugates were purified and characterized using RP-HPLC and MALDI-TOF mass spectrometry, respectively. Table 1 shows the chemical analysis and details of synthesized peptides and peptide–drug conjugates.

Table 1. The sequences of extra domain B of fibronectin (EDB-FN) targeting peptides and peptide–drug conjugates.

Peptide	Peptide Sequence	Chemical Formula	Exact Mass	Found (M/Z)
1	[CTVRTSADC]	$C_{35}H_{62}N_{13}O_{14}S_2$	952.3980	952.4786, $[M + H]^+$
5	Hydrz^b-Glutrt^b-GG-FK-C_6^{a}-[CTVRTSADC]	$C_{65}H_{111}N_{21}O_{21}S_2$	1585.7704	1585.5270, $[M +H]^+$
6	Hydrz^b-Glutrt^b-GG–VCita–C_6^{a}-[CTVRTSADC]	$C_{61}H_{109}N_{22}O_{22}S_2$	1565.7527	1565.5920, $[M + H]^+$
8	[KTVRTSADE]	$C_{40}H_{71}N_{14}O_{15}$	987.5223	987.2699, $[M + H]^+$
9	CGFLG-C_6^{a}-[KTVRTSADE]	$C_{68}H_{113}N_{20}O_{21}S$	1577.8110	1577.6815, $[M]^+$
10	Dox-s-s-Pyridine	$C_{36}H_{40}N_3O_{11}S_2$	754.2104	754.4053, $[M + H]^+$
11	Doce-βA-SMCC	$C_{58}H_{72}N_3O_{18}$	1098.4811	1098.1675, $[M + H]^+$
12	Dox-hydrz^c-Glutrt^b–GG–VCit^a–C_6^{a}-[CTVRTSADC]	$C_{88}H_{136}N_{23}O_{32}S_2$	2090.9003	2090.7130, $[M + H]^+$
13	Dox-s-s-CGFLG-C_6^{a}-[KTVRTSADE]	$C_{99}H_{150}N_{22}O_{32}S_2$	2223.0227	2222.3398, $[M +H]^+$
14	Doce-βA-thioether-CGFLG-C_6^{a}-[KTVRTSADE]	$C_{126}H_{184}N_{23}O_{39}S$	2675.2842	2675.8490, $[M + H]^+$
15	GFLG-C_6^{a}-[CTVRTSADC]	$C_{60}H_{100}N_{18}O_{19}S_2$	1440.6615	1440.5944, $[M + H]^+$
16	GFLG-C_6^{a}-[KTVRTSADE]	$C_{65}H_{109}N_{19}O_{20}$	1475.8096	1475.8738, $[M + H]^+$
17	FAM-GFLG-C_6^{a}-[CTVRTSADC]	$C_{81}H_{111}N_{18}O_{25}S_2$	1799.7249	1799.928, $[M + H]^+$
18	FAM-GFLG-C_6^{a}-[KTVRTSADE]	$C_{86}H_{118}N_{19}O_{26}$	1832.8495	1832.3740, $[M + H]^+$
19	CGFLG-C_6^{a}-[CTVRTSADC]	$C_{63}H_{106}N_{19}O_{20}S_3$	1544.6863	1544.1794, $[M + H]^+$

Cyclic peptides are shown using the [] bracket. [a]Unnatural amino acids denominated by βA, C_6, and Cit, which represent beta-alanine, 6-aminohexanoic acid, and citrulline, respectively. FAM stands for 5(6)-carboxyfluorescein. [b]Hydrazine and glutarate linkers are abbreviated as Hydrz and Glutrt, respectively. [c]Hydrazone.

2.1.5. Conjugation of 5(6)-Carboxyfluorescein (FAM) Motif with Peptides

Peptides **15** and **16** containing linker and targeting ligand (**1** and **8**, respectively) were conjugated with a carboxyfluorescein dye to determine whether ligand **8** showed higher localization and targeting toward EDB-FN as compared to its disulfide cyclized counterpart **1**. A 5(6)-carboxyfluorescein diisobutyrate (CFDI) was coupled to *N*-terminus of glycine in both peptides (**15** and **16**) in the presence of PyAOP, HOAt, and DIPEA using anhydrous DMF. Fluorescent-tagged peptides (**17** and **18**) were purified and characterized using RP-HPLC and MALDI mass spectrometry, respectively (Scheme 6).

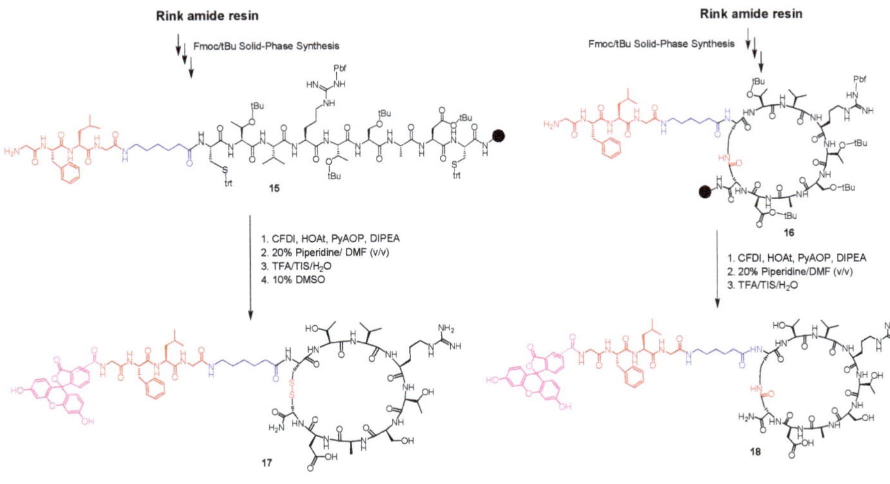

Scheme 6. Conjugation of FAM with the peptide targeting moiety through a hydrolyzable linker.

2.1.6. Stability and Hydrolysis

The stability of selected synthesized doxorubicin-peptide conjugate **13** was evaluated under different physiological conditions to mimic the in vivo environment, such as different pH, under redox condition, and in the human serum as depicted in the Figures 1–4. Conjugate **13** showed high stability in PBS with a half-life of ~10 h in pH values ranging from 6.5 to 7.4 (Figure 2). It also appears that the dox peptide conjugate **13** is stable at pH tested from 6.5 to 7.4. No drastic differences observed between stability in three different physiological pHs. Furthermore, conjugate **13** were treated with redox condition using dithiothreitol (DTT) to confirm the release of Dox from conjugate **13** that allows the free drug becomes bioactive. Figure 3 shows the hydrolysis of conjugate **13** within 2.6 min when incubated with 12 equivalents of DTT that confirms the release of Dox from disulfide linkage due to redox conditions. Similarly, the targeting peptides which consist of disulfide cyclization (peptide **1**) and lactam cyclization (peptide **8**) were incubated in the redox condition. In the presence of DTT, the targeting peptide ligand **1** behaved as expected and released reduced disulfide linkage quickly. During HPLC analysis of the ligand, it was found to be too difficult to separate the linear and cyclic forms of CTVRTSADC, so MALDI mass spectrometry was used to obtain a qualitative insight. Peptide **1** was reduced to its linear form that has a slightly higher mass because of the addition of two hydrogen atoms to the reduced disulfide bond into two sulfhydryl groups thereby increasing its mass by 2 Da in high-resolution MALDI-TOF mass spectroscopy (Figure 5). However, peptide **8** was stable under reducing conditions as expected (Figure 4).

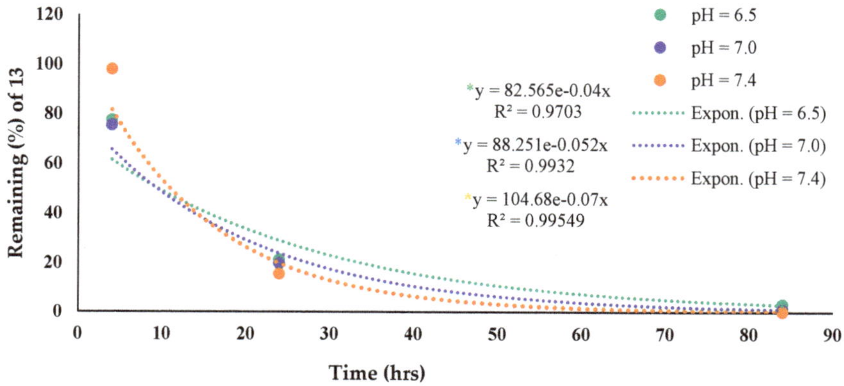

Figure 2. Conjugate **13** in PBS at different pH. The percent remaining of conjugate **13** was quantified using the area under the curve (AUC) in analytical HPLC.

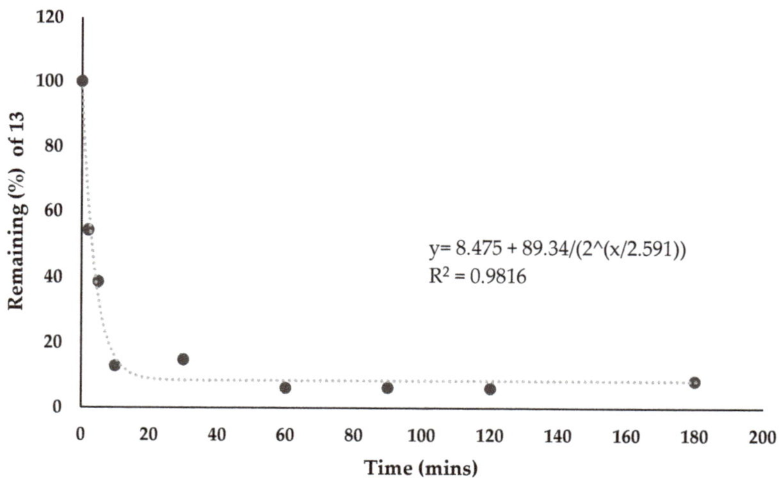

Figure 3. Stability analysis of conjugate **13** under reducing conditions using dithiothreitol (DTT). The experiment was done in PBS at pH 7.4. The percent remaining was quantified using the AUC, and the trendline equation gives the half-life of 2.6 min with an R^2 of 0.98. Fitting was done online using MyCurveFit.

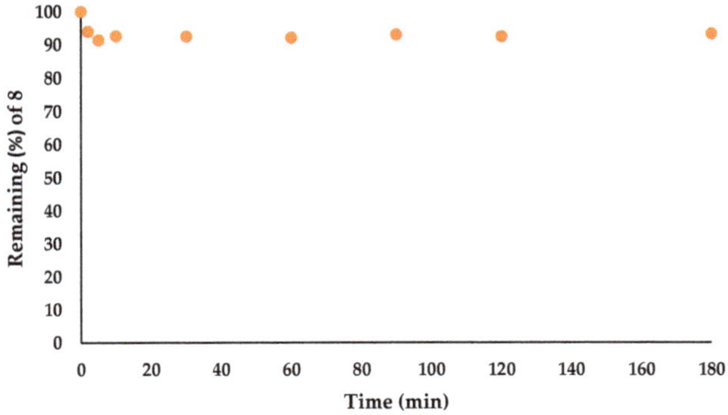

Figure 4. Stability of peptide **8** ([KTVRTSADE]) in pH 7.4 PBS with DTT. The percent remaining was calculated from the AUC.

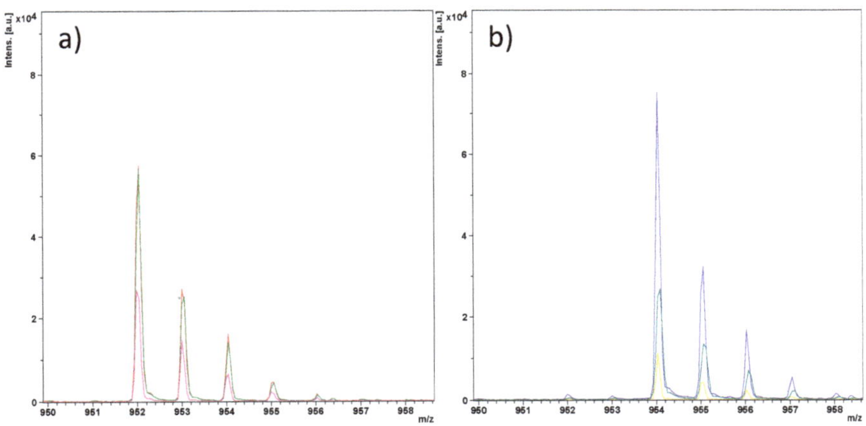

Figure 5. Peptide **1** before (**a**) and after 5 min (**b**) incubation with DTT. The higher molecular weight shows a full reduction of peptide **1** to the linear counterpart of (CTVRTSADC) within 5 min.

Dox–peptide conjugate **13** showed a quick degradation in human serum at 37 °C (Figure 6). The data shows that conjugate **13** has a half-life of approximately 11.8 min in 25% human serum. There was no visible difference in the shape of the percentage remaining vs. time plot when performed using a higher concentration of conjugate **13**. Thus, it was speculated that the rate of the reaction increased proportionally when a higher concentration (100 μM) of Dox–peptide was used. This indicates that the reaction was likely to be first-order with regard to Dox–peptide **13** under these conditions. Usually, peptides and peptide–drug conjugates have a very wide range of serum stabilities ranging from less than a min in human serum to over 200 min leading to an undetectable amount of degradation [27]. Conjugate **13** showed a half-life of 11.8 min, which displays that this peptide–drug conjugate, although does not have outstanding stability, it may have time to yield an effect on cancer cells in vivo.

Furthermore, the hydrolysis of peptide–drug conjugate **13** containing a cathepsin B cleavable linker was evaluated using in vitro assay as reported [28]. Figure 7 shows that cathepsin B enzyme recognizes GFLG linker in the conjugate **13** and cleaves it within 5 min of incubation time. Dox–peptide conjugate **13** showed a peak with the retention time of 34 min. After incubation with serum, the compound was hydrolyzed and showed different peaks. The peak with a retention time around 30.5 min was found to be dox conjugated with a couple of amino acid residues. The free doxorubicin was found with the retention time of 30 min (Supporting information, Figure S1). The peak at 30.5

min showed absorption at 490 nm and with some at 214 nm, which demonstrates amide linkage for conjugated amino acids. Therefore, the peak at 30.5 min was likely to be with Dox conjugated with a couple of amino acids. This enzymatic reaction occurred too fast to assess the kinetic nature of the reaction. However, it demonstrated the utility of the cathepsin B-sensitive linker in conjugate design.

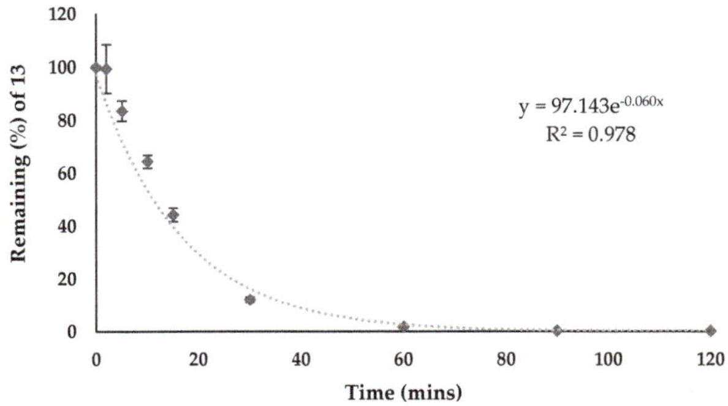

Figure 6. Stability of conjugate **13** (125 μM) in 25% human serum. Note the trendline gives the half-life of 11.8 min with an R^2 of 0.978. Each point represents an average of three readings.

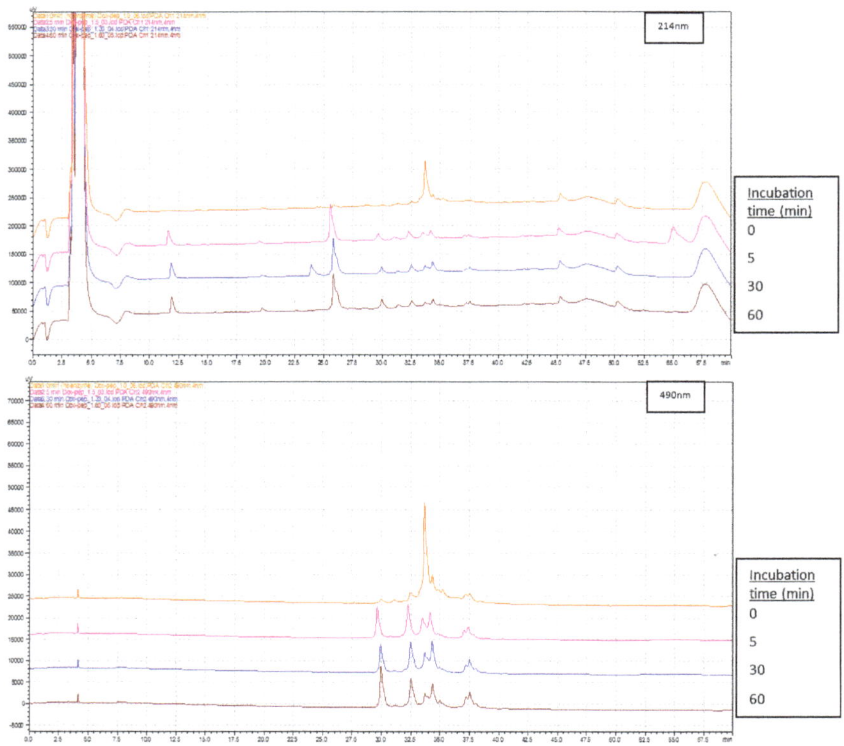

Figure 7. Analytical HPLC chromatogram over time for Dox-peptide conjugate **13** in human cathepsin B with absorbance detection at 214 nm (**top**) and 490 nm (**bottom**). Dox-peptide **13** retention time = 34 min. Dox alone retention time = 30 min. Note the similar shape of the 5, 30, and 60 min incubations. Also, the peak at ~30.5 min showed strong absorbance at both 214 nm and 490 nm.

2.2. Cell-Based Assays

2.2.1. Peptide Cytotoxicity

To evaluate the cytotoxicity of the synthesized compounds, we selected both a normal prostate cell line (prostate epithelial cells transformed by HPV, RWPE-1) and cancerous prostate cell lines (LNCaP, PC3, DU-145, and C4-2) [29]. LNCaP is an indolent form. DU-145 is a moderatetly metastatic cell form as compared to PC3, which is an aggressive form of PCa with a high metastatic potential [30]. C4-2 cells were originally derived from LNCaP xenografts grown in castrated mice exhibiting androgen-independence and were found to have elevated expression of EDB-FN, which would help explore the targeting of peptide conjugate [31]. Evaluation of the cytotoxicity of synthesized targeting peptides hydrazine-glutarate-GG-FK-C_6-[CTVRTSADC] **5** and hydrazine-glutarate-GG–VCit–C_6-[CTVRTSADC] **6** as compared to peptide C-GFLG-C_6-[KTVRTSADE] **9** were performed on different timelines (24, 48, and 72 h) using a panel of cell lines such as RWPE-1, LNCaP, PC3, DU-145, and C4-2 as depicted in Figure S2. These peptides were used in the conjugation of anticancer drugs connecting through three different types of peptidic cathepsin linkers such as phenylalanine lysine (FK), valine citrulline (VCit), and glycine phenylalanine leucine glycine (GFLG). These peptides were used to perform a cytotoxicity assay at a selected concentration of 5 μM as used previously in screening peptide–drug conjugates in different cancer cell lines [8]. Figure S2 (supporting information) showed that peptides **5, 6, 9** did not show any significant cytotoxicity to normal human prostate epithelial cells (RWPE-1), and cancerous prostate cell lines LNCaP, C4-2, PC3, and DU-145 after 72 h. These results are not surprising since the targeting peptides are not expected to have any intrinsic antiproliferative activity.

The determination of cytotoxicity of peptide C-GFLG-C_6-[KTVRTSADE] **9** and selected peptide–drug conjugate Dox-s-s-CGFLG-C_6-[KTVRTSADE] **13** were performed in PC3 cell line after 2 h of incubation using positive control (doxorubicin) and negative control (5% DMSO). Different concentrations (0.01, 0.1, 1, 10, and 100 μM) of Dox, peptide **9**, and conjugate **13** were prepared in 5% DMSO and used in the screening in the PC3 cell line after incubation for 2 h to observe the cytotoxicity of targeting peptide **9** in comparison to parent drug Dox and its conjugate **13** (Figure 8). The conjugate **13**, peptide **9**, and Dox showed no significant toxicity at or below 10 μM, possibly due to the shorter incubation time of 2 h. Dox usually requires 24 to 72 h incubation time to show any cytotoxicity at a lower concentration. The targeting peptide **9** did not show any toxicity up to the concentration of 100 μM.

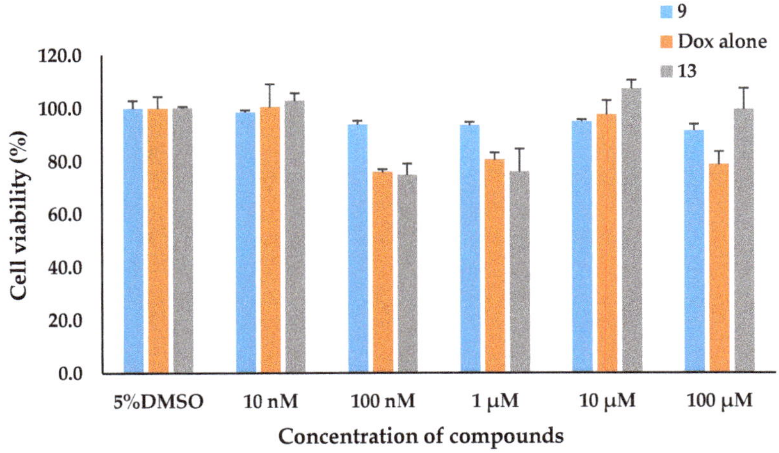

Figure 8. Cytotoxicity of peptide **9**, conjugate **13** (Dox-s-s-CGFLG-C6-[KTVRTSADE]) and Dox on a PC3 prostate cancer cell line after two hours of incubation at various concentrations.

The conjugates **13** and **14** were tested at 5 μM concentration using both positive and negative controls in all the four cell lines after 72 h incubation. Figure 9 depicted that Dox conjugate **13** was moderately toxic with a reduced cell proliferation to a range of 25–35% as compared to Dox which reduced cell proliferation in the range of 20–34% for all selected four cell lines. However, it was interesting to observe that Doce conjugate **14** was almost nontoxic (cell proliferation within the range of 89–96%) in all the cell lines as compared to Doce alone which reduced the cell proliferation in the range of 54–61% (Figure 9). The comparable potency of Dox conjugate **13** versus Dox suggests the possible reduction of disulfide linkage, and release of an active form of Dox from the conjugate after the cellular uptake. As described above in the stability studies, compound **13** was found to be unstable under redox conditions, and in the presence of cathepsin B. On the other hand, the inactivity of conjugate **14** is possibly due to lack of hydrolysis to get an active drug, Doce. Conjugate **14** does not have disulfide linkage and it is possible that ester bond of the secondary alcohol and amide bond do not undergo fast hydrolysis.

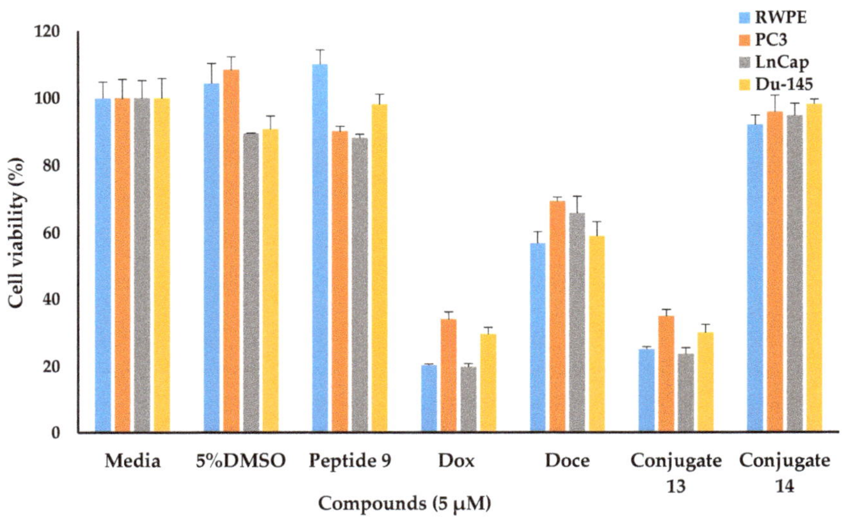

Figure 9. Cell viability of peptide **9**, Dox, Doce, Dox–peptide conjugate **13**, and Doce–peptide conjugate **14** at 5 μM on prostate epithelial and PCa cell lines (RWPE-1, PC3, LNCap, Du-145) after 72 h using MTS assay.

Furthermore, to evaluate the effect of selectivity of synthesized conjugates containing targeting ligand, the cells were treated with TGF-β for 3 days to induce overexpression of EDB-FN as reported and assayed using cell viability assay for 24–72 h [32]. Also, an equimolar amount of the physical mixture of drug and peptides were evaluated to determine the potency of these compounds in TGF-β positive treated cells. C4-2 cell line was used with and without TGF-β treatment (Figure 10a,b) as it was reported to show overexpression of EDB-FN [31]. C4-2 cells were derived from human prostate adenocarcinoma LNCaP cells. Figure 10a showed an increase in the cytotoxicity of Dox and Dox/peptide **9** physical mixture as compared to the conjugate **13** over an incubation period of 24 h to 72 h. Conjugates **13** and **14** were found to be less cytotoxic as compared to drug alone in 24–72 h. These cells were not treated with TGF-β, so very minimal or no overexpression of EDB-FN. Figure 10b showed the effect of overexpression of EDB-FN in the cell viability. There was no observed effect of TGF-β treatment for the cytotoxicity of Dox and physical mixture of Dox/peptide **9** on the cell viability as compared to the TGF-β untreated cell lines. However, conjugate **13** showed a decrease in cell viability by 17% after 72 h as compared to untreated cell lines. Similarly, Doce and Doce conjugate

14 showed decrease in cell viability by 16 and 10%, respectively, after 72 h. The physical mixtures of Doce/peptide **9** showed a decrease in cell viability by 16% as compared to untreated cells.

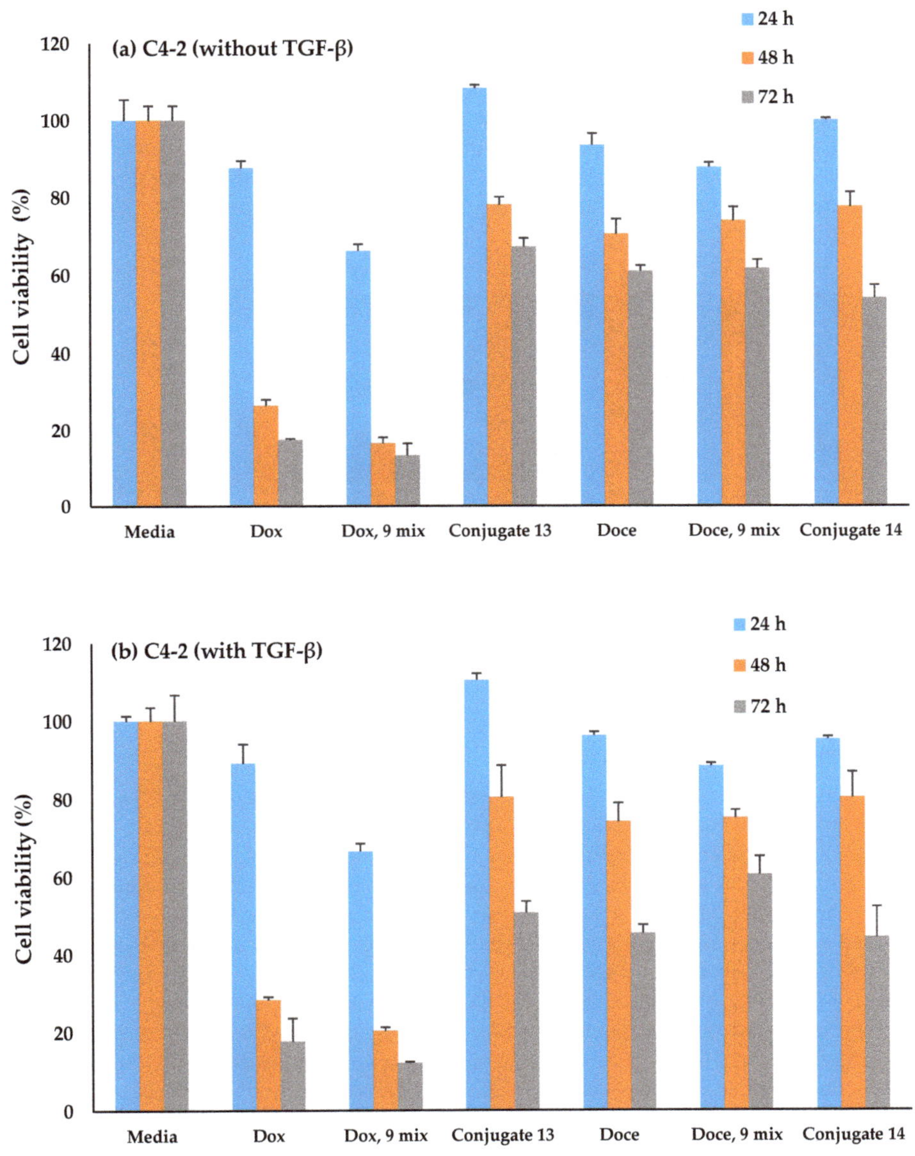

Figure 10. Time-dependent (24, 48, 72 h) cell viability studies of drugs, peptide 9, and drug–peptide conjugates **13**, **14**, and physical mixtures of drug and peptides using MTS on (**a**) prostate C4-2 cell line, a subline of LNCaP without TGF-β, (**b**) with TGF-β treatment.

2.2.2. Confocal Microscopy of the Fluorescently Tagged Compounds **17** and **18**

To evaluate the comparative EDB-FN binding capability of disulfide cyclized peptide **1** and its more stable analog **8**, we prepared the corresponding FAM-conjugated fluorescent tagged derivatives, namely peptide **17** and **18**, respectively. C4-2 cells were used for this assay due to overexpression of EDB-FN [31]. Expression of EDB-FN was confirmed using PCR. The C4-2 cells were incubated with the FAM-conjugated peptide **17** and **18** dissolved in the cell culture media for 12 h. Figure 11 shows the confocal images of EDB-FN stained cells without any peptide (a, control), incubated with peptide **17** (b) and peptide **18** (c) respectively at 60× magnification. As it is evident from Figure 11, peptide **18** showed notably higher binding and colocalization for EDB-FN as compared to peptide **17**. A binary contrast quantification showed a significant increase of 48% in the peptide-stained area in cells incubated with peptide **18** compared to that of peptide **17**. These images reflect binding of peptide **17** and **18** to EDB-FN, which includes linkers and spacer, which presumably did not change the affinity of ligands for EDB-FN. The enhanced binding of peptide **18** over peptide **17** was speculated to be due to the higher stability of ligand 18.

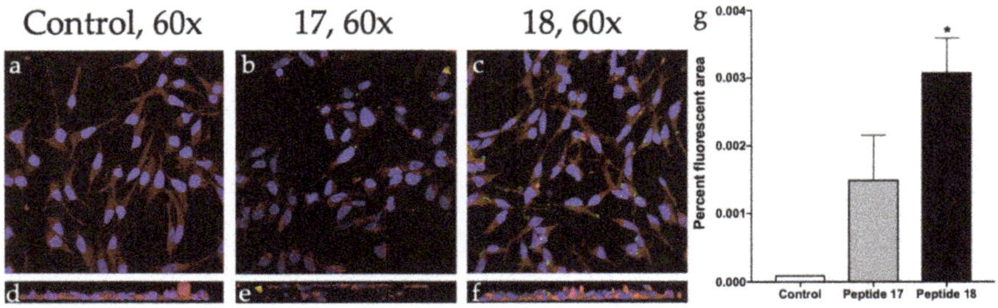

Figure 11. Comparative confocal microscopy of Fluorescent tagged peptide **17** and **18**. Control (**a**) shows no treatment; (**b**) shows FAM-linked disulfide cyclized targeting moiety in the peptide **17**; (**c**) shows FAM linked amide cyclized targeting moiety in the peptide **18**; (**d**), (**e**), and (**f**) are Z-slices of images in (**a**), (**b**), and (**c**); (**g**) shows the unpaired *t*-statistic reflects peptide **18** has higher binding to EDB-FN as compared to peptide **17**.

3. Materials and Methods

3.1. Chemistry

All amino acids and resins were purchased from AAPPTec LLC (Louisville, KY, USA), Chem-Impex International Inc. (Wood Dale, IL, USA), and CEM Corporation (Matthews, NC, USA). All organic solvents were purchased from Millipore Sigma Corporation (St. Louis, MO, USA), Fischer Scientific (Pittsburgh, PA, USA), and Gyros Protein Technologies, Inc (Tucson, AZ, USA). Anticancer agents Doxorubicin (Dox) and Docetaxel (Doce) were purchased from LC laboratories (Woburn, MA, USA). Masses of intermediate and final products were confirmed by high-resolution matrix-assisted laser desorption/ ionization time-of-flight (MALDI-TOF) mass spectrometer from Bruker Inc. (GT 0264, Billerica, MA, USA) or Applied Biosystems (4800 MALDI TOF/TOF Analyzer, Foster City, CA, USA). Intermediate and final compounds were purified by reversed-phase high-performance liquid chromatography (RP-HPLC) from Shimadzu (Prominence, Columbia, MD, USA) using a gradient system of acetonitrile and water with 0.1% trifluoroacetic acid using reverse phase C18 column (XBridge BEH130 Prep C18), from Waters Corporation (Milford, MA, USA). 5(6)-Carboxyfluorescein diisobutyrate (CFDI) was used to synthesize fluorescently-label peptide (USBiological Life Science, Swampscott, MA, USA).

3.1.1. Synthesis of Peptide 1 ([CTVRTSADC])

The linear peptide containing nine amino acids (CTVRTSADC) was synthesized by Fmoc/tBu solid-phase peptide synthesis. Rink Amide resin (0.3 mmol, 526 mg, 0.57 mmol/g) was swollen in DMF. The Fmoc group was deprotected using 20% piperidine/DMF under nitrogen for two times (20 min × 2). Side chain-protected amino acids (0.9 mmol) were coupled using HBTU (341 mg, 0.9 mmol, 3 equiv) and

DIPEA (313 μL, 1.8 mmol) in DMF as coupling and activating reagent for 90 min. The Fmoc group was deprotected using 20% piperidine/DMF under nitrogen twice (20 min × 2). The peptide was assembled on the resin following coupling of amino acid in the order, e.g., Fmoc-Cys(Trt)-OH (527 mg, 0.9 mmol), Fmoc-Asp(OtBu)-OH (370 mg, 0.9 mmol), Fmoc-Ala-OH (280 mg, 0.9 mmol), Fmoc-Ser(tBu)-OH (345 mg, 0.9 mmol), Fmoc-Thr(Trt)-OH (357 mg, 0.9 mmol), Fmoc-Arg(Pbf)-OH (583 mg, 0.9 mmol), Fmoc-Val-OH (305 mg, 0.9 mmol), Fmoc-Thr(Trt)-OH (357 mg, 0.9 mmol), and Fmoc-Cys(Trt)-OH (527 mg, 0.9 mmol). Coupling and deprotection cycles were repeated to assemble the sequence of the linear protected peptide. Then, the side-chain deprotection and cleavage from the resin were carried out by a freshly prepared cleavage cocktail, trifluoroacetic acid:triisopropylsilane:water (TFA:TIS:H_2O; 95:2.5:2.5, *v/v/v*, 13.5 mL:750 μL:750 μL,15 mL) for 2 h. The crude peptide was precipitated by the addition of cold diethyl ether (30 mL × 3 times, Et_2O) and purified by RP- HPLC on a water XBridge BEH130 Prep C18 OBD 10 μm ODS reversed-phase column (2.1 cm × 25 cm) using a gradient system. The crude peptide was purified at a flow rate of 10.0 mL/min using a gradient of 0–100% acetonitrile (0.1% TFA) and water (0.1% TFA) over 60 min on RP-HPLC and then was lyophilized to obtain the linear peptide. Linear peptide (CTVRTSADC): MALDI-TOF (*m/z*): $C_{35}H_{64}N_{13}O_{14}S_2$, calcd. $[M + H]^+$ 954.4137; found 954.5512. Cyclization of the linear peptide was carried out by exposing the peptide in 10% DMSO to air at pH 7.4, followed by purification using RP-HPLC. Peptide **1** ([CTVRTSADC]); MALDI-TOF (*m/z*): $C_{35}H_{62}N_{13}O_{14}S_2$, calcd. $[M + H]^+$ 952.3980; found 952.4786.

3.1.2. Synthesis of Peptide 8 ([KTVRTSADE])

The linear peptide containing sequence (KTVRTSADE) was synthesized by Fmoc/tBu solid-phase peptide synthesis. Rink amide resin (0.3 mmol, 526 mg, 0.57 mmol/g) was swelled in DMF. The Fmoc group was deprotected using 20% piperidine/DMF under nitrogen for two times (20 min × 2). Amino acids were coupled using HBTU (341 mg, 0.9 mmol, 3 equiv) and DIPEA (313 μL, 1.8 mmol) in DMF as coupling and activating reagents, respectively. The peptide was assembled on the resin following coupling of amino acid in the order, e.g., Fmoc-Glu(Phipr)-OH (438 mg, 0.9 mmol), Fmoc-Asp(OtBu)-OH (370 mg, 0.9 mmol), Fmoc-Ala-OH (280 mg, 0.9 mmol), Fmoc-Ser(tBu)-OH (345 mg, 0.9 mmol), Fmoc-Thr(Trt)-OH (357 mg, 0.9 mmol), Fmoc-Arg(Pbf)-OH (583 mg, 0.9 mmol), Fmoc-Val-OH (305 mg, 0.9 mmol), Fmoc-Thr(Trt)-OH (357 mg, 0.9 mmol), and Fmoc-Lys(Dde)-OH (479 mg, 0.9 mmol). Coupling and deprotection cycles were repeated to assemble the sequence of the linear protected peptide. Then, selective side-chain deprotection were performed in order to remove Phipr (1% TFA/DMF, *v/v*, 15 min × 3 times) on glutamic acid and then Dde (2% hydrazine monohydrate/DMF, 20 min × 2 times) to achieve carboxylic group on the side chain of glutamic acid and amine group at the side chain of lysine side chain amine group. Cyclization of carboxylic acid and the amino group in the side chain was performed using 1-[bis(dimethylamino)methylene]-1H-1,2,3-triazolo[4,5-b]pyridinium 3-oxid hexafluorophosphate (HATU; 344 mg, 0.9 mmol), ethyl cyano(hydroxyimino)acetate (Oxyma Pure; 136 mg, 0.96 mmol), and DIPEA (313 μL, 1.8 mmol) in DMF for 3 h. After cyclization was confirmed by MALDI mass, the cleavage of peptide from the resin was carried out by agitating to a freshly prepared cleavage cocktail, TFA:TIS:H_2O (95:2.5:2.5, *v/v/v*, 13.5 mL:750 μL:750 μL,15 mL) for 2 h. The crude peptide was precipitated by the addition of cold diethyl ether (30 mL × 3 times, Et_2O). Peptide **8** ([KTVRTSADE]): MALDI-TOF (*m/z*): $C_{40}H_{71}N_{14}O_{15}$, calcd. $[M + H]^+$ 987.5223; found 987.2699.

3.1.3. Synthesis of Peptide 5 (Hydrazine-Glutarate-GG-FK-C_6-[CTVRTSADC])

Peptide **1** on solid support from step 3.1.1 was used before final cleavage to attach a spacer Fmoc-6-Ahx-OH (C_6) (157 mg, 0.9 mmol, 3 equiv) in the presence of HBTU (341 mg, 0.9 mmol, 3 equiv) and DIPEA (313 μL, 1.8 mmol, 6 equiv) in DMF as coupling and activating reagents for 3 days. The resin was washed, and Fmoc-Lys(Boc)-OH (421 mg, 0.9 mmol) and Fmoc-Phe-OH (348 mg, 0.9 mmol) were conjugated similarly. Fmoc-Gly-OH (267 mg, 0.9 mmol) was coupled twice and then reacted with glutaric anhydride (102 mg, 0.9 mmol) to provide a spacer and a carboxylic acid functional group for further functionalization, respectively. After washing the resin, Boc-hydrazine (396 mg, 3 mmol),

HOBt (405 mg, 3 mmol), EDC (465 mg, 3 mmol), and DIPEA (1.045 mL, 6 mmol) in dry DMF coupled for 48 h to provide hydrazine functional group at the *N* terminal. The resin was washed and cleaved using cleavage cocktail, TFA:TIS:H_2O (95:2.5:2.5, *v/v/v*, 13.5 mL:750 μL:750 μL,15 mL) for 2 h to afford linear peptide **5**, which after RP-HPLC purification and cyclization in 10% DMSO under air afforded cyclic peptide **5**. MALDI-TOF (*m/z*): $C_{65}H_{111}N_{21}O_{21}S_2$ calcd. $[M + H]^+$ 1585.7704; found 1585.5270.

3.1.4. Synthesis of Peptide 6 (Hydrazine-Glutarate-GG-VCit-C_6-[CTVRTSADC])

The synthesis of peptide **6** was followed by using a similar procedure described in 3.1.3 upto the coupling of Fmoc-6-Ahx-OH, and then Fmoc-L-citrulline-OH (Cit; 476 mg, 1.2 mmol) and Fmoc-Val-OH (305 mg, 0.9 mmol) were coupled on the solid phase. After washing and deprotecting Fmoc group, the resin was further reacted with Fmoc-Gly-OH (twice), glutaric anhydride, and then Boc-hydrazine as mentioned in the 3.1.3. The cleavage of linear peptide with cleavage cocktail, TFA:TIS:H_2O (95:2.5:2.5, *v/v/v*, 13.5mL:750μL:750μL, 15 mL), purification by RP-HPLC, and cyclization under 10% DMSO under air afforded peptide **6**. MALDI-TOF (*m/z*): $C_{61}H_{109}N_{22}O_{22}S_2$, calcd. $[M + H]^+$ 1565.7527; found 1565.5920.

3.1.5. Synthesis of Peptides 19 (C-GFLG-C_6-[CTVRTSADC]) and 9 (C-GFLG-C_6-[KTVRTSADE])

Both the peptidyl resin from synthesis of peptide **1** (from step 3.1.1) and peptide **8** (from step 3.1.2) were further reacted to attach a Fmoc-6-Ahx-OH (157 mg, 0.9 mmol), Fmoc-Gly-OH (267 mg, 0.9 mmol), Fmoc-Leu-OH (318 mg, 0.9 mmol), Fmoc-Phe-OH (348 mg, 0.9 mmol), Fmoc-Gly-OH (267 mg, 0.9 mmol), and Fmoc-Cys(Trt)-OH (527 mg, 0.9 mmol) amino acid residues in the separate peptide synthesis vessel. Coupling and deprotection cycles were repeated to assemble the sequence of the linear protected peptide followed by cleavage with cleavage cocktail, TFA:TIS:H_2O (95:2.5:2.5, *v/v/v*, 13.5 mL:750 μL:750 μL,15 mL) for 2 h to afford complete form of peptide **9** and linear form of peptide **19** which further cyclized using 10% DMSO/H_2O. Peptides were purified using RP-HPLC and characterized using MALDI TOF. Peptide **19** (C-GFLG-C_6-[CTVRTSADC]); MALDI-TOF (*m/z*): $C_{63}H_{106}N_{19}O_{20}S_2$, calcd. $[M + H]^+$ 1544.6863; found 1544.1794; Peptide **9** (C-GFLG-C_6-[KTVRTSADE]); MALDI-TOF (*m/z*): $C_{68}H_{113}N_{20}O_{21}S$, calcd. $[M]^+$ 1577.8110; found 1577.6815.

3.1.6. Synthesis of Conjugate 12 (Dox-hydrazone-glutarate-GG-VCit-C_6-[CTVRTSADC])

A solution of dox (16.3 mg, 0.03 mmol) was prepared after dissolving in anhydrous methanol (2 mL) and was added dropwise to the solution of peptide **5** (47.3 mg, 0.03 mmol) in anhydrous methanol (1 mL) followed by addition of TFA (0.1%, 3 μL) to acidify the reaction mixture with continued stirring overnight at room temperature in 5 mL amber glass vial. The solvent was removed under reduced pressure and conjugate was purified by RP-HPLC. Conjugate **12** (Dox-hydrazone-glutarate-GG-VCit-C_6-[CTVRTSADC]); MALDI (*m/z*): $C_{88}H_{136}N_{23}O_{32}S_2$, calcd. $[M + H]^+$ 2090.9003; found: 2090.7130.

3.1.7. Synthesis of Conjugate 13 (Dox-s-s-CGFLG-C_6-[KTVRTSADE])

First, dox thiol (Dox-SH) was prepared by reacting Dox (25 mg, 43 μmol) with 2-iminothiolane (2.5 equiv, 107 μmol) in methanol (25 mL) and triethylamine (TEA) (5 equiv, 30 μL) stirred at room temperature (7 h). The reaction mixture was precipitated by ether (15 mL), and upon decantation, the precipitant was evaporated using rotary evaporator. The sulfhydryl group of Dox was activated with dithiodipyridine by reacting Dox-SH (1 equiv, 27 mg, 43 μmol) with 2,2′-dithiodipyridine (2.5 equiv, 50 mg, 107 μmol) in methanol (0.1% acetic acid, *v/v*) with stirring at room temperature for overnight. The resulting precipitate was washed again with cold ether and dried to afford compound (Dox-S-S-Pyridine). A single peak was detected in the analytical HPLC. Activated Dox compound **10** was coupled with peptide **9** (47 mg, 0.03 mmol) via disulfide bridge using overnight stirring in the degassed water (5 mL). The solvent was evaporated, and peptide **13** was purified using RP-HPLC.

Conjugate **13** (Dox-s-s-CGFLG-C_6-[KTVRTSADE]); MALDI-TOF (m/z): $C_{99}H_{150}N_{22}O_{32}S_2$, calcd. [M + H]$^+$ 2223.0227; found 2222.3398.

3.1.8. Synthesis of Conjugate 14 (Doce-βA-thioether-CGFLG-C_6-[KTVRTSADE])

Doce-βA-NH_2 was first prepared with the stirring of Docetaxel (5 mg, 0.012 mmol), Fmoc-β-alanine (5.5 mg, 0.062 mmol), using 2-(6-chloro-1-H-benzotriazole-1-yl)-1,1,3,3-tetramethylaminium hexafluorophosphate (HCTU) (14.9 mg, 36 μmol, 3 equiv) and DIPEA (12 μL, 72 μmol, 6 equiv) in anhydrous DMF (1.5 mL), at room temperature for 90 min. Fmoc was removed by treating with 40% piperidine in DMF (*v/v*, 10 mL, 20 min), and the resulting peptide was precipitated by ether. Purification was performed using RP-HPLC followed by lyophilization. Doce-βA-NH_2 (10.8 mg, 10 μmol) was reacted with sulfosuccinimidyl-4-(N-maleimidomethyl)cyclohexane-1-carboxylate (sulfo-SMCC) (6.5 mg, 15 μmol, 1.5 equiv) in *N,N′*-diisopropylcarbodiimide (DIC, 8 μL, 5 equiv) and TEA (8 μL, 6 equiv) using methanol (1 mL), overnight. Docetaxel-βA-SMCC (**11**) was purified using RP-HPLC. The Docetaxel-βA-SMCC (7.6 mg, 7 μmol) was dissolved in DMF (5 mL), reacted with peptide **9** (C-GFLG-C_6-[KTVRTSADE]) (12 mg, 76 μmol, 1.08 equiv) in PBS with 20 mM EDTA at pH 8 (20 mL) for 17 h at room temperature. The formation of conjugated compound **14** (Doce-*β*A-thioether-C-GFLG-C_6-[KTVRTSADE]) was confirmed by mass analysis using MALDI; MALDI-TOF (*m/z*): $C_{126}H_{184}N_{23}O_{39}S$, calcd. [M + H]$^+$ 2675.2842; found 2675.8490.

3.1.9. Synthesis of FAM Conjugate Peptide 17 (FAM-GFLG-C_6-[CTVRTSADC])

First, the linear protected peptide **15** (NH_2-GFLG-C_6-C(Trt)T(tBu)VR(Pbf)T(tBu)S(tBu)AD(OtBu)C (Trt)) on Rink amide resin was synthesized as described above at 0.2 mmol (350 mg) scale using Rink amide resin. The peptidyl-resin was washed in DMF (15 mL, 3 × 2 min). The coupling of 5(6)-carboxyfluorescein diisobutyrate (CFDI, 2.5 equiv, 258 mg), in the presence of HOAt (2.5 equiv, 68 mg), (7-azabenzotriazol-1-yloxy)tripyrrolidinophosphonium hexafluorophosphate (PyAOP, 2.5 equiv, 260 mg), and DIPEA (5 equiv, 174 μL) in anhydrous DMF (12 mL) was carried out for 4.5 h followed by washing with DMF. The deprotection of isobutyrate protection in CFDI was carried out by agitating the peptidyl resin using piperidine (20%, *v/v*, 25 mL, 2 × 15 min). The resin was finally washed with DMF (3 × 3 mL), DCM (3 × 3 mL), and MeOH (3 × 3 mL) and finally dried in vacuum for 30 min. The fluorescent-labeled peptide was cleaved using TFA:TIS:H_2O (95:2.5:2.5, *v/v/v*, 6.75 mL:375 μL:375 μL) for 2 h. The crude peptide was precipitated with cold diethyl ether (10 mL) and centrifuged (4500 rpm, 6 min). The solvent was decanted to obtain the precipitant, dried under nitrogen which was further dissolved in 10% DMSO (100 mL) under dark to form disulfide bridge between cysteine's thiol residues under air for overnight. The solvent was evaporated under reduced pressure to afford viscous peptide precipitate. The precipitate was dissolved in 30% ACN/H_2O (*v/v*, 4 mL), purified by using RP-HPLC, and lyophilized. The mass was confirmed for peptide **17** with MALDI-TOF (*m/z*); $C_{81}H_1N_{18}O_{25}S_2$, calcd. [M + H]$^+$ 1799.7249; found 1799.9280.

3.1.10. Synthesis of FAM Conjugate Peptide 18 (FAM-GFLG-C_6-[KTVRTSADE])

The cyclized protected peptide [KT(tBu)VR(Pbf)T(tBu)S(tBu)AD(OtBu)E] was assembled on Rink amide resin (0.2 mmol, 350 mg) using steps in Section 3.1.2. The peptidyl resin was washed in DMF (15 mL, 3 × 2 min) and then solvents were filtered off. Then, Fmoc-Ahx-OH, Fmoc-Gly-OH, Fmoc-Leu-OH, Fmoc-Phe-OH, and Fmoc-Gly-OH were coupled as mentioned above, followed by resin washing with DMF (15 mL, 3 × 2 min). The deprotection of *N*-terminal Fmoc was carried out using 20% piperidine in DMF (*v/v*, 2 × 10 min) and then washed with DMF (9 mL, 3 × 2 min) to afford intermediate peptide **16** (NH_2-C(Trt)GFLG-C_6-[KT(tBu)VR(Pbf)T(tBu)S(tBu)AD(OtBu)E]-Rink amide resin). The coupling of 5(6)-carboxyfluorescein diisobutyrate (CFDI, 2.5 equiv, 258 mg), in the presence of HOAt (2.5 equiv, 68 mg), (7-azabenzotriazol-1-yloxy)tripyrrolidinophosphonium hexafluorophosphate (PyAOP, 2.5 equiv, 260 mg), and DIPEA (5 equiv, 174 μL) in anhydrous DMF (12 mL) was carried out to the peptidyl resin for 4.5 h followed by washing with DMF. The deprotection of isobutyrate protection in CFDI was carried out by agitating the peptidyl resin using 20% piperidine/DMF (*v/v*, 25 mL, 2 × 15 min). The resin was finally washed with DMF (3 × 3 mL), DCM (3 × 3 mL), and MeOH (3 × 3 mL) and finally dried under vacuum for 30 min. The fluorescent-labeled peptide was cleaved using TFA:TIS:H_2O (95:2.5:2.5, *v/v/v*, 6.75 mL:375 μL:375 μL) for 2 h. The crude peptide was precipitated with cold diethyl ether (10 mL), centrifuged (4500 rpm, 6 min). The solvent was decanted to obtain the precipitant to afford viscous peptide precipitate. The precipitate was dissolved in ~30% ACN/H_2O (*v/v*, 4 mL), purified by using RP-HPLC, and lyophilized. The mass was confirmed for peptide **18** with MALDI-TOF (m/z); $C_{86}H_{118}N_{19}O_{26}$, calcd. $[M + H]^+$ 1832.8495; found 1832.3740.

3.2. Stability in PBS and Dithiothreitol (DTT)

The stability of peptide–drug conjugate **13** was evaluated in the phosphate buffered saline (PBS). The pH of PBS was adjusted to 6.8, 7.0, 7.4 using HCl and NaOH as needed. In brief, 250 μL of 1 × PBS was mixed with 45 μL of 0.25 mM Dox-peptide conjugate **13**. The mixture was incubated at 37 °C in a water bath covered with aluminum foil. An aliquot (40 μL) was removed at different time intervals, such as 4, 24, and 84 h and analyzed using analytical RP-HPLC at 490 nm. The percentage of the remained peptide was calculated using the area under the curve (AUC) by integrating the chromatogram in analytical HPLC. For evaluation of stability in DTT, a solution of DTT (0.120 mM) was prepared using PBS (pH 7.4). A solution of compounds (**1, 8**, and **13** at 1 mM) was prepared in PBS (pH = 7.4). A volume of 45 μL of the peptide was added to 2.250 mL of PBS in a UV-blocking glass vial. The glass vials were put under nitrogen and stirred using a small magnetic bead. A volume of DTT (222 mg, 12 equiv) was added to the solution, which was incubated with each peptide sample to obtain a working concentration of 0.12 mM DTT and 0.010 mM peptide. Aliquots of 200 μL were collected at different time intervals and analyzed using analytical HPLC vials (0 min was used as controls without DTT) after quenching the DTT with 10 μL of H_2O (pH = 2). The aliquots were run on analytical HPLC with detection at 214 nm.

3.3. Stability in Human Serum

The stability was performed using 25% human serum. 325 μL of RPMI-1640 media was incubated with 125 μL of human serum at 37 °C in a water bath. Then 50 μL of 125 μM of conjugate **13** was added to the RPMI and serum. The control experiment was performed without conjugate using an aqueous 5% DMSO. Several aliquots (55 μL) were removed at 2, 5, 10, 15, 30, 60, 90, and 120 min. The aliquots were added to 120 μL of cold methanol to precipitate the serum proteins with vigorous mixing using vortex and kept at ice for 10 min. Then, the samples were centrifuged at 17× *g* (13,300 rpm) for 10 min to get a pellet of the insoluble peptide. The supernatant was collected and analyzed on analytical HPLC, as mentioned before.

3.4. Hydrolysis Studies with Cathepsin Enzyme

Human cathepsin B (CTSB, Lot # CB2016-02 from Athens Research, Athens, GA, USA) was used to hydrolyze the cathepsin linker GFLG in conjugated peptide **13**. A stock solution of cathepsin B enzyme (0.434 μg/μL) was prepared using 50 mM acetate (pH 5.0) buffer with 1 mM EDTA. The activation buffer was separately prepared using 30 mM DTT and 15 mM EDTA in H_2O. An aliquot of 2.25 μL of cathepsin B stock was activated with 5 μL of activation buffer for 15 min at room temperature then diluted with 1.185 mL of acetate buffer and aliquoted into different tubes with 200 μL each. Further 8 μL of 1 mM of conjugated peptide **13** was added to aliquots and incubated for various time intervals. Then, the reaction was quenched by boiling to denature cathepsin B for 10 min. A 400 μL of cold methanol was added to precipitate the proteins followed by centrifugation of sample 17× *g* (13,300 rpm) for 10 min. A supernatant (500 μL) was collected, evaporated, and reconstituted with 300 μL of HPLC solvent A (H_2O with 0.1% TFA). The eluent fraction from HPLC was further collected and analyzed using MALDI-TOF mass spectrometry to confirm the hydrolysis of the conjugate.

3.5. Cell Culture

Human epithelial prostate cell line (RWPE-1, ATCC No. CRL-11610), human prostate cancer cell line (Human prostate carcinoma DU-145, ATCC No. HTB-81), human adenocarcinoma PC3, ATCC No. CRL-1435), androgen-sensitive human prostate adenocarcinoma cells (LNCaP, ATCC No. CRL-1740), and castration-resistant LNCaP-derived C4-2 (ATCC No. CRL-1595) were purchased from American Type Culture Collection (ATCC), Manassas, VA, USA. All the media, serum, and antibiotics were bought from American Type Culture Collection (ATCC) (Manassas, VA, USA) and Sigma life-science (St. Louis, MO, USA). Keratinocytes SFM (1×) with EGF + BPE was bought from Gibco life technologies (Thermo Fisher Scientific), Gaithersburg, MD, USA. CellTiter 96 was bought from Promega, Madison, WI, USA. Dithiothreitol (DTT) was bought from Thermo Fisher scientific, Carlsbad, CA, USA, and TGF-β was bought from Abcam, Boston, MA, USA.

3.5.1. Cell Viability Assays using MTS

Cell-proliferation assay using MTS reagent was conducted against five cell lines (RWPE-1, PC3, DU-145, LNCaP, and, C4-2). Cells were seeded into 96-well plates (5×10^3 cells for all the cells) and incubated with 100 μL of complete medium (RPMI-1640 for RWPE-1, DU-145, LNCaP, and C4-2; DMEM for PC3) overnight at 37 °C with 5% CO_2. Various concentrations (0, 0.06, 0.6, 6, 60, 600 μM) of the peptide solution (20 μL) were added to cells to yield the final concentrations of peptide (0, 0.01, 0.1, 1, 10, 100 μM). The cells were kept in an incubator (37 °C, 5% CO_2) for 24, 48, or 72 h accordingly for different treatment time responses. Then, a CellTiter 96 aqueous solution (20 μL) was added to each well and incubated for 4 h under the same condition. The absorbance was obtained at 490 nm using SpectraMax M2 microplate reader to detect the formazan product. Wells containing cells in the absence of any peptide were used as a control. The percentage of cell viability was calculated by [(OD value of cells treated with the test mixture of compounds) − (OD value of culture medium)] / [(OD value of control cells) − (OD value of culture medium)] × 100%. For other time-dependent experiments, compounds (5 μM) were added to each well in triplicate and kept for 24, 48, and 72 h at 37 °C in a 5% CO_2 incubator. Cell viability was determined by measuring the absorbance at 490 nm using SpectraMax M2 microplate spectrophotometer, as mentioned above.

3.5.2. Overexpression of EDB-FN using TGF-β

To compare the cytotoxicity of compounds using treatment of TGF-β, the cells were culture, and 8000 cells (RPWE-1, PC3, DU-145, LNCap, and C4-2) were added in 120 μL per well with TGF-β treatment. Cells were exposed to TGF-β at 10 ng/mL for 3 days. Then cells were seeded in the corresponding medium (DMEM containing 10% FBS and 1% penicillin–streptomycin for DU-145, LNCaP, and PC3; keratinocyte containing 0.1% FBS without antibiotics for RWPE-1), 24 h prior to the experiment. Then, MTS assay was followed up, as mentioned in 3.5.1.

3.5.3. Confocal Microscopy

Prostate cancer cell line (C4-2) was grown in RPMI medium containing 10% fetal bovine serum and 1% penicillin/streptomycin at the 37 °C in a 5% CO_2 incubator. The culture medium was changed every 2–3 days until confluency was reached to 80%. Each cells group was treated with 5 μg of fluorescence-tagged peptides (17 and 18) for 12 h. A set of control groups was also prepared, which was not treated with peptides. Then all cells were fixed with 4% paraformaldehyde for 10 min and washed for 10 min (3 times with PBS). The cells were blocked in 1% FBS/PBS for 10 min and washed for 10 min (3 times with PBS). To target EDB-FN on the cell surface, the cells were incubated with a mouse monoclonal anti-fibronectin antibody (Abcam, Cambridge, MA, USA) (1:1000 dilution using 1% FBS/PBS) overnight at the 4 °C. The cells were washed for 10 min (3 times with PBS), and a Texas red-conjugated anti-mouse secondary antibody was used in 1% FBS/PBS (5:1000 dilution) at the room temperature for 1 h. Then cells were washed for 10 min (3 times with PBS), and at the end, nuclei were stained using DAPI and covered with coverslips. Representative images of cells were captured using a standard confocal microscope (Nikon Eclipse Ti-E, Nikon Instruments Inc., Melville, NY, USA) at 60× magnification, and the images were processed for binary contrast quantification.

4. Conclusions

In conclusion, the synthesis and optimization of EDB-FN ligands were achieved using Fmoc/tBu solid phase synthesis. The lactam cyclized ligand [KTVRTSADE] **8** showed higher stability under redox condition (8 h) as compared to disulfide cyclized ligand [CTVRTSADC] **1** (2.6 min). A half-life of 11.8 min was detected for Dox conjugate **13** during treatment with 25% serum. A cathepsin enzyme based linkers were introduced in Dox/Doce conjugates (**13** and **14**), which releases the drug from conjugate within 5 min after treatment with cathepsin enzyme. The antiproliferative activity of Dox conjugate **13** exhibited comparable or the similar antiproliferative activity versus Dox alone or physical mixture in 72 h in all the cell lines (RWPE-1, PC3, DU-145, and LNCaP). The Doce conjugate **14** showed mild or no activity which might be due to the lack of hydrolysis of the drug from conjugate in all the cell lines. The TGF-β treatment in C4-2 cell lines demonstrates the effect of overexpression of EDB-FN on cell viability. Confocal microscopy demonstrated localization of ligands at EDB-FN in C4-2 cell line. The peptide **18** containing lactam cyclization showed higher targeting at EDB-FN in C4-2 cell lines as compared to peptide **17** containing disulfide cyclization. These studies provide a better understanding for designing peptide–drug conjugate to be used as a chemotherapeutic agent against PCa.

Supplementary Materials: Supplementary materials can be found at http://www.mdpi.com/1422-0067/20/13/3291/s1.

Author Contributions: R.K.T. conceived and designed the experiments; S.E.P. synthesized peptides, Doce intermediate and performed cellular studies, S.D. synthesized dox intermediate, T.A.K. performed hydrolysis and stability studies, M.B. assisted in cell-culture, provided PCa cell lines, and antiproliferative assay, K.S. performed confocal microscopy. S.E.P., T.A.K., M.B., K.P., and R.K.T. analyzed the data; S.E.P., K.P., and R.K.T. wrote the paper.

Funding: The authors acknowledge the financial support for this research from the Chapman University School of Pharmacy. RKT and SEP acknowledge support from FRDC and graduate program respectively for seed funding.

Acknowledgments: The authors acknowledge the financial support for this research from the Chapman University School of Pharmacy. R.K.T. and S.E.P. acknowledge support from FRDC and graduate program, respectively, for seed funding. We thank Ajay Sharma for help with Confocal microscopy.

Conflicts of Interest: The authors declare no conflict of interest.

References and Note

1. Siegel, R.L.; Miller, K.D.; Jemal, A. Cancer statistics, 2018. *CA Cancer J. Clin.* **2018**, *68*, 7–30. [CrossRef] [PubMed]
2. American Cancer Society Prostate Cancer. Available online: https://www.cancer.org/cancer/prostate-cancer.html (accessed on 16 February 2018).
3. Gravis, G. Systemic treatment for metastatic prostate cancer. *Asian J. Urol.* **2019**, *6*, 162–168. [CrossRef] [PubMed]
4. Evison, B.J.; Sleebs, B.E.; Watson, K.G.; Phillips, D.R.; Cutts, S.M. Mitoxantrone, More than Just Another Topoisomerase II Poison. *Med. Res. Rev.* **2016**, *36*, 248–299. [CrossRef] [PubMed]
5. Summers, N.; Vanderpuye-Orgle, J.; Reinhart, M.; Gallagher, M.; Sartor, O. Efficacy and safety of post-docetaxel therapies in metastatic castration-resistant prostate cancer: a systematic review of the literature. *Curr. Med. Res. Opin.* **2017**, *33*, 1995–2008. [CrossRef] [PubMed]
6. Alavi, M.; Hamidi, M. Passive and active targeting in cancer therapy by liposomes and lipid nanoparticles. *Drug Metab. Pers. Ther.* **2019**, *34*. [CrossRef]
7. Zhang, J.; Wang, L.; You, X.; Xian, T.; Wu, J.; Pang, J. Nanoparticle Therapy for Prostate Cancer: Overview and Perspectives. *Curr. Top. Med. Chem.* **2019**, *19*, 57–73. [CrossRef] [PubMed]
8. Nasrolahi Shirazi, A.; Tiwari, R.; Chhikara, B.S.; Mandal, D.; Parang, K. Design and biological evaluation of cell-penetrating peptide–doxorubicin conjugates as prodrugs. *Mol. Pharm.* **2013**, *10*, 488–499. [CrossRef]
9. Schally, A.V.; Nagy, A. Cancer chemotherapy based on targeting of cytotoxic peptide conjugates to their receptors on tumors. *Eur. J. Endocrinol.* **1999**, *141*, 1–14. [CrossRef]
10. Arap, W.; Pasqualini, R.; Ruoslahti, E. Cancer treatment by targeted drug delivery to tumor vasculature in a mouse model. *Science* **1998**, *279*, 377–380. [CrossRef]

11. Srinivasarao, M.; Low, P.S. Ligand-targeted drug delivery. *Chem. Rev.* **2017**, *117*, 12133–12164. [CrossRef]
12. Åkerfelt, M.; Härmä, V.; Nees, M. Advanced Models for Target Validation & Drug Discovery in Prostate Cancer. In *Prostate Cancer-From Bench to Bedside*; IntechOpen: Rijeka, Croatia, 2011.
13. Bonnans, C.; Chou, J.; Werb, Z. Remodelling the extracellular matrix in development and disease. *Nat. Rev. Mol. Cell Biol.* **2014**, *15*, 786. [CrossRef] [PubMed]
14. Kumra, H.; Reinhardt, D.P. Fibronectin-targeted drug delivery in cancer. *Adv. Drug Deliv. Rev.* **2016**, *97*, 101–110. [CrossRef] [PubMed]
15. Mani, S.A.; Guo, W.; Liao, M.-J.; Eaton, E.N.; Ayyanan, A.; Zhou, A.Y.; Brooks, M.; Reinhard, F.; Zhang, C.C.; Shipitsin, M.; et al. The Epithelial-Mesenchymal Transition Generates Cells with Properties of Stem Cells. *Cell* **2008**, *133*, 704–715. [CrossRef] [PubMed]
16. Albrecht, M.; Renneberg, H.; Wennemuth, G.; Möschler, O.; Janssen, M.; Aumüller, G.; Konrad, L. Fibronectin in human prostatic cells in vivo and in vitro: expression, distribution, and pathological significance. *Histochem. Cell Biol.* **1999**, *112*, 51–61. [CrossRef]
17. Kaspar, M.; Zardi, L.; Neri, D. Fibronectin as target for tumor therapy. *Int. J. Cancer* **2006**, *118*, 1331–1339. [CrossRef]
18. Han, Z.; Zhou, Z.; Shi, X.; Wang, J.; Wu, X.; Sun, D.; Chen, Y.; Zhu, H.; Magi-Galluzzi, C.; Lu, Z.-R. EDB Fibronectin Specific Peptide for Prostate Cancer Targeting. *Bioconjug. Chem.* **2015**, *26*, 830–838. [CrossRef]
19. Staudacher, A.H.; Brown, M.P. Antibody drug conjugates and bystander killing: is antigen-dependent internalisation required? *Br. J. Cancer* **2017**, *117*, 1736. [CrossRef]
20. Dubowchik, G.M.; Firestone, R.A.; Padilla, L.; Willner, D.; Hofstead, S.J.; Mosure, K.; Knipe, J.O.; Lasch, S.J.; Trail, P.A. Cathepsin B-Labile Dipeptide Linkers for Lysosomal Release of Doxorubicin from Internalizing Immunoconjugates: Model Studies of Enzymatic Drug Release and Antigen-Specific In Vitro Anticancer Activity. *Bioconjug. Chem.* **2002**, *13*, 855–869. [CrossRef]
21. Shao, L.-H.; Liu, S.-P.; Hou, J.-X.; Zhang, Y.-H.; Peng, C.-W.; Zhong, Y.-J.; Liu, X.; Liu, X.-L.; Hong, Y.-P.; Firestone, R.A.; et al. Cathepsin B cleavable novel prodrug Ac-Phe-Lys-PABC-ADM enhances efficacy at reduced toxicity in treating gastric cancer peritoneal carcinomatosis. *Cancer* **2012**, *118*, 2986–2996. [CrossRef]
22. Zhong, Y.; Shao, L.; Li, Y. Cathepsin B-cleavable doxorubicin prodrugs for targeted cancer therapy. *Int. J. Oncol.* **2013**, *42*, 373–383. [CrossRef]
23. Tsakalozou, E.; Eckman, A.M.; Bae, Y. Combination Effects of Docetaxel and Doxorubicin in Hormone-Refractory Prostate Cancer Cells. *Biochem. Res. Int.* **2012**, *2012*, 10. [CrossRef] [PubMed]
24. Wohl, A.R.; Michel, A.R.; Kalscheuer, S.; Macosko, C.W.; Panyam, J.; Hoye, T.R. Silicate esters of paclitaxel and docetaxel: Synthesis, hydrophobicity, hydrolytic stability, cytotoxicity, and prodrug potential. *J. Med. Chem.* **2014**, *57*, 2368–2379. [CrossRef] [PubMed]
25. Skwarczynski, M.; Hayashi, Y.; Kiso, Y. Paclitaxel prodrugs: toward smarter delivery of anticancer agents. *J. Med. Chem.* **2006**, *49*, 7253–7269. [CrossRef] [PubMed]
26. Kingston, D.G.I. Taxol, a molecule for all seasons. *Chem. Commun.* **2001**. [CrossRef]
27. Inoue, T.; Cavanaugh, P.G.; Steck, P.A.; Brünner, N.; Nicolson, G.L. Differences in transferrin response and numbers of transferrin receptors in rat and human mammary carcinoma lines of different metastatic potentials. *J. Cell. Physiol.* **1993**, *156*, 212–217. [CrossRef] [PubMed]
28. Pan, H.; Yang, J.; Kopeckova, P.; Kopecek, J. Backbone Degradable Multiblock N-(2-Hydroxypropyl) methacrylamide Copolymer Conjugates via Reversible Addition−Fragmentation Chain Transfer Polymerization and Thiol−ene Coupling Reaction. *Biomacromolecules* **2011**, *12*, 247–252. [CrossRef] [PubMed]
29. Cunningham, D.; You, Z. In vitro and in vivo model systems used in prostate cancer research. *J. Biol. Methods* **2015**. [CrossRef]
30. Tai, S.; Sun, Y.; Squires, J.M.; Zhang, H.; Oh, W.K.; Liang, C.Z.; Huang, J. PC3 is a cell line characteristic of prostatic small cell carcinoma. *Prostate* **2011**, *71*, 1668–1679. [CrossRef] [PubMed]
31. Bisoffi, M.; Klima, I.; Gresko, E.; Durfee, P.N.; Hines, W.C.; Griffith, J.K.; Studer, U.E.; Thalmann, G.N. Expression profiles of androgen independent bone metastatic prostate cancer cells indicate up-regulation of the putative serine-threonine kinase GS3955. *J. Urol.* **2004**, *172*, 1145–1150. [CrossRef]
32. Han, Z.; Lu, Z.-R. Targeting fibronectin for cancer imaging and therapy. *J. Mater. Chem. B* **2017**, *5*, 639–654. [CrossRef]

International Journal of
Molecular Sciences

Article

Inhibitory Effects of Peptide Lunasin in Colorectal Cancer HCT-116 Cells and Their Tumorsphere-Derived Subpopulation

Samuel Fernández-Tomé [1,2,†], Fei Xu [2], Yanhui Han [2], Blanca Hernández-Ledesma [1,*] and Hang Xiao [2,*]

1 Instituto de Investigación en Ciencias de la Alimentación (CIAL, CSIC-UAM CEI UAM+CSIC), Nicolás Cabrera, 9, 28049 Madrid, Spain; fernandeztome.samuel@gmail.com

2 Department of Food Science, University of Massachusetts, Amherst, MA 01003, USA; xufei5056@gmail.com (F.X.); yanhuihan@foodsci.umass.edu (Y.H.)

* Correspondence: b.hernandez@csic.es (B.H.-L.); hangxiao@foodsci.umass.edu (H.X.); Tel.: +34 910017970 (B.H.-L.); +1 413-545-2281 (H.X.)

† Current address: Hospital Universitario de La Princesa, Instituto de Investigación Sanitaria Princesa (IIS-IP), Centro de Investigación Biomédica en Red de Enfermedades Hepáticas y Digestivas (CIBEREHD), 28049 Madrid, Spain.

Received: 11 December 2019; Accepted: 13 January 2020; Published: 14 January 2020

Abstract: The involvement of cancer stem-like cells (CSC) in the tumor pathogenesis has profound implications for cancer therapy and chemoprevention. Lunasin is a bioactive peptide from soybean and other vegetal sources with proven protective activities against cancer and other chronic diseases. The present study focused on the cytotoxic effect of peptide lunasin in colorectal cancer HCT-116 cells, both the bulk tumor and the CSC subpopulations. Lunasin inhibited the proliferation and the tumorsphere-forming capacity of HCT-116 cells. Flow cytometry results demonstrated that the inhibitory effects were related to apoptosis induction and cell cycle-arrest at G1 phase. Moreover, lunasin caused an increase in the sub-GO/G1 phase of bulk tumor cells, linked to the apoptotic events found. Immunoblotting analysis further showed that lunasin induced apoptosis through activation of caspase-3 and cleavage of PARP, and could modulate cell cycle progress through the cyclin-dependent kinase inhibitor p21. Together, these results provide new evidence on the chemopreventive activity of peptide lunasin on colorectal cancer by modulating both the parental and the tumorsphere-derived subsets of HCT-116 cells.

Keywords: colorectal cancer; cancer stem cells; chemoprevention; bioactive peptide; lunasin

1. Introduction

Current statistics on colorectal cancer (CRC) have ranked this disease as the third most commonly diagnosed malignancy and the fourth leading cause of cancer death in the world [1]. In recent years, a great deal of research has been focused on CRC pathogenesis. Meanwhile, the existence of tumor-initiating cells or cancer stem-like cells (CSC) in this solid tumor has been established [2–4]. According to the CSC theory, a minor population of tumor cells is responsible for the driving of tumorigenesis [5]. These stem cells, like those in adult tissue, undergo unlimited proliferation and asymmetrically division into more differentiated cells leading to the neoplastic growth and maintenance [6]. In addition, it has been suggested that this CSC subpopulation might be potentially responsible for the tumor invasion, metastasis, recurrence, and resistance to therapy [7,8]. Therefore, the potential of preventive strategies needs to be evaluated not only against CRC cells, representing the bulk of the tumor mass (non-CSC), but also against colon CSC.

Accumulating evidence and epidemiological studies have revealed an inverse correlation between soybean consumption and the risk of CRC development [9,10], that can be in part due to the chemopreventive effects of the bioactive compounds described in this legume. Several soybean components including isoflavones [11], saponins [12], and bioactive proteins and peptides, such as lectins and protease inhibitors [13,14] have been shown to exert protective activities against the growth of CRC cells. Identified in soybean, lunasin is a bioactive peptide which chemopreventive properties have been recently reviewed [15]. It has been demonstrated that lunasin is able to cause cytotoxicity in four different human CRC cell lines, HCT-116, HT-29, KM12L4, and RKO, and their oxaliplatin-resistant variants [16]. Studies on the mechanisms of action involved in this antiproliferative activity have been mostly carried out in HT-29 and KM12L4 cells, in which Dia and de Mejia demonstrated lunasin´s effects on apoptosis-induction, cell cycle progression, and modulation of CRC-related biomarkers [16–18]. Moreover, García-Nebot and others reported the protective role played by lunasin in differentiated Caco-2 cells, as a model of human enterocytes, exposed to oxidizing agents through promoting cell viability and counteracting the rise in reactive oxygen species levels [19]. This notably antioxidant protection at intestinal level is also a noteworthy aspect, pointing lunasin as a promising chemopreventive agent against CRC.

The emergence of the CSC model has profound implications on cancer chemoprevention and the search of natural components targeting these cells has been markedly prompted [20]. Some dietary compounds and phytochemicals have been shown to potentially interact toward the pathways involved in the renewal and proliferation of CSC [21–23]. Despite the fact that food proteins and peptides have received increasing attention for their efficacy preventing the different stages of cancer, including initiation, promotion, and progression [24,25], their protective role against CSC has been scarcely studied. Accordingly, this study aimed to evaluate the cytotoxicity of peptide lunasin in human CRC HCT-116 cells by evaluating its inhibitory capacity on cell viability and CSC-related tumorsphere forming activity, as well as its effects on apoptosis induction, cell cycle progression, and carcinogenesis-related protein biomarkers.

2. Results and Discussion

2.1. Inhibitory Effect of Lunasin on Cell Viability and Tumorsphere Formation

In this study, the human HCT-116 cell line was grown in monolayer as parental CRC cells (Figure 1A) and used for the enrichment of tumor-derived colon-spheres (Figure 1B). We first examined the growth of adherent HCT-116 cells exposed to lunasin. HCT-116 cells were treated with serial concentrations of synthetic lunasin (5–160 μM) for 72 h and the number of viable cells was assessed by the 3-(4,5-dimethylthiazol-2-yl)-2,5-diphenyl tetrazolium bromide (MTT) assay. As shown in Figure 1C, lunasin showed cell proliferation inhibitory properties with increasing effects at higher doses. Hence, treatment with 10 μM lunasin was able to induce a significant reduction on cellular growth (12.9%, $p < 0.01$) compared to control cells. The cytotoxic effect increased up to the highest concentration used (64.1%, $p < 0.001$). The IC_{50} value, expressed as the peptide concentration needed to inhibit 50% of cell number, was 107.5 ± 1.9 μM. It had been previously demonstrated that lunasin is able to induce cytotoxicity in colon cancer HCT-116, HT-29, KM12L4, and RKO cells, with IC_{50} values of 26.3, 61.7, 13.0, and 21.6 μM, respectively, while it was no toxic for colon fibroblasts CCD-33Co [16]. While these authors used purified lunasin (~90%) from defatted soybean flour, in our study we have assessed the effects of synthetic lunasin. The higher IC_{50} value found in our study might be due to differences in the secondary and tertiary structures between plant-purified lunasin and the synthetic peptide. Additionally, other compounds present in the natural preparation could be responsible for the observed change in the inhibitory potency. In this regard, synthetic lunasin has been shown to suppress the growth of breast cancer MDA-MB-231 cells with a reported IC_{50} value of 181.0 μM [26].

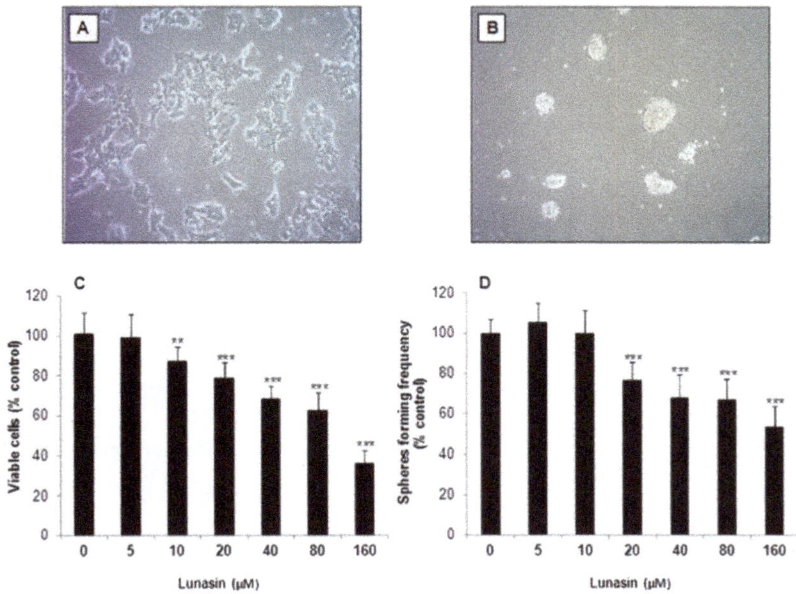

Figure 1. Cell culture and cytotoxic effect of lunasin on colorectal cancer (CRC) cells. Representative images of (**A**) HCT-116 cells in adherent conditions and (**B**) enrichment culture of tumor-derived colon-spheres formed from the parental HCT-116 cell line under anchorage-independent conditions. (**C**) HCT-116 cells were treated with lunasin (5–160 µM) for 72 h, and cell viability was determined by the MTT assay. (**D**) Colon tumorspheres were treated with lunasin (5–160 µM) for 10 days, stained with crystal violet solution and counted. Results, expressed as percentage of control cells, are means ± standard deviation (SD) of the replicates of experiments carried out. ** ($p < 0.01$), *** ($p < 0.001$) significantly different from control.

Since colon-sphere subpopulations were demonstrated to exert a key role in the CRC pathogenesis, the culture of tumor-derived spheroids has been widely used for the evaluation of chemotherapy drugs and chemopreventive agents [3]. The sphere formation assay is extensively applied as in vitro method for the derivation and characterization of stem-like cancer cells with intrinsic self-renewal and tumorigenic properties [27]. To evaluate whether lunasin might prevent the formation of CRC-derived colon-spheres, we performed the colony formation assay as we did previously [28], following some modifications to model the enrichment of tumor-derived spheroids in culture. Hence, colon-tumorspheres were enriched from adherent HCT-116 cells, cultured as non-adherent spheres under anchorage-independent conditions, and treated with lunasin for 10 days (Figure 1B). As shown in Figure 1D, lunasin at the lowest range assessed (5–10 µM) was not able to suppress tumorsphere-forming capacity. Likewise, Pabona and others had reported that while isoflavone genistein (40 nM) reduced the number of mammosphere-forming units in malignant breast cancer MCF-7 cells, peptide lunasin (2 µM) was not able to recapitulate this inhibitory protection [29]. However, as represented in Figure 1D, the peptide in the range of 20–160 µM, exerted a significant inhibitory effect ($p < 0.001$). Evidence supports that colon-spheres formed by culture in ultra-low attachment conditions in supplemented-serum-free medium presented more stem-like cell properties [30]. Following this culture, spheroid formation of DLD-1 and SW480 CRC cells with protein and mRNA expression of CSC markers including CD133, CD44, ALDHA1, Oct-4 and Nanog, was recently inhibited by (−)-epigallocatechin-3-gallate [31]. However, the characterization of these markers was not performed in the present study. The calculated IC_{50} value for HCT-116-derived spheres in our study was 161.0 ± 2.4 µM, indicating that colony-forming cells are less sensitive to peptide lunasin than parental

HCT-116 cells (107.5 ± 1.9 μM). These results are in agreement with the reported higher resistance of CSC to other anti-cancer therapies [7,32]. Similarly, Yang and others have shown that docosahexaenoic acid (DHA) exerts higher antiproliferative potency on adherent CRC SW620 cells than on their tumorspheres-derived CSC subpopulation [33]. Nevertheless, in the study of McConnell and others, it was found that peptide lunasin presented a higher anti-proliferative activity against non-small cell lung cancer cells when they were assessed under anchorage-independent growth conditions, compared to anchorage-dependent conditions [34]. In this line, detailed studies on soybean lunasin effects against melanoma CSC have been recently published [35,36]. These authors found that lunasin specifically targeted the cancer-initiating subset of melanoma cancer cells, suppressing not only their oncosphere formation capacity but also the expression of the CSC-markers aldehyde dehydrogenase and Nanog, while also inducing the expression of melanocyte-associated differentiation markers tyrosinase and microphthalmia-associated transcription factor. Interestingly, the functional domain arginine-glycine-aspartic acid (RGD) of lunasin sequence was found to be crucial in the interaction with integrins, cell internalization, inhibition of histone acetylation and anticancer-stem activity [36]. Therefore, lunasin´s modulatory chemoprevention might notably depend on the lunasin´s preparation and origin, as well as on the culture conditions and the cell line used. Inhibitory effects of lunasin over colon-spheres derived from other CRC cell lines apart from HCT-116 cells might be different and thus should be evaluated in future studies with different types of CRC.

2.2. Apoptosis Analysis of Lunasin-Treated CRC Cells

Tumor cell populations expand in number through several molecular processes such as the capability of evading programmed cell death by presenting an elevated apoptotic threshold [37]. In order to determine whether the inhibitory effect of lunasin on HCT-116 cells was through interacting with the apoptotic pathways, adherent and colon-spheres-derived cells were incubated with lunasin, and apoptosis detection was assessed by flow cytometry-based Annexin V/propidium iodide (PI) assay. Annexin V has high affinity for membrane phospholipid phosphatidylserine translocated to the outer cellular environment as one of the earliest processes during apoptosis. Phospholipid phosphatidylserine is exposed before the loss of membrane integrity, which can be revealed in later stages of cell apoptosis or necrosis by the viability dye PI. Based on the lunasin´s inhibitory effects on HCT-116 cell viability and colon-sphere forming-frequency, the range of 20–80 μM for this peptide was then chosen as the optimal treatment concentration for subsequent experiments.

Figure 2 presents the apoptotic state of adherent HCT-116 cells under control and lunasin-treated conditions for 72 h. The apoptotic populations of cells treated with the peptide were significantly increased (Figure 2A). Lunasin at 20, 40 and 80 μM induced 1.3, 1.7 and 1.8-fold increase of total apoptotic cells, respectively, compared to control. In the case of lunasin at 40 and 80 μM, cells both in the early and late apoptotic stages were significantly enhanced. The apoptosis-involved inhibitory role of lunasin against HCT-116 cells was further addressed by the immunoblotting study of the molecular proteins PARP and caspase-3. PARP is responsible for the regulation of many cellular functions, such as key events supporting cell viability and DNA repair [38]. PARP degradation has been shown to facilitate cellular disassembly, and serve as a marker of cells undergoing apoptosis, with this protein being the main cleavage target on the activity of the apoptotic trigger caspase-3 [39]. As shown in Figure 2B, lunasin activated the cleavage of caspase-3 and, consequently, the protein level of full-length PARP was decreased in lunasin-treated cells. This might be accompanied to increased expression of cleaved PARP, a hallmark of apoptosis, as we found previously [40]. In this line, Dia and de Mejia found that lunasin was able to activate the apoptotic mitochondrial pathway in HT-29 and KM12L4 cells, as evidenced by the modulation of Bcl-2/Bax family of proteins, nuclear clusterin, cytochrome c, and caspases-activity [16,17]. Similar apoptosis-related properties have been reported for this peptide against the growth of leukemia L1210 cells [41], and breast cancer MCF-7 and MDA-MB-231 cells [29,42].

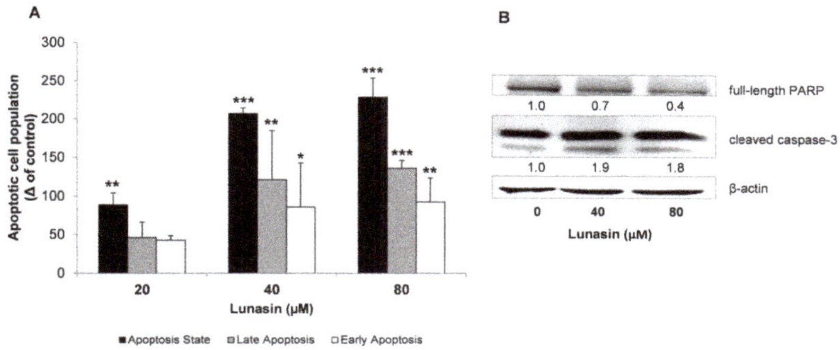

Figure 2. Effect of lunasin on the apoptosis state of HCT-116 cells. Cells were treated with lunasin at the indicated concentrations (20, 40, and 80 μM) for 72 h, and harvested for apoptosis analysis and Western immunoblotting. (**A**) Flow cytometry-based Annexin V/PI double labeling of apoptotic cells. Total apoptotic cells were identified as Annexin V-positive cells (apoptosis state), being Annexin V-positive/PI-negative and Annexin V-positive/PI-positive cells identified as early apoptotic and late apoptotic cells, respectively. Results, presented as the increased number in apoptotic cell populations compared to control cells, are means ± standard deviation (SD) of the replicates of experiments carried out. * ($p < 0.05$), ** ($p < 0.01$), *** ($p < 0.001$) significantly different from control. (**B**) Expression of full length PARP and cleaved caspase-3 proteins determined by Western Blot. The numbers underneath the blots represent band intensity that was normalized to β-actin and measured by Image J software (means of duplicates, and standard deviations within ± 15% of the means were not shown). β-actin was served as an equal loading control for cytosolic fraction.

We next aimed to determine whether the apoptosis-inducing property was also involved in the suppression of the spheroid-forming capacity of HCT-116 cells. Colon-spheres were treated with lunasin for 7 days and apoptosis detection was examined as shown in Figure 3. Results from the flow cytometry study after staining with Annexin-V/PI showed that lunasin led to induction in the cellular apoptotic state (Figure 3A,B). The raise in the number of apoptotic cells was not significantly promoted at lunasin 20 μM. However, lunasin both at 40 and 80 μM exerted a 2.0-fold apoptosis-induction effect, mostly in the late apoptotic cellular subset, independently of the dose. As shown in Figure 3C, the implication of the mechanism responsible for the inhibitory effect of lunasin peptide against the expansion of the HCT-116-derived colon-spheres was further demonstrated by immunoblotting. Again, cleaved caspase-3 activity was induced by lunasin treatment, with this activation being companied by a decrease of PARP protein levels. Our results suggested a dose-dependent trend in the down-regulated levels of PARP protein after lunasin treatment. However, this was not the case for cleaved caspase-3, which is more related to the inhibitory effects shown in the MTT assay and, mostly, over the spheroid-forming capacity where peptide lunasin displayed a ca. 30–40% inhibitory effect for the dose range of 20–160 μM. Therefore, in the present study, it has been suggested that lunasin has similar effects in the apoptosis-induction of both populations of CRC HCT-116 cells. In this sense, other food/natural compounds and phytochemicals have demonstrated to exert similar inhibitory effects through apoptosis induction against the expansion of the CSC subpopulation not only in CRC [33,43,44], but also in pancreatic and prostate cancer cells [45,46].

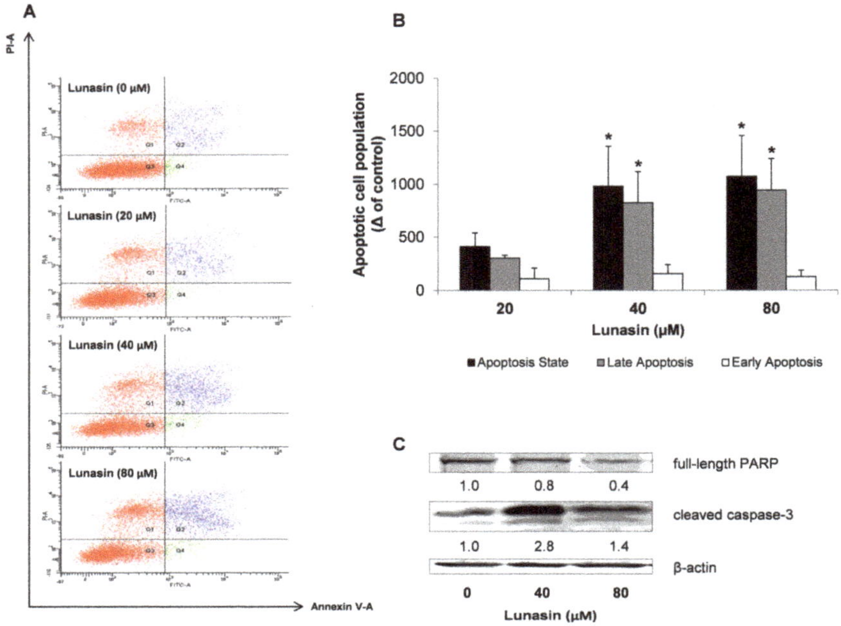

Figure 3. Effect of lunasin on the apoptosis state of colon tumorspheres. Cells were treated with lunasin at the indicated concentrations for 7 days, and harvested for apoptosis analyses and Western immunoblotting. (**A**) Flow cytometry-based Annexin V/PI double labeling of apoptotic cells. (**B**) Total apoptotic cells were identified as Annexin V-positive cells (apoptosis state), being Annexin V-positive/PI-negative and Annexin V-positive/PI-positive cells identified as early apoptotic and late apoptotic cells, respectively. Results, presented as the increased number in apoptotic cell populations compared to control cells, are means ± standard deviation (SD) of the replicates of experiments carried out. * ($p < 0.05$) significantly different from control. (**C**) Expression of full length PARP and cleaved caspase-3 proteins determined by Western Blot. The numbers underneath the blots represent band intensity that was normalized to β-actin and measured by Image J software (means of duplicates, and standard deviations within ± 15% of the means were not shown). β-actin was served as an equal loading control for cytosolic fraction.

2.3. Effect of Lunasin on Cell Cycle Progression of CRC Cells

To provide further insights into the growth inhibitory effects exerted by lunasin in HCT-116 cells, analyses on cell cycle distribution were performed on both adherent cells and colon-spheres after treatment with lunasin for 72 h and 7 days, respectively. Deregulation of cell cycle control and potential to replicate without limit are one of the hallmarks of cancer, with all these events being highly regulated by internal checkpoints that ensure the proper cellular division [35]. As shown in Figure 4A, control adherent HCT-116 cells were found to significantly increase their G1 phase (66.5 ± 1.7%) after lunasin´s treatment (20 µM lunasin, G1: 70.2 ± 0.3%, $p < 0.05$; 40 µM lunasin, 70.5 ± 0.7%, $P < 0.05$; 80 µM lunasin, 72.0 ± 1.2%, $p < 0.01$). Interestingly, as represented in Figure 4B, lunasin-treated cells also resulted in a marked accumulation of the sub-G0/G1 cell population, compared to control cells. Cells at the sub-G0/G1 fraction contain less amount of DNA than G1 cells, suggesting DNA degradation potentially caused by apoptotic events [47]. This effect had also been demonstrated for peptide lunasin in leukemia L1210 cells [41] and is in agreement with our results on apoptosis-induction in HCT-116 cells (Figure 2). On the other hand, our findings differ with other studies showing the capability of this peptide to arrest cell cycle at S-phase in breast cancer MDA-MB-231 cells [42], and at G2-phase in leukemia L1210 cells [41] and CRC HT-29 and KM12L4 cells [16,17]. However, other

RGD-motif-containing peptides have been also reported to result in a G0/G1-phase arrest in cancer cells [48]. Noteworthy, different cancer cells might respond differently to lunasin peptide accordingly to their diverse tumor phenotype. Moreover, regarding to the colon tumorspheres (Figure 5A), lunasin at 80 μM also led to an enhancement of G1-arrest (74.0 ± 0.6%, $p < 0.001$), accompanied with a reduction in the S-cellular subset (14.3 ± 0.9%, $p < 0.05$), compared to control cells (G1: 69.4 ± 0.6%; S: 16.4 ± 0.9%). This effect might be related to the antiproliferative and pro-apoptotic activities above indicated. However, 20–40 μM-treated colon-spheres showed a similar trend but in a weaker manner, lacking statistical significance in this dose range.

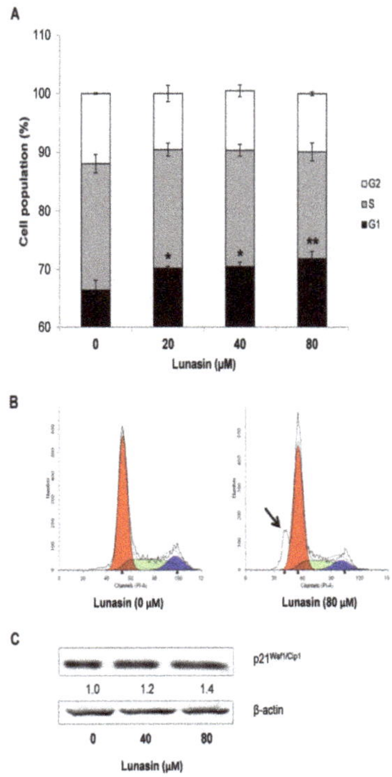

Figure 4. Effect of lunasin on cell cycle progression of HCT-116 cells. Cells were treated with lunasin at the indicated concentrations for 72 h, and harvested for cell cycle analysis and Western immunoblotting. (**A**) Cell cycle distribution was assessed by flow cytometry using PI staining. Results, presented as percentage of cells in G1, S, and G2 phases, are means ± standard deviation (SD) of the replicates of experiments carried out. * ($p < 0.05$), ** ($p < 0.01$) significantly different from control. (**B**) Representative images of lunasin-induced increase in the sub-GO/G1 cell population (black arrow). (**C**) Expression of p21$^{Waf1/Cip1}$ protein determined by Western Blot. The numbers underneath the blots represent band intensity that was normalized to β-actin and measured by Image J software (means of duplicates, and standard deviations within ± 15% of the means were not shown). β-actin was served as an equal loading control for cytosolic fraction.

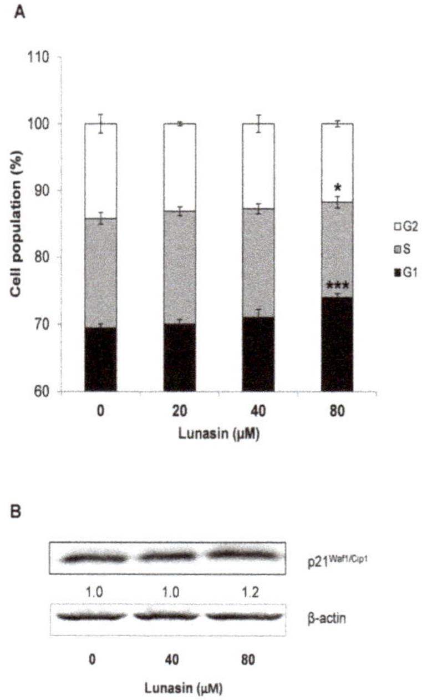

Figure 5. Effect of lunasin on cell cycle progression of colon tumorspheres. Cells were treated with lunasin at the indicated concentrations for 7 days, and harvested for cell cycle analyses and Western immunoblotting. (**A**) Cell cycle distribution was assessed by flow cytometry using PI staining. Results, presented as percentage of cells in G1, S, and G2 phases, are means ± standard deviation (SD) of the replicates of experiments carried out. * ($p < 0.05$), *** ($p < 0.001$) significantly different from control. (**B**) Expression of p21Waf1/Cip1 protein determined by Western Blot. The numbers underneath the blots represent band intensity that was normalized to β-actin and measured by Image J software (means of duplicates, and standard deviations within ± 15% of the means were not shown). β-actin was served as an equal loading control for cytosolic fraction.

To further explain lunasin´s effect on cell cycle progression, evaluation of the expression of the cyclin-dependent kinase (CDK) inhibitors $p21^{Waf1/Cip1}$ and $p27^{Kip1}$ was performed by Western Blot. Treatment of CRC cells with lunasin showed no effect on the level of p27 (data not shown), while it slightly increased the molecular expression of p21 protein up to 140% and 120% in adherent HCT-116 cells (Figure 4C) and colon-spheres CSC (Figure 5B), respectively. A consistent role of lunasin over these molecules cannot be thus extracted from our results. CDK-inhibitors p21 and p27 are two important cell cycle regulators at the G1-phase known to be usually co-regulated, although they have shown paradoxical roles in the literature [49]. In non-small cell lung H661 cancer cells, expressing a mutated form of p53 and thus low non-inducible levels of p21, McConnell and co-workers recently found that peptide lunasin also blocked cell cycle at the G1/S-phase through CDK-inhibitor p27 as well as by disrupted phosphorylation of the retinoblastoma protein [34]. Regarding studies on CRC, Dia and de Mejia reported lunasin´s capability to induce the expression of the CDK-inhibitor p21 in HT-29 and KM12L4 cells, and linked this effect with a decreased cell proliferation, cell cycle arrest, and up-regulation of the pro-apoptotic markers caspase-3 and nuclear clusterin isoform [16,17]. In these studies, CDK-inhibitor p27 induction was also demonstrated in KM12L4 cells [16] although it was not evaluated in the HT-29 cell line [17].

In order to provide more evidence on the cancer-preventive role of bioactive peptide lunasin, specifically against the CRC malignancy, some studies have been carried out. In the highly metastatic KM12L4 cell line, Dia and de Mejia demonstrated that lunasin is able to internalize into the cell and sit within the nucleus, to modify the expression of human extracellular matrix and cell adhesion genes by binding to $\alpha_5\beta_1$ integrin, and also to inhibit the FAK/ERK/NF-κB signaling pathway [16,18]. Indeed, peptides containing the RGD-motif can bind integrins and block their signaling pathways involved in cell adhesion, invasion and extracellular matrix components, mechanisms by which lunasin has recently shown inhibitory and anti-metastatic effects in some cancer models [15,50]. Studies have found that lunasin inhibits non small cell lung cancer cell proliferation acting as antagonist of αv integrin and histone acetylation modulatory agent [32,51]. Similarly, lunasin inhibited the migration and invasion properties of breast cancer cells via integrin-mediated FAK/Akt/ERK and nuclear factor (NF)-κβ pathways, and suppression of matrix metalloproteinases 2 and 9 [52]. The in vivo effect of this peptide was suggested in the CRC liver metastasis mice model by Dia and de Mejia [53], although disagreements between intraperitoneally- and orally-administered findings made it hard to establish a definitive lunasin´s role on preventing the CRC liver metastasis. Regarding the in vivo efficacy of lunasin against CSC, lunasin impaired the tumor growth initiated by CSC in a melanoma xenograft mouse model [35] and also suppressed the ability of these cancer-initiating cells to invade and proliferate in the lung of an experimental model of melanoma metastasis using B16-F10 cells [36].

3. Materials and Methods

3.1. Materials

Peptide lunasin was synthesized by Chengdu KaiJie Biopharm Co., Ltd. (Chengdu, China). Its purity (>95%) was confirmed by liquid chromatography (HPLC) coupled to mass spectrometry (HPLC-MS).

3.2. Cell Lines

The human CRC cell line HCT-116 was obtained from American Type Cell Collection (ATCC, Manassas, VA, USA), and maintained in RPMI medium (ATCC) supplemented with 5% heat inactivated fetal bovine serum (FBS; Mediatech, Herndon, VA, USA), 100 units/mL penicillin, and 0.1 mg/mL streptomycin (Sigma-Aldrich, St. Louis, MO, USA). Cells were grown in a humidified incubator containing 5% CO_2 and 95% air at 37 °C, kept sub-confluent, and medium was changed every other day. All cells were assayed within 5–25 passages. Enrichment culture of tumor-derived colon-spheres was performed by incubating parental HCT-116 cells in serum-free medium (SFM) composed of DMEM/F-12 medium supplemented with 2% B-27 supplement, 20 ng/mL recombinant human epidermal growth factor, 10 ng/mL fibroblast growth factor-basic (Life Technologies, Grand Island, NY, USA), 100 units/mL penicillin, 0.1 mg/mL streptomycin, and 10 µg/mL insulin (Sigma-Aldrich) in ultra low-attachment plates (Corning, Lowell, MA, USA) at 37 °C. Plated under these anchorage-independent conditions in supplemented-SFM, tumor cells form floating spheres reported to represent the growth of CSC [27,31,54].

3.3. Cell Proliferation Assay

HCT-116 cells were seeded in 96-well plates (1.1×10^4 cells/mL). After 24 h incubation, cells were treated with different concentrations of lunasin ranging from 5 to 160 µM. After 72 h treatment, cell viability was determined by the MTT assay. Treatment medium was replaced by 200 µL of fresh medium containing 0.5 mg/mL MTT (Sigma-Aldrich). After 1 h incubation at 37 °C, MTT-containing medium was removed and the reduced formazan dye was solubilized by adding 100 µL of dimethyl sulfoxide to each well. After gently mixing, the absorbance was read at 570 nm using a microplate reader (Elx800TM, BioTek Instrument, Winooski, VT, USA). The results were expressed as percentage of the control, considered as 100%. Experiments were carried out in triplicate with at least three replicates per concentration.

3.4. Tumorsphere Formation Assay

To examine the effect of lunasin on the formation of tumorspheres derived from CRC HCT-116 cells, cells were grown in SFM and plated as single cells in ultra low-attachment 24-well plates (6×10^3 cells/mL). Right after seeding, cells were treated with different concentrations of lunasin ranging from 5 to 160 μM and incubated at 37 °C for 10 days. After that time, tumorspheres were formed and transferred to 6-well dishes in differentiating medium (RPMI supplemented with 5% FBS and 1% antibiotics). Under these conditions, tumorspheres were adhered after 24 h incubation. Then, cells were stained with crystal violet solution (0.2% crystal violet in 2% ethanol) for 20 min at room temperature, photographed and counted. Results were presented as percentage of tumorspheres forming cells compared to control, considered as 100%. Analyses were performed in triplicate with at least three replicates per concentration.

3.5. Detection of Apoptosis

Apoptotic cells were quantified by Annexin V/PI double staining using an apoptotic detection kit (BioVision, Mountain View, CA, USA) according to manufacturer's instruction, followed by flow cytometry. HCT-116 cells (4×10^4 cells/mL) and colon tumorspheres (3×10^3 cells/mL) were seeded onto 6-well plates and treated (20–80 μM lunasin) as described above. After 72 h treatment, HCT-116 cells were collected as described by Qiu and others [55]. In the case of colon tumorspheres, after 7 days treatment, floating cells in medium were collected in ice-cold flow cytometry tubes. After centrifugation (2000 × g, 2 min), single-cell suspensions were generated by incubation with 0.5 mL trypsin (0.25% trypsin-ethylenediaminetetraacetic acid, EDTA, Sigma-Aldrich) and 1 mL medium for 5 min at 37 °C, and gentle pippeting. Afterwards, in both cell cultures, cell suspensions were centrifuged (2000 × g, 2 min) and washed twice with 0.5 mL ice-cold phosphate buffer saline (PBS). Then, cells were suspended in 0.3 mL binding buffer containing Annexin V and PI, and incubated for 15 min at room temperature in the dark. Total apoptotic cells were identified using a BD LSR II cell analyzer (BD Biosciences, San Jose, CA, USA) as Annexin V-positive cells (apoptosis state), being further identified based on PI staining as early apopototic cells (Annexin V-positive/PI-negative) or late apoptotic-necrotic cells (Annexin V-positive/PI-positive). At least 10,000 events were recorded to assess the percentage of apoptotic cells. Analyses were performed in duplicate with at least three replicates per concentration, and results were presented as the increased number in apoptotic cell populations, compared to control cells.

3.6. Cell Cycle Analyses

HCT-116 cells and colon tumorspheres were treated as described for the apoptosis detection assay. After 72 h treatment, HCT-116 cells were collected as described by Qiu and others [55]. In the case of colon tumorspheres, after 7 days treatment, cells were collected as described for apoptosis detection assay with some modifications. Briefly, floating tumorspheres in medium were collected, centrifuged, and single-cell suspensions were generated, washed with ice-cold PBS, and then fixed in 1 mL of 70% ethanol and kept at −20 °C overnight. After centrifugation (2000× g, 2 min), cells were washed with 0.5 mL PBS, and incubated with 0.3 mL PBS solution containing RNAse (10%; Sigma-Aldrich) and PI (1%; BioVision) for 25 min at room temperature in the dark. Cell cycle distribution was analyzed with at least 8000 events recorded using a BD LSR II cell analyzer (BD Biosciences), and data were processed using ModFit LT software. Analyses were performed in duplicate with at least three replicates per concentration, and results were presented as percentage of cells in G1, S, and G2-phases.

3.7. Immunoblotting

HCT-116 cells (3.5×10^4 cells/mL) were seeded in 10 cm cell culture dishes. Colon tumorspheres were seeded exactly same as described for apoptosis assay. After 72 h treatment (20–80 μM lunasin), HCT-116 cells were collected and whole-cell lysates were prepared as previously described [53]. In the case of colon tumorspheres, after 7 days-treatment (20–80 μM lunasin), cells were collected following the same procedure with some modifications. Briefly, floating tumorspheres in medium were collected, centrifuged, and washed with ice-cold PBS. Then, cells were incubated on ice for 30 min in RIPA lysis buffer containing a protease inhibitor cocktail (Boston BioProducts, Ashland, MA, USA), and processed as previously described [56]. Supernatants were collected and protein content was quantified by the bicinchoninic acid method (Pierce, Rockford, IL, USA), using bovine serum albumin as standard protein. Equal amount of proteins (50–70 μg) were resolved over 12% SDS-polyacrylamide gel electrophoresis and transferred to nitrocellulose membranes. After blocking, membranes were incubated with different monoclonal primary antibodies overnight at 4 °C, according to manufacturer's instructions. Primary antibodies for cleaved caspase-3 (Asp175), full-length PARP, $p21^{Waf1/Cip1}$, and $p27^{Kip1}$ were from Cell Signaling Technology (Beverly, MA, USA). β-actin was used as a loading control of cytosolic fraction, and its antibody was from Sigma-Aldrich. After 1 h incubation with the appropriate secondary antibodies (goat anti-mouse IgG, and goat anti-rabbit IgG IRDye (LI-COR Biosciences, Lincoln, NE, USA)), proteins of interest were visualized using enhanced chemiluminescence (Boston Bioproducts), processed with Image J Software and analyzed as we previously described [40].

3.8. Statistical Analysis

Data were evaluated using one-way ANOVA followed by Bonferroni post hoc test and expressed as the mean ± standard variation (SD) of the different experiments carried out. GraphPad Prism 5.0 software (San Diego, CA, USA) was used to perform statistical analyses. Differences with a p value < 0.05 (*), p value < 0.01 (**) or p value < 0.001 (***) were considered significant.

4. Conclusions

In the present study, our cellular model allowed us to approach the study of peptide lunasin towards the ideal evaluation of cancer-preventive agents by targeting both the parental and the stem-like tumorigenic populations. The protective mechanisms on lunasin-treated cells can be postulated in terms of inhibition of cell growth and tumorsphere-forming activity, induction of apoptosis, and regulation of cell cycle progression. The recent CSC hypothesis has supposed a challenge on the search of chemotherapeutic agents that efficiently target fast diving cancer cells as well as CSC responsible for the growth and maintenance of the tumorigenic bulk mass. To the best of our knowledge, this is the first study that suggests a protective role of lunasin against the formation of colon-spheres derived from CRC cells, specifically the HCT-116 cell line. The potential of bioactive peptides against the CSC subpopulation deserves additional studies characterizing CSC markers in more cellular models. Before concluding on lunasin´s effects over CSC, the promising results of this work clearly need to be further addressed to elucidate the molecular basis of the tumorsphere-inhibitory activity, to study its potential on stem-related markers and signaling pathways, such as Wnt/β-catenin, Hedgehog and Notch, and to confirm this role by using in vivo models of CSC self-renewal.

Author Contributions: Conceptualization, B.H.-L., and H.X.; methodology, S.F.-T., F.X., and Y.H.; writing—original draft preparation, S.F-T.; writing–review and editing, S.F.-T., B.H.-L., and H.X.; funding acquisition, B.H.-L., and H.X. All authors have read and agreed to the published version of the manuscript.

Funding: This work has received financial support from Ministry of Economy and Competitiveness (MINECO, Spain) through the project AGL2015-66886-R, and United States Department of Agriculture. SFT is currently funded by the Instituto de Salud Carlos III (Sara Borrell fellowship CD17/00014).

Conflicts of Interest: The authors declare no conflict of interest.

Abbreviations

ATCC	American Type Cell Collection
CDK	Cyclin-dependent kinase
CRC	Colorectal cancer
CSC	Cancer stem-like cells
DHA	Docosahexaenoic acid
EDTA	Ethylenediaminitetraacetic acid
FBS	Fetal bovine serum
MTT	3-(4,5-dimethylthiazol-2-yl)-2,5-diphenyl tetrazolium bromide
PBS	Phosphate buffer saline
PI	Propidium iodide
SFM	Serum-free medium

References

1. Arnold, M.; Sierra, M.S.; Laversanne, M.; Soerjomataram, I.; Jemal, A.; Bray, F. Global patterns and trends in colorectal cancer incidence and mortality. *Gut* **2017**, *66*, 683–691. [CrossRef] [PubMed]
2. O'Brien, C.A.; Pollett, A.; Gallinger, S.; Dick, J.E. A human colon cancer cell capable of initiating tumor growth in immunodeficient mice. *Nature* **2007**, *445*, 106–110. [CrossRef] [PubMed]
3. Anderson, E.C.; Hessman, C.; Levin, T.G.; Monroe, M.M.; Wong, M.H. The role of colorectal cancer stem cells in metastatic disease and therapeutic response. *Cancer* **2011**, *3*, 319–339. [CrossRef] [PubMed]
4. Munro, M.J.; Wickremesekera, S.K.; Peng, L.; Tan, S.T.; Itinteang, T. Cancer stem cells in colorectal cancer: A review. *J. Clin. Pathol.* **2018**, *71*, 110–116. [CrossRef]
5. Reya, T.; Morrison, S.J.; Clarke, M.F.; Weissman, I.L. Stem cells, cancer, and cancer stem cells. *Nature* **2001**, *414*, 105–111. [CrossRef]
6. Visvader, J.E.; Lindeman, G.J. Cancer stem cells in solid tumors: Accumulating evidence and unresolved questions. *Nat. Rev. Cancer* **2008**, *8*, 755–768. [CrossRef]
7. Zhou, B.B.S.; Zhang, H.Y.; Damelin, M.; Geles, K.G.; Grindley, J.C.; Dirks, P.B. Tumor-initiating cells: Challenges and opportunities for anticancer drug discovery. *Nat. Rev. Drug Discov.* **2009**, *8*, 806–823. [CrossRef]
8. Clevers, H. The cancer stem cell: Premises, promises and challenges. *Nat. Med.* **2011**, *17*, 313–319. [CrossRef]
9. Spector, D.; Anthony, M.; Alexander, D.; Arab, L. Soy consumption and colorectal cancer. *Nutr. Cancer* **2003**, *47*, 1–12. [CrossRef]
10. Yang, G.; Shu, X.O.; Li, H.L.; Chow, W.H.; Cai, H.; Zhang, X.L.; Gao, Y.T.; Zheng, W. Prospective cohort study of soy food intake and colorectal cancer risk in women. *Am. J. Clin. Nutr.* **2009**, *89*, 577–583. [CrossRef]
11. Kim, Y.S.; Farrar, W.; Colburn, N.H.; Milner, J.A. Cancer stem cells: Potential target for bioactive food components. *J. Nutr. Biochem.* **2012**, *23*, 691–698. [CrossRef] [PubMed]
12. Tsai, C.Y.; Chen, Y.H.; Chien, Y.W.; Huang, W.H.; Lin, S.H. Effect of soy saponin on the growth of human colon cancer cells. *World J. Gastroenterol.* **2010**, *16*, 3371–3376. [CrossRef] [PubMed]
13. De Mejia, E.G.; Bradford, T.; Hasler, C. The anticarcinogenic potential of soybean lectin and lunasin. *Nutr. Rev.* **2003**, *61*, 239–246. [CrossRef] [PubMed]
14. Clemente, A.; Moreno, F.J.; Marín-Manzano, M.C.; Jiménez, E.; Domoney, C. The cytotoxic effect of Bowman-Birk isoinhibitors, IBB1 and IBBD2, from soybean (*Glycine max*) on HT29 human colorectal cancer cells is related to their intrinsic ability to inhibit serine proteases. *Mol. Nutr. Food Res.* **2010**, *54*, 396–405. [CrossRef] [PubMed]
15. Fernández-Tomé, S.; Hernández-Ledesma, B. Current state of art after twenty years of the discovery of bioactive peptide lunasin. *Food Res. Int.* **2019**, *116*, 71–78. [CrossRef]
16. Dia, V.P.; de Mejia, E.G. Lunasin induces apoptosis and modifies the expression of genes associated with extracellular matrix and cell adhesion in human metastatic colon cancer cells. *Mol. Nutr. Food Res.* **2011**, *55*, 623–634. [CrossRef]
17. Dia, V.P.; de Mejia, E.G. Lunasin promotes apoptosis in human colon cancer cells by mitochondrial pathway activation and induction of nuclear clusterin expression. *Cancer Lett.* **2010**, *295*, 44–53. [CrossRef]

18. Dia, V.P.; de Mejia, E.G. Lunasin potentiates the effect of oxaliplatin preventing outgrowth of colon cancer metastasis, binds to $\alpha_5\beta_1$ integrin and suppresses FAK/ERK/NF-κB signaling. *Cancer Lett.* **2011**, *313*, 167–180. [CrossRef]
19. García-Nebot, M.J.; Recio, I.; Hernández-Ledesma, B. Antioxidant activity and protective effects of peptide lunasin against oxidative stress in intestinal Caco-2 cells. *Food Chem. Toxicol.* **2014**, *65*, 155–161. [CrossRef]
20. Kawasaki, B.T.; Hurt, E.M.; Mistree, T.; Farrar, W.L. Targeting cancer stem cells with phytochemicals. *Mol. Interv.* **2008**, *8*, 174–184. [CrossRef]
21. Li, Y.; Wicha, M.S.; Schwartz, S.J.; Sun, D. Implications of cancer stem cell theory for cancer chemoprevention by natural dietary compounds. *J. Nutr. Biochem.* **2011**, *22*, 799–806. [CrossRef] [PubMed]
22. Kim, G.N.; Song, J.H.; Kim, E.S.; Choi, H.T.; Jang, H.D. Isoflavone content and apoptotic effect in HT-29 cancer cells of a soy germ extract. *Food Chem.* **2012**, *130*, 404–407. [CrossRef]
23. Oh, J.; Hlatky, L.; Jeong, Y.S.; Kim, D. Therapeutic effectiveness of anticancer phytochemicals on cancer stem cells. *Toxins* **2016**, *8*, 199. [CrossRef] [PubMed]
24. De Mejia, E.G.; Dia, V.P. The role of nutraceutical proteins and peptides in apoptosis, angiogenesis, and metastasis of cancer cells. *Cancer Metastasis Rev.* **2010**, *29*, 511–528. [CrossRef] [PubMed]
25. Hernández-Ledesma, B.; Hsieh, C.-C. Chemopreventive role of food-derived proteins and peptides: A review. *Crit. Rev. Food Sci.* **2017**, *57*, 2358–2376. [CrossRef] [PubMed]
26. Hernández-Ledesma, B.; Hsieh, C.-C.; de Lumen, B.O. Relationship between lunasin's sequence and its inhibitory activity of histones H3 and H4 acetylation. *Mol. Nutr. Food Res.* **2011**, *55*, 989–998. [CrossRef] [PubMed]
27. Kanwar, S.S.; Yu, Y.J.; Nautiyal, J.; Patel, B.B.; Majumdar, A.P.N. The Wnt/beta-catenin pathway regulates growth and maintenance of colonospheres. *Mol. Cancer* **2010**, *9*, 212. [CrossRef]
28. Qiu, P.; Guan, H.; Dong, P.; Guo, S.; Zheng, J.; Li, S.; Chen, Y.; Ho, C.-T.; Pan, M.-H.; McClements, D.J.; et al. The inhibitory effects of 5-hydroxy-3, 6, 7, 8, 3′, 4′-hexamethoxyflavone on human colon cancer cells. *Mol. Nutr. Food Res.* **2011**, *55*, 1523–1532. [CrossRef]
29. Pabona, J.M.P.; Dave, B.; Su, Y.; Montales, M.T.E.; de Lumen, B.O.; de Mejia, E.G.; Rahal, O.M.; Simmen, R.C.M. The soybean peptide lunasin promotes apoptosis of mammary epithelial cells via induction of tumor suppressor PTEN: Similarities and distinct actions from soy isoflavone genistein. *Genes Nutr.* **2013**, *8*, 79–90. [CrossRef]
30. Yu, Y.; Nangia-Makker, P.; Farhana, L.; Rajendra, G.S.; Levi, E.; Majumdar, A.P. miR-21 and miR-145 cooperation in regulation of colon cancer stem cells. *Mol. Cancer* **2015**, *14*, 98. [CrossRef]
31. Chen, Y.; Wang, X.-Q.; Zhang, Q.; Zhu, J.-Y.; Li, Y.; Xie, C.-F.; Li, X.-T.; Wu, J.-S.; Geng, S.-S.; Zhong, C.-Y.; et al. (-)-epigallocatechin-3-gallate inhibits colorectal cancer stem cells by suppressing Wnt/β-catenin pathway. *Nutrients* **2017**, *9*, 572. [CrossRef] [PubMed]
32. Kim, T.I. Chemopreventive drugs: Mechanisms via inhibition of cancer stem cells in colorectal cancer. *World J. Gastroenterol.* **2014**, *20*, 3835–3846. [CrossRef]
33. Yang, T.; Fang, S.; Zhang, H.X.; Xu, L.X.; Zhang, Z.Q.; Yuan, K.T.; Xue, C.L.; Yu, H.L.; Zhang, S.; Li, Y.F.; et al. N-3 PUFAs have antiproliferative and apoptotic effects on human colorectal cancer stem-like cells in vitro. *J. Nutr. Biochem.* **2013**, *24*, 744–753. [CrossRef] [PubMed]
34. McConnell, E.J.; Devapatla, B.; Yaddanapudi, K.; Davis, K.R. The soybean-derived peptide lunasin inhibits non-small cell lung cancer cell proliferation by suppressing phosphorylation of the retinoblastoma protein. *Oncotarget* **2015**, *6*, 4649–4662. [CrossRef] [PubMed]
35. Shidal, C.; Al-Rayyan, N.; Yaddanapudi, K.; Davis, K.R. Lunasin is a novel therapeutic agent for targeting melanoma cancer stem cells. *Oncotarget* **2016**, *7*, 84128–84141. [CrossRef] [PubMed]
36. Shidal, C.; Inaba, J.-I.; Yaddanapudi, K.; Davis, K.R. The soy-derived peptide Lunasin inhibits potential of melanoma initiating cells. *Oncotarget* **2017**, *8*, 25525–25541. [CrossRef] [PubMed]
37. Hanahan, D.; Weinberg, R.A. Hallmarks of cancer: The next generation. *Cell* **2011**, *144*, 646–674. [CrossRef]
38. Satoh, M.S.; Lindahl, T. Role of poly (ADP-ribose) formation in DNA repair. *Nature* **1992**, *356*, 356–358. [CrossRef]
39. Oliver, F.J.; de la Rubia, G.; Rolli, V.; Ruiz-Ruiz, M.C.; de Murcia, G.; Menissier-de Murcia, J. Importance of poly(ADP-ribose) polymerase and its cleavage in apoptosis—Lesson from an uncleavable mutant. *J. Biol. Chem.* **1998**, *273*, 33533–33539. [CrossRef]

40. Charoensinphon, N.; Qiu, P.; Dong, P.; Zheng, J.; Ngauv, P.; Cao, Y.; Li, S.; Ho, C.-T.; Xiao, H. 5-demethyltangeretin inhibits human non-small lung cancer cell growth by inducing G2/M cell cycle arrest and apoptosis. *Mol. Nutr. Food Res.* **2013**, *57*, 2103–2111. [CrossRef]
41. De Mejia, E.G.; Wang, W.; Dia, V.P. Lunasin, with an arginine-glycine-aspartic acid motif, causes apoptosis to L1210 leukemia cells by activation of caspase-3. *Mol. Nutr. Food Res.* **2010**, *54*, 406–414. [CrossRef] [PubMed]
42. Hsieh, C.-C.; Hernández-Ledesma, B.; de Lumen, B.O. Cell proliferation inhibitory and apoptosis-inducing properties of anacardic acid and lunasin in human breast cancer MDA-MB-231 cells. *Food Chem.* **2011**, *125*, 630–636. [CrossRef]
43. Lin, L.; Liu, Y.; Li, H.; Li, P.K.; Fuchs, J.; Shibata, H.; Iwabuchi, Y.; Lin, J. Targeting colon cancer stem cells using a new curcumin analogue, GO-Y030. *Br. J. Cancer* **2011**, *105*, 212–220. [CrossRef] [PubMed]
44. Massey, A.R.; Reddivari, L.; Vanamala, J. The dermal layer of sweet sorghum (*Shorgum bicolor*) stalk, a byproduct of biofuel production and source of unique 3-deoxyanthocyanidins, has more antiproliferative and proapoptotic activity than the pith in p53 variants of HCT116 and colon cancer stem cells. *J. Agric. Food Chem.* **2014**, *62*, 3150–3159. [PubMed]
45. Ottinger, S.; Klöppel, A.; Rausch, V.; Liu, L.; Kallifatidis, G.; Gross, W.; Gebhard, M.M.; Brümmer, F.; Herr, I. Targeting of pancreatic and prostatic cancer stem cell characteristics by *Crambe crambe* marine sponge extract. *Int. J. Cancer* **2012**, *130*, 1671–1681. [CrossRef] [PubMed]
46. Soner, B.C.; Aktug, H.; Acikgoz, E.; Duzagac, F.; Guven, U.; Ayla, S.; Cal, G.; Oktem, G. Induced growth inhibition, cell cycle arrest and apoptosis in $CD133^{+}/CD44^{+}$ prostate cancer stem cells by flavopiridol. *Int. J. Mol. Med.* **2014**, *34*, 1249–1256. [CrossRef]
47. Nagata, S.; Nagase, H.; Kawane, K.; Mukae, N.; Fukuyama, H. Degradation of chromosomal DNA during apoptosis. *Cell Death Differ.* **2003**, *10*, 108–116. [CrossRef]
48. Yang, W.; Meng, L.; Wang, H.; Chen, R.; Wang, R.; Ma, X.; Xu, G.; Zhou, J.; Wang, Y.; Lu, Y.; et al. Inhibition of proliferative and invasive capacities of breast cancer cells by arginine-glycine-aspartic acid peptide in vitro. *Oncol. Rep.* **2006**, *15*, 113–117. [CrossRef]
49. Abukhdeir, A.M.; Park, B.H. p21 and p27: Roles in carcinogenesis and drug resistance. *Expert Rev. Mol. Med.* **2009**, *10*, e19. [CrossRef]
50. Vuyyuri, S.B.; Shidal, C.; David, K.R. Development of the plant-derived peptide lunasin as an anticancer agent. *Curr. Opin. Pharmacol.* **2018**, *41*, 27–33. [CrossRef]
51. Inaba, J.; McConnell, E.J.; Davis, K.R. Lunasin sensitivity in non-small lung cancer cells is linked to suppression of integrin signaling and changes in histone acetylation. *Int. J. Mol. Sci.* **2014**, *15*, 23705–23724. [CrossRef] [PubMed]
52. Jiang, Q.; Pan, Y.; Cheng, Y.; Li, H.; Liu, D. Lunasin suppresses the migration and invasion of breast cancer cells by inhibiting matrix metalloproteinase-2/-9 via the FAK/Akt/ERK and NF-κB signaling pathways. *Oncol. Rep.* **2016**, *36*, 253–262. [CrossRef] [PubMed]
53. Dia, V.P.; de Mejia, E.G. Potential of lunasin orally-administered in comparison to intraperitoneal injection to inhibit colon cancer metastasis in vivo. *J. Cancer Ther.* **2013**, *4*, 34–43. [CrossRef]
54. O'Brien, C.A.; Kreso, A.; Ryan, P.; Hermans, K.G.; Gibson, L.; Wang, Y.D.; Tsatsanis, A.; Gallinger, S.; Dick, J.E. ID1 and ID3 regulate the self-renewal capacity of human colon cancer-initiating cells through p21. *Cancer Cell* **2012**, *21*, 777–792. [CrossRef]
55. Qiu, P.; Dong, P.; Guan, H.; Li, S.; Ho, C.T.; Pan, M.H.; McClements, D.J.; Xiao, H. Inhibitory effects of 5-hydroxy polymethoxyflavones on colon cancer cells. *Mol. Nutr. Food Res.* **2010**, *54*, S244–S252. [CrossRef]
56. Xiao, H.; Yang, C.S.; Li, S.; Jin, H.; Ho, C.T.; Patel, T. Monodemethylated polymethoxyflavones from sweet orange (*Citrus sinensis*) peel inhibit growth of human lung cancer cells by apoptosis. *Mol. Nutr. Food Res.* **2009**, *53*, 398–406. [CrossRef]

International Journal of *Molecular Sciences*

MDPI

Article

Multifunctional Peptides from Spanish Dry-Cured Pork Ham: Endothelial Responses and Molecular Modeling Studies

Sara María Martínez-Sánchez [1,2], Horacio Pérez-Sánchez [3], José Antonio Gabaldón [2], José Abellán-Alemán [4] and Silvia Montoro-García [1,4,*]

1 Laboratorio de Cultivo Celular, Facultad de Ciencias de la Salud, UCAM Universidad Católica San Antonio de Murcia, Campus de los Jerónimos s/n, Guadalupe 30107, Murcia, Spain
2 Departamento Tecnología de la Alimentación y Nutrición, UCAM Universidad Católica San Antonio de Murcia, Campus de los Jerónimos s/n, Guadalupe 30107, Murcia, Spain
3 Bioinformatics and High Performance Computing Research Group (BIO-HPC), Computer Engineering Department, Universidad Católica de Murcia (UCAM), Guadalupe 30107, Murcia, Spain
4 Cátedra de Riesgo Cardiovascular, UCAM Universidad Católica San Antonio de Murcia, Campus de los Jerónimos s/n, Guadalupe 30107, Murcia, Spain
* Correspondence: smontoro@ucam.edu

Received: 24 July 2019; Accepted: 26 August 2019; Published: 28 August 2019

Abstract: Food peptides contain a very wide range of diversified structures, which explains their diverse range of functional activities. Proatherogenic endothelium is related to vasoconstriction, inflammation, and oxidative stress. In this line, four synthetic bioactive peptides from dry-cured pork ham, previously identified according to their Angiotensin I Converting Enzyme (ACE) inhibitory capacity and high bioavailability, were tested. Among them, KPVAAP displayed an estimated IC_{50} of 59.22 μM for human ACE inhibition, and docking simulations demonstrated the consistency of the noncompetitive binding with the protein. The addition of synthetic peptides to human endothelial cells significantly prevents the expression of genes related to endothelial dysfunction and inflammation (eNOS, ICAM-1, VCAM-1, IL-6) and lowers NF-κB activation (all $p < 0.05$). In silico dockings showed that the four bioactive peptides interact with the regulatory subunit NEMO of the NF-κB transcription factor at the same site as other characterized inhibitors (CC2-LZ region). This is the first study linking experimental and computational approaches that shows NF-κB to be the target of biopeptides of food origin. These multifunctional peptides from dry-cured pork ham make them good candidates for further research into their therapeutic or preventive use to attenuate the inflammatory atherosclerotic process.

Keywords: bioactive peptides; inflammation; NF-κB; dry-cured pork ham; angiotensin I converting enzyme; endothelial dysfunction; molecular blind docking

1. Introduction

The endothelium is an important regulator of vascular homeostasis and an essential part of the cardiovascular (CV) system, participating in all aspects of pathophysiological processes such as muscular tone, inflammation, thrombosis, or vascular wall remodeling [1]. Tumor necrosis factor-α (TNF-α) is a potent proinflammatory cytokine that triggers the canonical activation of the transcription nuclear factor-κB (NF-κB) and the endothelial gene expression of adhesion molecules, cytokines, and fibrinolytic proteins among others [2]. NF-κB is crucially involved in the pathogenesis of inflammatory diseases and represents a target for treatment. The canonical pathway is strictly regulated and involves various steps including the phosphorylation, ubiquitination, and degradation of the IκB kinase (IKK) complex, which leads to the nuclear translocation of the p50 and p65 subunits of NF-κB [3]. Small molecules such as peptides have been shown to bind with high affinity protein kinases, phosphatases [4] and inhibit ubiquitination which could explain their potential as NF-κB inhibitors [5].

Levels of inflammatory biomarkers are increased in common pathological conditions such as hypertension, hyperlipidemia, physical inactivity, among others [6]. In this last respect, beneficial dietary habits, which include anti-inflammatory and antioxidant foods, have gained in popularity for the prevention of cardiovascular diseases (CVD). Many experimental studies have documented that small sequence peptides are released during gastrointestinal digestion, food processing, and the microbial proteolysis of proteins [7]. Depending on the sequence of amino acids, these bioactive peptides (BP) can exhibit different biological activities [8], mainly through pathways that are still not clearly understood. Bioactive peptides have attracted a lot of scientific interest due to their wide range of biofunctional properties [9]. The inhibition of Angiotensin I-converting enzyme (ACE) is a widely studied effect of BP from pork meat [10], and many of these BP show high bioavailability [11]. However, few mechanistic studies have looked at possible functional properties other than ACE inhibition. In a previous clinical study, our group suggested a link between the regular consumption of dry-cured pork ham with its characteristic BP and an improvement in the inflammatory status [12].

The present study aims to confirm a robust cause and effect relationship between BP from dry-cured pork ham and beneficial physiological effects related to CV health in humans. For that purpose, changes in human ACE activity, endothelial dysfunction gene expression, NF-κB activation, oxidative and apoptotic markers were tested. In addition, molecular modelling was used to establish novel peptide-NF-κB interactions at the molecular level, which has not been attempted before. The current findings suggest that BP from dry-cured pork ham bind to the NEMO subunit of the IKK complex and might suppress the NF-κB-dependent gene expression in vitro.

2. Results

2.1. Peptides with Human in Vitro ACE Inhibitory Activity

The inhibitory activity of human overexpressed endothelial ACE was assayed in transfected cells (Figure S1), with the results expressed as IC_{50} (Table 1). By HPLC it was revealed that BP1 (KPVAAP) is a potent inhibitor of ACE activity with an estimated IC_{50} of 59.22 μM, making it 16 times more effective than BP2 and BP4 with an estimated IC_{50} > 1000 μM. BP3 also displayed a high IC_{50} of 485 μM. The initial linear dose-response pointed to the higher ACE inhibitory activity of BP1 and BP3 but was not directly comparable to the inhibitory effect of captopril on a weight-basis (IC_{50} below 10 μM).

Table 1. Angiotensin I-converting enzyme (ACE) inhibitory peptides derived from dry-cured pork ham: sequence, estimated IC_{50} value and interaction score. Captopril (10 μM) inhibition was used to represent 100% inhibition of ACE activity under the assay conditions.

Bioactive Peptide	Sequence	Source of Protein	IC_{50} (μM)	Interaction Score
PEPTIDE 1 (BP1)	KPVAAP	Myosin-XV	59.22 ± 3.8	−8.1
PEPTIDE 2 (BP2)	KAAAATP	PR domain Zinc Finger Protein 2	>1000	−7.1
PEPTIDE 3 (BP3)	KPGRP	Titin	485.50 ± 43.47	−8.2
PEPTIDE 4 (BP4)	AAATP	PR domain Zinc Finger Protein 2	>1000	−7.1

Data from three independent experiments are expressed as mean ± SD. Specific ACE activity was expressed as μmoles of substrate HHL converted to the product HA per unit of time and normalized for protein content (units per microgram of protein).

2.2. Peptides Penetrate in the Catalytic Active Site of ACE

Docking simulations were carried out (Figure 1) to provide insight at atomic level into the interactions established between the different peptides, captopril, and the active site of human ACE (PDB: 4APJ) [13]. The four peptides (BP1, BP2, BP3, BP4) and captopril were able to bind to the protease with interaction scores of −8.1, −7.1, −8.2, −7.1, and −8.2 kcal/mol, respectively (Table 1). The area of the hydrophobic interactions and/or the establishment of hydrogen bonds (Figure 1) depend on their distinct sequences and resulting interaction patterns. Figure 1F shows that BP1 occupied the S1 and S2 subsites (near the active site), while the commercial antihypertensive drug, captopril, was found deeper inside the active site of ACE.

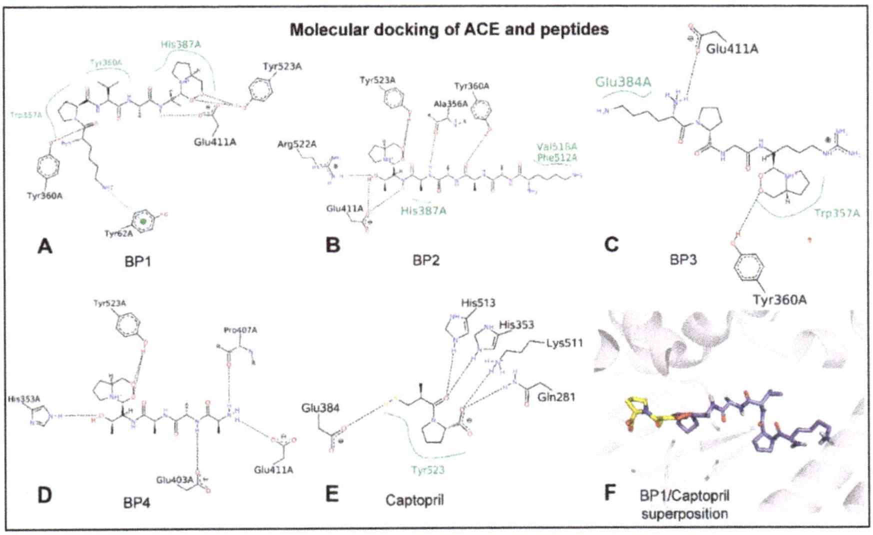

Figure 1. Depiction (in 2D) of the molecular docking between catalytic residues from the active site of ACE (PDB: 4APJ) and the bioactive peptides. (**A**) Peptide 1 (BP1) KPVAAP, (**B**) Peptide 2 (BP2) KAAAATP, (**C**) Peptide 3 (BP3) KPGRP, and (**D**) Peptide 4 (BP4) AAATP. (**E**) Captopril. (**F**) The superposition (in 3D) of BP1 (in purple) and captopril (in yellow) in stick representation. Continuous green lines represent hydrophobic interactions, while black dashed lines show hydrogen bonds. The absence of hydrophobic stabilization in the case of BP4 (**D**) might explain its high IC_{50} value. Moreover, BP2 (**B**) showed small hydrophobic interaction areas compared with BP1 (**A**) and BP3 (**C**), which contained voluminous hydrophobic groups.

2.3. Peptide Effects in Inflammatory Conditions

2.3.1. Peptides Affect Gene Expression in Inflammatory Conditions

To define the stimulatory conditions in which endothelial cells express high levels of adhesion and inflammatory markers, EA.hy926 cells, a well-established endothelial model of large vessel endothelium, were treated with 10–200 ng/mL of the prototypic inflammatory cytokine TNF-α [14,15]. To develop an optimal inflammatory cell model with TNF-α-stimulated cells and investigate its response to anti-inflammatory BP, doses (10, 50, and 100 ng/mL) and durations (6 h and 12 h) of TNF-α treatment were optimized. Since the increase in protein expression reached a maximum at 6 h, further experiments were performed in EA.hy926 cells after 6 h of treatment with 100 ng/mL TNF-α. Incubation with TNF-α rapidly increased the transcription of ICAM-1, VCAM-1, and IL-6 genes and decreased eNOS expression compared with control conditions (all $p < 0.05$) (Figure 2). However, the addition of 300 μM synthetic peptides significantly prevented ICAM-1 mRNA overexpression in

TNF-α activated cells ($p < 0.05$) (Figure 2A). In addition, BP2, BP3, and BP4 also downregulated the expression of VCAM-1 mRNA after stimulation with TNF-α ($p = 0.01$; $p = 0.03$; $p = 0.02$, respectively) (Figure 2B). Treatment with the four peptides produced a similar effect on IL-6 mRNA expression (all $p < 0.05$) (Figure 2C), while eNOS mRNA expression only recovered after treatment with BP1 and BP2 ($p = 0.0003$ and $p = 0.04$, respectively) (Figure 2D). Importantly, the treatment using 300 μM synthetic peptides alone—without stimulation—did not significantly alter the cell expression of these four genes or their viability compared to control conditions (Figures S2 and S3A, respectively).

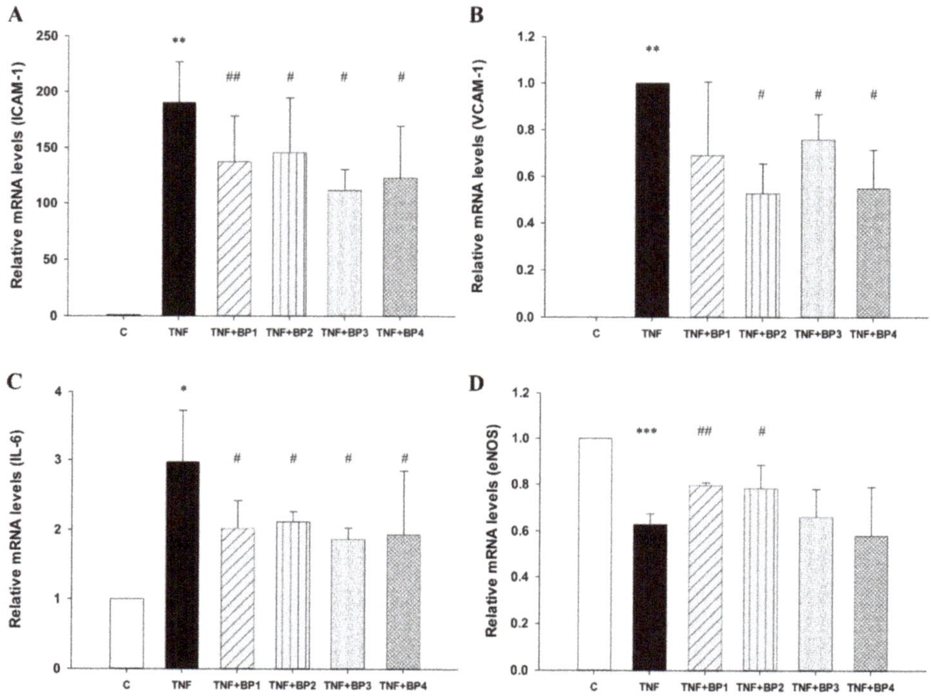

Figure 2. Relative mRNA expression of (**A**) ICAM-1, (**B**) VCAM-1, (**C**) IL-6, (**D**) eNOS in Ea.hy926 cells after treatment with 100 ng/mL TNF-α and 300 μM synthetic peptides. Data shown represent averaged values of three independent experiments. The asterisks *, **, and *** indicate statistically significant differences compared with unstimulated cells ($p < 0.05$, 0.01, or 0.001, respectively). #, ##, or ### indicate statistically significant differences compared with stimulated cells ($p < 0.05$, 0.01, or 0.001, respectively).

2.3.2. Peptides Affect Protein Expression in Inflammatory Conditions

The corresponding flow cytometry (FACS) analysis pointed to an increase in ICAM-1 mean fluorescence intensity (MFI) on the cell surface after TNF-α stimulation ($p = 0.0003$) (Table 2). The coincubation of synthetic peptides (BP1, BP2, and BP4) and TNF-α resulted in a reduced ICAM-1 surface density compared to TNF-α activated cells (all $p < 0.05$) (Table 2, Figure S4B). Despite the higher number of VCAM-1 positive cells, the surface staining of VCAM-1 was hardly detectable even after TNF-α stimulation and no effect was found after the addition of synthetic peptides to the activated cells (Figure S4C). Peptide treatment alone did not impair the MFI of adhesion molecules compared to control conditions.

Table 2. Mean fluorescence intensity (MFI) of ICAM-1 surface expression in inflammatory conditions measured by flow cytometry.

	ICAM-1 MFI	*p* Value
Control	71.36 ± 8.31	
TNFα	1271.91 ± 158.54	0.0003
TNFα + BP1	1151.37 ± 59.53	0.03
TNFα + BP2	1064.42 ± 32.02	0.009
TNFα + BP3	1203.01 ± 279. 89	0.09
TNFα + BP4	926.01 ± 152.46	0.04

ICAM-1: Intercellular Adhesion Molecule-1; VCAM-1 Vascular Adhesion Molecule-1; TNF-α: Tumor Necrosis Factor α; BP: Bioactive peptide. Data from six independent experiments are expressed as mean ± SD. TNF-α stimulated cells are compared to control conditions, and preincubations with 300 μM synthetic peptides are compared to 100 ng/mL TNF-α alone.

The intracellular protein expression of ICAM-1 in the same conditions as above was confirmed by western blot (Figure S4D). Treatment with three of the four peptides (BP2, BP3 and BP4) consistently resulted in a lower production of ICAM-1 protein, compared with that produced after TNF-α stimulation ($p < 0.01$) (Figure S4D). No intracellular VCAM-1 was detected in these cells using the same approach.

2.4. *Effect of Peptides in Oxidative Conditions*

2.4.1. Peptides Do Not Affect Cell Viability and Apoptosis after Treatment with H_2O_2

To explore the role of synthetic peptides in oxidative conditions, the endothelial cell function was impaired by means of high H_2O_2 concentrations, as it has been previously reported [16,17]. Treatment with synthetic peptides did not improve cell viability after the H2O_2 treatment (Figure S3B). Moreover, 300 μM H_2O_2 caused 10% apoptosis (AnV+, PI−cells) and 18% necrosis (AnV+, PI+ cells), and BPs were unable to improve this cytotoxic effect (Table S1).

2.4.2. Peptides Slightly Affect the Oxidative Status

The mRNA expression of the redox enzymes, IL-6 and BAX, were analyzed by RT-PCR, after 16 h of preincubation with synthetic peptides followed by 24 h with 300 μM H_2O_2 (Figure S5). In the presence of H_2O_2, the expression of eNOS increased 3.3-fold ($p = 0.01$). Pretreatment with synthetic peptides did not affect the expression of the redox enzymes in oxidative conditions. Consistent with this, H_2O_2 upregulated IL-6 mRNA expression 1.84-fold ($p = 0.001$), independently of the presence of synthetic peptides. The proapoptotic gene BAX showed a 1.5-fold increase in expression after H_2O_2 treatment ($p = 0.02$) and synthetic peptides did not modify the previously obtained expression levels.

Subsequently, the effect of synthetic peptides on protein carbonylation was determined by immunoblotting assay, which showed that the protein carbonylation levels significantly increased after only 30 min of H_2O_2 treatment. In the presence of BP1 and BP3, the basal oxidative status was maintained (Figure S6) (all $p < 0.05$). The overall protein carbonylation level of unstimulated cells was like that seen in the presence of peptides alone, except in the presence of BP3, when it was even lower (Figure S6).

2.5. *Peptides Decrease NF-κB Activity*

To evaluate whether the synthetic peptides could modulate the NF-κB biological activity, a luciferase reporter assay was carried out, using the Renilla reporter as a specificity control and indicator of cell proliferation [18]. In this study, NF-B activity was induced at a lower concentration of TNF-α (50 ng/mL) because of the high sensitivity of the approach. The normalized firefly/Renilla luciferase activity was reduced by around 25% of the activity when cells were pretreated with the BP1 ($p = 0.002$), BP2 ($p = 0.003$), and BP3 ($p = 0.003$) synthetic peptides (Figure 3). The NF-κB luciferase activity was also repressed by BP4 to a lesser degree ($p = 0.012$). The basal activity for NF-κB was not inhibited by synthetic peptides without TNF-α stimulation (data not shown). These results suggest that synthetic peptides may target events needed for NF-κB activation, rather than other transcription factor activations.

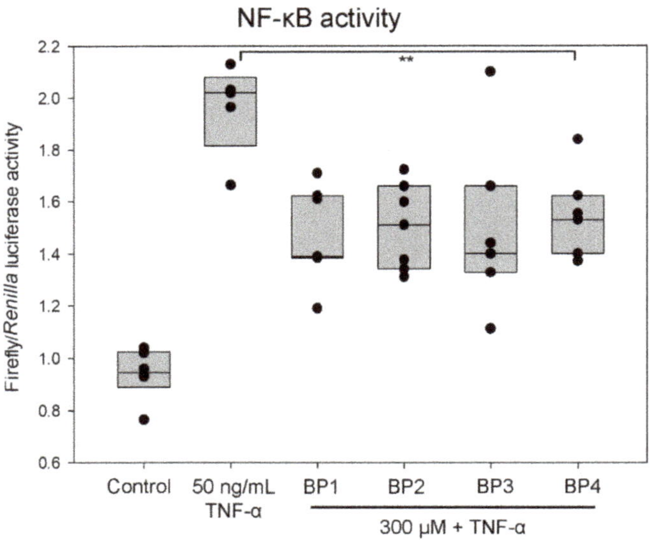

Figure 3. Activation of NF-κB after treatment with 50 ng/mL TNF-α and 300 µM synthetic peptides. Ea.hy926 cells were transfected with NF-κB:Luc together with the pRL-CMV (10:1) reporter vectors. The normalized luciferase activity (firefly/Renilla) was measured using the Dual-Luciferase Reporter Assay in extracts from control, TNF-α, and synthetic peptide-treated cells. Upper and lower bars of box plots represent the 25th and 75th percentiles, respectively. Median is represented as the bar inside the box plot. The data shown represent values of three independent experiments (dots). The asterisks ** indicate statistically significant differences compared with stimulated cells ($p < 0.015$). The BP treatment prevented TNF-α-induced proinflammatory NF-kB activation by 25%.

2.6. Peptides Bind to the Subunit NEMO

Blind docking approaches were used to model peptide binding interactions with the regulatory subunit of NF-κB, NEMO. In silico docking calculations support the potential binding of the four distinct peptides around the residue Glu315, which is crucial for functional assessment and could impair IKK recruitment through competition with the Lys63-linked poly-Ub (Table 3) [19]. The theoretical interactions of these BP and the NF-κB inhibitors are shown (Table 3A,B). Interestingly, BP1–4 efficiently bound to the same site as the anthraquinone derivative of the natural emodin (iNUB) (Table 3A,B). Indeed, the UBI peptide and BP1–4 were all found in the ubiquitin binding motifs of 4BNW and 3JSV (Table 3A,B, respectively and Figure 4). When the CCL-Z2 region displayed Lys63-linked poly-Ub (PDB: 3JSV), peptide binding appeared to be weaker due to the absence of hydrogen bonds, except in the case of BP2 (Table 3B).

Table 3. Blind docking simulation results showing the main residues involved in the interactions with the docked BP and compounds. (**A**) Docking to 4BWN PDB. (**B**) Docking to 3JSV PDB. Common residues are marked in bold.

A	Ligand	PDB Code	Hydrogen Bonds	Hydrophobic
	BP1	4BWN	ALA314A, GLU315A, GLN317A	ALA314A, ALA318A
	BP2	4BWN	ASP311A, GLN313A, ALA314A, GLU315A, GLN317A	PHE312B
	BP3	4BWN	ASP311A, GLU315A, LYS326B, GLU327B	ALA314A, LYS325A
	BP4	4BWN	ALA314A, GLN317A	
	iNUB	4BWN	LYS326B, GLU327B	LYS321A, ALA323B
	UBI peptide	4BWN	GLU315B, GLN317A, ARG319A, GLU320A, LYS321A,	GLN317A, ALA318A, GLU320A, LYS321A, ALA323A
B	**Ligand**	**PDB Code**	**Hydrogen Bonds**	**Hydrophobic**
	BP1	3JSV	THR55A, SER57A, ASP58A, ASN60A	
	BP2	3JSV	ASP39A, GLY76A	PRO37A
	BP3	3JSV	LYS63B, SER65B	
	BP4	3JSV	GLU18B, LYS63B, ARG74A, ARG312D	
	iNUB	3JSV	ARG312D	ALA311C, LEU315C, VAL316D, LYS319D
	UBI peptide	3JSV	GLU24A, ASN25A, ASP39A, ARG42A, LEU50A, ASP52A, GLY53A, GLN62B, GLU64B	THR22A, GLU 24A, PRO38A, GLY53A, THR55A, GLU64B

iNUB is an anthraquinone derivative (8-hydroxy-9,10-dioxo-9,10-dihydro-1-anthracenyl 2-phenylcyclopropanecarboxylate). UBI peptide: LKAQADIYKARFQAERHAREK (21 residues).

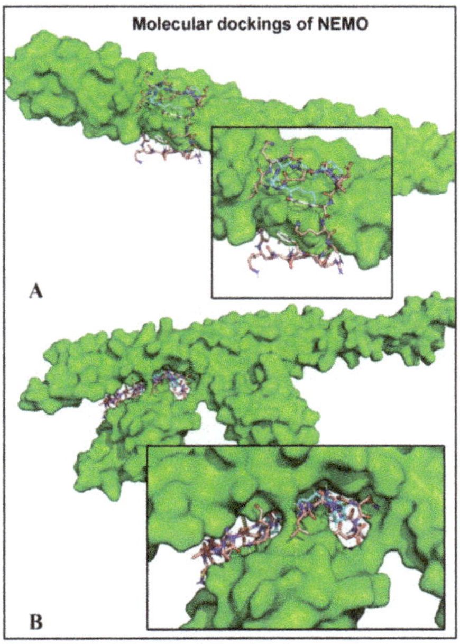

Figure 4. Blind docking of the UBI peptide and BP1 in the CCL-Z2 binding domain of NEMO. (**A**) 4BNW without linked ubiquitins. (**B**) 3JSV complexed with Lys 63-linked poly-ubiquitin. The superposition of UBI peptide and BP1 (in blue) is shown. In silico docking calculations support the potential binding of the four distinct BP, which could impair IKK recruitment through competition with the Lys63-linked poly-ubiquitin.

3. Discussion

Numerous reports have identified stable peptides with high ACE-inhibitory activity from pork dry-cured ham after gastrointestinal digestion [11,20]. Four of them were synthetized chemically (KPVAAP, KAAAATP, KPGRP, and AAATP) for the current experimental studies with human endothelial cells. The synthetic peptides were much more effective ACE inhibitors when purified rabbit ACE was used in vitro: KPVAAP (BP1, 12.37 μM), KAAAATP (BP2, 25.64 μM), KPGRP (BP3, 67.02 μM), and AAATP (BP4, 100 μM) [21], an inconsistency that might be due to the presence of other proteins in the endothelium lysate extract. In both assays, BP1 was the most potent peptide and BP4 the least potent.

Docking simulations were carried out with human ACE and the different BP in order to explain the respective IC_{50} values. Molecular modelling showed that BP1, BP2, and BP3 are stabilized by both hydrogen and hydrophobic interactions with at least one residue of the three pockets, consistent with the typical competitive inhibition model. BP1 and BP2, particularly, established hydrophobic interactions with HIS387 and hydrogen bonds with TYR360, TYR523, and GLU411, while BP3 reached hydrophobic stability with GLU384 and TRP357 and formed hydrogen bonds with TYR360 and GLU411. Moreover, BP4 only established hydrogen bonds with GLU411, HIS353, and TYR523. The absence of hydrophobic stabilization in the case of BP4 might explain its high IC_{50} value. BP2 showed small hydrophobic interaction areas compared with BP1 and BP3, both of which contained voluminous hydrophobic groups. Therefore, the stability of BP2 seems to be comparable to that of BP4. Besides, captopril formed up to five H-bonds with key residues within the active site (GLU384, TYR523, and HIS353 residues). BP1 was found to occupy the S1 and S2 subsites (near the active site) but was not located inside the active site of ACE, in contrast to captopril. Therefore, we concluded that, structurally, BP are noncompetitive inhibitors of ACE so their IC_{50} were much higher compared to captopril. The present data are of special interest for predicting and understanding mechanisms of action of BP and may also be of help for predicting new biomolecules before the relevant assays are carried out.

Apart from its ACE inhibitory capacity, food-derived peptides have been shown to display a wide range of functional activities over the CV system. In fact, multifunctional peptides interfere with more than one biological pathway, such as NO production, oxidative stress, and inflammation. The Ea.hy926 cell line demonstrates highly differentiated functions of human vascular endothelium, such as expression of inflammation cytokines (e.g., IL-6) and adhesive markers (ICAM-1 and VCAM-1) [14]. Our interest in the use of an in vitro model of endothelial dysfunction was to further understand the regulatory effects of the BP treatment in inflammatory conditions. In fact, endothelial NOS was found slightly overexpressed in the presence of these synthetic peptides, which could also be beneficial for attenuating endothelial dysfunction. Similarly, a decrease in inflammatory markers (IL-6 and eNOS) has also been reported with the milk-derived peptides VPP and IPP in spontaneous hypertensive rats using DNA microarray [22]. Egg-derived IRW peptide also inhibited the TNF-α-induced increase of adhesive molecules [23]. Moreover, BP from milk impaired human endothelial–monocyte interactions by inhibiting the expression of VCAM-1, ICAM-1, and E-selectin [24]. Regarding the surface expression of proteins, their measurement strongly depends on the detachment method. The overall surface expression of ICAM-1 was found to be trypsin-resistant, while VCAM-1 was much more sensitive to trypsin degradation [25]. This fact could explain the different results on VCAM-1 gene and protein expression here reported. No mechanisms that might regulate these inhibitions have been identified, but some studies suggest that NF-kB is the main target [24]. These results are in line with a previous clinical study carried out in humans, where the regular intake of dry-cured ham containing the currently described BP lowered plasmatic IL-6, P-selectin, and MCP-1 levels but did not alter plasmatic VCAM-1 [12]. All this evidence supports the hypothesis that the dual ability of these BP to modulate adhesive markers and eNOS might be due to the NF-κB interaction.

On the other hand, oxidative stress and inflammation are closely related pathophysiological processes, one of which can be easily induced by another. Likewise, oxidative stress can activate the NF-κB pathway [26]. In the current study, the simultaneous use of synthetic peptides and H_2O_2 was ineffective at reducing cell death, indicating that these peptides do not target apoptotic mechanisms of action (for instance, caspases or BAX proteins). Besides, the in vitro antioxidant capacity of these peptides [27] was not supported by the current results since the redox enzymes remained unaltered and the carbonylation protein approach was not sensitive enough to confirm the reduction of reactive oxygen species.

Nonetheless, it is apparent from the current data that TNF-α-induced NF-κB activation is sharply attenuated in human endothelial cells in the presence of these peptides from dry-cured pork ham. Several studies have also demonstrated the anti-inflammatory role of specific peptides [28]. However, the research efforts were limited to the biological effects and the action mechanisms were not deduced [29]. In our case, blind docking approaches were used to model peptide binding interactions with the regulatory subunit of NF-κB, NEMO. Recent studies have shown that direct binding of NEMO to linear polyubiquitin (poly-Ub) chains in the TNF-α signaling pathway is crucial for kinase (IKKα/β) recruitment and further NF-κB activation [30,31].

Furthermore, optineurin and small molecules (anthraquinone derivatives and peptides) have previously been suggested to negatively regulate TNF-α-induced NF-κB activation by competing with this CC2-LZ region for Lys63-linked poly-Ub [32,33]. We further found that the BP used experimentally could bind to the same region as a competitive peptide, the UBI peptide, specifically designed to interfere with the coiled interfaces of NEMO [33]. The targeting of IKKβ by dry-cured pork ham peptides was unexpected and suggests the potential regulatory role of the canonical pathway of NF-κB activation identified in the in vitro approaches.

The CCL-Z2 region interacts with Lys63-linked poly-Ub chains with relatively low affinity, which could facilitate the disruption by small molecular compounds, such as peptides [19]. Therefore, the possibility that the BP under study have a biological effect due to their interaction with NEMO before and/or after polyubiquitin binding (PDB: 4BWN and 3JSV, respectively) is very likely. Since both interactions are possible, our data do not determine whether these peptides impair ubiquitin binding by competing at the site or modulating it after TNF-α stimulation. Nonetheless, these data must be interpreted with caution because the inhibitory concentration and bioavailability of these multifunctional BP are still unknown, thereby limiting clinical anti-inflammatory therapeutic strategies. At this point, it is important to mention that this study does not attempt to use attainable eating levels of bioactive peptides (300 μM) but provides the conceptual and operational tools for investigating the sites of action of BP in the context of inflammatory pathological mechanisms. The study may lead to better understanding of the effects of food-derived BP as ACE-inhibitors, the findings being of indirect clinical relevance.

4. Material and Methods

4.1. Peptides

Spanish dry-cured ham has been reported as a good source of bioactive peptides with potent ACE inhibitory activity in vitro [21]. Four of these identified peptides were synthesized chemically by GenScript Corporation (Piscataway, NJ, USA) at the highest purity certified using liquid-chromatography mass spectrometry (LC-MS) analysis for the current experimental approaches with human endothelial cells. The sequence and protein origin of the peptides are shown in Table 1.

4.2. Cell Culture

EA.hy926 cells, the hybrid human umbilical vein endothelial cell line, were obtained from the American Type Culture Collection (ATCC® CRL2922™, Rockville, MD, USA). Cells were cultured in high glucose Dulbecco's Modified Eagle's Medium (DMEM), containing 10% heat-inactivated fetal bovine serum (FBS, Biowest, Riverside, CA, USA) and 50 U/mL of penicillin and 50 μg/mL streptomycin (Sigma Aldrich Chemical Co., Saint Louis, MO, USA). Cells were grown in 5% CO_2 in a humidified air incubator at 37 °C. Subculture was performed when 90% confluence was reached.

4.3. Human ACE Inhibition Assay

To overexpress the ACE enzyme, EAhy926 cells were transiently transfected with a human Ace ORF mammalian expression plasmid (Sinobiological, Beijing, China) using trans-it X2 reagent (Myrus®, Madison, WI, USA). The resultant ACE had a terminal peptide Myc and the efficiency of transfection was checked by immunofluorescence, using an antimyc antibody (Sigma Aldrich Chemical Co., Saint Louis, MO, USA) (Figure S1).

Cells were lysed using 200 μL M-PER® Mammalian Protein Extraction Reagent (ThermoFisher, Waltham, MA, USA). Human ACE inhibitory activity of the four chemically synthetized peptides was tested in vitro. Briefly, 20 μg of lysate (protein) was incubated for one hour at 37 °C with peptides concentrations (50–1000 μM) containing 5 mM Hippuryl Histidyl Leucine (HHL) (Sigma Aldrich Chemical Co., Saint Louis, USA) as substrate. Captopril (10 μM) was used as the positive control of inhibition for the assay conditions. The HHL was transformed into hippuric acid (HA), which was detected by high-performance liquid chromatography (HPLC, Shimadzu, Kioto, Japan) in a C18 column (Teknokroma, Madrid, Spain). The mobile phase was composed of solvent A, 0.05% trifluoroacetic in water; and solvent B, 100% acetonitrile. The ratio of solvent A/solvent B was 7/3, and the flow rate was 1.5 mL/min. The elute was analyzed at a wavelength of 214 nm, which is the maximum absorbance of HA, and the column temperature was maintained at 25 °C. The IC_{50} value (the concentration of inhibitor resulting in a 50% reduction of ACE activity) was calculated by regression analysis from the ACE inhibition curve obtained with increasing amounts of synthetic peptides.

4.4. TNF-α and H_2O_2 Stimulation

EA.hy926 were seeded in six-well plates at 0.25×10^6 cells/well in DMEM supplemented with 5% FBS. Cells were treated with 300 μM of each peptide for 16 h. Then, 100 ng/mL TNF-α or 300 μM H_2O_2 were added for an additional 6 h or 24 h, respectively.

4.5. Quantitative RT-PCR

Total RNA was extracted from EA.hy926 cells using 300 μL Trisure™ (Bioline, Taunton, MA, USA) reagent and Direct-zol™ RNA MiniPrep (Zymo Research Irvine, Irvine, CA, USA) according to the manufacturer's protocol. Total RNA was reverse-transcribed into complementary DNA (Sensifast cDNATM Synthesis kit, Bioline, Taunton, MA, USA). The mRNA levels of the target genes were quantified by RT-PCR using SensiFAST SYBER Hi-ROX Kit (Bioline, Taunton, MA, USA) with StepOnePlus Real-Time PCR System (Applied Biosystems, Foster City, CA, USA). Briefly, 5 μL of 1:5 diluted cDNA was added to the qPCR reaction containing 10 μL 2X SensiFAST Mix and 400 nM of each primer in a total volume of 20 μL.

Specific and validated primers for human glyceraldehyde-3-phosphate dehydrogenase (GADPH), intercellular adhesion marker-1 (ICAM-1), vascular cell adhesion marker-1 (VCAM-1), endothelial nitric oxide synthase (eNOS), interleukin 6 (IL-6), catalase, SOD (Super Oxide Dismutase), NADPH (Nicotinamide Adenine Dinucleotide Phosphate oxidase) and Bax (proapoptosis regulator) genes were used (Sigma-Aldrich Chemical Co., Saint Louis, MO, USA).

The relative mRNA expression of the genes of interest was represented by:

$$2\hat{}(-\Delta\Delta CT) = [CT_{(\text{gene of interest})} - CT_{(\text{GADPH})}]\text{test} - [CT_{(\text{gene interest})} - CT_{(\text{GADPH})}]\text{control}.$$

The relative quantification of gene expression was determined by the comparative fold change 2^ΔΔCT method [34]. An average value of each target gene after GAPDH normalization at the time point showing highest expression was used as a calibrator to determine the relative levels in the rest of the experimental conditions. All the assays were performed in triplicate. Each qPCR reaction had three replicates.

4.6. Flow Cytometry

To measure adhesion molecule content in human endothelial cells by flow cytometry, the cell monolayer was detached and fixed before immunofluorescence labelling [25]. The following antibodies were used per sample: 20 μL anti-ICAM-1/CD54-phycoerythrin (PE) (clone HA58, BD Biosciences, Franklin Lakes, NJ, USA) and 5 μL anti-VCAM-1/CD106-PerCP-Cy5.5 (clone 51–10C9; BD Biosciences, San Jose, CA, USA), according to the manufacturer's protocol. Stained cells (10,000 events) were examined by flow cytometry (FACS Calibur, Becton Dickinson, Mountain View, CA, USA) in a simultaneous two-color analysis with FL2 (PE) and FL3 (PerCP-Cy5.5) channels. Markers were set according to the negative controls to quantify the percentage of positively stained cells. Mean Fluorescence Intensity (MFI) was calculated for each antigen (MFI of total stained cells-MFI of negative control cells).

ApoScreen Annexin V Apoptosis Kit-FITC (Southern Biotech, Birmingham, AL, USA) was used to detect the apoptosis rate. After the co-incubation of synthetic peptides and 300 μM H_2O_2, 0.4×10^6 cells were detached and washed in cold PBS, then resuspended in 100 μL Annexin V binding buffer and treated following the manufacturer's protocol. Annexin V-fluorescein isothiocyanate (FITC) binding was assessed in FL1 channel simultaneously with propidium iodide (FL2 channel). This test discriminates among intact cells (FITC−/PI−), apoptotic cells (FITC+/PI−), and necrotic cells (FITC+/PI+). Assays were performed in triplicate (Table S1).

4.7. NF-κB Activity

The dual luciferase assay has been widely used in cell lines to rapidly and accurately determine the activity of the promoter of NF-κB. The transfected vector, pNF-κB:Luc, carries the luciferase gene under the control of three synthetic copies of the κB consensus of the immunoglobulin κ-chain promoter cloned in the BamHI site located upstream of the conalbumin transcription start site. This construct, together with the pRL-CMV, were kindly provided by Dr. Cayuela-Fuentes (Hospital Universitario Virgen de la Arrixaca, Murcia, Spain) [18].

For luciferase reporter assays, 80,000 cells/well were seeded in 24-well plates overnight, followed by cotransfection with 0.5 μg pNF-κB:Luc/pRL-CMV at a ratio of 10:1 using 1.5 μL trans-it X2 reagent (Myrus®, Madison, WI, USA). The transfection media was changed 6 h after transfection by complete growth medium. The following day, cells were treated with the synthetic peptides for 16 h and then with 50 ng/mL TNF-α for 6 h. NF-κB-dependent firefly luciferase activity and NF-κB-independent Renilla luciferase activity were assessed using Dual-Luciferase® Reporter Assay System (Promega, Madison, WI, USA) in a Luminometer Optocomp I (MGM Instruments). Data were normalized to the amounts of Renilla luciferase activities, according to the manufacturer´s protocol.

4.8. Molecular Modelling

In order to obtain detailed information at atomic level about the interactions between the different peptide molecules, inhibitors and the human proteins, molecular modelling studies were carried out.

A representative X-ray crystal structure for human ACE (PDB code 4APJ) was chosen, and its full atom model for the docking simulation was prepared. Docking simulations were chosen as the most adequate molecular modelling technique for the ACE study, since they efficiently predict at a reasonable computing cost electrostatic, van der Waals, hydrogen bond and hydrophobic interactions between interacting ligands and protein [35]. Partial charges and hydrogens were added with Autodock Tools [36]. The studied peptide molecules and characterized inhibitors were built manually using Pymol [37].

Two representative X-ray crystal structures of the NF-κB essential modulator, NEMO, were retrieved from the PDB database (PDB: 4BWN and 3JSV), and their full atom models for the docking simulations were prepared. The sequence for the UBI peptide was extracted from the work of Chiaravalli et al. [32] and the structure from the anthraquinone derivative, iNUB, was retrieved from the work of Vincendeau et al. [38]. In order to determine in which part of the CC2-LZ region of NEMO the different ligands interact, a blind docking approach [39] where the researcher does not define any preferred interaction spot [40] was followed, where multiple docking runs started around the geometric centers of all residues. A histogram with the resulting distribution of binding energies and their structural clusters of poses was generated. Each individual docking simulation was performed with the Autodock Vina software AUTODOCK using default configuration parameters [41]. The size of the grid box was set to extend 120 Å in each direction from the geometric center of each individual docking simulation. The docking score produced by Autodock Vina was taken as the predicted value of the ligand binding energy. Only the top-ranked poses were used for structural and energy analyses. The scoring function from Vina considers the Lennard-Jones term (LJ), hydrogen bonds (H-bonds), electrostatic interactions, hydrophobic stabilization, entropic penalty due to the number of rotatable bonds, and the internal energy of the ligand.

4.9. Statistical Analyses

Data were expressed as mean ± standard deviation (SD) of three determinations. A Student's *t*-test was used to compare the differences between the mean of two groups. Statistical analyses were performed with SPSS 21.0. Statistical significance was considered at $p < 0.05$.

5. Conclusions

The overall goal of the study was to identify which specific mechanisms of dry-cured pork ham peptides are operative in inflammatory pathways and, as a result, to identify the most promising targets for functional food development. Reported molecular modelling results explained the rationale of their anti-inflammatory activity for the first time. The amino acid sequence of these inhibitory peptides may also form the basis for the design of analogues with therapeutic potential. Further, the multifunctional peptides characterized here may herald an important avenue in food and pharmacological research.

Supplementary Materials: Supplementary materials can be found at http://www.mdpi.com/1422-0067/20/17/4204/s1.

Author Contributions: S.M.M.-S.: investigation and methodology. H.P.-S. and J.A.G.: molecular modeling calculations. J.A.-A.: conceptualization. S.M.-G.: writing-review and editing the drafted manuscript.

Funding: S.M.M.-S. belongs to the "Programa de Doctorado en Ciencias de la Salud. Universidad Católica de Murcia (UCAM)" and holds an FPU grant (FPU 15/03134) from the Ministry of Education (Spain). This work was supported by a grant from the Spanish Ministry of Economy and Competitiveness (CTQ2017-87974-R), partially supported by the e-infrastructure programme of the Research Council of Norway, and the supercomputer center of UiT - the Arctic University of Norway and by the supercomputing infrastructure of Poznan Supercomputing Center and by the Fundación Séneca with the projects "Jóvenes Líderes en Investigación 20646/JLI/18", from the Programa Regional de Fomento de la Investigación (Plan de Actuación 2018, Región de Murcia, Spain) and by "Ayudas a la realización de proyectos para el desarrollo de investigación científica y técnica por grupos competitivos 20988/PI/18".

Acknowledgments: The authors are grateful to the Instituto de Agroquímica y Tecnología de Alimentos de Valencia, (IATA, CSIC, Spain) for identifying the peptides from dry-cured pork ham.

Conflicts of Interest: The authors declare no conflict of interest.

Abbreviations

ACE	Angiotensin I-Converting Enzyme
BP	Bioactive Peptide
BSA	Bovine Serum Albumin
CV	Cardiovascular
CVD	Cardiovascular Disease
DMEM	Dulbecco Modified Eagle Medium
FBS	Fetal Bovine Serum
HA	Hippuric Acid
HHL	Hippuryl Histidyl Leucine
ICAM-1	Intercellular Adhesion Molecule-1
IKK	IκB kinase
IL-6	Interleukin-6
NF-κB	Nuclear Factor-κB
OD	Optic Density
PBS	Phosphate Buffer Saline
PVDF	Polyvinylidene Difluoride
VCAM-1	Vascular Adhesion Molecule-1
TNF- α	Tumor Necrosis Factor-α

References

1. Vanhoutte, P.M.; Shimokawa, H.; Feletou, M.; Tang, E.H. Endothelial dysfunction and vascular disease—A 30th anniversary update. *Acta Physiol. (Oxf.)* **2017**, *219*, 22–96. [CrossRef] [PubMed]
2. Sehnert, B.; Burkhardt, H.; Wessels, J.T.; Schröder, A.; May, M.J.; Vestweber, D.; Zwerina, J.; Warnatz, K.; Nimmerjahn, F.; Schett, G.; et al. NF-κB inhibitor targeted to activated endothelium demonstrates a critical role of endothelial NF-κB in immune-mediated diseases. *Proc. Natl. Acad. Sci. USA* **2013**, *110*, 16556–16561. [CrossRef] [PubMed]
3. Solt, L.A.; Madge, L.A.; May, M.J. NEMO-binding domains of both IKKalpha and IKKbeta regulate IkappaB kinase complex assembly and classical NF-kappaB activation. *J. Biol. Chem.* **2009**, *284*, 27596–27608. [CrossRef] [PubMed]
4. Yan, Z.; Deng, P.; Liu, Y. Recent Advances in Protein Kinase Activity Analysis Based on Nanomaterials. *Int. J. Mol. Sci.* **2019**, *20*, 1440. [CrossRef] [PubMed]
5. Xu, T.; Guo, Y.; Qi, X. Ubiquitination-Mediated Inflammasome Activation during Bacterial Infection. *Int. J. Mol. Sci.* **2019**, *20*, 2110. [CrossRef] [PubMed]
6. De Leo, F.; Panarese, S.; Gallerani, R.; Ceci, L.R. Angiotensin converting enzyme (ACE) inhibitory peptides: Production and implementation of functional food. *Curr. Pharm. Des.* **2009**, *15*, 3622–3643. [CrossRef]
7. Daliri, E.B.; Ofosu, F.K.; Chelliah, R.; Park, M.H.; Kim, J.H.; Oh, D.H. Development of a Soy Protein Hydrolysate with an Antihypertensive Effect. *Int. J. Mol. Sci.* **2019**, *20*, 1496. [CrossRef]
8. Ben Slama-Ben Salem, R.; Ktari, N.; Bkhairia, I.; Nasri, R.; Mora, L.; Kallel, R.; Hamdi, S.; Jamoussi, K.; Boudaouara, T.; El-Feki, A.; et al. In vitro and in vivo anti-diabetic and anti-hyperlipidemic effects of protein hydrolysates from Octopus vulgaris in alloxanic rats. *Food Res. Int.* **2018**, *106*, 952–963. [CrossRef]
9. Chakrabarti, S.; Guha, S.; Majumder, K. Food-Derived Bioactive Peptides in Human Health: Challenges and Opportunities. *Nutrients* **2018**, *10*, 1738. [CrossRef]
10. Escudero, E.; Aristoy, M.C.; Nishimura, H.; Arihara, K.; Toldrá, F. Antihypertensive effect and antioxidant activity of peptide fractions extracted from Spanish dry-cured ham. *Meat. Sci.* **2012**, *91*, 306–311. [CrossRef]
11. Gallego, M.; Grootaert, C.; Mora, L.; Aristoy, M.C.; Van Camp, J.; Toldrá, F. Transepithelial transport of dry-cured ham peptides with ACE inhibitory activity through a Caco-2 cell monolayer. *J. Funct. Foods* **2016**, *21*, 388–395. [CrossRef]
12. Martinez-Sanchez, S.M.; Minguela, A.; Prieto-Merino, D.; Zafrilla-Rentero, M.P.; Abellan-Aleman, J.; Montoro-Garcia, S. The Effect of Regular Intake of Dry-Cured Ham Rich in Bioactive Peptides on Inflammation, Platelet and Monocyte Activation Markers in Humans. *Nutrients* **2017**, *9*, 321. [CrossRef] [PubMed]

13. Masuyer, G.; Schwager, S.L.; Sturrock, E.D.; Isaac, R.E.; Acharya, K.R. Molecular recognition and regulation of human angiotensin-I converting enzyme (ACE) activity by natural inhibitory peptides. *Sci. Rep.* **2012**, *2*, 717. [CrossRef] [PubMed]
14. Seto, S.W.; Chang, D.; Ko, W.M.; Zhou, X.; Kiat, H.; Bensoussan, A.; Lee, S.M.; Hoi, M.P.; Steiner, G.Z.; Liu, J. Sailuotong Prevents Hydrogen Peroxide (H_2O_2)-Induced Injury in EA.hy926 Cells. *Int. J. Mol. Sci.* **2017**, *18*, 95. [CrossRef] [PubMed]
15. Chen, Y.; Zhang, H.; Mats, L.; Liu, R.; Deng, Z.; Mine, Y.; Tsao, R. Anti-inflammatory Effect and Cellular Uptake Mechanism of Peptides from Common Bean. *J. Agric. Food Chem* **2019**, *67*, 8370–8381. [CrossRef] [PubMed]
16. Wang, B.; Zhang, Q.; Yao, R.; Liu, X.; Qu, Z. 7,8-Dihydroxyflavone Protects an Endothelial Cell Line from H_2O_2 Damage. *PLoS ONE* **2015**, *10*, e0135345. [CrossRef] [PubMed]
17. Guo, S.; Long, M.; Li, X.; Zhu, S.; Zhang, M.; Yang, Z. Curcumin activates autophagy and attenuates oxidative damage in EA.hy926 cells via the Akt/mTOR pathway. *Mol. Med. Rep.* **2016**, *13*, 2187–2193. [CrossRef] [PubMed]
18. Alcaraz-Pérez, F.; Mulero, V.; Cayuela, M.L. Application of the dual-luciferase reporter assay to the analysis of promoter activity in Zebrafish embryos. *BMC Biotechnol.* **2008**, *8*, 81. [CrossRef]
19. Lo, Y.C.; Lin, S.C.; Rospigliosi, C.C.; Conze, D.B.; Wu, C.J.; Ashwell, J.D.; Eliezer, D.; Wu, H. Structural basis for recognition of diubiquitins by NEMO. *Mol. Cell* **2009**, *33*, 602–615. [CrossRef]
20. Escudero, E.; Mora, L.; Toldrá, F. Stability of ACE inhibitory ham peptides against heat treatment and in vitro digestion. *Food Chem.* **2014**, *161*, 305–311. [CrossRef]
21. Escudero, E.; Mora, L.; Fraser, P.D.; Aristoy, M.C.; Arihara, K.; Toldrá, F. Purification and Identification of antihypertensive peptides in Spanish dry-cured ham. *J. Proteomics* **2013**, *78*, 499–507. [CrossRef] [PubMed]
22. Yamaguchi, N.; Kawaguchi, K.; Yamamoto, N. Study of the mechanism of antihypertensive peptides VPP and IPP in spontaneously hypertensive rats by DNA microarray analysis. *Eur. J. Pharmacol.* **2009**, *620*, 71–77. [CrossRef] [PubMed]
23. Huang, W.; Chakrabarti, S.; Majumder, K.; Jiang, Y.; Davidge, S.T.; Wu, J. Egg-derived peptide IRW inhibits TNF-α-induced inflammatory response and oxidative stress in endothelial cells. *J. Agric. Food Chem.* **2010**, *58*, 10840–10846. [CrossRef] [PubMed]
24. Chakrabarti, S.; Wu, J. Milk-derived tripeptides IPP (Ile-Pro-Pro) and VPP (Val-Pro-Pro) promote adipocyte differentiation and inhibit inflammation in 3T3-F442A cells. *PLoS ONE* **2015**, *10*, e0117492. [CrossRef] [PubMed]
25. Gräbner, R.; Till, U.; Heller, R. Flow cytometric determination of E-selectin, vascular cell adhesion molecule-1, and intercellular cell adhesion molecule-1 in formaldehyde-fixed endothelial cell monolayers. *Cytometry* **2000**, *40*, 238–244. [CrossRef]
26. Janssen-Heininger, Y.M.; Poynter, M.E.; Baeuerle, P.A. Recent advances towards understanding redox mechanisms in the activation of nuclear factor kappaB. *Free Radic. Biol. Med.* **2000**, *28*, 1317–1327. [CrossRef]
27. Escudero, E.; Mora, L.; Fraser, P.D.; Aristoy, M.C.; Toldrá, F. Identification of novel antioxidant peptides generated in Spanish dry-cured ham. *Food Chem.* **2013**, *138*, 1282–1288. [CrossRef] [PubMed]
28. Meram, C.; Wu, J. Anti-inflammatory effects of egg yolk livetins (α, β, and γ-livetin) fraction and its enzymatic hydrolysates in lipopolysaccharide-induced RAW 264.7 macrophages. *Food Res. Int.* **2017**, *100*, 449–459. [CrossRef]
29. Wu, W.; Zhang, M.; Ren, Y.; Cai, X.; Yin, Z.; Zhang, X.; Min, T.; Wu, H. Characterization and Immunomodulatory Activity of a Novel Peptide, ECFSTA, from Wheat Germ Globulin. *J. Agric. Food Chem.* **2017**, *65*, 5561–5569. [CrossRef]
30. Ea, C.K.; Deng, L.; Xia, Z.P.; Pineda, G.; Chen, Z.J. Activation of IKK by TNFalpha requires site-specific ubiquitination of RIP1 and polyubiquitin binding by NEMO. *Mol. Cell* **2006**, *22*, 245–257. [CrossRef]
31. Tokunaga, F.; Sakata, S.; Saeki, Y.; Satomi, Y.; Kirisako, T.; Kamei, K.; Nakagawa, T.; Kato, M.; Murata, S.; Yamaoka, S.; et al. Involvement of linear polyubiquitylation of NEMO in NF-kappaB activation. *Nat. Cell Biol.* **2009**, *11*, 123–132. [CrossRef] [PubMed]
32. Chiaravalli, J.; Fontan, E.; Fsihi, H.; Coic, Y.M.; Baleux, F.; Véron, M.; Agou, F. Direct inhibition of NF-κB activation by peptide targeting the NOA ubiquitin binding domain of NEMO. *Biochem. Pharmacol.* **2011**, *82*, 1163–1174. [CrossRef] [PubMed]

33. Agou, F.; Courtois, G.; Chiaravalli, J.; Baleux, F.; Coïc, Y.M.; Traincard, F.; Israël, A.; Véron, M. Inhibition of NF-kappa B activation by peptides targeting NF-kappa B essential modulator (nemo) oligomerization. *J. Biol. Chem.* **2004**, *279*, 54248–54257. [CrossRef] [PubMed]
34. Livak, K.J.; Schmittgen, T.D. Analysis of relative gene expression data using real-time quantitative PCR and the 2(-Delta Delta C(T)) Method. *Methods* **2001**, *25*, 402–408. [CrossRef] [PubMed]
35. Hauser, A.S.; Windshügel, B. LEADS-PEP: A Benchmark Data Set for Assessment of Peptide Docking Performance. *J. Chem. Inf. Model.* **2016**, *56*, 188–200. [CrossRef] [PubMed]
36. Morris, G.M.; Huey, R.; Lindstrom, W.; Sanner, M.F.; Belew, R.K.; Goodsell, D.S.; Olson, A.J. AutoDock4 and AutoDockTools4: Automated docking with selective receptor flexibility. *J. Comput. Chem.* **2009**, *30*, 2785–2791. [CrossRef] [PubMed]
37. Delano, W.L.; LAM, J.W. PyMOL: A communications tool for computational models. *Am. Chem. Soc.* **2005**, *230*, U1371–U1372.
38. Vincendeau, M.; Hadian, K.; Messias, A.C.; Brenke, J.K.; Halander, J.; Griesbach, R.; Greczmiel, U.; Bertossi, A.; Stehle, R.; Nagel, D.; et al. Inhibition of Canonical NF-κB Signaling by a Small Molecule Targeting NEMO-Ubiquitin Interaction. *Sci. Rep.* **2016**, *6*, 18934. [CrossRef]
39. Hetényi, C.; van der Spoel, D. Blind docking of drug-sized compounds to proteins with up to a thousand residues. *FEBS Lett.* **2006**, *580*, 1447–1450. [CrossRef]
40. Tapia-Abellán, A.; Angosto-Bazarra, D.; Martínez-Banaclocha, H.; de Torre-Minguela, C.; Cerón-Carrasco, J.P.; Pérez-Sánchez, H.; Arostegui, J.I.; Pelegrin, P. MCC950 closes the active conformation of NLRP3 to an inactive state. *Nat. Chem. Biol.* **2019**, *15*, 560–564. [CrossRef]
41. Trott, O.; Olson, A.J. AutoDock Vina: improving the speed and accuracy of docking with a new scoring function, efficient optimization, and multithreading. *J. Comput. Chem.* **2010**, *31*, 455–461. [CrossRef] [PubMed]

International Journal of *Molecular Sciences*

MDPI

Article

Cytoprotective Effect of Antioxidant Pentapeptides from the Protein Hydrolysate of Swim Bladders of Miiuy Croaker (*Miichthys miiuy*) against H_2O_2-Mediated Human Umbilical Vein Endothelial Cell (HUVEC) Injury

Shi-Ying Cai [1], Yu-Mei Wang [2], Yu-Qin Zhao [2], Chang-Feng Chi [1,*] and Bin Wang [2,*]

1 National and Provincial Joint Laboratory of Exploration and Utilization of Marine Aquatic Genetic Resources, National Engineering Research Center of Marine Facilities Aquaculture, School of Marine Science and Technology, Zhejiang Ocean University, Zhoushan 316022, China; shiyingcai1996@outlook.com

2 Zhejiang Provincial Engineering Technology Research Center of Marine Biomedical Products, School of Food and Pharmacy, Zhejiang Ocean University, Zhoushan 316022, China; wangym731@126.com (Y.-M.W.); zhaoy@hotmail.com (Y.-Q.Z.)

* Correspondence: chicf@zjou.edu.cn (C.-F.C.); wangbin@zjou.edu.cn (B.W.)

Received: 2 October 2019; Accepted: 28 October 2019; Published: 31 October 2019

Abstract: In our previous research, ten antioxidant pentapeptides including FYKWP, FTGMD, GFEPY, YLPYA, FPPYERRQ, GFYAA, FSGLR, FPYLRH, VPDDD, and GIEWA were identified from the hydrolysate of miiuy croaker (*Miichthys miiuy*) swim bladder. In this work, their protective function on H_2O_2-induced oxidative damage to human umbilical vein endothelial cells (HUVECs) was studied. Results indicated that there was no significant difference in the HUVEC viability between the normal group and the treated groups with the 10 pentapeptides at the concentration of 100 μM for 24 h ($p < 0.05$). Furthermore, FPYLRH of 100 μg/mL extremely significantly ($p < 0.001$) increased the viability (80.58% ± 5.01%) of HUVECs with H_2O_2-induced oxidative damage compared with that of the model group. The protective mechanism indicated that FPYLRH could extremely significantly ($p < 0.001$) increase the levels of superoxide dismutase (SOD) (211.36 ± 8.29 U/mg prot) and GSH-Px (53.06 ± 2.34 U/mg prot) and decrease the contents of reactive oxygen species (ROS) (139.1 ± 11.8% of control), malondialdehyde (MDA) (13.66 ± 0.71 nM/mg), and nitric oxide (NO) (4.36 ± 0.32 μM/L) at the concentration of 100 μM in HUVECs with H_2O_2-induced oxidative damage compared with those of the model group. In addition, FPYLRH dose-dependently protected DNA in oxidative damage HUVECs model. These results suggested that FPYLRH could significantly attenuate the H_2O_2-induced stress injury in HUVECs and might be used as a potential natural antioxidant in the functional food industries.

Keywords: miiuy croaker (*Miichthys miiuy*); swim bladder; FPYLRH; antioxidant activity; cytoprotective effect

1. Introduction

Reactive oxygen species (ROS) are essential components of the cells in organisms and play a dual role in the physiological processes involved in cell signaling and homeostasis [1,2]. Excessive ROS can cause oxidative damage to biomolecules in cell membranes, which further initiates a series of chronic diseases, such as rheumatoid arthritis, hypertension, diabetes, cardiovascular disease, atherosclerosis, chronic fatigue syndrome, and neurodegenerative diseases [1,3,4]. Retarding the progress of oxidative stress is a key step towards alleviating stress-related diseases [5]. In living organisms, antioxidant enzymes (superoxide dismutase (SOD), catalase (CAT), glutathione peroxidase (GSH-Px), and glutathione reductase (GSH-Rx)) and non-enzymatic antioxidant agents (glutathione (GSH), ascorbic acid, and β-carotene) can maintain the oxidative balance by reducing the concentration of ROS to prevent oxidative stress [2,6]. Therefore, the elimination of ROS damages by antioxidants is an effective treatment of those chronic diseases and food deterioration [7,8].

Recently, antioxidant peptides have been isolated and identified from various seafood proteins, such as edible marine fishes, alga, shellfish, and their processing by-products [3,9]. These peptides showed excellent properties, including high bioactivity, easy absorption, low molecular weight (MW), and lower toxicity or side effects, which are important for their potential application in health products [3]. Collagen peptides from jellyfish could alleviate ultraviolet (UV)-induced abnormal changes of antioxidant defense systems such as SOD and GSH-Px [10]. Similarly, Chen et al. found that gelatin hydrolysate of Pacific cod skin could prevent UV radiation-induced skin damage by suppressing the depletion of an endogenous antioxidant enzyme and the expression of nuclear factor-κB (NF-κB) and pro-inflammatory cytokines [11]. MDLFTE and WPPD from protein hydrolysate of *Tergillarca granosa* exhibited strong radical scavenging activities and high inhibiting ability on lipid peroxidation. In addition, MDLFTE and WPPD were stable and could retain strong antioxidant activity at the temperatures lower 80 °C and acidic and weakly alkaline environments (pH < 9) [12]. GPA from gelatin hydrolysate of fish skin showed an approximately 2.5-fold increase in antioxidant response element (ARE)-luciferase activity and suppressed the H_2O_2-induced intracellular ROS production by dose-dependently activating the expression of ARE-driven antioxidant enzyme genes [13]. The antioxidant functions of these peptides were thought to be associated with their MW, amino acid compositions and sequences, and spatial structures [3,14]. Moreover, these results suggested that seafood proteins were high-quality raw materials for the preparation of antioxidant peptides for protecting human health and food quality by reducing oxidative stress.

Miiuy croaker (*Miichthys miiuy*) is mainly spread from the Western Japan Sea to the East China Sea and has been extensively farmed in China since the late 1990s due to its high nutrient content, delicious taste, and economic value [14,15]. Previous research indicated that the swim bladder of miiuy croaker showed positive curative effects on dozens of diseases, such as protective liver function, removing ROS, cure dizziness, and warding against inflammation and cancers [1,16]. In our previous work, ten antioxidant pentapeptides were prepared from hydrolysate of miiuy croaker swim bladder and determined as FYKWP (S1), FTGMD (S2), GFEPY (S3), YLPYA (S4), FPPYERRQ (S5), GFYAA (S6), FSGLR (S7), FPYLRH (S8), VPDDD (S9), and GIEWA (S10) [1]. Of these, FPYLRH (S8) could effectively inhibit lipid peroxidation and exhibit strong scavenging activities on hydroxyl radical (EC_{50} 0.68 mg/mL), 2,2-diphenyl-1-picrylhydrazyl (DPPH) radical (EC_{50} 0.51 mg/mL), and superoxide anion radical (EC_{50} 0.34 mg/mL). In this work, the protective functions of the antioxidant pentapeptides, especially FPYLRH, regarding oxidative damage to human umbilical vein endothelial cells (HUVECs) by H_2O_2 were studied for further elucidating their antioxidant mechanism.

2. Results and Discussion

2.1. *Effects of Antioxidant Pentapeptides (S1–S10) on the Viability of HUVEC*

HUVEC is the main type of endothelial cell and provides a classic model system to study many aspects of endothelial function and disease, such as cell and cardiovascular protection effects of bioactive molecules, cardiovascular-related complications associated with various diseases, hypoxia and inflammation-related pathways in endothelia under normal and pathological conditions, etc. [17].

As shown in Figure 1A, the viability of cells treated with FSGLR (S7) was 95.47% ± 4.20% at 100 μM for 24 h, which was lower than that of the other nine pentapeptides and normal control groups. The viability of cells treated with VPDDD (S9) was 108.20% ± 3.08%, which was higher than that of the other nine pentapeptides and normal control groups at the same concentration. However, there were no significant differences between the control group and peptides groups at 100 μM for 24 h ($p < 0.05$). Cell proliferation is a physiological process that occurs in almost all tissues and under many circumstances, and the balance between proliferation and programmed cell death (apoptosis) is maintained by regulating both processes to ensure the integrity of tissues and organs under normal conditions [18]. The cell viability measured by 3-(4,5-di methyl thiazol-2-yl)-2,5-diphenyltetrazolium bromide (MTT) method is an important index for evaluating the effects of pharmacological compounds on cell proliferation according to the response to stress stimuli, and it is often used to screen compounds for developing different pharmaceuticals [19]. Therefore, ten isolated pentapeptides (S1–S10) had the possibility for developing non-tumor functional products because of their insignificant effect on the normal proliferation of HUVECs.

Figure 1B indicated that the viability of HUVECs treated with H_2O_2 was negatively correlated with the concentrations ranging from 100 to 350 μM. In addition, the viability of HUVECs treated with H_2O_2 at the concentration of 200 μM was 51.66% ± 2.48%, which was significantly different from other groups ($p < 0.05$). Therefore, oxidative damage to HUVECs was established at the H_2O_2 concentration of 200 μM. Oxidative stress caused by ROS can activate apoptosis-related signaling pathways in vascular and cardiac endothelial cells, which further induce endothelial dysfunction to the initiation and development of cardiovascular diseases (CVDs) [20]. H_2O_2 can cause oxidative damage because it can be converted into hydroxyl radicals and oxygen radicals in liver cells [17]. Herein, a H_2O_2-induced HUVEC injury model was used to screen the antioxidant molecules and explore their mechanisms involved in the pathogenesis of ROS-induced oxidative stress.

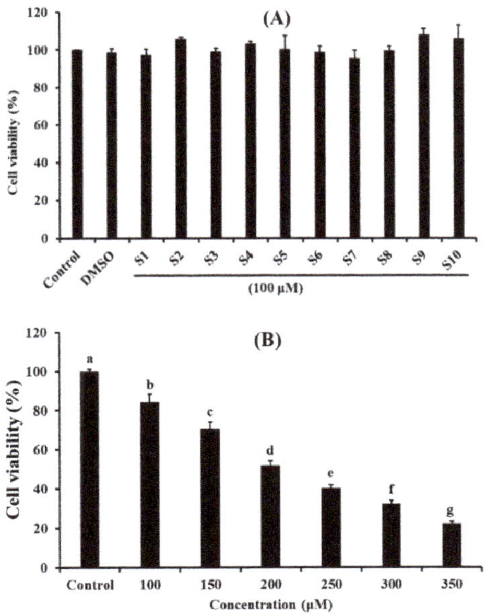

Figure 1. Effects of antioxidant pentapeptides S1–S10 at the concentration of 100 μM (**A**) and H_2O_2 with the concentration ranged from 100 to 350 μM (**B**) on cell viability of HUVECs for 24 h. Mean ± standard deviation (SD) ($n = 3$) is used to express the experiment data. $^{a–g}$ Values with same letter in Figure 1B indicate no significant difference of the cell viability of HUVECs treated with different H_2O_2 concentrations ($p > 0.05$).

2.2. Protective Effect of Antioxidant Pentapeptides (S1–S10) on the Oxidative Injury HUVEC by H_2O_2

Figure 2A showed the protective effects of the 10 antioxidant pentapeptides (S1–S10) on the H_2O_2-induced HUVEC injury model, and the cell viabilities of the S1–S10-treated groups were improved compared with the model group. At the concentration of 100 μM, the HUVEC viability of the FPYLRH (S8)-treated group increased to 80.58% ± 5.01%, which was extremely ($p < 0.001$) significantly higher than that of the model group; the HUVEC viability of the YLPYA (S4)- and GFYAA (S6)-treated groups increased to 70.75% ± 3.65% and 65.67% ± 1.99%, respectively, which were very ($p < 0.01$) significantly higher than that of model group ($p < 0.01$); FYKWP (S1) and GIEWA (S10) increased the HUVEC viability to 64.74% ± 2.02% and 63.99% ± 3.78%, respectively, which were significantly ($p < 0.05$) higher than that of model group. The present data indicated that FPYLRH (S8) has the strongest protective ability on H_2O_2-induced oxidative damage HUVEC among the 10 antioxidant pentapeptides (S1–S10) at the concentration of 100 μM.

Figure 2B shows that there was a positive correlation between the concentration of FPYLRH (S8) and its protective ability on the H_2O_2-induced HUVEC injury model. The FPYLRH (S8)-treated group increased the HUVEC viability to 80.58% ± 5.01% and 68.21% ± 3.59% at the concentrations of 50 and 100 μM, respectively, which were extremely ($p < 0.001$) and very ($p < 0.01$) significantly higher than that of model group. However, there was no difference between the FPYLRH (S8) group at the concentration of 10 μM and the H_2O_2-induced oxidative damage group ($p > 0.05$).

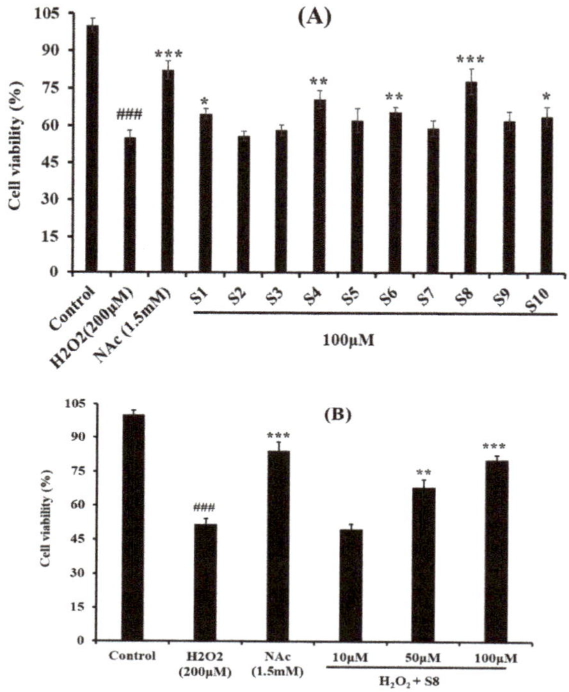

Figure 2. Protective effects of the 10 antioxidant pentapeptides (S1–S10) at the concentration of 100 μM (**A**) and FPYLRH (S8) at the concentrations of 10, 50, and 100 μM (**B**) on the H_2O_2-induced HUVEC injury model. Mean ± SD ($n = 3$) is used to express the experiment data. ### $p < 0.001$ vs. Control group; *** $p < 0.001$, ** $p < 0.01$, and * $p < 0.05$ vs. H_2O_2-induced HUVEC injury model.

Furthermore, images of Hoechst 33,342 staining HUVECs treated with H_2O_2, positive control of acetylcysteine (NAc), and FPYLRH (S8) at the concentrations of 100 μM for 24 h are shown in Figure 3. In the blank control group (Figure 3A), HUVECs were uniform in size, plump in shape, and presenting blue fluorescence. Compared with the blank control group, HUVECs in the model group (Figure 3B) showed a state of apoptosis because the number of HUVECs significantly decreased, the cells became smaller, and the fluorescence of most of the remaining adherent cells became bright. Figure 3C indicated that the most of HUVECs in the FPYLRH (S8) group adhered to the wall compared with the model group, and a small number of cells was washed away and another small number of cells showed blue fluorescence brightened compared with the blank control group. Those results indicated that FPYLRH (S8) had a strong protective effect on the oxidative damage HUVECs induced by H_2O_2, and the results were in line with those of Figure 2.

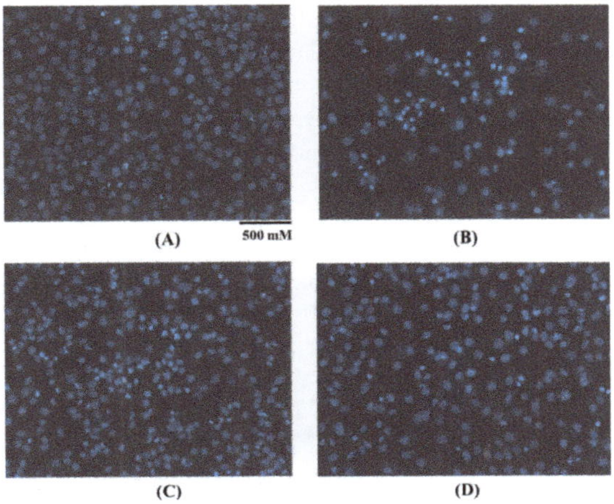

Figure 3. Apoptosis analysis of FPYLRH (S8) in the H_2O_2-induced HUVECs injury model at the concentration of 100 μM by Hoechst 33,342 staining assay. (**A**): Blank control; (**B**): Model (H_2O_2); (**C**): Positive control (acetylcysteine, NAc); (**D**): FPYLRH (S8).

2.3. *Effect of FPYLRH (S8) on the Levels of ROS in H_2O_2-Induced HUVEC Injury Model*

Excessive production of intracellular ROS will destroy key biological macromolecules, which further cause oxidative stress and serial chronic diseases [21,22]. Therefore, the effect of FPYLRH (S8) on the levels of ROS in HUVECs was measured. As shown in Figure 4, the level of ROS observed in the HUVECs exposed to H_2O_2 were 231.7% ± 13.5% of control, which was extremely ($p < 0.001$) significantly higher than those of the control group. As expected, the intracellular ROS levels (185.2% ± 12.4%, 155.3% ± 16.5%, and 139.1% ± 11.8% of control at the concentrations of 10, 50, and 100 μM) were significantly attenuated by FPYLRH (S8) pretreatment in a dose-effect manner.

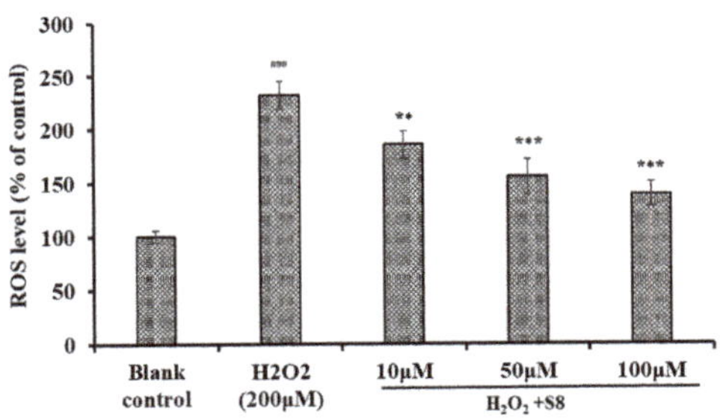

Figure 4. Effect of FPYLRH (S8) on ROS levels of H_2O_2-induced HUVEC injury model at the concentration of 10, 50, and 100 μM. Mean ± SD (n = 3) is used to express the experiment data. ### $p < 0.001$ vs. Control group; *** $p < 0.001$ and ** $p < 0.01$ vs. H_2O_2-induced HUVEC injury model.

2.4. *Effect of FPYLRH (S8) on the Levels of GSH-Px, SOD, Malondialdehyde (MDA) and Nitric Oxide (NO) in H_2O_2-Induced HUVEC Injury Model*

In living organisms, antioxidant enzymes (SOD, GSH-Px, CAT, and GSH-Rx) and non-enzymatic antioxidant agents (GSH, ascorbic acid, and β-carotene) make up a complex protective system, which can promote oxidative balance by reducing the concentration of ROS and forming less oxidation metabolites (MDA, NO, etc.) [6]. However, this protective system can be overwhelmed by uncontrolled generation of ROS, thus the system requires additional antioxidant agents to balance the oxidative status [6,19]. Therefore, effects on antioxidant enzymes and lipid oxidation metabolites in cells were usually applied to evaluate the antioxidant capacity of antioxidant molecules.

As shown in Figure 5, the effects of FPYLRH (S8) on the levels of SOD, GSH-Px, MDA, and NO in HUVECs were measured for illuminating its protection in HUVECs with H_2O_2-induced oxidative damage. Figures 5A and 5B indicated that the levels of GSH-Px and SOD observed in the HUVECs exposed to H_2O_2 were 20.38 ± 0.82 U/mg prot and 111.35 ± 3.47 mg prot, respectively, which were extremely significantly lower than those of the normal HUVECs control ($p < 0.001$). Moreover, the levels of GSH-Px (34.85 ± 1.65 U/mg prot and 51.06 ± 2.03 U/mg prot) and SOD (151.32 ± 4.52 U/mg prot and 212.56 ± 4.68 U/mg prot) of HUVECs incubated by FPYLRH (S8) at the concentrations of 50 and 100 μM were very ($p < 0.01$) and extremely ($p < 0.001$) significantly higher than those of the H_2O_2-damaged group.

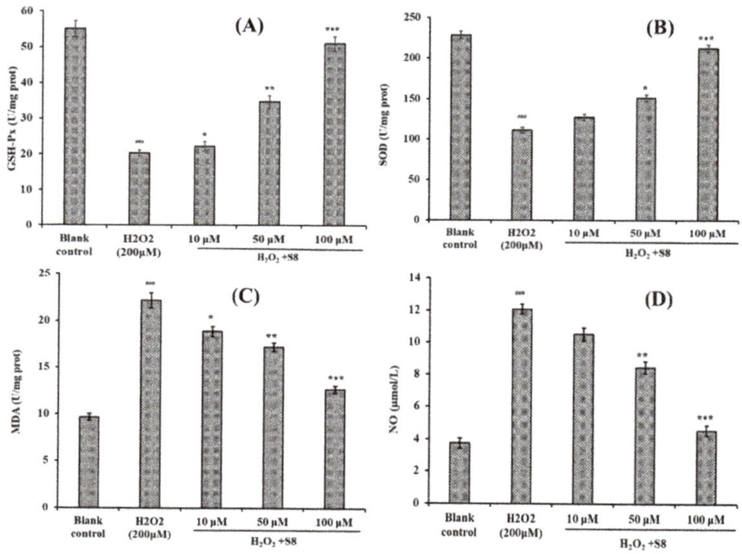

Figure 5. Effect of FPYLRH (S8) on GSH-Px (**A**), SOD (**B**), MDA (**C**), and NO (**D**) levels of the H_2O_2-induced HUVEC injury model at the concentrations of 10, 50, and 100 μM. Mean ± SD ($n = 3$) is used to express the experiment data. ### $p < 0.001$ vs. control group; *** $p < 0.001$, ** $p < 0.01$, and * $p < 0.05$ vs. H_2O_2-induced HUVEC injury model.

Figure 5C,D showed that the contents of MDA and NO in the HUVECs exposed to H_2O_2 were 22.15 ± 0.81 nM/mg prot and 12.06 ± 0.29 μM/L, respectively, which were extremely significantly lower than those of the normal HUVECs control ($p < 0.001$). The data indicated that H_2O_2 at the concentration of 200 μM could significantly damage the membrane of HUVECs. Compared with the model group, the contents of MDA of HUVECs incubated by FPYLRH (S8) at the concentrations of 10, 50, and 100 μM were 18.89 ± 0.54 nM/mg prot, 17.21 ± 0.46 nM/mg prot, and 12.64 ± 0.38 nM/mg prot, respectively, which were significantly ($p < 0.05$), very significantly ($p < 0.01$) and extremely significantly ($p < 0.001$) lower than those of the model group. The contents of NO of HUVECs incubated with FPYLRH (S8) at the concentrations of 50 and 100 μM were 8.42 ± 0.38 μM/L and 4.56 ± 0.31 μM/L, respectively, which were very ($p < 0.01$) and extremely ($p < 0.001$) significantly lower than those of model group.

At present, some protein hydrolysates, fractions, and peptides showed intracellular ROS scavenging activities through regulating antioxidant enzyme levels [3]. Corn gluten peptide fractions of CPF1 (MW < 1 kDa) and CPF2 (1 < MW < 3 kDa) exhibited cytoprotective effects and intracellular ROS scavenging activities through increasing the activity levels of SOD, CAT, and GR) as well as the total GSH levels in oxidized HepG2 cells [23]. Liang, Zhang, and Lin reported that the pulse electric field (PEF) could increase the antioxidant activity of QDHCH from pine nut (*Pinus koraiensis*) [24]. Compared with H_2O_2 damaged group, the PEF-treated QDHCH has a better protective oxidative stress inhibition of 74.22% ± 3.70% by significantly increasing the T-SOD, CAT, GSH-Px, and GSH-Rx activities and decreasing the MDA content in HepG2 cells. Zheng et al. reported that GPA from fish skin gelatin hydrolysate could activate the expression of antioxidant response element (ARE)-driven antioxidant enzyme genes in a dose-dependent manner, and subsequently suppressed the H_2O_2-induced intracellular ROS production in IPEC-J2 cells [13]. EAMAPK and AVPYPQ from stracchino hydrolysate by in vitro gastro-intestinal digestion showing antioxidant activities in a wide concentration range (5–150 µg/mL), involving ROS reduction, SOD expression increases and Nrf2 antioxidant response activation in intestinal epithelial cells (IEC-6) [25]. The present results suggested that FPYLRH (S8) had similar cytoprotective functions with QDHCH, GPA, EAMAPK, and AVPYPQ through enhancing endogenous antioxidant defense systems to reduce the H_2O_2 damage in cells.

2.5. Protective Activity of FPYLRH (S8) on Oxidative Damage DNA Induced by H2O2

2.5.1. Protective Activity on Plasmid DNA (pBR322 DNA)

The excessive production of ROS may cause a quantity of degenerative processes in organisms, such as cancer, premature aging, cardiovascular, and neurodegenerative diseases, while DNA damage is a key step in these ROS-induced effects [19,26]. Therefore, the protective activity of FPYLRH (S8) on plasmid DNA (pBR322 DNA) oxidatively damaged by H_2O_2 is presented in Figure 6. The plasmid DNA (pBR322 DNA) was mainly of the supercoiled (SC) form under normal conditions (Figure 6A). The damage of plasmid DNA results in a cleavage of one of the phosphodiester chains and produces a relaxed open circular (OC) form. Further cleavage near the first breakage leads to linear (LIN) double stranded DNA molecules. The OC formation of DNA is indicative of single-strand breaks and the LIN formation of DNA is indicative of double-strand breaks [16].

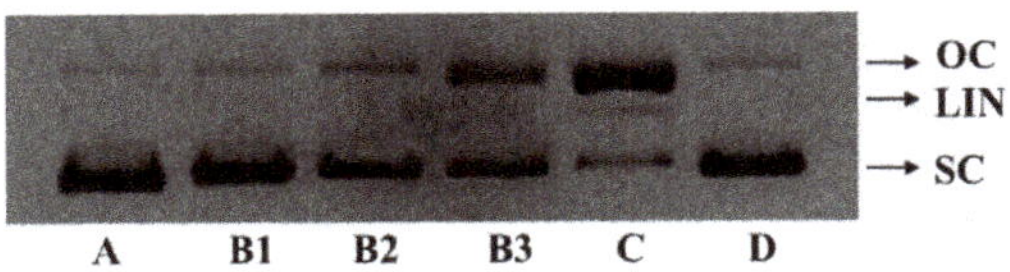

Figure 6. Agarose gel electrophoresis pattern of plasmid pBR322 DNA treated with H_2O_2 and FPYLRH (S8) at different conditions. A: pBR322 DNA; B1: pBR322 DNA + $FeSO_4$ + FPYLRH (S8) (2.0 mg/mL) + H_2O_2; B2: pBR322 DNA + $FeSO_4$ + FPYLRH (S8) (1.0 mg/mL) + H_2O_2; B3: pBR322 DNA + $FeSO_4$ + FPYLRH (S8) (0.5 mg/mL) + H_2O_2; C: pBR322 DNA +$FeSO_4$ + H_2O_2; D: pBR322 DNA + $FeSO_4$ + GSH (2.0 mg/mL) + H_2O_2.

In the experiment, hydroxyl radical was produced from iron-mediated decomposition of H_2O_2 when $FeSO_4$ and H_2O_2 were added into and reacted in the sample solution, and it subsequently broke the plasmid DNA (pBR322 DNA) and converted the supercoiled form into the OC form (Figure 6C). The linear form of DNA was observed in Figure 6C, which indicated that the excess hydroxyl radicals further broke a small amount of the double-strand of DNA. As shown in Figure 6B, FPYLRH (S8) positively influenced the contents of SC form of the plasmid DNA (pBR322 DNA) in a dose-dependent manner. Correspondingly, the contents of OC form of the plasmid DNA (pBR322 DNA) were decreased with increasing concentration of FPYLRH (S8). Moreover, FPYLRH (S8) at the concentration of 2 mg/mL showed similar protective effect on DNA damage with the positive control of GSH (Figure 6D).

Therefore, FPYLRH (S8) could protect the supercoiled plasmid DNA through preventing the reaction of Fe^{2+} with H_2O_2 and scavenge hydroxyl radical by donating a hydrogen-atom or electron. This finding was in line with the result that FPYLRH (S8) could effectively scavenge hydroxyl radical in radical scavenging assay in vitro [1].

2.5.2. Protective Activity on DNA in H_2O_2-Induced HUVEC Injury Model

The comet assay, known as single cell gel electrophoresis (SCRE), helps to determine whether there has been DNA damage to a single cell from apoptosis (cell death) or cytotoxicity and the extent of this damage [27]. Cells embedded in agarose on a microscope slide are lysed with detergent and high salt to form nucleoids containing supercoiled loops of DNA linked to the nuclear matrix. Electrophoresis results in structures that resemble comets at high pH and the intensity of the comet tail relative to the head that reflects the number of DNA breaks [28]. The comet assay has been devoted to testing novel chemicals for genotoxicity, monitoring environmental contamination with genotoxins, human biomonitoring and molecular epidemiology, and basic research in DNA damage and repair [28,29].

As shown in Figure 7A, the comet head is bright and almost no comet tail is found in the blank group, but the comet tail is long and has a large area in the model group (Figure 7B), which indicated that the damage model was successfully established. Compared with the model group, the comet tails on damaged HUVECs were gradually reduced with the increasing concentration of FPYLRH (S8) (Figure 7D). In addition, the length and area of comet tails in FPYLRH (S8) group with the concentration of 200 μM was close to that of positive control group (Figure 7C).

The comet tail length or comet length can intuitively reflect the experimental results, but they usually cause large errors. For a more accurate expression of the data of the comet assay, more indicators including head DNA (HDNA), tail DNA (TDNA), and torque class indicator (Olive tail moment, OTM) were employed (Table 1). The results indicated that the TDNA, tail length (TL), tail moment (TM), and OTM of model group were extremely significantly larger than that of the blank control group ($p < 0.001$), the comet length (CL) of model group was significantly larger than that of blank control group ($p < 0.05$), and HDNA was extremely significantly less than that of the blank control group ($p < 0.001$). With the decrease of the concentration of FPYLRH (S8), the HDNA of comet decreased gradually, while the TDNA, CL, TL, TM, and OTM increased gradually. Moreover, there were extremely significant differences of measured indicators at tested concentrations, except CL at 50 μM, between the sample and model groups ($p < 0.001$). Therefore, the comet assay indicated that FPYLRH (S8) showed a significant protective effect on DNA in H_2O_2-induced HUVEC injury model.

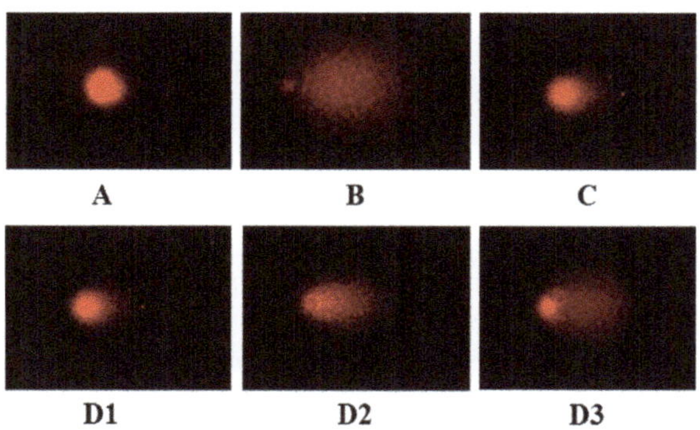

Figure 7. Typical comet images observed by PI (Propidium iodide) staining. (**A**): Blank control group; (**B**): Model (H_2O_2, 200 μM) group; (**C**): Positive control (NAc) group; (**D1**): H_2O_2 + FPYLRH (S8) (200 μM); (**D2**): H_2O_2 + FPYLRH (S8) (100 μM); (**D3**): H_2O_2 + FPYLRH (S8) (50 μM).

Table 1. Protective activity of FPYLRH (S8) on DNA in H_2O_2-induced HUVEC injury model in comet assay.

Group	Control	Model	NAc	FPYLRH (S8, μM)		
				200	100	50
Cell Number (n)	**104**	**117**	**121**	**111**	**107**	**118**
HDNA (%)	91.9 ± 4.2	18.2 ± 3.2 ###	80.7 ± 3.8 ***	78.4 ± 7.9 ***	53.2 ± 6.2 ***	37.5 ± 5.1 ***
TDNA (%)	8.1 ± 4.2	81.8 ± 3.2 ###	19.3 ± 4.8 *	21.6 ± 7.9 ***	46.8 ± 6.2 ***	62.5 ± 5.1 ***
CL (pix)	68.3 ± 3.3	77.3 ± 13.5 #	66.2 ± 3.3 ***	58.4 ± 5.7 ***	60.0 ± 7.3 ***	76.1 ± 10.6 *
TL (pix)	7.3 ± 0.33	59.3 ± 6.6 ###	27.2 ± 1.6 ***	23.4 ± 2.4 ***	25.0 ± 3.3 ***	45.1 ± 6.5 ***
TM	0.6 ± 0.03	46.4 ± 8.1 ###	5.2 ± 0.24 ***	5.0 ± 0.5 ***	11.7 ± 1.4 ***	28.1 ± 3.9 ***
OTM	2.3 ± 0.1	24.5 ± 4.3 ###	4.7 ± 0.2 ***	4.7 ± 0.5 ***	10.4 ± 1.2 ***	17.2 ± 2.4 ***

Head DNA, tail DNA, comet length, tail length, tail moment, and olive tail moment are referred to as HDNA, TDNA, CL, TL, TM, and OTM, respectively. All data are presented as the mean ± SD of triplicate results. # $p < 0.05$ vs. Control group, ### $p < 0.001$ vs. Control group; * $p < 0.05$ vs. model group, *** $p < 0.001$ vs. model group.

Oxidative damage DNA leads to pathological processes involved in the development of cancer, cardiovascular diseases, and ageing [30], and the damage degree of DNA can increase under conditions of oxidative stress, arising from exposure to a variety of physical or chemical insults [31,32]. Antioxidant peptides attract extensive attention due to their key roles in maintaining cellular function upon DNA insult. Zhao et al. indicated that a peptide fraction from croceine croaker swim bladder could have an anti-fatigue effect through inhibiting the oxidative reactions and DNA damage [16]. I/LNI/LCCN and WCTSVS from marine *Sepia brevimana* and *Loligo duvauceli*, respectively, exhibited significant protective effects on DNA damage and inhibition for the linoleic acid auto-oxidation in the model system due to hydroxyl radical induction [33]. YGDEY protected liver hepatocellular cells (HepG2 cells) from alcohol-induced injury by inhibiting oxidative stress including decreasing the amount of ethanol-induced DNA damage, and this may be associated with the Akt/NF-κB/mitogen-activated protein kinase (MAPK) signal transduction pathway [34]. In our previous research, FPYLRH (S8) exhibited strong radical scavenging activity and lipid peroxidation inhibition [1], and the presented results indicated FPYLRH could decrease the H_2O_2-induced stress injury in HUVECs by increasing the levels of SOD and GSH-Px, decreasing the contents of MDA and NO, and protecting DNA in oxidative damage. The antioxidant mechanism of FPYLRH might be related to activating the Nrf2-antioxidant response element signaling pathway.

2.6. Relationship Among the Molecular Size, Amino Acid Composition, and Antioxidant Activity

Molecular size and amino acid composition significantly affect the biological function of peptides [3,35]. Short peptides with 2–10 amino acid residues showed high radical scavenging and inhibition activities of lipid peroxidation because they were more easily accessible to active radicals to provide potential effects in a reaction mixture than their parent native proteins [36,37]. In the experiment, FPYLRH (S8) exhibited strong cytoprotective effect on H_2O_2-mediated HUVECs, which indicated that FPYLRH (S8) with 6 amino acid residues could easily enter into cells, interact with ROS, and inhibit oxidative stress reaction.

Hydrophobic amino acids with non-polar aliphatic groups, such as Tyr, Leu, Trp, Pro, Ile, and Val, had high reactivity to hydrophobic PUFAs and exert their significant effects on radical scavenging [1,38]. Aromatic amino acids (Phe, Trp, and Tyr) with aromatic residues can donate protons to electron-deficient radicals, keeping ROS stable during the radical scavenging process [14,17]. Sheih et al. reported that Tyr residues could remove free radicals, change them into more stable phenoxy radicals, and further inhibit the peroxidizing chain reaction mediated by free radicals [39]. Chen et al. found that the pyrrolidine ring of Pro can increase the flexibility of peptides and quench singlet oxygen due to its low ionization potential [40]. Therefore, the hydrophobic amino acid residues (Leu and Pro) and the aromatic residues (Phe and Tyr) in the sequences of FPYLRH (S8) should contribute to its radical-scavenging and lipid peroxidation inhibitory activities. In addition, polar amino acids are reported to play a critical role in the metal ion chelating and hydroxyl radical scavenging activities, which is related to the carboxyl and amino groups in their side chains [3,41,42]. Memarpoor-Yazdi et al. found that the basic (Arg) amino acid residues in the sequence of NTDGSTDYGILQINSR were of great importance to their antioxidant activities [43]. Therefore, the presence of Arg and His in FPYLRH (S8) should play a crucial role for their antioxidant capabilities.

3. Method and Materials

3.1. Materials

HUVECs were purchased from the Cell Bank of Type Culture Collection of the Chinese Academy of Sciences (Shanghai, China). Dulbecco's modified Eagle's medium (DMEM), phosphate buffered saline (PBS, pH 7.2), NAc, dimethyl sulfoxide (DMSO), 3-(4,5-dimethylthiazol-2-y1)-2,5-diphenyltetrazolium bromide (MTT) were purchased from Sigma-Aldrich (Shanghai) Trading Co., Ltd. (China). FYKWP (S1), FTGMD (S2), GFEPY (S3), YLPYA (S4), FPPYERRQ (S5), GFYAA (S6), FSGLR (S7), FPYLRH (S8), VPDDD (S9), and GIEWA (S10) with a purity higher than 98% were synthesized in China Peptides Co. (Suzhou, China).

3.2. Cell Culture and Viability Assay

The cell culture method and viability assay were performed according to the method of Lim et al. [44]. The HUVECs were cultured in DMEM containing 1% penicillin-streptomycin and 10% FBS at 37 °C and 5% CO_2 atmosphere. The MTT test was used to measure the cell viability. After 24 h incubation in a 96-well plate (7×10^3 cells/well), the HUVECs were cultured in 100 μM peptide solution for 12 h. After that, the wells were washed twice with PBS and the MTT with a final concentration of 0.5 mg/mL was added for an additional 4 h. After that, the formazan crystals formed by active cells were dissolved in 150 μL of DMSO. The absorbance at 570 nm was measured and the cell viability was calculated by the following equation:

$$\text{Cell viability} = (A_{sample}/A_{control}) \times 100\%. \quad (1)$$

3.3. Protection of FPYLRH on Oxidative Damage HUVECs by H_2O_2

The HUVECs were seeded on a 96-well plate with the density of 1.5×10^4 cells/well. After culturing for 24 h, the supernatant was aspirated and H_2O_2 (final concentration of 100, 200, 300, 400, and 500 mM) was added and sequentially incubated for 24 h. The optimal H_2O_2 concentration was confirmed when the cell viability was close to 50%.

The HUVECs with oxidative damage under the H_2O_2-induced optimal conditions were used to test the protective effects of FPYLRH on the damaged cells. The peptides were dissolved in the DMEM medium with the concentrations of 10, 50, and 100 μg/mL. The selected HUVECs were cultured (1.5×10^4 cells/well) in a 96-well plate for 24 h. Then the supernatant was aspirated and 100 μL of samples were added into the protection groups respectively incubating for 8 h. After removing samples, H_2O_2 was added into the damage and protection groups with the optimal concentration and sequentially incubated for 24 h. The positive control group used 100 μL of 1.5 mM NAc instead of 100 μL of peptide solution.

3.4. Hoechst 33,342 Staining Assay

The logarithmic growth HUVECs were treated by trypsinization and grown (2.0×10^5 cells/well) in a 6-well plate for 24 h. Then the supernatant was aspirated and 300 μL of FPYLRH and NAc solutions were separately added into the protection and positive control groups incubating for 2 h, respectively. After removing samples, 300 μL of H_2O_2 was separately added into the model, protection, and positive control groups which were incubated for 24 h, respectively. After that, HUVECs were exposed to 8 mg/mL Hoechst 33,342 solution for 30 min at 37 °C and 5% CO_2 atmosphere. After removing the Hoechst 33,342 solution and washing three times with serum-free DMEM, the morphology of HUVECs was observed using a fluorescence microscope (LSM710; Carl Zeiss Microscopy GmbH, Jena, Germany) with the excitation and emission wavelengths of 550 and 460 nm, respectively.

3.5. Determination of the Levels of ROS in H_2O_2-Induced HUVECs

Intracellular ROS accumulation in HUVECs was monitored according to the previous method described by Zheng et al. [13]. In brief, HUVECs were preincubated with samples at the concentrations of 10, 50, or 100 μM (B) for 12.0 h, and then incubated with 200 μM H_2O_2 for 2 h. After that, the cells were washed with PBS and incubated with 10 μM DCFH2-DA in fresh culture medium for 0.5 h. Intracellular ROS levels indicated by DCF fluorescence were quantified on a BD FACS Calibur flow cytometer (BD Biosciences, San Diego, C, USA) using excitation and emission filters of 488 and 530 nm, respectively. The data were expressed as % of control values.

3.6. Determination of the Levels of Antioxidant Enzymes in H_2O_2-Induced HUVECs

HUVECs were cultured in 6-well plates (1×10^6 cells/well). FPYLRH with final concentration of 10, 25, and 50 μg/mL was added into the protection groups, respectively. Last, the damage and protection groups were exposed to H_2O_2. Subsequently, 500 mL of cell lysis buffer was added into each well on ice lysed for 30 min and centrifuged at 12,000× *g* and 4 °C for 10 min. The resulting liquid supernatant was subsequently stored on cold standby at 4 °C (the indicators should be measured within 6 h). The levels of SOD, GSH-Px, NO, and MDA were measured using assay kits according to the protocols of the manufacturer (Nanjing Jiancheng Bioengineering Institute Co., Ltd., Nanjing, China), and protein concentrations were determined using the bicinchoninic acid (BCA) method to normalize their levels. The levels of T-SOD and GSH-Px were expressed as units of enzymatic activity per milligram of protein (U/mg prot).

3.7. Protective Effect of FPYLRH on pBR322 Plasmid DNA Damaged by H_2O_2

The ability of FPYLRH to protect supercoiled pBR322 plasmid DNA was measured according to the previous method [16]. The reaction mixtures (15 μL) containing 5 μL of PBS (10 mM, pH 7.4), 1μL of plasmid DNA (0.5 μg), 5 μL of FPYLRH, 2 μL of 1 mM $FeSO_4$, and 2 μL of 1 mM H_2O_2 were incubated at 37 °C for 30 min. After incubation, 2 μL of loading buffer (50% glycerol (*v*/*v*), 40 mM EDTA, and 0.05% bromophenol blue) was added to stop the reaction, and the reaction mixtures were electrophoresed on 1% agarose gel containing 0.5 μg/mL ethidium bromide in Tris/acetate/EDTA gel buffer for 50 min (60 V). The DNA in the gel was visualized and photographed under ultraviolet light. GSH was used as positive control.

3.8. DNA Comet Assay

The DNA comet assay was according to the previous method [45]. One hundred μL of 0.5% normal melting point agarose (NMA) preheated at 45 °C was spread on the frosted glass slides at 45 °C and covered with a clean cover glass. Subsequently, the slide was placed at 4 °C allowing the NMA to set. After 10 min, the slide was taken out, and the cover glass was gently peeled off. Then 75 μL melted 0.7% NMA mixed with 10 μL cells (1×10^6 cells/mL) were quickly dropped to the first layer of gel and immediately covered with another clean cover glass. Then, the slide was placed at 4 °C for 10 min allowing the NMA to set. After removing the cover glass slide, the glass slide was placed in the plate, the pre-cooled lysis buffer was added, and the plate was put into a refrigerator at 4 °C for 1 h. The glass slide was taken out, rinsed with PBS, and placed in an electrophoresis tank in cold 0.3 M NaOH, 1 mM EDTA for 20–60 min. After that, the electrophoresis was carried out for 25 min at a constant voltage of 25 V. After electrophoresis, the gel was neutralized with 0.4 mM Tris-HCl buffer (pH 7.5) at 4 °C for 10 min, and the neutralization process was repeated three times. Gels were stained with PI (propidium iodide) dye liquor for 10 min and measured with a fluorescence microscope at 515~560 nm wavelength. About 100 cells per gel were randomly selected for measuring the diameter and migration length of DNA.

3.9. Statistical Analysis

Data were presented as means ± standard deviation (SD) ($n = 3$). ANOVA test (SPSS 19.0 software) was applied to compare the mean value of each treatment. Significant differences were determined by using Duncan's multiple range Test ($p < 0.05$).

4. Conclusions

In this work, the protective effect of 10 antioxidant pentapeptides (S1–S10), especially FPYLRH (S8), from the hydrolysate of miiuy croaker swim bladder on oxidative damage of HUVECs by H_2O_2 was studied systematically. There was no significant difference between the normal HUVECs and the HUVECs treated with the 10 pentapeptides at 100 μM for 24 h ($p < 0.05$). Furthermore, FPYLRH could significantly inhibit H_2O_2-induced oxidative stress in HUVECs compared with that of the model group through increasing the levels of SOD and GSH-Px, decreasing the contents of ROS, MDA, and NO, and protecting DNA in HUVECs with oxidative damage compared with the model group. These results suggested that FPYLRH could be used as a potential natural antioxidant in functional food and other products.

Author Contributions: B.W. conceived and designed the experiments. S.-Y.C. and Y.-M.W. performed the experiments. Y.-M.W. and Y.-Q.Z. analyzed the data. B.W. and C.-F.C. contributed the reagents, materials, and wrote the paper.

Funding: This work was funded by the National Natural Science Foundation of China (NSFC) (No. 31872547), the Public Service Technology Application Research Project of Science and Technology Department of Zhejiang Province (No. LGN18D060002), and International S&T Cooperation Program of China (No. 2012DFA30600).

Conflicts of Interest: The authors declare no conflict of interest.

References

1. Zhao, W.H.; Luo, Q.B.; Pan, X.; Chi, C.F.; Sun, K.L.; Wang, B. Preparation, identification, and activity evaluation of ten antioxidant peptides from protein hydrolysate of swim bladders of miiuy croaker (*Miichthys miiuy*). *J. Funct. Foods* **2018**, *47*, 503–511. [CrossRef]
2. Wu, R.; Wu, C.; Liu, D.; Yang, X.; Huang, J.; Zhang, J.; Liao, B.; He, H. Antioxidant and anti-freezing peptides from salmon collagen hydrolysate prepared by bacterial extracellular protease. *Food Chem.* **2018**, *248*, 346–352. [CrossRef]
3. Sila, A.; Bougatef, A. Antioxidant peptides from marine by-products: Isolation, identification and application in food systems. *A review. J. Funct. Foods* **2016**, *21*, 10–26. [CrossRef]
4. Zhao, G.X.; Yang, X.R.; Wang, Y.M.; Zhao, Y.Q.; Chi, C.F.; Wang, B. Antioxidant peptides from the protein pydrolysate of spanish mackerel (*Scomberomorous niphonius*) muscle by in vitro gastrointestinal gigestion and their in vitro activities. *Mar. Drugs* **2019**, *17*, 531. [CrossRef]
5. Lu, X.; Zhang, L.; Sun, Q.; Song, G.; Huang, J. Extraction, identification and structure-activity relationship of antioxidant peptides from sesame (*Sesamum indicum* L.) protein hydrolysate. *Food Res. Int.* **2019**, *116*, 707–716. [CrossRef]
6. Lorenzo, J.M.; Munekata, P.E.S.; Gómez, B.; Barba, F.J.; Mora, L.; Pérez-Santaescolástica, C.; Toldrá, F. Bioactive peptides as natural antioxidants in food products—A review. *Trends Food Sci. Tech.* **2018**, *79*, 136–147. [CrossRef]
7. Wang, B.; Li, L.; Chi, C.F.; Ma, J.H.; Luo, H.Y.; Xu, Y.F. Purification and characterisation of a novel antioxidant peptide derived from blue mussel (*Mytilus edulis*) protein hydrolysate. *Food Chem.* **2013**, *138*, 1713–1719. [CrossRef]
8. Timón, M.L.; Andrés, A.I.; Otte, J.; Petrón, M.J. Antioxidant peptides (<3 kDa) identified on hard cow milk cheese with rennet from different origin. *Food Res. Int.* **2019**, *120*, 643–649.
9. Wong, F.C.; Xiao, J.; Ong, M.G.L.; Pang, M.J.; Wong, S.J.; Teh, L.K.; Chai, T.T. Identification and characterization of antioxidant peptides from hydrolysate of blue-spotted stingray and their stability against thermal, pH and simulated gastrointestinal digestion treatments. *Food Chem.* **2019**, *271*, 614–622. [CrossRef]

10. Zhuang, Y.; Hou, H.; Zhao, X.; Zhang, Z.; Li, B. Effects of collagen and collagen hydrolysate from jellyfish (*Rhopilema esculentum*) on mice skin photoaging induced by UV irradiation. *J. Food Sci.* **2009**, *74*, H183–H188. [CrossRef]
11. Chen, T.; Hou, H. Protective effect of gelatin polypeptides from Pacific cod (*Gadus macrocephalus*) against UV irradiation-induced damages by inhibiting inflammation and improving transforming growth Factor-β/Smad signaling pathway. *J. Photoch. Photobio. B* **2016**, *162*, 633–640. [CrossRef]
12. Yang, X.R.; Zhang, L.; Zhao, Y.Q.; Chi, C.F.; Wang, B. Purification and characterization of antioxidant peptides derived from protein hydrolysate of the marine bivalve mollusk Tergillarca granosa. *Mar. Drugs* **2019**, *17*, 251. [CrossRef]
13. Zheng, L.; Yu, H.; Wei, H.; Xing, Q.; Zou, Y.; Zhou, Y.; Peng, J. Antioxidative peptides of hydrolysate prepared from fish skin gelatin using ginger protease activate antioxidant response element-mediated gene transcription in IPEC-J2 cells. *J. Funct. Foods* **2018**, *51*, 104–112. [CrossRef]
14. He, Y.; Pan, X.; Chi, C.F.; Sun, K.L.; Wang, B. Ten new pentapeptides from protein hydrolysate of miiuy croaker (*Miichthys miiuy*) muscle: Preparation, identification, and antioxidant activity evaluation. *LWT* **2019**, *105*, 1–8. [CrossRef]
15. Che, R.; Sun, Y.; Sun, D.; Xu, T. Characterization of the Miiuy Croaker (*Miichthys miiuy*) Transcriptome and Development of Immune-Relevant Genes and Molecular Markers. *PLoS ONE* **2014**, *9*, e94046. [CrossRef]
16. Zhao, Y.Q.; Zeng, L.; Yang, Z.S.; Huang, F.F.; Ding, G.F.; Wang, B. Anti-fatigue effect by peptide fraction from protein hydrolysate of croceine croaker (*Pseudosciaena crocea*) swim bladder through inhibiting the oxidative reactions including DNA damage. *Mar. Drugs* **2016**, *14*, 221. [CrossRef]
17. Zhu, Z.; Shi, Z.; Xie, C.; Gong, W.; Hu, Z.; Peng, Y. A novel mechanism of Gamma-aminobutyric acid (GABA) protecting human umbilical vein endothelial cells (HUVECs) against H2O2-induced oxidative injury. Comp. *Biochem. Phy. C* **2019**, *217*, 68–75. [CrossRef]
18. Pan, X.; Zhao, Y.Q.; Hu, F.Y.; Wang, B. Preparation and identification of antioxidant peptides from protein hydrolysate of skate (*Raja porosa*) cartilage. *J. Funct. Foods* **2016**, *25*, 220–230. [CrossRef]
19. Tao, J.; Zhao, Y.Q.; Chi, C.F.; Wang, B. Bioactive peptides from cartilage protein hydrolysate of spotless smoothhound and their antioxidant activity in vitro. *Mar. Drugs* **2018**, *16*, 100. [CrossRef]
20. Yang, B.; Oo, T.N.; Rizzo, V. Lipid rafts mediate H_2O_2 prosurvival effects in cultured endothelial cells. *FASEB J.* **2006**, *20*, 1501–1503. [CrossRef]
21. Zhang, L.; Zhao, G.X.; Zhao, Y.Q.; Qiu, Y.T.; Chi, C.F.; Wang, B. Identification and active evaluation of antioxidant peptides from protein hydrolysates of skipjack tuna (*Katsuwonus pelamis*) head. *Antioxidants* **2019**, *8*, 318. [CrossRef] [PubMed]
22. León-López, A.; Fuentes-Jiménez, L.; Hernández-Fuentes, A.D.; Campos-Montiel, R.G.; Aguirre-Álvarez, G. Hydrolysed collagen from sheepskins as a source of functional peptides with antioxidant activity. *Int. J. Mol. Sci.* **2019**, *20*, 3931. [CrossRef] [PubMed]
23. Wang, L.; Ding, L.; Yu, Z.; Zhang, T.; Ma, S.; Liu, J. Intracellular ROS scavenging and antioxidant enzyme regulating capacities of corn gluten meal-derived antioxidant peptides in HepG2 cells. *Food Res. Int.* **2016**, *90*, 33–41. [CrossRef] [PubMed]
24. Liang, R.; Zhang, Z.; Lin, S. Effects of pulsed electric field on intracellular antioxidant activity and antioxidant enzyme regulating capacities of pine nut (*Pinus koraiensis*) peptide QDHCH in HepG2 cells. *Food Chem.* **2017**, *237*, 793–802. [CrossRef] [PubMed]
25. Pepe, G.; Sommella, E.; Ventre, G.; Scala, M.C.; Adesso, S.; Ostacolo, C.; Marzocco, S.; Novellino, E.; Campiglia, P. Antioxidant peptides released from gastrointestinal digestion of "Stracchino" soft cheese: Characterization, in vitro intestinal protection and bioavailability. *J. Funct. Foods* **2016**, *26*, 494–505. [CrossRef]
26. Limmongkon, A.; Pankam, J.; Somboon, T.; Wongshaya, P.; Nopprang, P. Evaluation of the DNA damage protective activity of the germinated peanut (*Arachis hypogaea*) in relation to antioxidant and anti-inflammatory activity. *LWT* **2019**, *101*, 259–268. [CrossRef]
27. Afanasieva, K.; Sivolob, A. Physical principles and new applications of comet assay. *Biophys. Chem.* **2018**, *238*, 1–7. [CrossRef]
28. Collins, A.R. The comet assay for DNA damage and repair: Principles, applications, and limitations. *Mol. Biotechnol.* **2004**, *26*, 249–261. [CrossRef]
29. Xue, Z.; Liu, Z.; Wu, M.; Zhuang, S.; Yu, W. Effect of rapeseed peptide on DNA damage and apoptosis in Hela cells. *Exp. Toxicol. Pathol.* **2010**, *62*, 519–523. [CrossRef]

30. Malik, Q.; Herbert, K.E. Oxidative and non-oxidative DNA damage and cardiovascular disease. *Free Radical Res.* **2012**, *46*, 554–564. [CrossRef]
31. Zaremba, T.; Oliński, R. Oxidative DNA damage—Analysis and clinical significance. *Postepy Biochem.* **2010**, *56*, 124–138. [PubMed]
32. Jackson, S.P.; Bartek, J. The DNA-damage response in human biology and disease. *Nature* **2009**, *461*, 1071–1078. [CrossRef] [PubMed]
33. Sudhakar, S.; Naz, R.A. Preparation of potent antioxidant peptide from edible part of shortclub cuttlefish against radical mediated lipid and DNA damage. *LWT* **2015**, *64*, 593–601. [CrossRef]
34. Chen, M.F.; Gong, F.; Zhang, Y.Y.; Li, C.; Zhou, C.; Hong, P.; Sun, S.; Qian, Z.J. Preventive effect of YGDEY from Tilapia fish skin gelatin hydrolysates against alcohol-induced damage in HepG2 cells through ROS-mediated signaling pathways. *Nutrients* **2019**, *11*, E392. [CrossRef]
35. Li, Z.; Wang, B.; Chi, C.; Gong, Y.; Luo, H.; Ding, G. Influence of average molecular weight on antioxidant and functional properties of cartilage collagen hydrolysates from *Sphyrna lewini*, *Dasyatis akjei* and *Raja porosa*. *Food Res. Int.* **2013**, *51*, 283–293. [CrossRef]
36. Chi, C.; Wang, B.; Li, Z.; Hu, F.; Luo, H. Influence of different hydrolysis conditions on physicochemical, antioxidant and functional properties of collagen hydrolysates from cartilages of *Sphyrna lewini*, *Dasyatis akjei* and *Raja porosa*. *J. Aquat. Food Prod. Technol.* **2016**, *25*, 616–632. [CrossRef]
37. Qiu, Y.T.; Wang, Y.M.; Yang, X.R.; Zhao, Y.Q.; Chi, C.F.; Wang, B. Gelatin and antioxidant peptides from gelatin hydrolysate of skipjack tuna (*Katsuwonus pelamis*) scales: Preparation, identification and activity evaluation. *Mar. Drugs* **2019**, *17*, 565. [CrossRef]
38. Zhang, J.B.; Zhao, Y.Q.; Wang, Y.M.; Chi, C.F.; Wang, B. Eight collagen peptides from hydrolysate fraction of spanish mackerel skins: Isolation, identification, and in vitro antioxidant activity evaluation. *Mar. Drugs* **2019**, *17*, 224. [CrossRef]
39. Sheih, I.C.; Wu, T.K.; Fang, T.J. Antioxidant properties of a new antioxidative peptide from algae protein waste hydrolysate in different oxidation systems. *Bioresour. Technol.* **2009**, *100*, 3419–3425. [CrossRef]
40. Chen, C.; Chi, Y.J.; Zhao, M.Y.; Lv, L. Purification and identification of antioxidant peptides from egg white protein hydrolysates. *Amino Acids* **2012**, *43*, 457–466. [CrossRef]
41. Gimenez, B.; Aleman, A.; Montero, P.; Gomez-Guillen, M.C. Antioxidant and functional properties of gelatin hydrolysates obtained from skin of sole and squid. *Food Chem.* **2009**, *114*, 976–983. [CrossRef]
42. Wang, B.; Wang, Y.; Chi, C.; Hu, F.; Deng, S.; Ma, J. Isolation and characterization of collagen and antioxidant collagen peptides from scales of croceine croaker (*Pseudosciaena crocea*). *Mar. Drugs* **2013**, *11*, 4641–4661. [CrossRef] [PubMed]
43. Memarpoor-Yazdi, M.; Asoodeh, A.; Chamani, J. A novel antioxidant and antimicrobial peptide from hen egg white lysozyme hydrolysates. *J. Funct. Foods* **2012**, *4*, 278–286. [CrossRef]
44. Lim, S.; Choi, A.H.; Kwon, M.; Joung, E.J.; Shin, T.; Lee, S.G.; Kim, N.G.; Kim, H.R. Evaluation of antioxidant activities of various solvent extract from *Sargassum serratifolium* and its major antioxidant components. *Food Chem.* **2019**, *278*, 178–184. [CrossRef] [PubMed]
45. Ghassemi-Barghi, N.; Varshosaz, J.; Etebari, M.; Dehkordi, A.J. Role of recombinant human erythropoietin loading chitosan-tripolyphosphate nanoparticles in busulfan-induced genotoxicity: Analysis of DNA fragmentation via comet assay in cultured HepG2 cells. *Toxicol. in Vitro* **2016**, *36*, 46–52. [CrossRef]

MDPI
St. Alban-Anlage 66
4052 Basel
Switzerland
Tel. +41 61 683 77 34
Fax +41 61 302 89 18
www.mdpi.com

International Journal of Molecular Sciences Editorial Office
E-mail: ijms@mdpi.com
www.mdpi.com/journal/ijms

www.ingramcontent.com/pod-product-compliance
Lightning Source LLC
Chambersburg PA
CBHW061610220926
51850CB00005B/474
9783039360802